STECK-VAUGHN

GED

Mathematics

PROGRAM CONSULTANTS

Myra K. Baum
New York, New York

Sheron Lee Chic
Community School South
East Palo Alto, California

June E. Dean
Parent and Adult Education
Detroit Public Schools
Detroit, Michigan

Ruth E. Derfler, M.Ed.
State GED Chief Examiner
Massachusetts Department of Education
Malden, Massachusetts

Joan S. Flanery
Adult Education and Literacy Program
Ashland Independent Board of Education
Ashland, Kentucky

Michael E. Snyder, Ed.S.
Tennessee Department of Corrections
Pikeville, Tennessee

Staff Credits

Executive Editor: Ellen Northcutt
Senior Editor: Tim Collins
Design Manager: John J. Harrison
Cover Design: Rhonda Childress
Photo Editor: Margie Foster

Editorial Development: McClanahan & Company, Inc.

Photography: Cover: (computer) © Ralph Mercer/Tony Stone Images. p.26 David Young-Wolff/PhotoEdit; p.194 © Tom Prettyman/PhotoEdit; p.242 Port Authority of New York and New Jersey.

ISBN 0-8114-7364-3

Printed in the United States of America.

1 2 3 4 5 6 7 8 9 BP 00 99 98 97 96 95

Contents

To the Learner

What Is the GED Test?

You are taking a very big step toward changing your life with your decision to take the GED test. By opening this book, you are taking your second important step: preparing for the test. You may feel nervous about what is ahead, which is only natural. Relax and read the following pages to find out the answers to your questions.

The GED test, the Test of General Educational Development, is given by the GED Testing Service of the American Council on Education for adults who did not graduate from high school. When you pass the GED test, you will receive a certificate that is regarded as being equivalent to a high school diploma. Employers in private industry and government, as well as admissions officers in colleges and universities, accept the GED certificate as they would a high school diploma.

The GED test covers the same subjects people study in high school. The five subject areas are: Writing Skills, Interpreting Literature and the Arts, Social Studies, Science, and Mathematics. You will not be required to know all the information that is usually taught in high school. You will, however, be tested on your ability to read and process information. Certain U.S. states also require a test on the U.S. Constitution or on state government. Check with your local adult education center to see if your state requires such a test.

Each year hundreds of thousands of adults pass the GED test. The *Steck-Vaughn GED Series* will help you develop and refine the reading and thinking skills you need to pass the GED test.

GED Scores

After you complete the GED test, you will get a score for each section and a total score. The total score is an average of all the other scores. The highest score possible on a single test is 80. The scores needed to pass the GED test vary depending on where you live. The chart on page 2 shows the minimum state requirements. A score of *40 or 45* means that the score for each test must be 40 or more, but if one or more scores is below 40, an average of at least 45 is required. A minimum score of *35 and 45* means that the score for each test must be 35 or more and an average of at least 45 is required.

GED Score Requirements

Area	Minimum Score on Each Test		Minimum Average on All Five Tests
UNITED STATES			
Alabama, Alaska, Arizona, Colorado, Connecticut, Georgia, Hawaii, Illinois, Indiana, Iowa, Kansas, Kentucky, Maine, Massachusetts, Michigan, Minnesota, Montana, Nevada, New Hampshire, North Carolina, Ohio, Pennsylvania, Rhode Island, South Carolina, Tennessee, Vermont, Virginia, Wyoming	35	and	45
Arkansas, California, Delaware, District of Columbia, Florida, Idaho, Maryland, Missouri, New York, Oklahoma, Oregon, South Dakota, Utah, Washington, West Virginia	40	and	45
Louisiana, Mississippi, Nebraska, Texas	40	or	45
New Jersey (42 is required on Test 1; 40 is required on Tests 2, 3, and 4; 45 is required on Test 5; and 225 is required as a minimum total test score.)			
New Mexico, North Dakota	40	or	50
Wisconsin	40	and	50
CANADA			
Alberta, British Columbia, Manitoba, New Brunswick (English and French), Northwest Territories, Nova Scotia, Prince Edward Island, Saskatchewan, Yukon Territory	45		—
Newfoundland	40	and	45
U.S. TERRITORIES & OTHERS			
Guam, Kwajalein, Micronesia, Puerto Rico, Virgin Islands	35	and	45
Panama Canal Area, Palau	40	and	45
Mariana Islands, Marshall Islands	40	or	45
American Samoa	40		—

This chart gives you information on the content, number of items, and time limit for each test. In some places you do not have to take all sections of the test on the same day. If you want to take all the test sections in one day, the GED test will last an entire day. Check with your local adult education center for the requirements in your area.

Test	Content Areas	Number of Items	Time Limit (minutes)
Writing Skills Part I	Sentence Structure Usage Mechanics	55	75
Writing Skills Part II	Essay	1	45
Social Studies	Geography U.S. History Economics Political Science Behavioral Science	64	85
Science	Biology Earth Science Physics Chemistry	66	95
Interpreting Literature and the Arts	Popular Literature Classical Literature Commentary	45	65
Mathematics	Arithmetic Algebra Geometry	56	90

Where Do You Go to Take the GED Test?

The GED test is offered year-round throughout the United States, its possessions, U.S. military bases worldwide, and in Canada. To find out when and where tests are held near you, contact the GED Hot Line at 1-800-62-MY-GED (1-800-626-9433) or one of these institutions in your area:

- An adult education center
- A continuing education center
- A local community college
- A public library
- A private business school or technical school
- A public board of education

In addition, the Hot Line and the institutions can give you information regarding necessary identification, testing fees, and writing implements. Schedules vary: some testing centers are open several days a week; others are open only on weekends.

Why Should You Take the GED Test?

A GED certificate can help you in the following ways:

Employment

People without high school diplomas or GED certificates have much more difficulty changing jobs or moving up in their present companies. In many cases employers will not hire someone who does not have a high school diploma or the equivalent.

Education

If you want to enroll in a technical school, a vocational school, or an apprenticeship program, you often must have a high school diploma or the equivalent. If you want to enter a college or university, you must have a high school diploma or the equivalent.

Personal

The most important thing is how you feel about yourself. You have the unique opportunity to turn back the clock by making something happen that did not happen in the past. You can attain a GED certificate that will help you in the future and make you feel better about yourself now.

How to Prepare for the GED Test

Classes for GED preparation are available to anyone who wants to take the GED. The choice of whether to take classes is up to you; they are not required. If you prefer to study by yourself, the *Steck-Vaughn GED Series* has been prepared to guide your study. *Steck-Vaughn GED Exercise Books* are also available to give you additional practice for each test.

Most GED preparation programs offer individualized instruction and tutors who can help you identify areas in which you may need help. Many adult education centers offer free day or night classes. The classes are usually informal and allow you to work at your own pace and with other adults who also are studying for the GED. In addition to working on specific skills, you will be able to take practice GED tests (like those in this book) in order to check your progress. For information about classes available near you, contact one of the institutions in the list on page 3.

What You Need to Know to Pass Test Five: **Mathematics**

The GED Mathematics Test focuses on the practical use of basic arithmetic, algebra, and geometry. You will be tested on your understanding of how to solve a problem and on your ability to do the math to find a solution.

The test takes 90 minutes and has 56 items. The items are divided into three areas: arithmetic, algebra, and geometry.

Within each area, some of the items require you to solve the problem and others ask you to show how you would solve (set up) the problem. These set-up items test how well you can find the correct approach to solving a problem.

Each area tests your ability to do the basic mathematics operations: addition, subtraction, multiplication, and division. In the set-up items, you will be asked to show which operation(s) you would use to solve a problem.

Basic mathematical concepts are also tested within each of the three areas. Some concepts, such as percent, ratio, and proportion, can be applied in arithmetic, algebra, and geometry.

Another concept tested in all three areas is the use of formulas to solve problems. You will be provided with a formula page when you take the GED Mathematics Test. The formula page includes all the formulas you will need to take the test. For some items, you will use the formulas to find solutions. In other items, you will use a formula to show the correct approach to solving a problem.

The following information summarizes the areas and concepts tested within each section of the GED Mathematics Test.

Arithmetic

Approximately fifty percent of the test items focus on arithmetic. You will be expected to work problems involving whole numbers, fractions, decimals, and percents. Some problems require an understanding of ratio and proportion or probability. There are three topics in the arithmetic section: measurement, number relations, and data analysis.

Measurement items test your ability to use basic math skills in items about length, perimeter, circumference, area, volume, and time. You will need to understand the standard measurement system and be able to make conversions. Some measurement items test less commonly used concepts such as square roots, exponents, and scientific notation.

Number relations items test your ability to compare numbers and draw conclusions. You may be asked to make an estimate or identify the range that the solution would fall within. In each number relations item, you are asked to show how your solution relates to other quantities given in the problem.

Data analysis items test your ability to use information presented in graphs, charts, and tables. You decide which pieces of information you need to solve the problem, locate the information, and work the problem. You may also be asked to find the mean (average) or median of a set of data.

Algebra

Approximately thirty percent of the items on the test focus on algebra. You will be tested on the use of algebraic symbols and expressions. Some items will test your ability to write equations and find the solutions.

Some items will require the use of the formulas page. You will be asked to show how to solve for any variable within a formula. Percent, ratio, and proportion are also applied in the algebra section.

A few items may include powers and roots, factoring, solving inequalities, graphing equations, and finding the slope of a line.

Geometry

Approximately twenty percent of the items on the test focus on geometry. The geometry concepts covered on the test are angles, triangles, quadrilaterals, and indirect measurement. Indirect measurement items require an understanding of similarity, congruence, and the Pythagorean Theorem. You will use the four basic operations to find the values of angles and line segments. The concepts of perimeter, circumference, area, and volume will also be tested and applied in indirect measurement problems.

The Mathematics Test Items

1. Which of the following expressions can be used to find the area of a rectangle that is 8 feet long and 5 feet wide:

   (1) $A = 2(8) - 2(5)$
   (2) $A = 2(8) + 2(5)$
   (3) $A = 8^2$
   (4) $A = \frac{1}{2}(8)(5)$
   (5) $A = (8)(5)$

Answer: **(5) $A = (8)(5)$**

Explanation: This item is an example of an arithmetic item that tests the topic of measurement. To solve this item, you must know which formula to select from the formulas page that accompanies the test. Then you must select the one option that uses the information from the problem correctly in the formula. Since the formula for finding the area of a rectangle is $A = lw$, where l = length and w = width, option (5) $A = (8)(5)$ is the correct answer.

2. The square root of 150 is between which of the following pairs of numbers?

   (1) 10 and 11
   (2) 11 and 12
   (3) 12 and 13
   (4) 13 and 14
   (5) 14 and 15

Answer: **(3) 12 and 13**

Explanation: This item is an example of an arithmetic item that tests the topic of number relationships. This item tests your understanding of square roots. Note that the item is not asking you to find the exact square root of 150, but rather is testing your ability to estimate the square root.

3. Approximately how many times as much money does the Cortez family spend on entertainment than it spends on utilities?

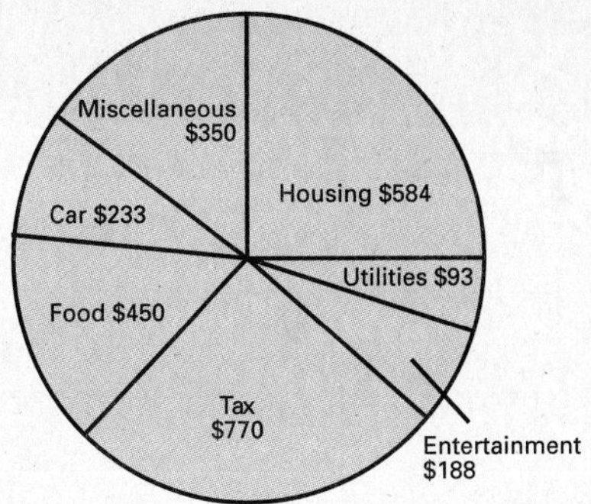

   (1) 2
   (2) 3
   (3) 5
   (4) 6
   (5) 8

Answer: **(1) 2**

Explanation: This is an example of an arithmetic item that tests the topic of data analysis. To find the information you need to solve the item, you must be able to read the circle graph. Then you can compare either the amounts from the graph or sectors on the graph. The entertainment sector is approximately twice as large as the utilities sector: \$188 is approximately twice \$93.

4. If cartons of strawberries are priced at 3 for $1.00, how much would 60 cartons of strawberries cost?

(1) $15
(2) $18
(3) $20
(4) $40
(5) $60

Answer: **(3) $20**

Explanation: This is an example of an algebra item. The item tests your ability to apply the concept of ratio and proportion. To find the correct answer, you must set up and solve an equation.

$$\frac{3}{\$1} = \frac{60}{x}$$

$$3x = \$60$$

$$x = \frac{\$60}{3}$$

$$x = \$20$$

5. If angles 1 and 2 in triangle *ABC* are each 60 degree angles, what is the measure of angle 3?

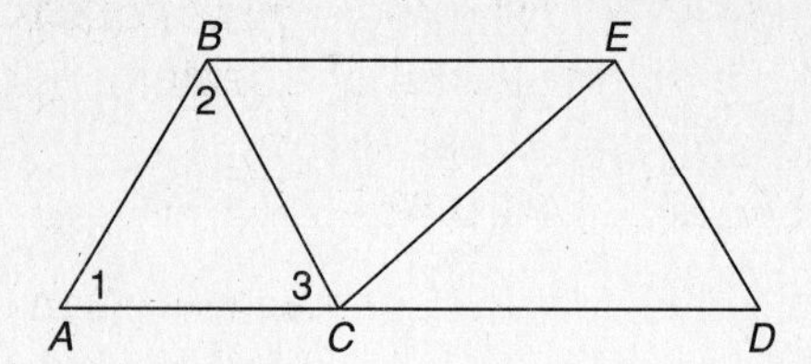

(1) 180 degrees
(2) 120 degrees
(3) 90 degrees
(4) 60 degrees
(5) 45 degrees

Answer: **(4) 60 degrees**

Explanation: This is an example of a geometry item. To answer this item, you need to know that there are 180 degrees in the angles of a triangle. Add the two known angles (60 + 60 = 120) and subtract from 180: 180 − 120 = 60 degrees.

Test-Taking Skills

The GED Test is not the kind of test you can cram for. There are, however, some ways that you can improve your performance on the test.

Answering the Test Items

- Never skim the directions. Read them carefully so that you know exactly what to do. If you are unsure, ask the test-giver if the directions can be explained.

- Read all of the answer options carefully, even if you think you know the right answer. Some of the answers may not seem wrong at first glance, but one answer will always be better than the others.

- Answer all the items. Wrong answers will not be subtracted from your score. If you cannot find the correct answer, reduce the number of possible answers by eliminating all the answers you know are wrong. If you still cannot decide, make your best guess.

- Fill in your answer sheet carefully. To record your answers, mark one numbered space on the answer sheet beside the number that corresponds to the item. Mark only one answer space for each item; multiple answers will be scored as incorrect.

- Remember that the GED is a timed test. When the test begins, write down the time you have to finish. Then keep an eye on the time. Do not take a long time on any one item. Answer each item as best you can and go on. If you are spending a lot of time on one item, skip it. If you finish before time is up, go back to the items you skipped or were unsure of, and give them more thought.

- Don't change an answer unless you are certain your answer was wrong. Usually the first answer you choose is the correct one.

- If you feel you are getting nervous, stop working for a moment. Take a few deep breaths and relax. Then begin working again.

Study Skills

Study Regularly

- If you can, set aside an hour to study every day. If you do not have time every day, set up a schedule of the days you can study. Be sure to pick times when you will be the most relaxed and least likely to be bothered by outside distractions.

- Let others know your study time. Ask them to leave you alone for that period. It helps if you explain to others why this is important.

- You should be relaxed when you study, so find an area that is comfortable for you. If you cannot study at home, go to the library. Most public libraries have areas for reading and studying. If there is a college or university near you, find out if you can use its library. All libraries have dictionaries, encyclopedias, and other resources you can use if you need more information while you're studying.

Organize Your Study Materials

- Be sure to have pens, sharp pencils, and paper for any notes you might want to take.

- Keep all of your books together. If you are taking an adult education class, you probably will be able to borrow some books or other study material.

- Make a notebook or folder for each subject you are studying. Folders with pockets are useful for storing loose papers.

- Keep all of your material in one place so you do not waste time looking for it each time you study.

Read Regularly

- Read the newspaper, read magazines, read books. Read whatever appeals to you—but read! Regular, daily reading is the best way to improve your reading skills.

- Use the library to find material you like to read. Check the magazine section for publications of interest to you. Most libraries subscribe to hundreds of magazines ranging in interest from news to cars to music to sewing to sports. If you are not familiar with the library, ask a librarian for help. Get a library card so that you can check out material to use at home.

Take Notes

- Take notes on things that interest you or things that you think might be useful.
- When you take notes, do not copy the words directly from the book. Restate the information in your own words.
- Take notes any way you want. You do not have to write in full sentences as long as you can understand your notes later.
- Use outlines, charts, or diagrams to help you organize information and make it easier to learn.
- You may want to take notes in a question-and-answer form, such as: *What is the main idea? The main idea is . . .*

Improve Your Vocabulary

- As you read, do not skip a word you do not know. Instead, try to figure out what the word means. First, omit it from the sentence. Read the sentence without the word and try to put another word in its place. Is the meaning of the sentence the same?
- Make a list of unfamiliar words, look them up in the dictionary, and write down the meanings.
- Since a word may have several meanings, it is best to look up the word while you have the passage with you. Then you can try out the different meanings in the context.
- When you read the definition of a word, restate it in your own words. Use the word in a sentence or two.
- Use the Glossary at the end of this book to review the meanings of the key terms. All of the words you see in **boldface** type are defined in the Glossary. In addition, definitions of other important words are included. Use this list to review important vocabulary for the content areas you are studying.

Make a List of Subject Areas that Give You Trouble

- As you go through this book, make a note whenever you do not understand something. Then ask your teacher or another person for help. Later go back and review the topic.

Taking the Test

Before the Test

- If you have never been to the test center, go there the day before the test. If you drive, find out where to park. This way you won't get lost the day of the test.

- Prepare the things you need for the test: your admission ticket (if necessary), acceptable identification, some sharpened No. 2 pencils with erasers, a watch, glasses, a jacket or sweater (in case the room is cold), and a snack to eat during breaks.

- You will do your best work if you are rested and alert. So do not cram before the test. In fact, if you prepared for the test, cramming should be unnecessary. Instead, eat a meal and get a good night's sleep. If the test is early in the morning, set the alarm.

The Day of the Test

- Eat a good breakfast. Wear comfortable clothing. Make sure that you have all of the materials you need.

- Try to arrive at the test center about twenty minutes early. This allows time if, for example, there is a last-minute change of room.

- If you are going to be at the test center all day, you might pack a lunch. If you have to find a restaurant or if you wait a long time to be served, you may be late for the rest of the test.

Using this Book

- Start with the Pretest. It is identical to the real test in format and length. It will give you an idea of what the GED test is like. Then use the Pretest Correlation Chart to figure out your areas of strength and the areas you need to review. The chart will tell you exact units and page numbers to study.

- As you study, use a copy of the Study Record Sheet to keep track of the times and pages you study. Use the GED Cumulative Review and the Performance Analysis chart at the end of each unit to find out if you need to review any lessons before continuing.

- After you complete your review, use the Posttest to decide if you are ready for the test. The Correlation Chart will tell you if you need additional review. Then use the Simulated Test and its Correlation Chart as a final check.

MATHEMATICS

Directions

The Mathematics Pretest consists of multiple-choice questions intended to measure your general mathematical skills and problem-solving ability. The questions are based on short readings which often include a graph, chart, or diagram.

You should spend no more than 90 minutes answering the 56 questions. Work carefully, but do not spend too much time on any one question. Do not skip any items. Make a reasonable guess when you are not sure of an answer. You will not be penalized for incorrect answers. When time is up, mark the last item you finished. This will tell you whether you can finish the real GED test in the time allowed. Then complete the test.

Formulas you may need are given on page 427. Only some of the questions will require you to use a formula. Not all the formulas given will be needed.

Some questions contain more information than you will need to solve the problem. Other questions do not give enough information to solve the problem. If the question does not give enough information to solve the problem, the correct answer choice is "Not enough information is given."

The use of calculators is not allowed.

Record your answers on a copy of the separate answer sheet provided on page 428. Be sure all required information is properly recorded on the answer sheet.

To record your answers, mark the numbered space on the answer sheet that corresponds to the answer you choose for each question on the test.

Example: If a grocery bill totaling $15.75 is paid with a $20.00 bill, how much change should be returned?

(1) $5.26
(2) $4.75
(3) $4.25
(4) $3.75
(5) $3.25

① ② ● ④ ⑤

The correct answer is $4.25; therefore, answer space 3 would be marked on the answer sheet.

Do not rest the point of your pencil on the answer sheet while you are considering your answer. Make no stray or unnecessary marks. If you change an answer, erase your first mark completely. Mark only one answer for each question; multiple answers will be scored as incorrect. Do not fold or crease your answer sheet.

When you finish the test, use the Correlation Chart on page 24 to determine whether you are ready to take the real GED Test, and, if not, which skill areas need additional review.

Adapted with permission of the American Council on Education.

Directions: Choose the best answer to each item.

1. Kevin assembles 35 valves in one hour and works 8 hours per day. Which expression represents how many valves he can assemble in 10 days?

   (1) $35(8) + 10(8)$
   (2) $35(8)(10)$
   (3) $\frac{35(10)}{8}$
   (4) $\frac{35(8)}{10}$
   (5) Not enough information is given.

2. The Mataeles pay 5.4 cents for each kilowatt-hour of electricity they use in their home. Their electric bill for one month is $40. About how many kilowatt-hours of electricity did the Mataeles use during the month?

   (1) fewer than 650
   (2) between 650 and 720
   (3) between 720 and 760
   (4) between 760 and 800
   (5) more than 800

3. Which of the following represents 273,815 written in scientific notation?

   (1) 2.73815×10^5
   (2) 2.73815×10^6
   (3) 27.3815×10^6
   (4) 273.815×10^5
   (5) 2738.15×10^3

4. Joel is competing in the City Bowling Tournament. In one round, he bowled games of 230, 212, and 257. Which expression represents Joel's mean score (average) for the round?

   (1) $\frac{230 + 212 + 257}{3}$
   (2) $3(230 + 212 + 257)$
   (3) $\frac{(230)(212)(257)}{3}$
   (4) $\frac{3(230 + 212)}{257}$
   (5) Not enough information is given.

Item 5 refers to the following information.

Computer Games	Price
A. Sky Trainer	$29.00
B. Castlequest	$21.95
C. Wordplay	$24.50
D. Webworks	$53.80
E. Flame Arrow	$13.95
F. Knighthawk	$22.45

5. Ky has $90 to spend on computer games. He has already chosen two games: Dictionary ($15.25) and Alphabet ($34.95). Which combination of two or more games from the list above can he also buy without spending more than a total of $90?

   (1) A and E
   (2) B and E
   (3) B and C
   (4) B and F
   (5) D only

50.20

Item 6 refers to the following diagram.

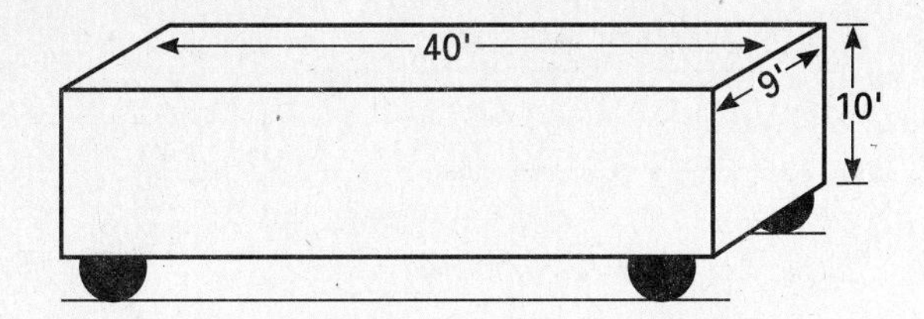

6. The inside dimensions of a rectangular freight car are shown in the diagram. What is the volume of the freight car in cubic feet?

(1) 1,800
(2) 2,400
(3) 3,600
(4) 18,000
(5) Not enough information is given.

Item 7 refers to the following table.

PAYROLL DEDUCTIONS FOR JUNE	
Federal Tax	$280.77
State Tax	$124.33
FICA	$206.82
Retirement	$36.18
Other	$87.12

7. The table shows the amounts subtracted from Masao's monthly earnings. His total earnings for June are $2,759. Approximately, what percent of his earnings are deducted for state taxes?

(1) between 3 and 4 percent
(2) between 4 and 5 percent
(3) between 5 and 6 percent
(4) between 6 and 7 percent
(5) Not enough information is given.

Item 8 refers to the following figures.

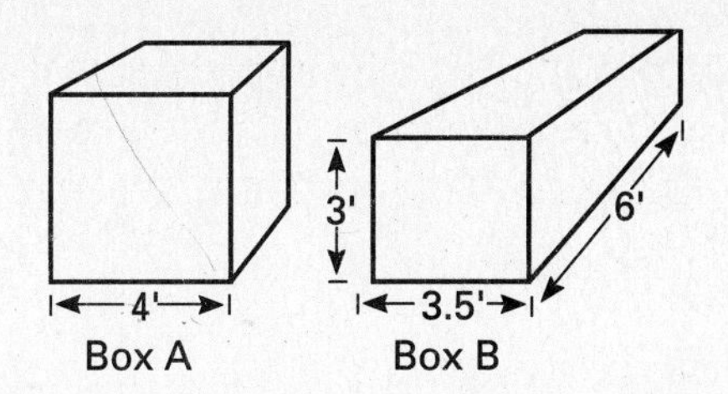

8. Box *A* is a cube. Box *B* is a rectangle. Which statement is true about the boxes?

(1) The volume of Box *A* is one cubic foot smaller than the volume of *B*.
(2) The volume of Box *B* is one cubic foot smaller than the volume of *A*.
(3) The volume of Box *A* is twice as much as Box *B*.
(4) The volume of Box *B* is twice as much as Box *A*.
(5) The volume of the boxes is equal.

Item 9 refers to the following drawing.

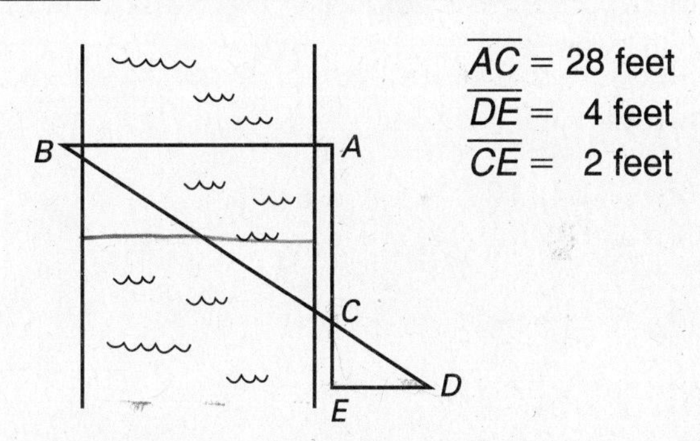

9. Lines *AB* and *DE* are parallel. What is the approximate width in feet of the river *(AB)*?

(1) 20
(2) 32
(3) 40
(4) 56
(5) Not enough information is given.

10. It takes $3\frac{3}{8}$ yards of fabric to make a pillowcase. How many pillowcases can you make from $13\frac{1}{2}$ yards of fabric?

(1) 3
(2) 4
(3) 5
(4) 6
(5) Not enough information is given.

Item 11 refers to the diagram below.

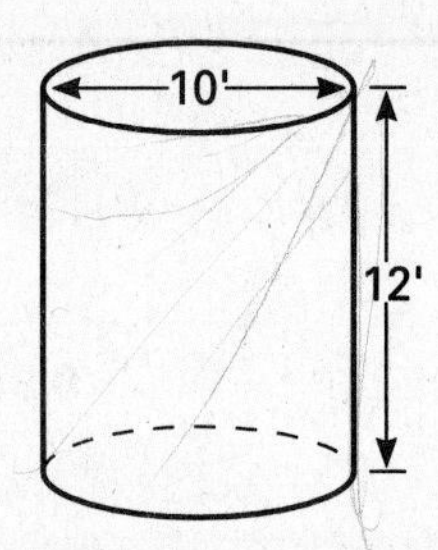

11. A cylindrical water tank is 10 feet across and 12 feet tall. Which expression shows the approximate volume of the water tank in cubic feet?

(1) $(3.14)(5)^2(12)$
(2) $(3.14)(10)^2(12)$
(3) $(3.14)(5)(12)^2$
(4) $\frac{(3.14)(12)}{5^2}$
(5) $(3.14)(10)(12)^2$

12. Out of every 400 television sets produced at a factory, 8 are defective. What is the probability that a television will be defective?

(1) $\frac{1}{8}$
(2) $\frac{1}{40}$
(3) $\frac{1}{50}$
(4) $\frac{1}{80}$
(5) $\frac{1}{150}$

Items 13 and 14 refer to the following diagram.

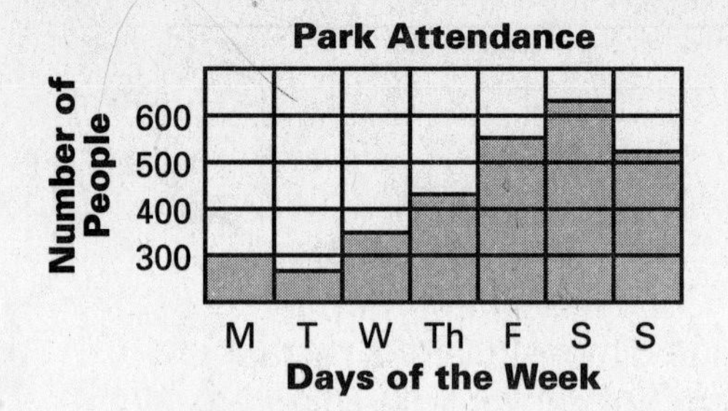

13. Approximately how many more people visited the park on Saturday than on Monday?

(1) 150–200
(2) 200–250
(3) 250–300
(4) 300–350
(5) Not enough information is given.

14. The park rules require that one security guard be on duty for every 50 people who visit the park. How many guards should be on duty Friday?

(1) 5
(2) 11
(3) 15
(4) 18
(5) 24

15. Ann is cooking for a soup kitchen. She uses $1\frac{1}{2}$ teaspoons of seasoning to make one quart of soup. If she has 12 teaspoons of seasoning on hand, how many quarts of soup can she make?

(1) 6
(2) 8
(3) 9
(4) 10
(5) 12

Item 16 refers to the following diagram.

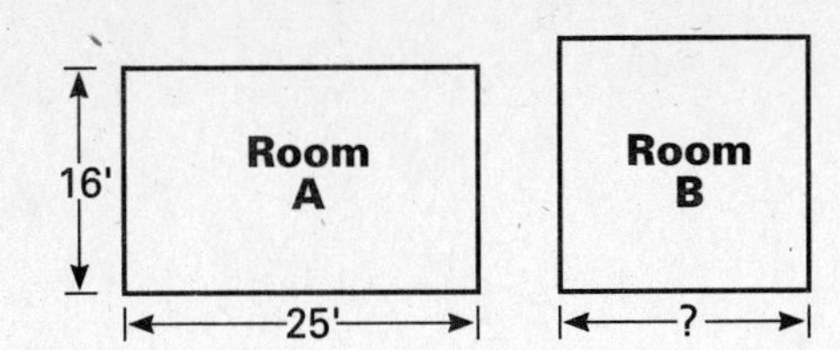

16. Room A and Room B have the same area. If Room B is square, what is the length, in feet, of each of its sides?

    (1) 16
    (2) 20
    (3) 24
    (4) 30
    (5) Not enough information is given.

17. A doctor tells a nurse to give a patient 5 grains of an allergy drug. Each allergy tablet contains $1\frac{1}{4}$ grains of the drug. How many allergy tablets should the nurse give the patient?

    (1) $1\frac{1}{2}$
    (2) $2\frac{1}{4}$
    (3) 3
    (4) $3\frac{1}{2}$
    (5) 4

18. Burger World opens at 10 A.M. and serves an average of 150 customers per hour. At this rate, how many customers will the restaurant serve by 3 P.M.?

    (1) 600
    (2) 750
    (3) 825
    (4) 975
    (5) Not enough information is given.

Item 19 is based on the following figure.

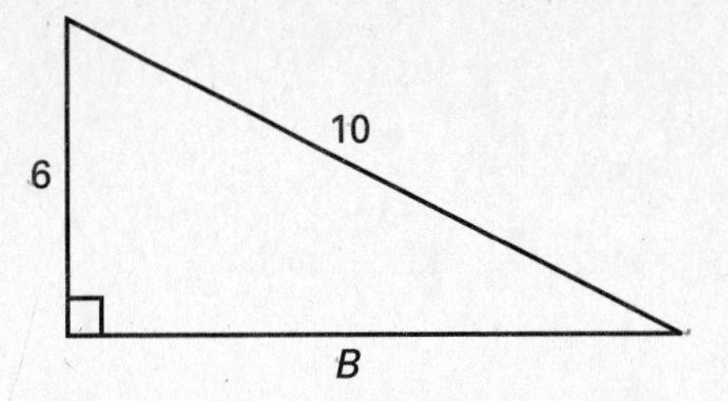

19. According to the Pythagorean Theorem, which statement is true for the triangle shown?

    (1) $6^2 + B^2 = 10^2$
    (2) $6^2 - B^2 = 10^2$
    (3) $6^2 - 10^2 = B^2$
    (4) $6^2 + 10^2 = B^2$
    (5) $6^2 + 10^2 = B$

Item 20 is based on the following figure.

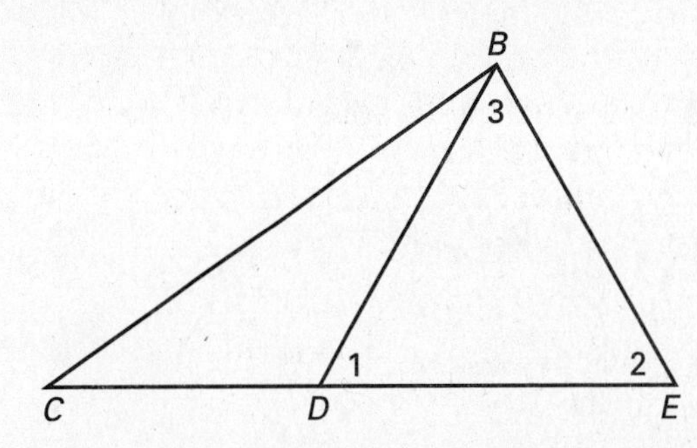

20. If angles 1 and 2 in triangle BDE are each 60-degree angles, what is the measure of angle 3?

    (1) 180 degrees
    (2) 120 degrees
    (3) 90 degrees
    (4) 60 degrees
    (5) 40 degrees

21. The square root of 125 is between which of the following pairs of numbers?

(1) 8 and 9
(2) 9 and 10
(3) 10 and 11
(4) 11 and 12
(5) 12 and 13

22. On a map, two cities are $5\frac{1}{2}$ inches apart. The map has a scale of 1 inch = 20 miles. What is the actual distance, in miles, between the two cities?

(1) 90
(2) 110
(3) 180
(4) 250
(5) 520

Item 23 is based on the following graph.

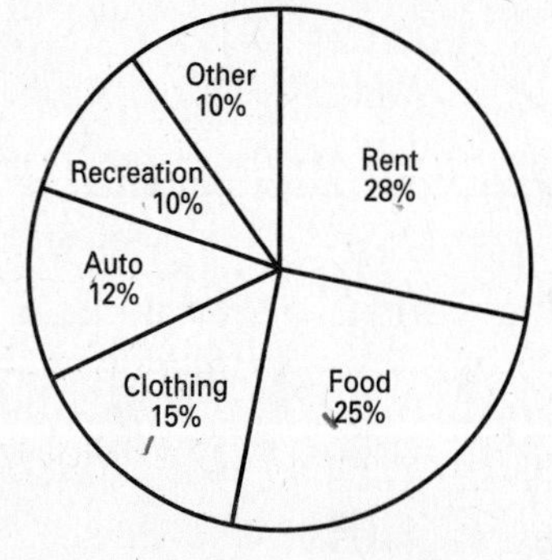

23. The graph shows the Enriquez family budget. According to the graph, what percent of their income is left after rent and food expenses?

(1) 8%
(2) 47%
(3) 53%
(4) 72%
(5) 112%

Items 24 and 25 refer to the following figure.

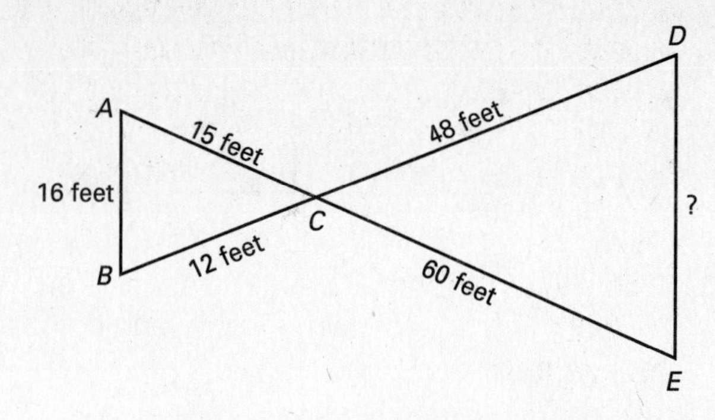

24. Triangles *ABC* and *CDE* are similar. Which side of triangle *ABC* corresponds to side $\overline{CE}$?

(1) side $\overline{BC}$
(2) side $\overline{AC}$
(3) side $\overline{CD}$
(4) side $\overline{AB}$
(5) Not enough information is given.

25. What is the length in feet of side $\overline{DE}$ in the figure?

(1) 49
(2) 52
(3) 64
(4) 88
(5) 120

26. Karen has a picture that is 24 inches wide and 36 inches long. She wants to reduce the picture on a copy machine so the width will be 16 inches. Which expression represents the new length, in inches, of the reduced picture?

(1) $\frac{24 + 36}{16}$
(2) $\frac{(16)(36)}{24}$
(3) (24)(16)(36)
(4) $\frac{(36 - 24)}{16}$
(5) $\frac{(36 - 16)}{24}$

27. Alicia borrowed $900 for two years. She paid $252 in interest. Which expression represents the annual rate of interest she was charged?

(1) $\frac{252}{900(2)}$

(2) $\frac{252(2)}{900}$

(3) $\frac{252(900)}{2}$

(4) $252(900)(2)$

(5) $\frac{(900)(2)}{252}$

28. The price of a gallon of gasoline *(x)* has tripled since 1970. If the current price is $1.36, which equation could be used to find the price of a gallon of gasoline in 1970?

(1) $\frac{x}{3} = \$1.36$

(2) $x - 3 = \$1.36$

(3) $3x = \$1.36$

(4) $x + 3 = \$1.36$

(5) $3(x + 3) = \$1.36$

29. Given the formula $c = 2a^2(b - 4)$, find c if $a = 3$ and $b = 6$.

(1) 10

(2) 12

(3) 15

(4) 24

(5) 36

30. The Perrones made a 25 percent down payment on a new car costing $11,960. How much do they still owe?

(1) $ 299

(2) $ 897

(3) $2,990

(4) $3,289

(5) $8,970

Item 31 refers to the following drawing.

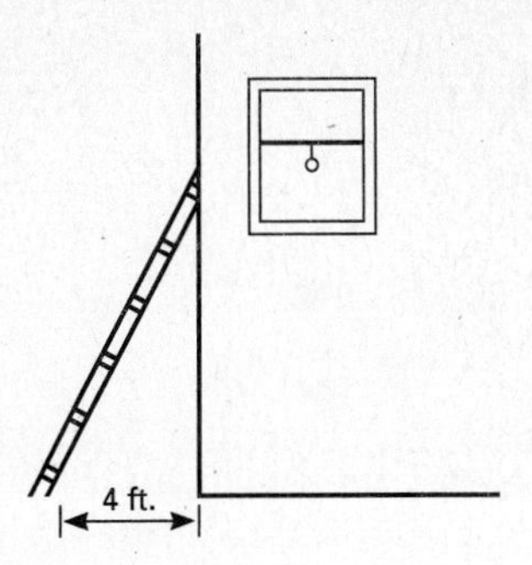

31. Approximately how high up on the wall does the 12-foot ladder in the drawing reach if the bottom of the ladder is 4 feet from the wall?

(1) between 8 and 9 feet

(2) between 9 and 10 feet

(3) between 10 and 11 feet

(4) between 11 and 12 feet

(5) Not enough information is given.

Item 32 refers to the following drawing.

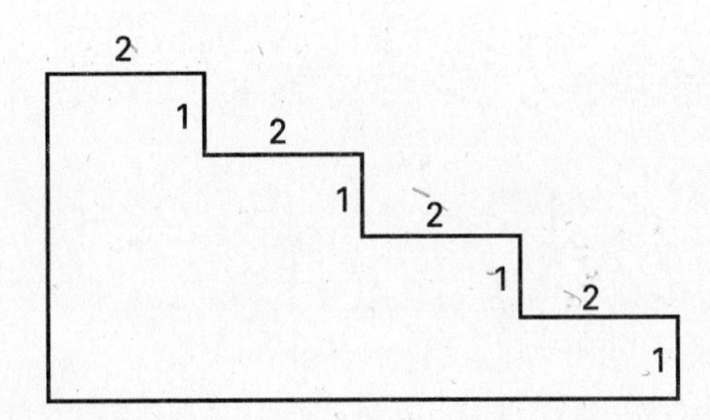

32. What is the perimeter of the figure?

(1) 10

(2) 12

(3) 20

(4) 24

(5) 30

33. Radio station KPRQ broadcasts at a frequency of 21.0×10^5 cycles per second. Station WRLO broadcasts at 4.2×10^6 cycles per second. Which statement is true?

(1) KPRQ's frequency is 5 times higher than WRLO's frequency.
(2) WRLO's frequency is 5 times higher than KPRQ's frequency.
(3) KPRQ's frequency is 2 times higher than WRLO's frequency.
(4) WRLO's frequency is 2 times higher than KPRQ's frequency.
(5) KPRQ's frequency is 14.8 times higher than WRLO's frequency.

34. Carolyn is planning a trip from San Francisco to Kansas City, a distance of 1,860 miles. What is the average (mean) number of miles Carolyn must drive each day to complete the trip in 5 days?

(1) 304
(2) 310
(3) 372
(4) 426
(5) 460

35. A city planner draws a circle with a 4-inch diameter on a map to show the area of the city that would suffer the most damage during an earthquake. Which expression represents the area in square miles of the city shown by the circle?

(1) $(3.14)(4)^2$
(2) $(3.14)(2)^2$
(3) $(3.14)(8)$
(4) $(3.14)^2(4)$
(5) Not enough information is given.

36. Specialty Graphics uses 3 boxes of copier paper every 5 days. Which equation could be used to find how many days *(d)* it will take the company to use 18 boxes of paper?

(1) $5(18) = 3d$
(2) $\frac{3(5)}{18} = d$
(3) $5(3d) = 18$
(4) $\frac{3(18)}{d} = 5$
(5) Not enough information is given.

37. A baseball team won 18 of the first 40 games it played. At the same rate, how many games will the team win during a 160-game season?

(1) 60
(2) 72
(3) 108
(4) 121
(5) Not enough information is given.

38. Kira is selling magazines door-to-door to raise money for a charity group. If 65% of the money she collects goes to the magazine company, how much will the charity get to keep if Kira collects $180?

(1) $ 17.00
(2) $ 63.00
(3) $ 94.00
(4) $117.00
(5) $132.00

Item 39 refers to the following number line.

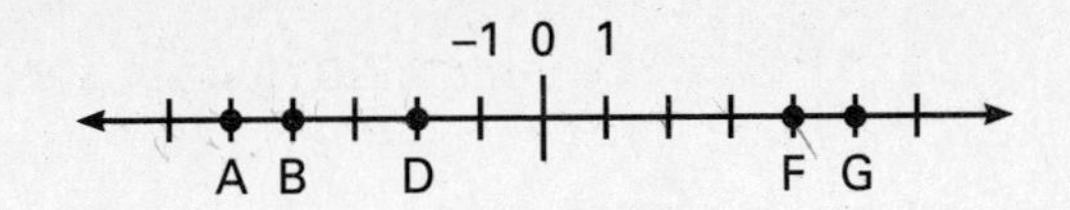

39. Which letter on the number line represents 4?

    (1) A
    (2) B
    (3) D
    (4) F
    (5) G

40. On Thursdays, Lynn is paid a 3 percent bonus on sales for the day. Last Thursday he sold three software programs priced at $725, $385, and $260. What was the amount of his bonus for that day?

    (1) $ 21.75
    (2) $ 33.30
    (3) $ 41.10
    (4) $ 645.00
    (5) $1,370.00

41. In 1995, 14,000 students enrolled at State University. If the student population increases at the rate of 1,500 students per year, in what year will the university reach its goal of 20,000 students?

    (1) 1996
    (2) 1997
    (3) 1998
    (4) 1999
    (5) 2000

42. John earned $14,500 this year as a receptionist. After receiving a raise, he will earn $15,080 next year. What percent raise did he get?

    (1) 2%
    (2) 3%
    (3) 4%
    (4) 5%
    (5) Not enough information is given.

Item 43 is based on the following figure.

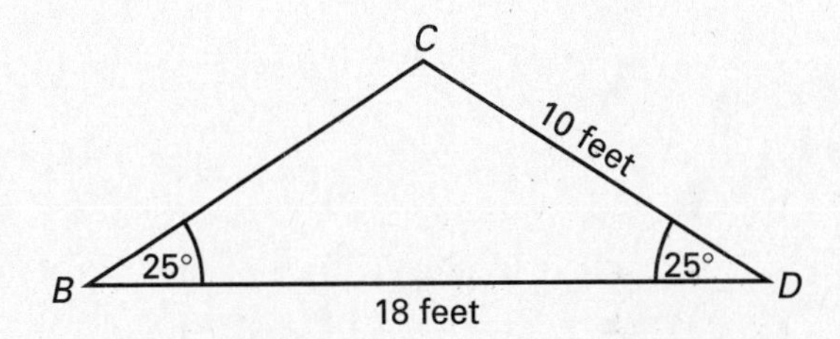

43. If triangle *BCD* is isosceles, what is the perimeter (in feet) of the triangle?

    (1) 10
    (2) 18
    (3) 29
    (4) 38
    (5) 46

44. A telephone pole casts a shadow of 36 feet, while a nearby 3-foot post casts a shadow of 4 feet. What is the height in feet of the telephone pole?

    (1) 108
    (2) 76
    (3) 58
    (4) 39
    (5) 27

45. 45% of the employees at La Plata Engineering responded to a company survey. If 126 employees responded, how many employees does the company have?

(1) 160
(2) 188
(3) 210
(4) 255
(5) 280

46. Ned takes his car in for repairs. The costs for the repairs are $875 for labor and $1,400 for new parts. Used parts would cost 60 percent less than new parts. What will be the total cost of the repairs if Ned asks for used parts?

(1) $ 840
(2) $1,435
(3) $1,715
(4) $1,750
(5) $1,925

47. What is the sum of $4x + y + 2x + y$?

(1) $4x + 2y$
(2) $6x + 2y$
(3) $x^4 + y^2$
(4) $x^6 + y^2$
(5) $8xy$

48. In an apartment complex, 85% of the 400 residents voted for a proposal to increase parking fees. How many residents voted against the proposal?

(1) 16
(2) 30
(3) 60
(4) 80
(5) 96

Item 49 refers to the following figure.

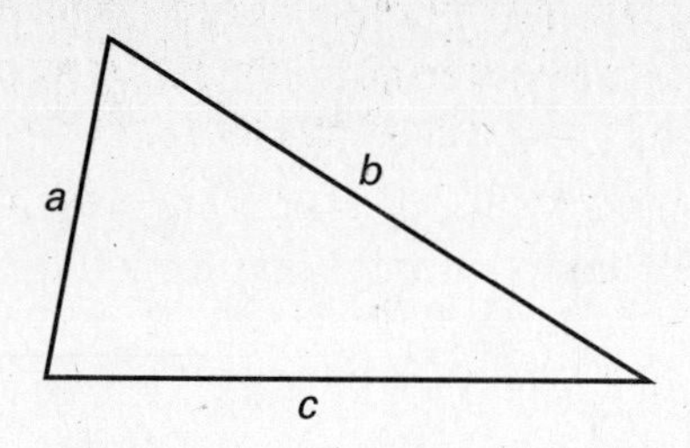

49. The perimeter of the triangle shown is 86 feet. What is the length of a if $b = 32$ feet and $c = 36$ feet?

(1) 12
(2) 16
(3) 18
(4) 24
(5) 30

50. Carmen bought a toaster oven. The store took $10.00 off the price because the cover was scratched. She also mailed in a coupon to receive a $5.50 rebate from the manufacturer. If p represents the original price, which expression represents the cost of the product to her?

(1) $p + \$10.00 + \$5.50$
(2) $p(\$10.00 - \$5.50)$
(3) $p + \$10.00 - \$5.50$
(4) $p - \$10.00 - \$5.50$
(5) $p - \$10.00 + \$5.50$

51. Evaluate $3(n^2 - z^4)$, if $n = 5$ and $z = 2$.

(1) 6
(2) 27
(3) 35
(4) 64
(5) 108

52. Denise earns $9.50 for each hour of overtime she works. How much overtime pay did she earn during the week if she worked 5 hours of overtime on Monday, 3 hours on Wednesday, and $1\frac{1}{2}$ hours on Friday?

(1) $70.75
(2) $75.00
(3) $80.75
(4) $85.50
(5) $90.25

53. The 48 employees of the Young Construction Company were given a choice of two different retirement plans. If Plan A was chosen by twice as many employees as Plan B, how many employees chose Plan A?

(1) 8
(2) 16
(3) 25
(4) 32
(5) 44

Item 54 refers to the following diagram.

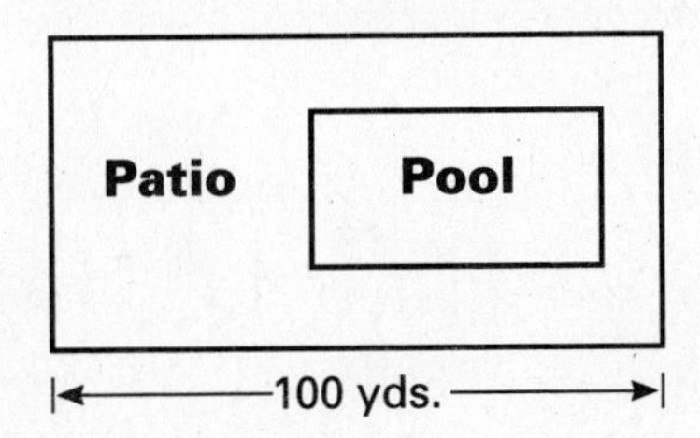

54. A fenced area encloses a pool with a patio. If the area of the patio is 4,500 square yards, and the area of the pool is 1,500 square yards, what is the width of the fenced area?

(1) 100 yards
(2) 60 yards
(3) 45 yards
(4) 30 yards
(5) Not enough information is given.

Item 55 is based on the following figure.

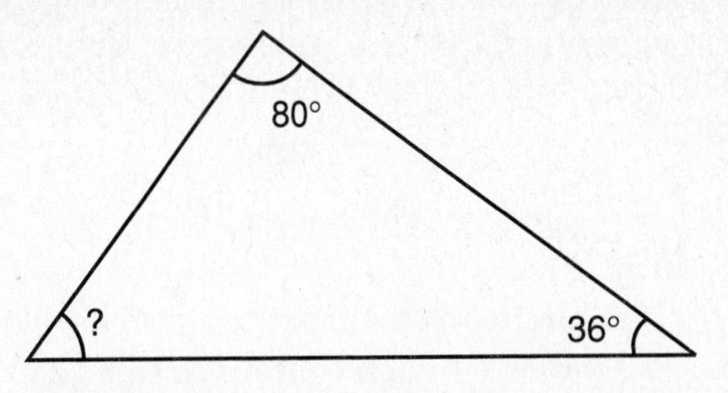

55. The triangle has three internal angles. One angle measures 36 degrees; the other measures 80 degrees. How many degrees does the third angle measure?

(1) 34°
(2) 64°
(3) 116°
(4) 244°
(5) Not enough information is given.

Item 56 is based on the following graph.

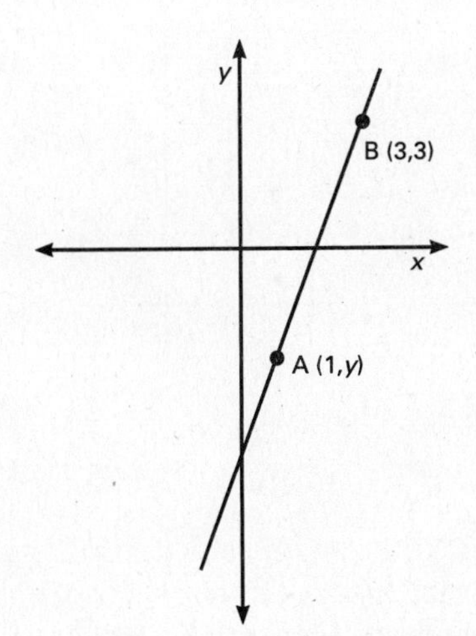

56. In the graph, what is the y-coordinate of point A if the slope of line AB is 3?

(1) 12
(2) 9
(3) 5
(4) −3
(5) −5

Answers are on page 333.

Pretest Correlation Chart: Mathematics

Name: ______________________ **Class:** ____________ **Date:** ____________

This chart can help you determine your strengths and weaknesses on the content and skill areas of the Mathematics GED Test. Use the Answer Key on pages 333–417 to check your answers to the test. Then circle on the chart the numbers of the test items you answered correctly. Put the total number correct for each content area and skill area in each row and column. Look at the total items correct in each column and row and decide which areas are difficult for you. Use the page references to study those areas. Use a copy of the Study Record Sheet on page 25 to guide your studying.

Content	Item Number	Total Correct	Page Ref.
Arithmetic *(pages 26–193)*			
Whole Number Operations	1, 18, 41	_____out of 3	26–57
Fractions	10, 15, 17	_____out of 3	58–87
Decimals	2, 5, 52	_____out of 3	88–113
Percent	7, 27, 30, 38, 40, 42, 45, 46, 48	_____out of 9	114–145
Ratio, Proportion, Mean, Median, Probability	4, 12, 22, 26, 34, 37, 53	_____out of 7	146–165
Graphs and Charts	13, 14, 23	_____out of 3	177–193
Algebra *(pages 194–241)*			
Equations	28, 29, 36, 39, 47, 50, 51	_____out of 7	194–211
Graphs, Slope, Line	56	_____out of 1	212–223
Powers and Roots	3, 21, 33	_____out of 3	224–233
Geometry *(pages 242–309)*			
Perimeter, Area, Volume	6, 8, 11, 16, 32, 35, 49, 54	_____out of 8	244–251 284–297
Angles	20, 55	_____out of 2	252–259
Triangles and Quadrilaterals	43	_____out of 1	260–265
Congruence and Similarity	9, 24, 25, 44	_____out of 4	266–275
Pythagorean Theorem	19, 31	_____out of 2	276–281

For additional help, see the Steck-Vaughn GED Mathematics Exercise Book.

Study Record Sheet

Name: ______________________ **Class:** ____________ **Date:** ____________

Use this chart to help you track your studies after you take the Pretest, Posttest, or Simulated Test. After each test, use the Mathematics Correlation Charts on pages 24, 320, and 332 to help you figure out the areas you need to study. Then, make a copy of the chart for you to complete. In the column on the left, write the content and skill areas you want to study. Then, after each session, write the date and the pages you studied. Use the sheet to review your study habits from time to time. Are you studying regularly? Are you reviewing the material you need to cover? Do you need to schedule more frequent sessions?

Area: ________ **Pages:** ________	Pages Date	Pages Date	Pages Date
Area: ________ **Pages:** ________	Pages Date	Pages Date	Pages Date
Area: ________ **Pages:** ________	Pages Date	Pages Date	Pages Date
Area: ________ **Pages:** ________	Pages Date	Pages Date	Pages Date
Area: ________ **Pages:** ________	Pages Date	Pages Date	Pages Date
Area: ________ **Pages:** ________	Pages Date	Pages Date	Pages Date

ARITHMETIC

Arithmetic is an important part of daily life.

In this part of the book you will learn how to use arithmetic facts to solve a variety of problems. Most problems involving **measurement, number relationships,** and the **analysis of data** can be solved with arithmetic. In the first part of this book you will sharpen your arithmetic skills and apply them in practical word problems. You will also learn how to estimate to narrow your options or check your answers.

place value
the value of a number determined by its position. The 5 in $589 has a value of 500; the 5 in 0.05 has a value of $\frac{5}{100}$

In lesson 1 you will learn about **place value** and how to use it to read, write, compare, and order numbers. In lesson 2 you will learn how to perform the four basic arithmetic operations: adding, subtracting, multiplying, and dividing.

A special section in each lesson titled "Strategies for Solving Word Problems" will help you know when to add, subtract, multiply, and divide. Lessons 3 and 4 discuss word problem strategies thoroughly. The concepts in these lessons will be built upon throughout the book.

fraction
part of a whole; examples are $\frac{1}{4}$, $\frac{1}{2}$, $\frac{3}{8}$

Lessons 5 through 8 cover the subject of **fractions.** First you will understand the concept of fractions. Then you will learn to perform the four arithmetic operations with fractions. Finally, you will apply the operations to practical problems to test your understanding.

Decimals are another way to write fractions. Perhaps the most common use of decimals is our system of dollars and cents. When you give a clerk $2.00 to pay for a $1.59 hamburger, you are comparing decimals. When you count your change, you are using basic operations. You will learn more about decimals in lessons 9 and 10. In lesson 11 you will learn how decimals and common fractions relate.

Percent is still another way we express a fraction. Just as 50 cents or $0.50 is one-half of a dollar, 50% is one-half of a whole. The whole or 100% can represent one dollar, one million dollars, a tank of gas, or a sales quota. Regardless, 50% means one-half.

In lessons 12 through 14 you will learn how to use percent in practical situations. Figuring out how much interest you will pay on a loan or how much you will earn on your savings account are two applications covered in lesson 14. You will also learn the basic parts of every percent problem and how to solve for a missing part.

Lesson 15 covers a variety of related topics: **ratio and proportion,** finding an average, and **probability.** These skills help you compare information and make decisions. Should you play the lottery? Is Brand X cheaper per ounce than Brand Y? If there is a 30% chance of rain, should you cancel your plans? These everyday decisions can be made more confidently using the skills covered in lesson 15.

Lessons 16 and 17 cover measurement concepts. You will learn about two common measurement systems: the standard and the metric systems. You will learn how to solve problems about distance, weight, time, and volume. You will also learn to find the **perimeter, area,** and **circumference** of basic shapes.

Lesson 18 shows you how to understand basic graphs, charts, and tables. A great deal of information can be presented on a graph. Deciding what information you need to solve a problem is an important skill. Lesson 18 will help you focus on the information you need.

Throughout this part of the book, you will be introduced to many useful problem-solving strategies. Using estimation and mental math are two strategies that can save you time as you work. Identifying needed information and weeding out unnecessary information are two other very important skills. Combine these strategies with your arithmetic skills and you will greatly improve your mathematic abilities.

decimal
a fraction that uses a place value system to show a part of 10, 100, 1000, and so on; examples are 3.25 and 0.9

percent
a fraction system based on the idea that there are 100 parts in a whole; examples are 75%, 10%, and so on

ratio
a way of comparing two numbers using division: 3 out of 4 dentists recommend.... A ratio can also be written 3:4 or $\frac{3}{4}$

probability
determining how likely it is that an event will happen

perimeter
the distance around a geometric figure

area
a measure of the surface of any flat object

circumference
the perimeter of a circle

SEE ALSO: Steck-Vaughn GED Mathematics Exercise Book, Unit 1: Arithmetic

Number Values and Facts

Place Value

Whole numbers are written using ten **digits: 0, 1, 2, 3, 4, 5, 6, 7, 8,** and **9.** The number **3,820** has four digits. **1,000,000** is a seven-digit number, although six of the digits are the same.

hundred billions	ten billions	billions	hundred millions	ten millions	millions	hundred thousands	ten thousands	thousands	hundreds	tens	ones
							3	0,	3	9	6

The **place value** of a digit is found by its position in a number. The chart shows the names of the first 12 whole number places. Commas are placed every three digits counting from the right. The number **30,396** is on the chart.

Each digit has a value determined by its place value. The sum of the values of the digits in each place equals the total value of the number.

Example:
What is the value of each digit in **215,073?**

2 is in the **hundred thousands** place. $2 \times 100{,}000 = \mathbf{200{,}000}$
1 is in the **ten thousands** place. $1 \times 10{,}000 = \mathbf{10{,}000}$
5 is in the **thousands** place. $5 \times 1{,}000 = \mathbf{5{,}000}$
0 is in the **hundreds** place. $0 \times 100 = \mathbf{0}$
7 is in the **tens** place. $7 \times 10 = \mathbf{70}$
3 is in the **ones** place. $3 \times 1 = \mathbf{3}$

Reading and Writing Whole Numbers

Numbers can also be expressed in words. The number **16,000,000** can be written **sixteen million.** It can also be written using numbers and words; **16 million.**

Numbers are read from left to right. To read a large number, read the first group of digits before the comma. Each group of digits is called a **period.** Then say the name of the period the comma represents. To read 984,356, say the digits before the comma, "nine hundred eighty-four," then say the period name, "thousand." Continue with the next period. Do not say the name of the **ones** period after the final period: "three hundred fifty-six". Also, do not say the word "and."

Zeros are sometimes used to fill up a three-place period, but they are not read. **062** is read "sixty-two." **008** is read "eight."

Example:
What is the word form of **13,023?**
Read the first period as you normally would, then say "thousand."
Read the final period as you normally would.
"thirteen thousand, twenty-three"

Comparing Numbers

Sometimes you need to compare numbers in order to find the highest or the lowest number in a group. Your knowledge of place value will help you when you compare numbers.

$=$	**equals**
$>$	**is greater than**
$<$	**is less than**

Use these three symbols to compare numbers:

$140 = 140$	One hundred forty **equals** one hundred forty.
$25 > 23$	Twenty-five **is greater than** twenty-three.
$4 < 5$	Four **is less than** five.

Think of the "greater than" and "less than" symbols as arrows that *always point to the smaller number.*

To compare numbers, follow these rules:

Rule 1: The whole number with the most digits is greater.
Example: $7{,}235 > 848$

Rule 2: For numbers with the same number of digits, compare digits from left to right.
Examples: $6{,}746 > 6{,}699$ because $700 > 600$
$588{,}210 < 589{,}429$ because $8{,}000 < 9{,}000$

Rounding

Rounding makes numbers easier to remember. It also makes difficult calculations easier. A number can be rounded when the exact amount is not needed.

Use these steps to round whole numbers:

Step 1: Find the digit you want to round to. It may help to circle this digit. **2④,902**
Step 2: Look at the digit immediately to the right of the circled digit. **2④,902**
Step 3: If the digit to the right is 5 or more, add 1 to the circled digit. If the digit to the right is less than 5, do not change the circled digit. Change all digits to the right of the circled digit to zeros. 2⑤,000

Examples: Round 3,499 to the nearest thousand—
③,499 rounds to 3,000
Round 125,193 to the nearest hundred—
125,①93 rounds to 125,200
Round 19,930 to the nearest thousand—
1⑨,930 rounds to 20,000

Lesson 1

Directions: Write the value of the underlined digit in words.

Example: $\underline{6}27$ six hundred

1. $5,\underline{5}17$ __________________
2. $\underline{3},742,691$ __________________
3. $2\underline{6},154$ __________________
4. $1\underline{8},063,500,000$ __________________
5. $4,\underline{7}00,510$ __________________
6. $9\underline{6}4,211$ __________________
7. $1\underline{8}6,502,110$ __________________

8. What is the value of each digit in **679,308**?

6 __________________

7 __________________

9 __________________

3 __________________

0 __________________

8 __________________

Directions: Match each number in Column A with its word form in Column B.

Column A

________ 9. 8,416

________ 10. 8,420,106

________ 11. 84,200,160

________ 12. 842,016

Column B

a. Eighty-four million, two hundred thousand, one hundred sixty
b. Eight thousand, four hundred sixteen
c. Eight hundred forty-two thousand, sixteen
d. Eight million, four hundred twenty thousand, one hundred six

Column A

________ 13. 10,250,900

________ 14. 12,509

________ 15. 1,259

________ 16. 125,090

Column B

a. One hundred twenty-five thousand, ninety
b. Twelve thousand, five hundred nine
c. Ten million, two hundred fifty thousand, nine hundred
d. One thousand, two hundred fifty-nine

Directions: Compare the following numbers. Write >, <, or = between the numbers.

17. 1,305   1,503
18. 34,000   29,989
19. 102,667   102,657
20. 5,690,185   5,690,185,100
21. 7,650,300   7,649,950
22. 875,438   875,438
23. 3,492,012,558   3,492,012,558
24. 75,390,000   75,391,540
25. 9,500,000   9,500,000,000
26. 45,100   45,099
27. 7,456,795   7,500,000
28. 319,002,110   319,002,110

29. Round **8,621** to the hundreds place. ____________

30. Round **675,924** to the nearest hundred thousand. ____________

31. Round **5,099,620** to the nearest million. ____________

32. Round **49,962,750** to the millions place. ____________

33. Round **10,562** to the nearest thousand. ____________

34. Round **46,055** to the nearest ten thousand. ____________

35. Round **64,000,000** to the nearest ten million. ____________

36. Round **73,895** to the nearest thousand. ____________

Items 37 to 41 refer to the following chart.

Daily Sales Totals Week Ending March 4	
Monday	$18,756
Tuesday	12,316
Wednesday	13,940
Thursday	13,772
Friday	21,592
Saturday	28,795

37. The manager of the store must report his daily sales to his supervisor. The supervisor wants the sales reported to the nearest thousand dollars. What amount will the manager report for each day shown?

Monday ____________

Tuesday ____________

Wednesday ____________

Thursday ____________

Friday ____________

Saturday ____________

38. Which day had the greater sales, Wednesday or Thursday? ____________

39. Which day of the week had the lowest sales? ____________

40. Which day had the highest sales? ____________

41. Arrange the sales on the chart in order from lowest to highest.

____

____

____

Answers are on page 335.

Strategies for Solving Word Problems

Choose the Operation

Many problems are written out in words. Most word problems can be solved using addition, subtraction, multiplication, and division, but you must decide which operation to use.

Read the problem carefully before you choose the operation that you need to solve it. Identify what you are trying to find out. Think of how the information in the problem can help you find the answer.

Here are some guidelines:

You	When You Need To
Add	Combine quantities. Find a total.
Subtract	Find a difference. Take away a quantity. Compare to find "how many more," "how much less," or "how much is left."
Multiply	Put together a number of equal amounts to find a total. Add the same number repeatedly.
Divide	Split a quantity into equal parts.

Examples:

1. Vic needs to pay a telephone bill of \$54.12 and an electric bill of \$79.48. Which operation shows the total amount of the bills?

   (1) $\$79.48 + \$54.12$
   (2) $\$79.48 - \$54.12$
   (3) $\$79.48 \times \$54.12$
   (4) $\$79.48 \div \$54.12$

   The correct answer is (1). You **add** to find the total.

2. Paul, Rita, and Tanya are sharing equally the \$126 profit from their yard sale. Which operation shows how much each person will receive?

   (1) $\$126 + 3$
   (2) $\$126 - 3$
   (3) $\$126 \times 3$
   (4) $\$126 \div 3$

   The correct answer is (4). You **divide** to split \$126 into 3 equal parts.

Solving Word Problems

Directions: Choose the best answer to each item.

1. Carl's prescription medicine costs \$28. He pays with two twenty-dollar bills. Which operation shows how much money he should get back?

   (1) \$28 + \$40
   (2) \$40 − \$28
   (3) \$28 × \$40
   (4) \$40 ÷ \$28

2. The Rios pay \$269 a month for their car loan. Which operation shows how much they will spend on the car loan in a 12-month period?

   (1) \$269 + 12
   (2) \$269 − 12
   (3) \$269 × 12
   (4) \$269 ÷ 12

3. Last month the Lees paid \$137 to heat their home. This month they paid \$124. Which operation shows the total cost of heating their home for the 2 months?

   (1) \$137 + \$124
   (2) \$137 − \$124
   (3) \$137 × \$124
   (4) \$137 ÷ \$124

4. Mark has \$827 in his checking account. He writes a check for \$189. Which operation shows how much will be left in the account?

   (1) \$827 + \$189
   (2) \$827 − \$189
   (3) \$827 × \$189
   (4) \$827 ÷ \$189

5. Kim is reviewing a 348-page book for a test. The test will be in 6 weeks. Which operation shows how many pages Kim should review each week?

   (1) 348 + 6
   (2) 348 − 6
   (3) 348 × 6
   (4) 348 ÷ 6

6. Alex is buying carpet for his home. One room to be carpeted has 168 square feet. The other room has 154 square feet. Which operation shows the total square feet of carpet Alex needs?

   (1) 168 + 154
   (2) 168 − 154
   (3) 168 × 154
   (4) 168 ÷ 154

7. Six friends commute to work together Monday through Friday. Each week gas, parking, and tolls cost \$114. Which operation shows how much each friend pays each week?

   (1) \$114 + 6
   (2) \$114 − 6
   (3) \$114 × 6
   (4) \$114 ÷ 6

8. Ella worked 68 hours in 2 weeks. She earns \$7 per hour. Which operation shows how much she earned in the 2 weeks?

   (1) 68 + \$7
   (2) 68 − \$7
   (3) 68 × \$7
   (4) 68 ÷ \$7

Answers are on page 336.

Whole Number Operations

There are four whole number operations: addition, subtraction, multiplication, and division. One key to success in mathematics is knowing how to do these four operations accurately.

Addition

You can use addition to combine quantities or find a total. Use these steps:

Step 1: Line up the numbers so that each column has digits with the same place value.

Step 2: Starting with the ones column and working to the left, add the numbers in each column.

Step 3: When a column of digits has a sum greater than 9, you will need to carry to the next column on the left.

Example: Find the total of 169 and 683.

$$\begin{array}{r} {}^{1}\\ 169 \\ +\ 683 \\ \hline 2 \end{array} \qquad \begin{array}{r} {}^{1\,1}\\ 169 \\ +\ 683 \\ \hline 852 \end{array}$$

Add the ones column (9 + 3 = 12). Write the 2 in the ones column and carry the 1 to the tens column. Continue adding the remaining columns.

Subtraction

Use subtraction to find the difference between numbers. Use these steps:

Step 1: Write the lower number below the higher number, lining up the place value columns.

Step 2: Starting with the ones column and working to the left, subtract each column.

Step 3: When a digit in the bottom number is greater than the digit in the same column above it, you must borrow to subtract the column.

Step 4: Check your answer by adding the result and the bottom number.

Example: Subtract 2,438 from 5,025.

Step 1:

$$\begin{array}{r} {}^{1\,1}\\ 5{,}025 \\ -\ 2{,}438 \\ \hline 7 \end{array}$$

Borrow one ten from the tens column to subtract the ones column (15 − 8 = 7).

Step 2:

$$\begin{array}{r} {}^{4\ \ 1\,1\,1}\\ 5{,}025 \\ -\ 2{,}438 \\ \hline 7 \end{array}$$

To subtract the tens column, you need to borrow again. Since there are no hundreds, borrow from the thousands column. Now there are ten hundreds in the hundreds column.

Step 3:

```
     9
   4 111
   5,025
 − 2,438
   2,587
```

Finally, borrow from the hundreds column. Now the subtraction can be completed.

Step 4: Check using addition.

Check:

```
   2,587
 + 2,438
   5,025
```

Multiplication

Use multiplication when you need to add the same number repeatedly. Follow these steps to multiply:

Step 1: Work from the right-hand digit of the number on the bottom. Multiply the ones column and continue to the left. Multiply the top number by each digit in the bottom number.

Step 2: Line up the result under the digit you multiplied by. Add the results of each multiplication.

Example: Multiply 912 by 43.

Step 1:

```
    912
  ×  43
  2 736
```

Multiply 912 by 3. The result, 2,736, is lined up under the ones column.

Step 2:

```
  36 48
 39,216
```

Multiply 912 by 4. The result, 3,648, is lined up under the tens column. Add the results of each multiplication.

Division

You can use division to find out how many times one number goes into another number. The example will help you understand the steps in long division.

Example: Divide 7,310 by 18.

Step 1:

```
        4
 18) 7,310
     7 2
       11
```

How many times will 18 divide 73? You may not be sure whether it is 3 or 4. Multiply 18 by 3 and by 4 to find which answer is closer to 73 without going over 73. Write the 4 in the answer, multiply, and subtract. Bring down the next digit.

```
   18      18
 × 3     × 4
   54      72
            ↑
   closer to 73
```

Step 2:

```
      406r2
 18) 7,310
     7 2
       110
       108
         2
```

18 will not divide 11. Write 0 above the tens place in the answer and bring down the next digit. Continue dividing. Use the letter **r** to show the **remainder.**

Step 3: Check using multiplication.

Check:

```
    406
  ×  18
   3248
   406
   7308
  +   2
  7,310
```

ged Practice Lesson 2

Directions: Solve.

1. $\begin{array}{r} 305 \\ +\ 463 \\ \hline \end{array}$

2. $\begin{array}{r} 4{,}172 \\ +\ 4{,}510 \\ \hline \end{array}$

3. $\begin{array}{r} 6{,}795 \\ +\ 132 \\ \hline \end{array}$

4. $\begin{array}{r} 193 \\ 317 \\ +\ 629 \\ \hline \end{array}$

5. $\begin{array}{r} 56{,}439 \\ +\ 4{,}796 \\ \hline \end{array}$

6. $\begin{array}{r} 36{,}075 \\ 1{,}936 \\ 189{,}006 \\ +\ 17{,}950 \\ \hline \end{array}$

7. $\begin{array}{r} 19{,}067 \\ +\ 35{,}196 \\ \hline \end{array}$

8. $\begin{array}{r} 65{,}196 \\ 6{,}725 \\ 114{,}021 \\ +\ 27{,}716 \\ \hline \end{array}$

---

9. 81,427 + 3,584 + 24,625 =

10. Add 76 and 58.

11. What is the total of 36, 9, 74, 48, 6, and 15?

12. 588,394 + 61,042 + 109,014 =

13. Add 176 and 54,095.

14. 35,100 + 49,257 + 7,566 =

15. What is the total of 950, 308, 77, 29, and 50?

16. 6,019 + 85,200 + 116,896 =

---

17. $\begin{array}{r} 86 \\ -\ 51 \\ \hline \end{array}$

18. $\begin{array}{r} 51{,}964 \\ -\ 20{,}651 \\ \hline \end{array}$

19. $\begin{array}{r} 494 \\ -\ 167 \\ \hline \end{array}$

20. $\begin{array}{r} 1{,}258 \\ -\ 295 \\ \hline \end{array}$

21. $\begin{array}{r} 680 \\ -\ 268 \\ \hline \end{array}$

22. $\begin{array}{r} 3{,}205 \\ -\ 2{,}276 \\ \hline \end{array}$

23. $\begin{array}{r} 800 \\ -\ 219 \\ \hline \end{array}$

24. $\begin{array}{r} 5{,}067 \\ -\ 3{,}795 \\ \hline \end{array}$

---

25. 10,508 − 3,679 =

26. Subtract 3,695 from 5,000.

27. 419,003 − 12,018 =

28. Find the difference of 10,000 and 8,975.

29. Subtract 16,567 from 20,000.

30. 375,000 − 186,425 =

31. Subtract 4,768 from 510,000.

32. Find the difference of 100,000 and 75,510.

---

33. $\begin{array}{r} 432 \\ \times\ 2 \\ \hline \end{array}$

34. $\begin{array}{r} 746 \\ \times\ 5 \\ \hline \end{array}$

35. $\begin{array}{r} 15{,}663 \\ \times\ 8 \\ \hline \end{array}$

36. $\begin{array}{r} 30{,}409 \\ \times\ 9 \\ \hline \end{array}$

37. $\begin{array}{r} 36 \\ \times\ 23 \\ \hline \end{array}$

38. $\begin{array}{r} 5{,}084 \\ \times\ 76 \\ \hline \end{array}$

39. $\begin{array}{r} 1{,}193 \\ \times\ 45 \\ \hline \end{array}$

40. $\begin{array}{r} 3{,}276 \\ \times\ 24 \\ \hline \end{array}$

41. $2{,}584 \times 2{,}700 =$

45. $65 \times 885 =$

42. $25{,}097 \times 60 =$

46. $190 \times 2{,}186 =$

43. $8{,}050 \times 509 =$

47. $775 \times 775 =$

44. $1{,}247 \times 3{,}014 =$

48. $3{,}056 \times 2{,}500 =$

---

49. $7\overline{)3{,}206}$

52. $16\overline{)130{,}112}$

55. $13\overline{)100{,}000}$

50. $4\overline{)23{,}984}$

53. $6\overline{)3{,}502}$

56. $24\overline{)219{,}315}$

51. $6\overline{)254{,}178}$

54. $12\overline{)76{,}402}$

57. $35\overline{)38{,}430}$

---

58. Divide 30,321 by 46.

62. $606{,}450 \div 15 =$

59. $360{,}537 \div 68 =$

63. Divide 712,000 by 25.

60. $139{,}400 \div 205 =$

64. $419{,}357 \div 163 =$

61. Divide 159,035 by 155.

65. Divide 508,320 by 120.

Answers are on page 336.

Strategies for Solving Word Problems

In Lesson 1 you learned how to choose the correct operation to solve a word problem. In this lesson you will calculate the answer.

Solving word problems requires an organized approach. These ideas will help you organize your work.

Read: Make sure you understand the problem. Identify what it is that you need to find out. Think of how the information in the problem can help you find the answer.
Plan: Choose the operation (addition, subtraction, multiplication, or division) you will use to solve the problem.
Solve: You may think you can solve the problem in your head. This may be possible, but you will make fewer errors if you write the problem out.
Check: Make sure your answer makes sense. Does it answer the question posed by the problem? Does it seem reasonable? Check your answer.

These ideas can be stated in 4 words to help you remember them.

Read
Plan
Solve
Check

Apply this strategy to the following problem.

Example: Roberto drove 357 miles on Monday and 509 miles on Tuesday. How much farther did he drive on Tuesday than on Monday?

Read: You are trying to find out the difference between the miles driven on Monday and Tuesday.

Plan: Since Tuesday's mileage (509) is greater than Monday's (357), subtract 357 from 509.

Solve:

$$\begin{array}{r} 509 \\ -\ 357 \\ \hline 152 \end{array}$$

Check: Is the answer reasonable?
Check subtraction using addition.

$$\begin{array}{r} 152 \\ +\ 357 \\ \hline 509 \end{array}$$

The answer is **152 miles.**

Solving Word Problems

Directions: Choose the best answer to each item.

1. Carl's prescription medicine costs $28. He pays with 2 twenty-dollar bills. How much money should he get back?

   (1) $11
   (2) $12
   (3) $22
   (4) $28
   (5) $68

2. The Rios pay $269 a month for their car loan. How much will they spend on the car loan in a 12-month period?

   (1) $ 807
   (2) $3,128
   (3) $3,188
   (4) $3,218
   (5) $3,228

3. Last month the Lees paid $137 to heat their home. This month they paid $124. What is the total cost of heating their home for the two months?

   (1) $161
   (2) $251
   (3) $260
   (4) $261
   (5) $262

4. Mark has $827 in his checking account. He writes a check for $189. How much will be left in the account?

   (1) $1,016
   (2) $ 762
   (3) $ 738
   (4) $ 648
   (5) $ 638

5. Kim is reviewing a 348-page book for a test. The test will be in 6 weeks. How many pages should Kim review each week?

   (1) 50.8
   (2) 57
   (3) 58
   (4) 342
   (5) 418

6. Alex is buying carpet for his home. One room to be carpeted has 168 square feet. The other room has 154 square feet. What is the total square feet of carpet Alex needs?

   (1) 322
   (2) 320
   (3) 312
   (4) 222
   (5) 214

7. Six friends commute to work together Monday through Friday. Each week gas, parking, and tolls cost $114. How much does each friend pay each week?

   (1) $ 17
   (2) $ 18
   (3) $ 19
   (4) $108
   (5) $109

8. Ella worked 68 hours in 2 weeks. She earns $7 per hour. How much did she earn in the two weeks?

   (1) $416
   (2) $435
   (3) $466
   (4) $468
   (5) $476

Answers are on page 338.

Lesson 3 Steps to Solving Word Problems

Estimation

To **estimate** means to find an approximate solution to a problem. Use estimation to solve problems that do not require an exact solution.

Example: The city of Southampton has 11,968 registered voters. In a recent election, only 4,787 of the people registered actually voted. Approximately how many registered voters did not vote?

(1) between 4,500 and 5,500

(2) between 5,500 and 6,500

(3) between 6,500 and 7,500

(4) between 7,500 and 8,500

(5) between 8,500 and 9,500

To compare the number of registered voters with the number of people who voted you need to subtract. However, the problem does not require an exact solution. The word approximately tells you to estimate the answer. To estimate, round the numbers to the nearest thousand and subtract.

$$\begin{array}{r} 12{,}000 \\ -\ 5{,}000 \\ \hline 7{,}000 \end{array} \quad \text{ESTIMATED SOLUTION} \qquad \begin{array}{r} 11{,}968 \\ -\ 4{,}787 \\ \hline 7{,}181 \end{array} \quad \text{EXACT SOLUTION}$$

The correct answer is **(3) between 6,500 and 7,500.**

Identifying Necessary Information

Some problems may not include all the information you need to solve them. Other problems may include information that is not necessary to solve them. When you solve problems, think carefully about the situation. Be sure you have the information you need and are considering only the necessary information.

Example: A stockroom manager needs to send a truck to the warehouse to get 200 cartons of housewares, 150 cartons of men's clothes, and 75 cartons of bedding. How many trips must the truck make to transport all cartons to the stockroom?

(1) 3

(2) 17

(3) 25

(4) 425

(5) Not enough information is given.

Do you see that something is missing? You need to know how many cartons the truck can hold. It is impossible to solve the problem with only the information from the problem. The correct answer is

(5) Not enough information is given.

A common error is performing an operation using the numbers from the problem even though there may not be a reason for using them. For instance, if you added the numbers in the problem, 200 + 150 + 75, you could find the answer 425, which is among the choices, but it does not make sense. Does it seem reasonable that a truck would need to make 425 trips to transport the cartons? Adding gives the total number of cartons to be shipped, not the number of trips.

For some problems, a table or chart of information is given. You may be asked several questions about the information. Before solving each problem, identify the information you need.

Example: The McCaffy Public Library kept records on book circulation from January through April.

January	February	March	April
10,256	7,542	7,625	9,436

How many more books were borrowed in April than in February?

(1) 83

(2) 820

(3) 1,811

(4) 1,894

(5) 16,978

To solve the problem, you need **only** the numbers for February and April. The problem asks you to compare two numbers and find the difference. Using subtraction, the correct answer is **(4) 1,894.**

$$9{,}436 - 7{,}542 = 1{,}894$$

Since this is a subtraction problem, you can check the answer using addition. The answer is correct.

$$\begin{array}{r} 1{,}894 \\ +\ 7{,}542 \\ \hline 9{,}436 \end{array}$$

ged Practice Lesson 3

Directions: Choose the best answer to each item.

1. Sarah's bank account balance is $715. If she makes a deposit of $375, what will her balance be?

   (1) $ 340
   (2) $ 880
   (3) $1,090
   (4) $1,465
   (5) Not enough information is given.

2. The Coast Cafe has 12 tables. How many people can be seated in the restaurant during rush hour?

   (1) 40
   (2) 48
   (3) 60
   (4) 72
   (5) Not enough information is given.

3. Marshall is eating 2,400 calories per day in an effort to gain weight. Today he has eaten 685 calories. Approximately how many more calories must he eat today?

   (1) between 1,450 and 1,550
   (2) between 1,550 and 1,650
   (3) between 1,650 and 1,750
   (4) between 1,750 and 1,850
   (5) between 1,850 and 1,950

4. Ahmed wants to save $6,000 per year. If he saves the same amount each month, how much should that amount be?

   (1) $ 50
   (2) $ 72
   (3) $500
   (4) $720
   (5) $800

5. Ricardo's weekly gross pay is $440. If $146 is taken out for taxes and other deductions, what is the amount of his take-home pay?

   (1) $194
   (2) $294
   (3) $364
   (4) $584
   (5) Not enough information is given.

6. Meg needs to have her car radiator repaired. Used parts will cost $85 and labor on the job will be $125. How much will it cost to get her car fixed if new parts are used?

   (1) $ 40
   (2) $160
   (3) $364
   (4) $584
   (5) Not enough information is given.

7. Today's sales of the 20-inch HiTech television total $5,760. If each set sells for $320, how many were sold?

   (1) 15
   (2) 16
   (3) 17
   (4) 18
   (5) 19

8. This week Selma earned $236 in commissions on her sales. This is $48 more than she earned last week. How much commission did she earn last week?

   (1) $472
   (2) $284
   (3) $236
   (4) $188
   (5) Not enough information is given.

9. The Melendez family drove 167 miles on the first day of a two-day trip. What was the total number of miles driven for the two days?

(1) 84
(2) 247
(3) 334
(4) 453
(5) Not enough information is given.

10. Quan Le sells hot dogs at the park seven days a week. He uses 15 bags of buns each day. There are 6 buns in a bag. How many buns does he use a day?

(1) 90
(2) 105
(3) 180
(4) 450
(5) 630

11. Gail needs $223 to pay the cost of an evening class. She has $37 and plans to borrow the rest from her brother. How much does she need to borrow?

(1) $150
(2) $186
(3) $209
(4) $223
(5) $260

12. Roy is collecting aluminum cans to raise money for the blood bank. He collects four bags of cans in May and five bags in June. How many pounds of cans did Roy collect in May and June?

(1) 9
(2) 15
(3) 18
(4) 20
(5) Not enough information is given.

13. A company gave each of its 412 employees a $500 year-end bonus. What was the total amount spent by the company on year-end bonuses?

(1) $194,000
(2) $200,000
(3) $206,000
(4) $212,000
(5) Not enough information is given.

14. A six-floor building contains 112 large offices. If two people are assigned to each office, how many people could work in the offices?

(1) 120
(2) 224
(3) 264
(4) 448
(5) 672

15. Lotte works 40 hours per week. She earns $8 per hour and spends $2 per hour for child care. How much of her hourly wage is left after child care?

(1) $6
(2) $11
(3) $26
(4) $29
(5) Not enough information is given.

16. A hotel has 350 rooms. The custodial staff can clean one room per hour. How many rooms are assigned to each staff member?

(1) 25
(2) 35
(3) 50
(4) 70
(5) Not enough information is given.

Items 17 to 22 refer to the following information.

Department	City Budget
Police	$180,500
Fire protection	92,750
Maintenance	225,125
Recreation	18,725
Office	51,350

17. What is the total amount of the budget shown?

(1) $387,900
(2) $492,000
(3) $517,100
(4) $568,450
(5) Not enough information is given.

18. What is the approximate difference between the largest and smallest amounts in the budget?

(1) between $120,000 and $150,000
(2) between $150,000 and $180,000
(3) between $180,000 and $200,000
(4) between $200,000 and $220,000
(5) between $220,000 and $240,000

19. A citizens' group wants to double the police budget. What amount are they asking the city to spend on the police department?

(1) $ 90,250
(2) $199,000
(3) $361,000
(4) $405,600
(5) Not enough information is given.

20. The recreation budget will be spent equally on five city parks. How much will be spent on each park?

(1) $ 3,600
(2) $ 3,745
(3) $ 7,515
(4) $10,270
(5) $18,550

21. Last year's office budget was $6,500 less than the amount on the chart. What was last year's office budget?

(1) $41,350
(2) $44,850
(3) $52,850
(4) $59,350
(5) Not enough information is given.

22. After some cuts, the final budget for twelve months was $541,908. How much will this cost the city per month?

(1) $ 45,159
(2) $ 58,526
(3) $ 90,318
(4) $127,950
(5) $173,458

23. Mr. and Mrs. Pierotti are remodeling their kitchen. They spent $548 for new cabinets and $618 on plumbing. The tiles for the kitchen floor will cost $3 each. If they need 240 tiles, how much will the tiles cost?

(1) $ 80
(2) $ 720
(3) $1,166
(4) $1,644
(5) $1,886

Items 24 to 26 are based on the following information.

Marina Outboard Motors

	Work Quotas	
	Goal	Actual
Monday	100	103
Tuesday	95	136
Wednesday	120	117
Thursday	110	122
Friday	90	180

24. What was the total goal for the week?

(1) 500
(2) 515
(3) 555
(4) 620
(5) 658

25. The workers missed the quota one day during the week. Which day was it?

(1) Monday
(2) Tuesday
(3) Wednesday
(4) Thursday
(5) Friday

26. How many more motors were built on Friday than on Thursday?

(1) 12
(2) 58
(3) 70
(4) 90
(5) Not enough information is given.

27. Maya buys a stereo system on credit and agrees to pay $45 per month. What is the total amount she will pay?

(1) $270
(2) $315
(3) $540
(4) $675
(5) Not enough information is given.

28. Bowling 226 and 241, Bea easily won the first two games of the Riverside Lanes Tournament. What was her total score for the three-game series?

(1) 467
(2) 598
(3) 693
(4) 708
(5) Not enough information is given.

29. Wrightway Shoes has 1,192 employees working at 32 branches in Washington and Oregon. If 372 employees are laid off, how many employees will remain?

(1) 404
(2) 409
(3) 788
(4) 820
(5) 1,160

30. Arcadia Coliseum seats 26,500. The East-West football game has been held at the coliseum for the past 8 years. This year the tickets for the football game sold for $6. If the game was sold out, how much did the coliseum earn in ticket sales?

(1) $ 53,000
(2) $106,000
(3) $159,000
(4) $185,000
(5) $212,000

Answers are on page 339.

Steps for Solving Word Problems

In Lesson 3 you learned a basic strategy for solving word problems. You learned how to handle problems that have too little or too much information. The same strategies and techniques apply to solving multi-step problems.

A multi-step problem requires more than one calculation to find the solution. The key to solving multi-step problems is to identify each step and the operation needed for each step **before** working the problem.

Example:

Carpet Sale ....$16 per square yard
Carpet Pad Special ....$3 per square yard

Jana wants to take advantage of the sale prices in this ad. She measures her room and finds that it is 12 feet wide and 18 feet long. How much will it cost to buy the pad and carpet for her room?

(1) $72
(2) $216
(3) $384
(4) $456
(5) Not enough information is given.

There are several steps to solving this problem.

Step 1: Find the area of the room in square feet by **multiplying** the length and width of her room.

Step 2: Change the area from square feet to square yards by **dividing** by 9, the number of square feet in 1 square yard.

Step 3: Find the cost of the carpet by **multiplying** the number of square yards (from Step 2) by $16, the price of 1 square yard of carpet.

Step 4: Find the cost of the carpet pad by **multiplying** the number of square yards (from Step 2) by $3, the price of 1 square yard of carpet padding.

Step 5: Find the total cost by **adding** the cost of the carpet (from Step 3) and the cost of the carpet pad (from Step 4).

Now solve the problem.

Step 1:

$$\begin{array}{r} 12 \\ \times\ 18 \\ \hline 96 \\ 12 \\ \hline 216 \end{array}$$ **square feet**

Find the area of Jana's room in square feet (feet × feet = square feet).

Step 2:

$$\begin{array}{r} 24 \\ 9\overline{)216} \\ 18 \\ \hline 36 \\ 36 \\ \hline 0 \end{array}$$ **square yards**

Find the area of her room in square yards (square feet ÷ feet/sq yard = square yards).

Step 3:

$$\begin{array}{r} 24 \\ \times\ \$16 \\ \hline 144 \\ 24 \\ \hline \$384 \end{array}$$ **carpet cost**

Find the cost of the carpet.

Step 4:

$$\begin{array}{r} 24 \\ \times\ \$3 \\ \hline \$72 \end{array}$$ **pad cost**

Find the cost of the pad.

Step 5:

$$\begin{array}{r} \$384 \text{ carpet cost} \\ +\ \ 72 \text{ pad cost} \\ \hline \$456 \text{ total cost} \end{array}$$

Find the total cost.

The correct answer option is **(4) $456.**

Notice that some of the other step results are found among the answer choices. Option (1), **$72,** was the cost of the pad. Option (2), **$216,** was not correct because it should not be money. It was actually the number of square feet in the room. Option (3), **$384,** was the cost of the carpet alone. Do not stop working when you find a result that matches one of the answer choices. Always decide the steps you will need first; then work every step.

Organizing Information

Some problems contain a large amount of data. Creating lists and tables are two effective ways of organizing information.

The following problem asks you to count many items. Keeping a tally may be the best way to organize the information.

Example: The employee union at Berczi Manufacturing recently held an election for new officers. Reynaldo, the election chairperson, is in charge of counting the votes. He has the following ballots.

Which choice below lists the new officers using this order: president, vice president, and secretary?

(1) Tom, Ellen, and Joan

(2) Tom, Joan, and Art

(3) Pete, Ellen, and Art

(4) Pete, Tom, and Joan

(5) Pete, Ellen, and Joan

The problem can best be solved by making a table using tally marks. List the names of the nominees down the side and the offices across the top. Then make marks to record the votes.

	President	*Vice President*	*Secretary*
Tom	卌 I	卌	III
Art		II	卌 II
Joan	III	卌	卌
Ellen	IIII	卌 I	III
Pete	卌 II	II	II

Using the table, it is easy to see that the correct answer is **(3) Pete, Ellen, and Art.** If you had to answer more questions about the election, you would already have the information organized.

In the following example, the data are organized for you. Decide what information you need to solve the problem. Remember to label your work.

Example:

United Solar Energy Sales Report (in Units)			
	January	February	March
Eastern Regional Office	735	650	866
Western Regional Office	1,104	1,062	1,200
Southern Regional Office	865	727	1,129
Northern Regional Office	935	600	1,163

The sales manager wants to give awards to the two regional offices with the greatest increase in sales from January to March. Which two offices will win the awards?

(1) the Eastern and Western regional offices
(2) the Eastern and Southern regional offices
(3) the Western and Southern regional offices
(4) the Southern and Northern regional offices
(5) Not enough information is given.

To solve this problem subtract the January sales figure from the March figure for each regional office. Compare the results.

Eastern: 866 − 735 = 131
Western: 1,200 − 1,104 = 96
Southern: 1,129 − 865 = 264 **Top office**
Northern: 1,163 − 935 = 228 **Second highest office**

The correct answer is **(4) the Southern and Northern regional offices.**

Example: How much greater was the increase of the top office from January to March than the increase of the office with the smallest improvement?

(1) 11
(2) 24
(3) 64
(4) 75
(5) 168

From the previous calculations, you can quickly find the offices with the highest, 264, and lowest, 96. Subtract to solve the problem. The correct answer is **(5) 168,** because 264 − 96 = 168.

ged Practice Lesson 4

Directions: Choose the best answer to each item.

Items 1 and 2 refer to the following information.

Felicia and Jay Oser buy a new refrigerator. They must make eight payments of $115 each and one final payment of $162.

1. Which combination of operations would you need to find the total amount the Osers will spend on the refrigerator?

   (1) addition and subtraction
   (2) addition and division
   (3) subtraction and division
   (4) multiplication and addition
   (5) multiplication and subtraction

2. How much will the Osers spend to buy the refrigerator?

   (1) $ 477
   (2) $ 889
   (3) $ 920
   (4) $ 998
   (5) $1,082

3. Hugh Coles pays $312 every six months for insurance for his car. Brenda Coles pays $866 per year for insurance for her car. How much do the Coles pay for car insurance in one year?

   (1) $ 624
   (2) $1,178
   (3) $1,490
   (4) $1,732
   (5) Not enough information is given.

Items 4 and 5 refer to the following information.

An electronics company advertises a 24-inch stereo color television set for $460 cash or $45 per month for 12 months on credit.

4. Which combination of operations would you need to find how much a customer will save by paying cash?

   (1) addition and subtraction
   (2) subtraction and division
   (3) multiplication and addition
   (4) multiplication and subtraction
   (5) division only

5. How much will a customer save by paying cash instead of buying on credit?

   (1) $40
   (2) $45
   (3) $80
   (4) $85
   (5) $90

6. Susan Landow paid $3,794 for medical insurance, hospital and doctor's bills last year. If her insurance company refunded $2,382 of her money, how much did she spend in medical bills last year?

   (1) $ 908
   (2) $1,294
   (3) $1,412
   (4) $2,786
   (5) $3,290

Items 7 and 8 refer to the following information.

The tenants in two apartment buildings agree to hire a private trash collector to serve their buildings. The total yearly cost of the service will be $3,096. There are 18 apartments in one building and 25 apartments in the other.

7. Which combination of operations would you need to find out how much each apartment will pay per year for the service if each apartment pays an equal share of the cost?

   (1) addition and multiplication
   (2) addition and division
   (3) subtraction and division
   (4) multiplication and subtraction
   (5) multiplication only

8. How much is each apartment's share of the annual cost?

   (1) $43
   (2) $56
   (3) $65
   (4) $68
   (5) $72

Items 9 to 13 refer to the following chart.

Employee	Hourly Wage	Hours Worked S	M	T	W	T	F	S
F. Blau	$7		8	7	8	8	6	3
M. Bodine	8		8	8	9	8	8	8
T. Ortiz	8		6	6	8	9	8	
R. Perez	6		8	8	8	8	8	

9. How much money did F. Blau earn for the week?

   (1) $280
   (2) $301
   (3) $320
   (4) $329
   (5) Not enough information is given.

10. How many more hours did M. Bodine work than T. Ortiz?

   (1) 9
   (2) 12
   (3) 16
   (4) 49
   (5) 86

11. R. Perez earned $240 this week. How many more hours would he have had to work to earn $252?

   (1) 2
   (2) 3
   (3) 8
   (4) 12
   (5) 40

12. If T. Ortiz were given a $2 hourly raise, how much would he earn for the same number of hours worked?

   (1) $ 74
   (2) $296
   (3) $370
   (4) $490
   (5) Not enough information is given.

13. The following week, F. Blau worked 8 hours each day, Monday through Saturday. How much more money did he earn than the week before?

   (1) $ 48
   (2) $ 56
   (3) $280
   (4) $336
   (5) Not enough information is given.

Items 14 to 16 refer to the following table.

Item	Price	Sale Price
T-shirt	$8	$7
Jersey	10	8
Pants	18	15

14. On Friday, the store sold 24 pairs of pants at the sale price. How much money did the store take in on pants?

    (1) $270
    (2) $360
    (3) $405
    (4) $432
    (5) $518

15. Two parents are buying school clothes for their five boys. They buy 12 T-shirts, 5 jerseys, and 10 pairs of pants. How much do they save buying at the sale prices instead of the regular prices?

    (1) $28
    (2) $30
    (3) $36
    (4) $48
    (5) $52

16. After the sale, the store has only 12 pairs of pants in stock. If they take an additional $3 off the sale price, what would be the cost of 4 pairs of pants?

    (1) $36
    (2) $48
    (3) $60
    (4) $72
    (5) Not enough information is given.

Items 17 to 19 refer to the following table.

Daily Production Report	
Cynthia Bonales	56
Sadie Johns	50
Ariste Sanchez	48

17. Ms. Bonales earns $1 for each item produced. If she continues to produce the same number for each of the next 5 days, how much will she earn for the 5 days?

    (1) $280
    (2) $300
    (3) $336
    (4) $356
    (5) Not enough information is given.

18. Al Butler, a new worker not shown on the chart, produced fewer items than Ariste Sanchez on the same day. How much did Mr. Butler earn for the day?

    (1) $132
    (2) $176
    (3) $204
    (4) $352
    (5) Not enough information is given.

19. Ms. Johns worked only 5 hours during the day shown on the chart. On the average, how many items did she produce per hour?

    (1) 10
    (2) 13
    (3) 14
    (4) 15
    (5) Not enough information is given.

Items 20 to 22 refer to the following information.

ABC Preschool called in this order for disposable diapers:

Quantity	Diapers	Number in Box
15 boxes	small	66
18 boxes	medium	42
24 boxes	large	24

20. How many more small diapers were ordered than large diapers?

(1) 234
(2) 414
(3) 514
(4) 594
(5) Not enough information is given.

21. If the preschool uses about 60 diapers a day, approximately how long will these diapers last?

(1) between 25 and 30 days
(2) between 30 and 35 days
(3) between 35 and 40 days
(4) between 40 and 45 days
(5) Not enough information is given.

22. Regardless of size, each box of diapers costs the preschool $11. How much will the entire order cost?

(1) $ 57
(2) $264
(3) $462
(4) $627
(5) $726

Items 23 and 24 refer to the following information.

Baseball Game Attendance		
Day	**Day Game**	**Night Game**
S	47,551	-------
M	**No Game Played**	
T	34,196	-------
W	-------	44,640
T	-------	42,079
F	-------	50,302
S	49,653	-------

23. How many fans attended games on the weekdays (Monday through Friday)?

(1) 123,666
(2) 137,021
(3) 171,217
(4) 220,870
(5) 268,421

24. Approximately how many more fans attended night games than day games during the week?

(1) between 3,500 and 5,000
(2) between 5,000 and 6,500
(3) between 6,500 and 8,000
(4) between 8,000 and 9,500
(5) Not enough information is given.

25. The total enrollment for four quarters at Claremont College of Business in 1989 was 2,829 students. If 650 students enrolled for winter quarter, 624 students enrolled for spring quarter, and 705 students enrolled for summer quarter, how many students enrolled for fall quarter?

(1) 569
(2) 707
(3) 840
(4) 850
(5) 943

Answers are on page 340.

Lessons 1–4

Directions: Choose the best answer to each item.

1. Grace charges $15 for a haircut and $45 for a permanent. If she does 10 haircuts and 3 permanents on Tuesday, how much money will she earn that day from haircuts?

   (1) $15 + $45
   (2) 10 − 3
   (3) $15 × 10
   (4) $15 × 3
   (5) $45 × 10

2. Westside Car Rental held an awards banquet for its 65 employees. The company paid a hotel $125 for the use of a conference room and $9 per person for food. How much did the company pay for the room and food?

   (1) $ 199
   (2) $ 398
   (3) $ 585
   (4) $ 710
   (5) $1,125

3. Carmel, a California city, had a population of 5,724 in 1980. In 1990, the population was 10,054. Approximately how many more people lived there in 1990 than in 1980?

   (1) between 2,500 and 3,500
   (2) between 3,500 and 4,500
   (3) between 4,500 and 5,500
   (4) between 5,500 and 6,500
   (5) between 6,500 and 7,500

4. The Conrad Electronics factory can produce 500 calculators per week. How many calculators per year would that be?

   (1) 6,000
   (2) 15,000
   (3) 26,000
   (4) 30,000
   (5) 60,000

Items 5 and 6 refer to the following table.

Final Medal Standings 1992 Summer Olympics	
Unified Team	112
United States	108
Germany	87
China	54
Cuba	31
Hungary	30
South Korea	29

5. The United States and China together won more medals than the Unified Team. How many more?

   (1) 4
   (2) 25
   (3) 50
   (4) 162
   (5) Not enough information is given.

6. How many more medals did the Unified Team win than Cuba, South Korea, and Hungary combined?

   (1) 22
   (2) 31
   (3) 90
   (4) 112
   (5) 202

7. Art works in an auto parts factory. He makes 12 parts per hour and earns $2 for each part. How much will Art earn each hour?

   (1) 12 + $2
   (2) 12 − $2
   (3) 12 × $2
   (4) 12 ÷ $2
   (5) Not enough information is given.

Items 8 to 10 refer to the following information.

Tape Rentals at Video Wonderland	
January	12,820
February	6,521
March	13,042
April	25,907
May	26,095

8. What is the total number of tapes rented in March, April, and May?

(1) 32,383
(2) 58,290
(3) 65,044
(4) 84,385
(5) Not enough information given.

9. How many more tape rentals were there in the month with the most rentals compared to the month with the fewest rentals?

(1) 26,095 + 6,521
(2) 25,907 + 6,521
(3) 25,907 − 6,521
(4) 26,095 − 6,521
(5) 26,095 ÷ 6,521

10. May's rentals are about how many times higher than February's rentals?

(1) about 2 times higher
(2) about 3 times higher
(3) about 4 times higher
(4) about 5 times higher
(5) about 6 times higher

11. Manuel works 3 hours a day, 5 days a week, at a car wash. How many hours does he work in 4 weeks?

(1) 11
(2) 12
(3) 15
(4) 20
(5) 60

12. Teresa had $150 to spend on her college books. She bought 2 textbooks costing $35 each, 3 textbooks for $18 each, and a notebook for $3. How much money did she have left after her purchase?

(1) $ 6
(2) $ 17
(3) $ 23
(4) $ 56
(5) $127

13. An office supply store advertises typewriter ribbons for $4 each. How much will a box of ribbons cost?

(1) $24
(2) $40
(3) $48
(4) $80
(5) Not enough information is given.

14. Dave can drive 160 miles on a tank of gas. Approximately how many tanks of gas will he need to drive 880 miles?

(1) between 4 and 5
(2) between 5 and 6
(3) between 6 and 7
(4) between 7 and 8
(5) Not enough information is given.

15. A parking garage has spaces for 70 cars and charges each driver $6 per day for parking. If the garage is full, how much more would the garage owner make if he charged $8 per day?

(1) $ 14
(2) $ 48
(3) $105
(4) $140
(5) $420

Items 16 to 18 refer to the following information.

The Clothes Closet		
Item	**Original Price**	**Sale Price**
Wool Sweaters	$45	$36
Dress Shirts	$21	$17
Neckties	$19	$15

16. Tim buys 1 wool sweater, 1 dress shirt, and 1 necktie, each at the sale price. How much money does he save on the 3 items by buying them at the sale price instead of the original price?

(1) $ 4
(2) $16
(3) $17
(4) $18
(5) $68

17. Phil buys 2 dress shirts and 2 neckties at the sale price. What is the total cost of the items?

(1) $32
(2) $49
(3) $64
(4) $74
(5) $80

18. How much would Chet save by buying 2 sweaters at the sale price instead of the original price?

(1) $ 9
(2) $18
(3) $22
(4) $81
(5) $90

19. Greg drove for 3 hours on Monday at an average speed of 55 miles per hour. On Tuesday he drove for 4 hours at an average speed of 50 mph. How many miles did he drive in those 7 hours?

(1) 165
(2) 200
(3) 365
(4) 565
(5) 730

20. Luisa needs 900 feet of rope. How many packages of rope does she need to buy if each package contains 50 feet of rope?

(1) 18
(2) 19
(3) 190
(4) 4,500
(5) 45,000

21. Marcia buys a box of 1,000 writing pads for office meetings. She gives out 188 pads on Tuesday and 234 pads on Wednesday. How many writing pads does she have left?

(1) 422
(2) 578
(3) 766
(4) 812
(5) 1,422

22. Eddie's Electronics shipped 72 boxes that weighed 1,800 pounds altogether. If each box weighed the same, how many pounds did 1 box weigh?

(1) 25
(2) 40
(3) 250
(4) 400
(5) 650

23. Jaime is taking inventory at the hardware store. He counts the number of $\frac{3}{8}$-inch bolts in stock and finds he has 3,325 on hand. His records show he received 21 boxes of $\frac{3}{8}$-inch bolts on his last order. If each box contained about 506 bolts, approximately how many bolts did Jaime receive on his last order?

(1) between 9,000 and 11,000
(2) between 11,000 and 13,000
(3) between 13,000 and 15,000
(4) between 15,000 and 17,000
(5) Not enough information is given.

24. Barbara discovered she was supposed to pay only $4,516 in federal income taxes last year. She actually paid $5,752. How much did she overpay?

(1) $ 988
(2) $1,056
(3) $1,137
(4) $1,236
(5) Not enough information is given.

25. In a town election, a candidate for council representative won by 47 votes over the other candidate. The winner had 2,035 votes. How many votes did the other candidate have?

(1) 1,978
(2) 1,988
(3) 2,012
(4) 2,082
(5) 2,088

26. Kenneth sold 2 cameras for $175 each and 3 cameras for $150 each. If he earns a $35 commission for each sale, how much did he earn in commission?

(1) $ 35
(2) $ 70
(3) $ 75
(4) $150
(5) $175

27. If 50,302 fans attended the Friday night game and 34,196 attended the game on Tuesday, how many more fans saw the Friday night game?

(1) 15,601
(2) 16,106
(3) 17,114
(4) 26,421
(5) 84,498

28. Janet has a $1,800 dental bill for herself and her children. If she pays $150 each month, how many months will it take her to pay off the bill?

(1) 12
(2) 13
(3) 120
(4) 1,650
(5) 1,950

29. The Parkville School District had 18,596 children enrolled in its schools last year. 20,132 children are enrolled in its schools this year. By how many children did the enrollment grow?

(1) 1,536
(2) 2,464
(3) 2,536
(4) 20,132
(5) 38,728

Answers are on page 341.

Lesson 5 Introduction to Fractions

- Ultralite margarine contains $\frac{1}{4}$ the fat of butter.
- Five out of six doctors recommend NoPain aspirin.
- Going Out of Business Sale! All merchandise $\frac{1}{2}$ off!

Fractions are all around us. Fractions measure part of something. They can be used to show a part of one thing: "Add $\frac{1}{2}$ cup of milk to the mixture." They can also be used to show part of one group: "One-third of our employees have used up all their sick leave."

In this lesson you will cover the basic facts about fractions. You will learn to recognize these types of fractions: **proper fractions, improper fractions,** and **mixed numbers.** You will also learn to change between improper fractions and mixed numbers.

Lesson 6 explains how to compare fractions. You will learn how to tell when two fractions are equal. You will also learn how to reduce a fraction to lower terms and how to build equal fractions with higher terms.

The four operations you used with whole numbers are also used to solve problems involving fractions. Lesson 7 covers addition and subtraction of fractions. Lesson 8 shows how to multiply and divide fractions.

Facts About Fractions

To count, you need whole numbers. To show part of something, you need fractions. Fractions show part of a whole.

Every **fraction** has a **numerator** and a **denominator.** The denominator is the bottom number. It tells the number of equal parts that are in the whole object or group. The numerator is the top number. It tells the number of equal parts or things you are referring to. The numerator and denominator are the **terms** of the fractions.

$$\frac{2}{3} \begin{array}{l} \leftarrow \text{numerator} \\ \leftarrow \text{denominator} \end{array}$$

Fractions have three special uses.

1. Fractions show a part of one whole thing.

one whole object

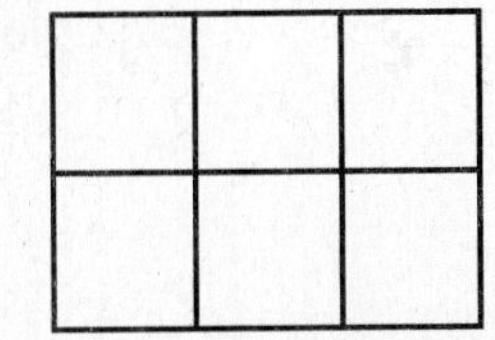

The object is divided into 6 equal parts.

One whole object represents 6 parts out of 6, or $\frac{6}{6}$. Any fraction with the same numerator and denominator equals 1.

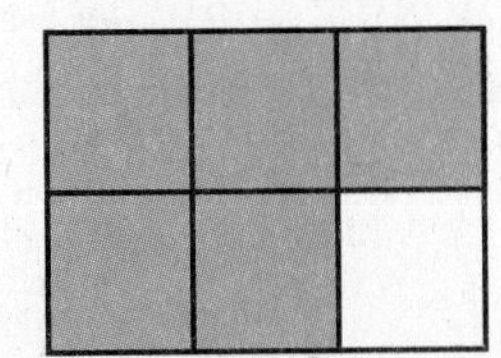

5 of the 6 parts are shaded.
The fraction $\frac{5}{6}$ represents the shaded part.

5 ← number of shaded parts
6 ← number of equal parts in the whole

Example: At a restaurant, a chef divides a cheesecake into 8 equal pieces. By the end of the evening, 5 of the pieces are sold. What part of the cheesecake is sold?

5 ← number of pieces sold
8 ← number of equal parts in the whole cheesecake

2. Fractions show part of a group.

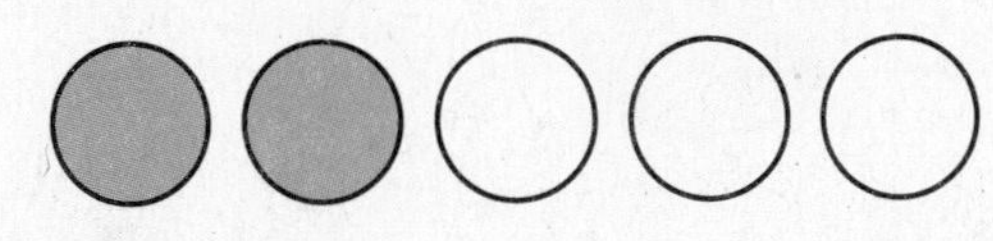

The picture shows a group of objects. There are 5 objects in the group. 2 of the objects are shaded.

The fraction $\frac{2}{5}$ represents the shaded parts.

2 ← number of shaded parts
5 ← number of objects in the whole group

Example: At the shop where Nicco works, 18 orders were received one day. There were 11 phone orders. What fraction of the orders came by phone?

11 ← number of orders that came by phone
18 ← number of orders received in one day

3. Fractions also mean division.
The fraction $\frac{12}{3}$ means $12 \div 3$ or $3\overline{)12}$.

$\frac{12}{3} = 4$

Proper Fractions

A **proper fraction** shows a quantity that is less than 1. The numerator of a proper fraction is always less than the denominator.

Example:

$\frac{3}{4}$ is shaded

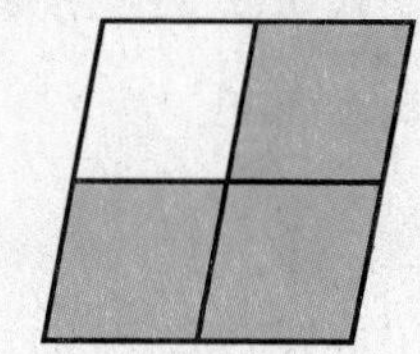

The numerator, 3, is less than the denominator, 4. $\frac{3}{4}$ represents a quantity less than 1.

Improper Fractions

An **improper fraction** shows a quantity equal to or greater than 1. The numerator of an improper fraction is equal to or greater than the denominator.

Examples:

$\frac{2}{2}$ is shaded

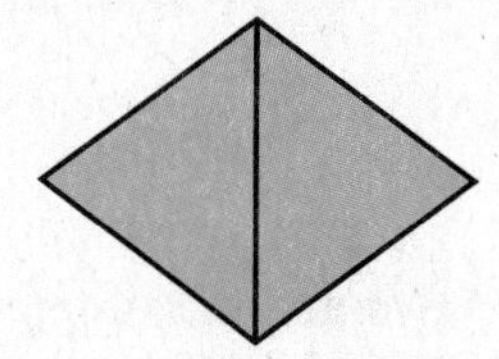

The figure is divided into 2 equal parts, and both parts are shaded. $\frac{2}{2}$ is an improper fraction because it is equal to 1 whole and the numerator and the denominator are equal.

$\frac{7}{5}$ is shaded

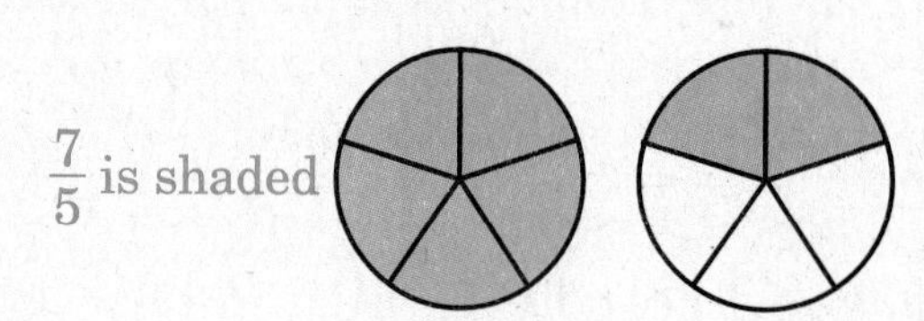

Each circle is divided into 5 equal parts. 7 parts are shaded. Because the numerator is greater than the denominator, $\frac{7}{5}$ is an improper fraction. This fraction shows a quantity greater than 1.

Mixed Numbers

A **mixed number** is another way to show a fraction that is greater than 1. A mixed number is a whole number and a proper fraction. Although the plus sign (+) is not written, the whole number is added to the fraction part of the mixed number.

Example: $2\frac{1}{4}$ means "two and one-fourth" or $2 + \frac{1}{4}$.

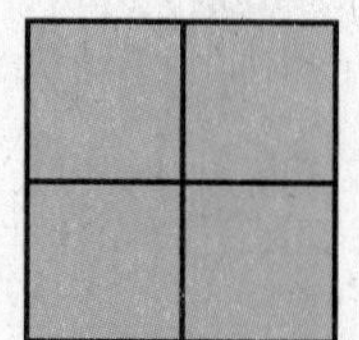

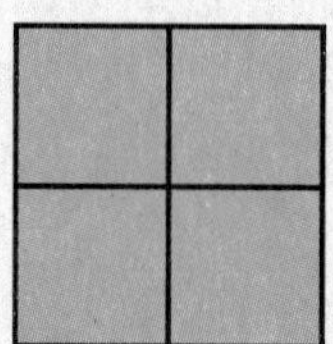

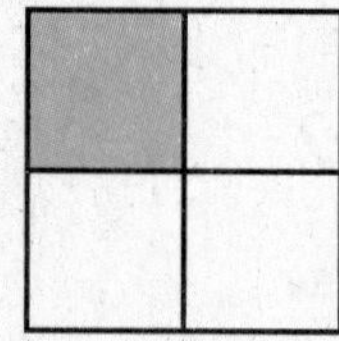

Each figure is divided into 4 equal parts. Two figures are completely shaded. One-fourth of the last figure is shaded.

Changing Improper Fractions and Mixed Numbers

When you perform basic operations with fractions, you need to know how to change back and forth between improper fractions and mixed numbers.

An improper fraction can be changed to a mixed number or a whole number. Remember that a fraction is a division problem.

Follow these steps:

Step 1: Divide the numerator of the improper fraction by the denominator. The whole number answer becomes the whole number part of the mixed number.

Step 2: Write the remainder over the original denominator. This is the fraction part of the mixed number.

Example: Change $\frac{14}{3}$ to a mixed number.

$\frac{14}{3}$ means $14 \div 3$.

Divide:

$$\begin{array}{r} 4 \\ 3\overline{)14} \\ \underline{12} \\ 2 \end{array}$$

denominator → 3; 4 ← whole number; 2 ← numerator

Write as a mixed number. $4\frac{2}{3}$.

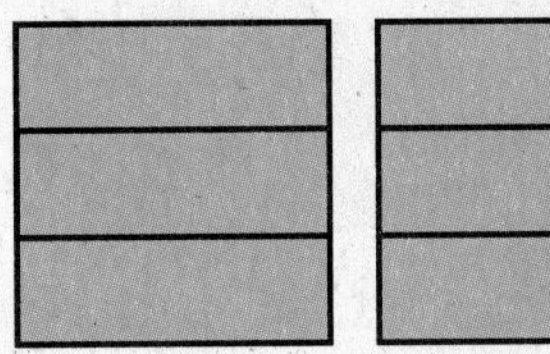
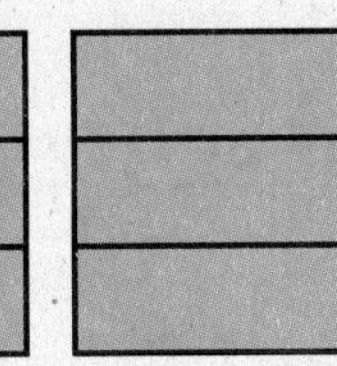
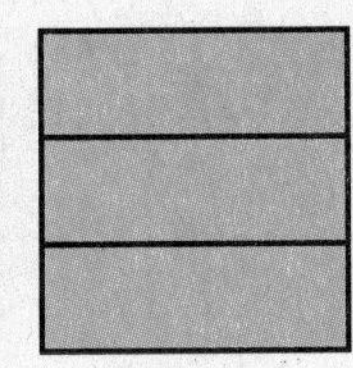
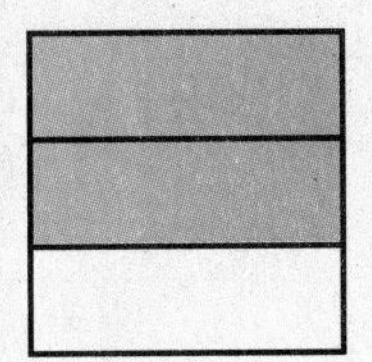

$\frac{14}{3} = 4\frac{2}{3}$

A mixed number can also be changed to an improper fraction. Follow these steps:

Step 1: Multiply the denominator of the fraction by the whole number part of the mixed number.

Step 2: Add the numerator of the fraction part.

Step 3: Write the result over the original denominator.

Example: Change $2\frac{3}{8}$ to an improper fraction.

Multiply 8 by 2. $8 \times 2 = 16$

Add the numerator, 3. $16 + 3 = 19$

Write over the denominator, 8. $\frac{19}{8}$

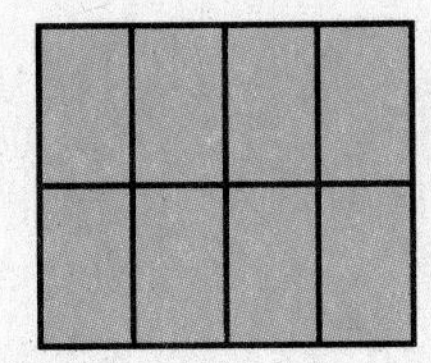
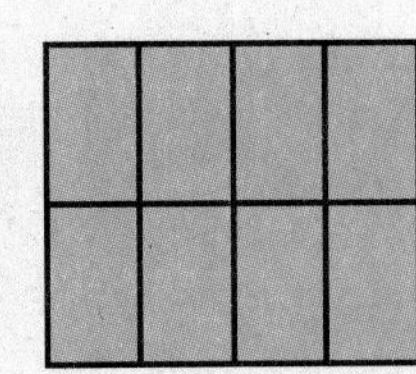
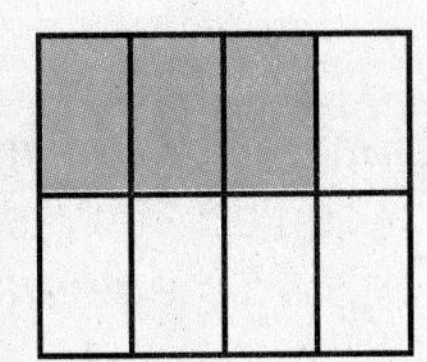

$2\frac{3}{8} = \frac{19}{8}$

Lesson 5

Directions: Write the fraction for the shaded portion.

1. 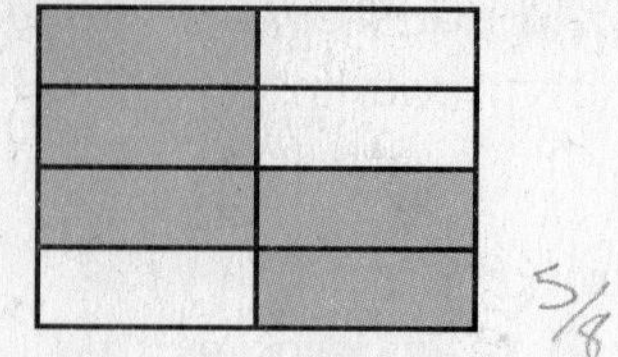

2. 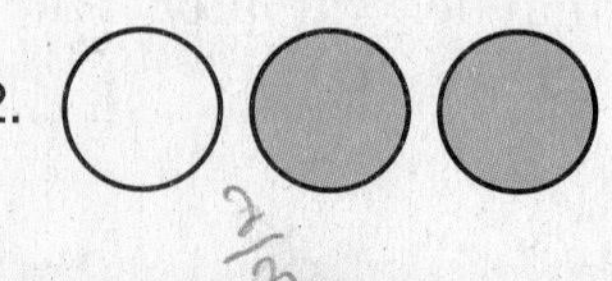

3. 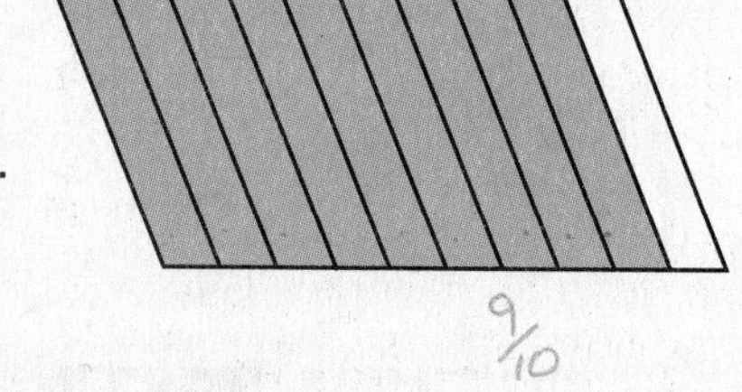

4. 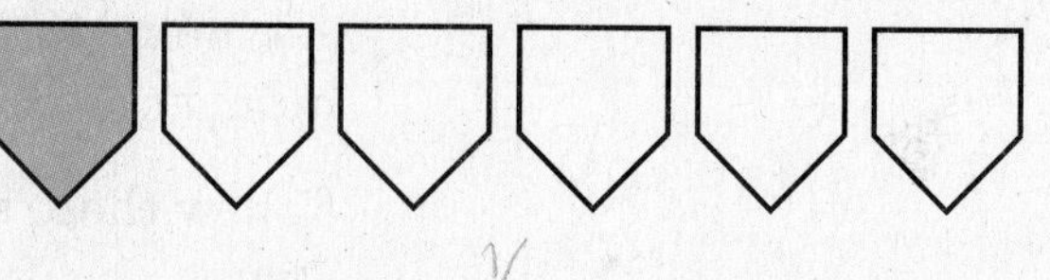

5. 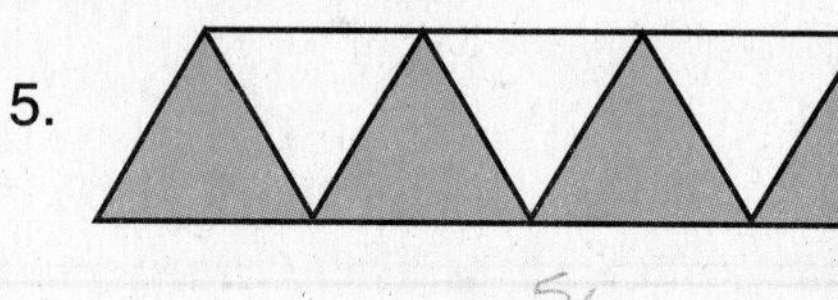

Directions: Solve.

6. On Tuesday, 9 of Steve's 16 customers paid using credit cards. What fraction of his customers used credit cards?

7. Out of the 10 people in Carla's office, 7 regularly watch the evening news on TV. What fraction of the office workers watch the evening news?

8. In the election for union representative, Carlos received 19 votes. What fraction of the union's 45 members voted for Carlos?

9. Joy Chan arrived over half an hour early on 3 of the last 50 workdays. What fraction of the time has she been over half an hour early?

10. Out of 70 cases delivered yesterday, 9 were damaged. What fraction of the cases were damaged?

11. A video rental store has 500 movies in stock. Only 43 of the movies are for children. What fraction of the movies in stock are children's movies?

Directions: Write the improper fraction for the shaded portion.

12. 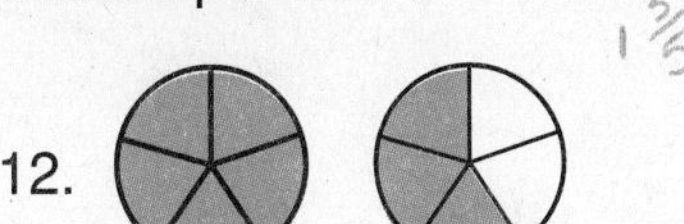

13. 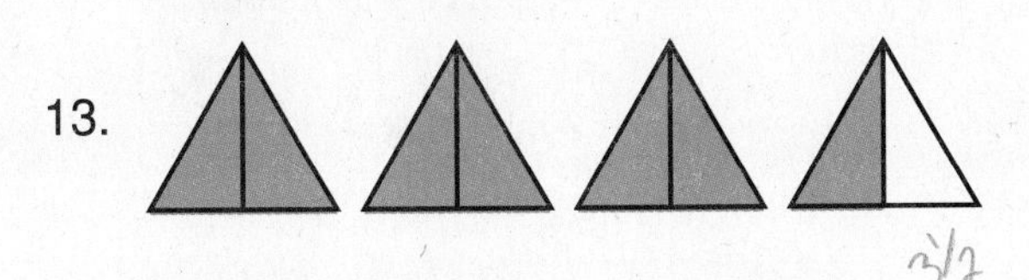

14. 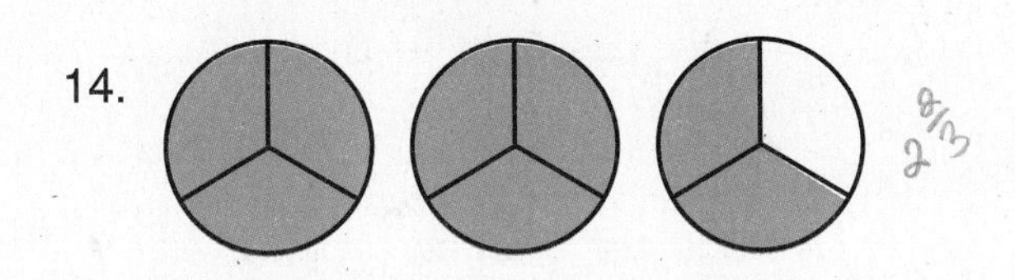

Directions: Write the mixed number for the shaded portion.

15. 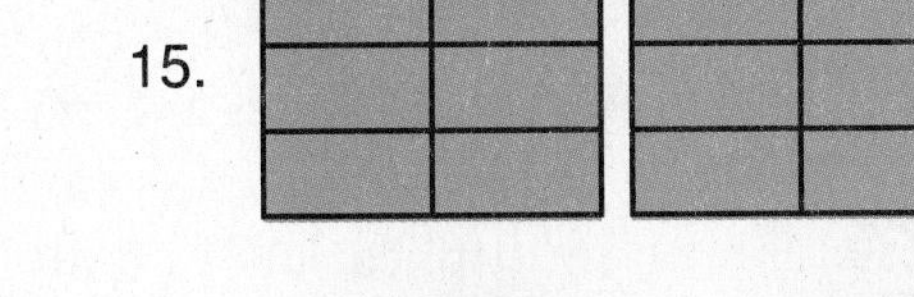

16. 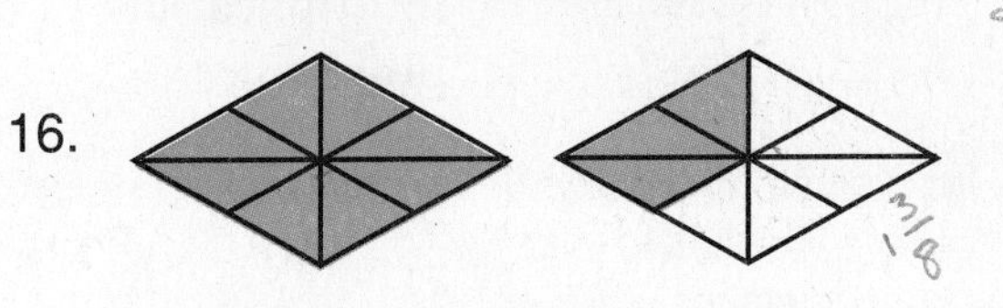

17.

Directions: Change each improper fraction to a whole or a mixed number.

18. $\frac{7}{2} =$

19. $\frac{8}{3} =$

20. $\frac{57}{10} =$

21. $\frac{15}{5} =$

22. $\frac{10}{3} =$

23. $\frac{77}{9} =$

24. $\frac{19}{4} =$

25. $\frac{31}{8} =$

26. $\frac{17}{6} =$

27. $\frac{9}{2} =$

28. $\frac{42}{6} =$

29. $\frac{41}{12} =$

Directions: Change each mixed number to an improper fraction.

30. $6\frac{1}{2} =$

31. $8\frac{1}{2} =$

32. $5\frac{3}{8} =$

33. $3\frac{5}{6} =$

34. $2\frac{1}{5} =$

35. $6\frac{4}{7} =$

36. $1\frac{5}{8} =$

37. $11\frac{3}{4} =$

38. $10\frac{1}{2} =$

39. $4\frac{2}{9} =$

40. $5\frac{1}{12} =$

41. $9\frac{2}{5} =$

Answers are on page 343.

Strategies for Solving Word Problems

Choose the Operation

Fraction problems can also be written in words. The same operations you used to solve whole number problems—addition, subtraction, multiplication, and division—can also be used to solve fraction problems.

Read the problem carefully before you choose the operation you need to solve it. Identify what you are trying to find out.

Review these guidelines for choosing the operation:

You	When You Need To
Add	Combine quantities. Find a total.
Subtract	Find a difference. Take away a quantity. Compare to find "how many more," "how much less," or "how much is left."
Multiply	Put together a number of equal amounts to find a total. Add the same number repeatedly. Find a part "of" a whole item or group
Divide	Split a quantity into equal parts. Find how many equal parts are in a whole item or group.

Examples:

1. A worker at a coffee shop mixed $10\frac{3}{8}$ pounds of gourmet coffee with $6\frac{3}{4}$ pounds of regular grind coffee. How many more pounds of the gourmet coffee than the regular grind coffee were used?

(1) $10\frac{3}{8} + 6\frac{3}{4}$
(2) $10\frac{3}{8} - 6\frac{3}{4}$
(3) $10\frac{3}{8} \times 6\frac{3}{4}$
(4) $10\frac{3}{8} \div 6\frac{3}{4}$

The correct answer is (2). You subtract because you need to compare to find "how many more."

2. Of the 128 shipping room employees $\frac{3}{8}$ volunteered to work on Saturday. How many employees volunteered to work Saturday?

(1) $128 + \frac{3}{8}$
(2) $128 - \frac{3}{8}$
(3) $128 \times \frac{3}{8}$
(4) $128 \div \frac{3}{8}$

The correct answer is (3). You need to find a part "of" a group.

Solving Word Problems

Directions: Circle the operation you would use to solve each problem.

1. A recipe calls for $5\frac{3}{4}$ cups of flour. There are only $2\frac{3}{8}$ cups left in the bag. How much more is needed?

   (1) $5\frac{3}{4} + 2\frac{3}{8}$
   (2) $5\frac{3}{4} - 2\frac{3}{8}$
   (3) $5\frac{3}{4} \times 2\frac{3}{8}$
   (4) $5\frac{3}{4} \div 2\frac{3}{8}$

2. The Wilderness Club hiked $8\frac{1}{8}$ miles on Saturday and $4\frac{4}{5}$ miles on Sunday. What was the combined mileage for the two days?

   (1) $8\frac{1}{8} + 4\frac{4}{5}$
   (2) $8\frac{1}{8} - 4\frac{4}{5}$
   (3) $8\frac{1}{8} \times 4\frac{4}{5}$
   (4) $8\frac{1}{8} \div 4\frac{4}{5}$

3. A restaurant ordered $15\frac{3}{4}$ pounds of almonds. The nut supplier sent only $\frac{1}{2}$ the order. How many pounds of almonds did the supplier send?

   (1) $15\frac{3}{4} + \frac{1}{2}$
   (2) $15\frac{3}{4} - \frac{1}{2}$
   (3) $15\frac{3}{4} \times \frac{1}{2}$
   (4) $15\frac{3}{4} \div \frac{1}{2}$

4. A class at the driving school lasts $\frac{3}{4}$ of an hour. How many classes can Kim teach in a $7\frac{1}{2}$ hour day?

   (1) $7\frac{1}{2} + \frac{3}{4}$
   (2) $7\frac{1}{2} - \frac{3}{4}$
   (3) $7\frac{1}{2} \times \frac{3}{4}$
   (4) $7\frac{1}{2} \div \frac{3}{4}$

5. Phil rode his motorcycle $18\frac{1}{10}$ miles on Friday and $12\frac{2}{5}$ miles on Saturday. How many total miles did he ride?

   (1) $18\frac{1}{10} + 12\frac{2}{5}$
   (2) $18\frac{1}{10} - 12\frac{2}{5}$
   (3) $18\frac{1}{10} \times 12\frac{2}{5}$
   (4) $18\frac{1}{10} \div 12\frac{2}{5}$

6. At Howard Brothers Manufacturing, $\frac{5}{8}$ of the employees take the bus to work. Of these, $\frac{2}{5}$ take the express bus. What fraction of the employees take the express bus?

   (1) $\frac{5}{8} + \frac{2}{5}$
   (2) $\frac{5}{8} - \frac{2}{5}$
   (3) $\frac{5}{8} \times \frac{2}{5}$
   (4) $\frac{5}{8} \div \frac{2}{5}$

7. A hiking trail in a national park is $6\frac{1}{4}$ miles long. If Steve averages $2\frac{1}{2}$ miles per hour, how many hours will it take him to reach the end of the trail?

   (1) $6\frac{1}{4} + 2\frac{1}{2}$
   (2) $6\frac{1}{4} - 2\frac{1}{2}$
   (3) $6\frac{1}{4} \times 2\frac{1}{2}$
   (4) $6\frac{1}{4} \div 2\frac{1}{2}$

8. A flight from San Francisco to Seattle took $3\frac{1}{4}$ hours. The return flight took only $2\frac{4}{5}$ hours. How much shorter was the return flight?

   (1) $3\frac{1}{4} + 2\frac{4}{5}$
   (2) $3\frac{1}{4} - 2\frac{4}{5}$
   (3) $3\frac{1}{4} \times 2\frac{4}{5}$
   (4) $3\frac{1}{4} \div 2\frac{4}{5}$

Answers are on page 344.

Lesson 6 Comparing Fractions

Equal Fractions

You know from experience that different fractions can have the same value.

There are 100 pennies in a dollar.
20 pennies out of 100 = $\frac{20}{100}$.

There are 20 nickels in a dollar.
4 nickels out of 20 = $\frac{4}{20}$.

Both $\frac{20}{100}$ and $\frac{4}{20}$ are worth 20 cents or $\frac{1}{5}$ of a dollar.

$$\frac{20}{100} = \frac{4}{20} = \frac{1}{5}$$

Fractions that have the same value are called **equal fractions.**

You can tell if two fractions are equal by **cross multiplying.** If the products are equal, the fractions are equal.

Example: Are $\frac{4}{8}$ and $\frac{3}{6}$ equal fractions?

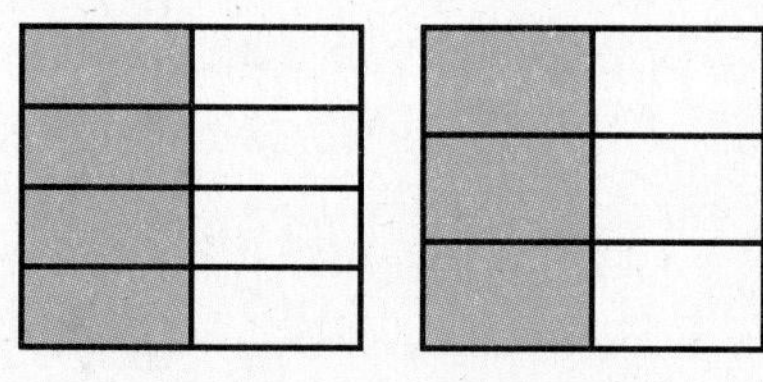

Multiply: $\frac{4}{8} \times \frac{3}{6}$ $4 \times 6 = 24$ $8 \times 3 = 24$

$$\frac{4}{8} = \frac{3}{6}$$

Because both products are the same, $\frac{4}{8}$ and $\frac{3}{6}$ are equal.

Reducing Fractions

Reducing a fraction means finding an equal fraction with a smaller numerator and denominator. To reduce a fraction, divide the numerator and denominator by the same number. A fraction has been reduced to **lowest terms** when there is no number other than 1 that will divide evenly into both the numerator and the denominator.

Example: Reduce $\frac{6}{10}$ to its lowest terms.

$$\frac{6}{10} = \frac{6 \div 2}{10 \div 2} = \frac{3}{5}$$

The fraction $\frac{3}{5}$ is $\frac{6}{10}$ reduced to its lowest terms. There is no number other than 1 that will divide evenly into 3 and 5.

Building Equal Fractions

Sometimes you need to find an equal fraction with higher terms (numerator and denominator) than those of the original. You can build or raise a fraction by multiplying both the numerator and the denominator by the same number (except 0).

$\frac{5}{8} = \frac{5 \times 4}{8 \times 4} = \frac{20}{32}$ $\quad \frac{5}{8}$ and $\frac{20}{32}$ are equal fractions.

Often you need to find an equal fraction with a given denominator.

Example: $\frac{3}{4} = \frac{?}{24}$

Because $4 \times \mathbf{6} = 24$,
multiply the numerator 3 by **6.** $\quad \frac{3 \times 6 = 18}{4 \times 6 = 24}$

The fractions $\frac{3}{4}$ and $\frac{18}{24}$ are equal fractions.

Comparing Fractions

When fractions have the same number as the denominator, they are said to have a **common denominator** and the fractions are called **like fractions.** When comparing two like fractions, the fraction with the greater numerator is the greater fraction.

Example: Which fraction is greater, $\frac{3}{5}$ or $\frac{4}{5}$?
The fractions $\frac{3}{5}$ and $\frac{4}{5}$ are like fractions because they have a common denominator, 5. Since 4 is greater than 3, $\frac{4}{5}$ is greater than $\frac{3}{5}$.

To compare **unlike fractions**—fractions with different denominators—you must change them to fractions with a common denominator. The common denominator must be a multiple of both of the original denominators.

Follow these steps to compare unlike fractions:

Step 1: Think of a common denominator.
Step 2: Build equal fractions with a common denominator.
Step 3: Compare the numerators.
Example: Which is greater, $\frac{5}{6}$ or $\frac{3}{4}$?
The number 24 is a multiple of both 6 and 4 ($6 \times 4 = 24$). However, 12 is the smallest multiple of 6 and 4. Using 12, the **lowest common denominator,** build equal fractions.

$\frac{5 \times 2}{6 \times 2} = \frac{10}{12}$

$\frac{3 \times 3}{4 \times 3} = \frac{9}{12}$

Compare the like fractions.
$\frac{10}{12}$ is greater than (>) $\frac{9}{12}$,
so $\frac{5}{6}$ is greater than $\frac{3}{4}$.

ged Practice Lesson 6

Directions: Reduce each fraction to lowest terms.

1. $\frac{2}{4} =$
2. $\frac{6}{9} =$
3. $\frac{10}{25} =$
4. $\frac{6}{8} =$
5. $\frac{6}{15} =$
6. $\frac{18}{27} =$
7. $\frac{5}{20} =$
8. $\frac{12}{48} =$
9. $\frac{8}{10} =$
10. $\frac{12}{30} =$
11. $\frac{7}{42} =$
12. $\frac{24}{36} =$

Directions: Find an equal fraction with the given denominator.

13. $\frac{1}{2} = \frac{?}{8}$
14. $\frac{2}{3} = \frac{?}{12}$
15. $\frac{2}{7} = \frac{?}{21}$
16. $\frac{4}{5} = \frac{?}{25}$
17. $\frac{5}{8} = \frac{?}{32}$
18. $\frac{7}{9} = \frac{?}{63}$
19. $\frac{3}{10} = \frac{?}{120}$
20. $\frac{3}{8} = \frac{?}{96}$
21. $\frac{3}{4} = \frac{?}{36}$
22. $\frac{4}{9} = \frac{?}{81}$
23. $\frac{9}{50} = \frac{?}{150}$
24. $\frac{12}{25} = \frac{?}{100}$

Directions: Compare the fractions. Write >, <, or = between the fractions.

25. $\frac{1}{3} \quad \frac{1}{4}$
26. $\frac{3}{4} \quad \frac{7}{8}$
27. $\frac{3}{9} \quad \frac{1}{3}$
28. $\frac{2}{3} \quad \frac{1}{2}$
29. $\frac{1}{3} \quad \frac{2}{5}$
30. $\frac{5}{6} \quad \frac{15}{18}$
31. $\frac{9}{12} \quad \frac{3}{4}$
32. $\frac{7}{10} \quad \frac{2}{3}$
33. $\frac{7}{15} \quad \frac{2}{5}$
34. $\frac{3}{4} \quad \frac{17}{20}$
35. $\frac{8}{9} \quad \frac{16}{18}$
36. $\frac{5}{6} \quad \frac{3}{4}$

Directions: Choose the best answer to each item.

37. At 10 A.M., $\frac{9}{10}$ of the shirts manufactured at Crown Shirt passed inspection. At 5 P.M., $\frac{7}{8}$ passed inspection. Which statement is true?

    (1) A greater fraction of shirts passed inspection at 10 A.M.
    (2) A greater fraction of shirts passed inspection at 5 P.M.
    (3) The same fraction of shirts passed at each time.

38. Morley's Paint Store ran an ad on television. On Thursday, 4 out of 12 customers said they saw the ad. On Friday, 6 out of 18 customers said they saw it. Which statement is true?

    (1) A greater fraction of customers saw the ad on Thursday.
    (2) A greater fraction of customers saw the ad on Friday.
    (3) The same fraction of customers saw the ad each day.

Items 39 to 43 refer to the following information.

Five teams are polling voters before an election. With two weeks to go, they compare their work.

Election Polling Results				
Team A	**Team B**	**Team C**	**Team D**	**Team E**
$\frac{2}{5}$ of goal	$\frac{1}{2}$ of goal	$\frac{1}{4}$ of goal	$\frac{3}{5}$ of goal	$\frac{3}{4}$ of goal

39. Which team completed the smallest part of its goal?

    (1) Team A
    (2) Team B
    (3) Team C
    (4) Team D
    (5) Team E

40. Which team completed the greatest part of its goal?

    (1) Team A
    (2) Team B
    (3) Team C
    (4) Team D
    (5) Team E

41. Which two teams completed more than $\frac{1}{2}$ of their goals?

    (1) Teams A and B
    (2) Teams B and D
    (3) Teams B and E
    (4) Teams C and D
    (5) Teams D and E

42. Which teams completed *at least* $\frac{1}{2}$ of their goals?

    (1) Teams A, D, and E
    (2) Teams B, D, and E
    (3) Teams C, D, and E
    (4) Teams D and E
    (5) Teams A, B, and C

43. An additional team, Team F, completed $\frac{2}{3}$ of its goal. How does Team F's work compare with the work of Team E?

    (1) Team F completed less than Team E.
    (2) Team F completed the same fraction of its goal as Team E.
    (3) Team F completed more than Team E.

Answers are on page 344.

Estimation and Fractions

Use Estimation

Estimating means to find an approximate amount. Rounding fractions to the nearest whole number is a good way to estimate answers to problems that involve fractions.

To round a fraction to the nearest whole number, compare the fraction to $\frac{1}{2}$.

Rule: If the fraction is less than $\frac{1}{2}$, the whole number stays the same.

Example: Round $5\frac{1}{3}$ to the nearest whole number. Compare $\frac{1}{3}$ and $\frac{1}{2}$. Change to equal fractions with a common denominator of 6.

$$\frac{1}{3} = \frac{2}{6} \quad \frac{1}{2} = \frac{3}{6}$$

The fraction $\frac{1}{3}$ is less than $\frac{1}{2}$ because $\frac{2}{6}$ is less than $\frac{3}{6}$. The mixed number $5\frac{1}{3}$ rounds to the whole number 5.

Rule: If the fraction is $\frac{1}{2}$ or more, add 1 to the whole number.

Example: Round $8\frac{5}{8}$ to the nearest whole number.

Compare $\frac{5}{8}$ and $\frac{1}{2}$. Since 8 is a multiple of 2, you can change $\frac{1}{2}$ to an equal fraction with a denominator of 8.

$$\frac{1}{2} = \frac{4}{8} \qquad \text{Compare. } \frac{5}{8} > \frac{4}{8}$$

The fraction $\frac{5}{8}$ is greater than $\frac{1}{2}$ because $\frac{5}{8}$ is greater than $\frac{4}{8}$. The mixed number $8\frac{5}{8}$ rounds to the whole number 9.

Estimation with fractions is most accurate in addition and subtraction problems.

Example: Kikko and Maria each jog two times a week. Kikko jogged $4\frac{3}{4}$ miles on Monday and $3\frac{1}{6}$ miles on Tuesday. Maria jogged $3\frac{1}{4}$ miles on Tuesday and $4\frac{1}{8}$ miles on Thursday. Who jogged farther?

Estimate Kikko's total:

$4\frac{3}{4}$ is about 5

$+\ 3\frac{1}{6}$ is about 3

8 miles rounded

Estimate Maria's total:

$3\frac{1}{4}$ is about 3

$+\ 4\frac{1}{8}$ is about 4

7 miles rounded

Kikko jogged approximately 1 mile farther than Maria.

Note: Estimation in fraction multiplication and division problems can help you decide if your answer is reasonable, but your estimate may not be very close to the exact solution.

GED Practice — Estimation and Fractions

Directions: Compare the fractions to one half. Write $>$, $<$, or $=$ between the fractions.

1. $\frac{1}{3}$ $\frac{1}{2}$
2. $\frac{5}{6}$ $\frac{1}{2}$
3. $\frac{2}{3}$ $\frac{1}{2}$
4. $\frac{1}{4}$ $\frac{1}{2}$
5. $\frac{5}{8}$ $\frac{1}{2}$
6. $\frac{4}{5}$ $\frac{1}{2}$
7. $\frac{1}{8}$ $\frac{1}{2}$
8. $\frac{3}{6}$ $\frac{1}{2}$
9. $\frac{3}{4}$ $\frac{1}{2}$
10. $\frac{2}{7}$ $\frac{1}{2}$
11. $\frac{2}{5}$ $\frac{1}{2}$
12. $\frac{7}{10}$ $\frac{1}{2}$
13. $\frac{4}{9}$ $\frac{1}{2}$
14. $\frac{3}{7}$ $\frac{1}{2}$
15. $\frac{3}{5}$ $\frac{1}{2}$

Directions: Round the mixed numbers or fractions to the nearest whole number.

16. $3\frac{3}{4}$

Nearest whole number:

17. $5\frac{1}{3}$

Nearest whole number:

18. $\frac{2}{3}$

Nearest whole number:

19. $\frac{1}{4}$

Nearest whole number:

20. $5\frac{3}{5}$

Nearest whole number:

21. $2\frac{1}{6}$

Nearest whole number:

22. $7\frac{7}{8}$

Nearest whole number:

23. $1\frac{1}{2}$

Nearest whole number:

24. $4\frac{2}{5}$

Nearest whole number:

25. $6\frac{7}{10}$

Nearest whole number:

26. $\frac{11}{20}$

Nearest whole number:

27. $3\frac{3}{7}$

Nearest whole number:

28. $10\frac{5}{6}$

Nearest whole number:

29. $8\frac{3}{8}$

Nearest whole number:

30. $\frac{7}{15}$

Nearest whole number:

Items 31 to 33 refer to the following information.

Tom Ortez buys supplies for Superior Construction. Last week he bought:

Paint: $14\frac{1}{3}$ gallons red paint
$6\frac{3}{4}$ gallons green paint
$9\frac{1}{4}$ gallons yellow paint

Hardware: $12\frac{1}{6}$ pounds of nails

Lumber: $9\frac{5}{8}$ feet of 2-by-4 boards
$27\frac{1}{4}$ feet of 2-by-8 boards
$4\frac{2}{3}$ feet of 1-by-4 boards
$36\frac{3}{8}$ feet of 1-by-6 boards

31. Approximately how many gallons of paint did Tom buy?

(1) 28
(2) 29
(3) 30
(4) 31
(5) 32

32. Approximately how many feet of lumber did Tom buy?

(1) 75
(2) 76
(3) 77
(4) 78
(5) 79

33. Tom has $10\frac{2}{5}$ pounds of nails on hand. After this purchase, approximately how many pounds of nails will he have?

(1) 21
(2) 22
(3) 24
(4) 25
(5) 26

Items 34 to 36 refer to the following information.

Walton Nut Company sells two kinds of nut mixtures.

Mix A: $2\frac{2}{3}$ pounds of cashews,
$2\frac{3}{8}$ pounds of peanuts,
$3\frac{1}{2}$ pounds of salted walnuts, and
$2\frac{1}{8}$ pounds of Brazil nuts.

Mix B: $6\frac{1}{2}$ pounds of almonds,
$3\frac{7}{8}$ pounds of black walnuts, and
$4\frac{1}{5}$ pounds of peanuts.

34. Approximately how many more pounds of peanuts are in Mix B than in Mix A?

(1) 1
(2) 2
(3) 3
(4) 4
(5) 5

35. Estimate the number of pounds of cashews and Brazil nuts in Mix A.

(1) 2
(2) 3
(3) 4
(4) 5
(5) 6

36. Mix B contains approximately how many more pounds of nuts than Mix A?

(1) 1
(2) 2
(3) 3
(4) 4
(5) 5

37. Lorna has a $5\frac{1}{3}$-foot wood board. She needs to cut $1\frac{7}{8}$ feet from the board. After she cuts the length she needs, approximately how many feet of board will Lorna have left?

(1) 3
(2) 4
(3) 5
(4) 6
(5) 7

38. Paul prepared $2\frac{2}{3}$ cups of salad dressing. He used $\frac{3}{4}$ cup of the dressing. About how many cups of salad dressing are left?

(1) 1
(2) 2
(3) 3
(4) 4
(5) 5

39. John is working at his pizza shop. There are 3 whole pies, $\frac{1}{4}$ of a pie, and $\frac{3}{4}$ of a pie already made. About how many pies are already made?

(1) 1
(2) 2
(3) 3
(4) 4
(5) 5

40. Mark needs $1\frac{1}{3}$ cups of milk for the bran muffin mix and $1\frac{1}{8}$ cups of milk for the corn muffin mix. Approximately how many cups of milk does he need?

(1) 1
(2) 2
(3) 3
(4) 4
(5) 5

41. Tara walked $2\frac{1}{4}$ miles from her home to the office. She walked $1\frac{7}{8}$ miles from the office to the store. Then she walked $2\frac{1}{4}$ miles home. Approximately what is the total number of miles she walked?

(1) 4
(2) 5
(3) 6
(4) 7
(5) 8

42. Ella has $1\frac{1}{3}$ yards of fabric. She needs $4\frac{7}{8}$ yards to recover the pillows in her living room. About how many more yards of fabric does she need?

(1) 2
(2) 3
(3) 4
(4) 5
(5) 6

43. Sandy read these driving directions to the doctor's office: "Go straight on Main for $3\frac{3}{4}$ miles. Make a right onto Green. Take Green for $1\frac{1}{3}$ miles to the office, 190 Green." About how many miles is it to the doctor's office?

(1) 2
(2) 3
(3) 4
(4) 5
(5) 6

44. Elena needs $1\frac{3}{4}$ yards of silk to make a blouse and $2\frac{1}{8}$ yards to make matching pants. Approximately how many yards of silk does she need?

(1) 1
(2) 2
(3) 3
(4) 4
(5) 5

Answers are on page 345.

Adding and Subtracting Fractions

Like Fractions

Like fractions have the same or **common denominators.** You can add or subtract like fractions by adding or subtracting the numerator and writing the answer over the common denominator. If necessary, reduce the answer to lowest terms.

Example: Subtract $\frac{2}{12}$ from $\frac{11}{12}$.

$\frac{11}{12}$ Since the common denominator is 12, $\frac{11}{12}$ and $\frac{2}{12}$ are like fractions.

$-\frac{2}{12}$ Subtract the numerators $(11 - 2 = 9)$.

$\frac{9}{12}$ Write the answer using the common denominator.

Reduce the answer to lowest terms.

$\frac{9 \div 3}{12 \div 3} = \frac{3}{4}$; $\frac{11}{12} - \frac{2}{12} = \frac{9}{12}$ or $\frac{3}{4}$

Unlike Fractions

Unlike fractions have different denominators. Use these steps to add or subtract unlike fractions:

Step 1: Find the least common denominator and change one or both of the fractions to make like fractions.

Step 2: Add or subtract the like fractions.

Step 3: Reduce the answer if necessary.

Example: Add $\frac{1}{2}$ and $\frac{2}{8}$.

Step 1: Find the least common denominator and change to like fractions. The least common denominator is 8, so you do not need to convert $\frac{2}{8}$.

$$\frac{1 \times 4}{2 \times 4} = \frac{4}{8}$$

Step 2: Add: $\frac{1}{2} = \frac{4}{8}$

$+\frac{2}{8} = \frac{2}{8}$

$\frac{6}{8}$

Step 3: Reduce to lowest terms: $\frac{6 \div 2}{8 \div 2} = \frac{3}{4}$; $\frac{1}{2} + \frac{2}{8} = \frac{6}{8}$ or $\frac{3}{4}$

Adding and Subtracting Mixed Numbers

Follow these steps to add or subtract mixed numbers.

Step 1: Write the fractions with common denominators.

Step 2: Add or subtract the fractions, then add or subtract the whole numbers.

Step 3: If the sum of the fractions is an improper fraction, change it to a mixed number and combine it with the whole number.

Step 4: Reduce to lowest terms if necessary.

Example: Add $3\frac{3}{4} + 6\frac{1}{3}$.

Step 1: Write the fractions with common denominators.

$$3\frac{3}{4} = 3\frac{3 \times 3}{4 \times 3} = 3\frac{9}{12}$$
$$+\ 6\frac{1}{3} = 6\frac{1 \times 4}{3 \times 4} = 6\frac{4}{12}$$

Step 2: Add the fractions, then the whole numbers.

$$3\frac{3}{4} = 3\frac{9}{12}$$
$$+\ 6\frac{1}{3} = 6\frac{4}{12}$$

Answer contains an improper fraction. $9\frac{13}{12}$

Step 3: Change $\frac{13}{12}$ to a mixed number: $\frac{13}{12} = 1\frac{1}{12}$

Add this to the whole number answer: $9 + 1\frac{1}{12} = 10\frac{1}{12}$

So, $3\frac{3}{4} + 6\frac{1}{3} = 10\frac{1}{12}$

Step 4: Make sure the fraction is reduced to lowest terms. In this case, $\frac{1}{12}$ cannot be reduced.

Sometimes the number you are subtracting from does not have the larger fraction. In this case, you will need to borrow 1 from the whole number.

Example: Subtract $3\frac{2}{3}$ from 6.

The whole number 6 needs a fraction part in order to subtract. Borrow 1 from the whole number and write it as the fraction. Remember, a fraction where the numerator and denominator are the same equals 1. So, 6 can be written as $5\frac{3}{3}$.

Subtract.

$$6 = 5\frac{3}{3}$$
$$-\ 3\frac{2}{3} = 3\frac{2}{3}$$
$$2\frac{1}{3}$$

Example: Subtract $2\frac{5}{6}$ from $5\frac{1}{3}$.

Find the common denominator.

$$5\frac{1}{3} = 5\frac{2}{6}$$
$$-\ 2\frac{5}{6} = 2\frac{5}{6}$$

Because $\frac{2}{6}$ is less than $\frac{5}{6}$, you need to borrow. Borrow 1 from the whole number 5 and write it as a fraction using the common denominator of 6. Add $\frac{6}{6}$ to the fraction part.

$$5\frac{2}{6} = 4\frac{6}{6} + \frac{2}{6} = 4\frac{8}{6}$$
$$-\ 2\frac{5}{6} \qquad\qquad -\ 2\frac{5}{6}$$

Subtract.

$$5\frac{1}{3} = 5\frac{2}{6} = 4\frac{8}{6}$$
$$-\ 2\frac{5}{6} = 2\frac{5}{6} = 2\frac{5}{6}$$

Reduce to lowest terms. $2\frac{3}{6} = 2\frac{1}{2}$

GED Practice Lesson 7

Directions: Add. Reduce answers to lowest terms.

1. $\frac{2}{3} + \frac{1}{3}$

2. $\frac{1}{6} + \frac{3}{6}$

3. $\frac{3}{4} + \frac{1}{8}$

4. $\frac{1}{6} + \frac{1}{2}$

5. $\frac{3}{8} + \frac{1}{5}$

6. $\frac{2}{5} + \frac{1}{3}$

7. $2\frac{1}{4} + 5\frac{5}{8}$

8. $3\frac{5}{6} + 10\frac{2}{3}$

9. $2\frac{4}{5} + 8\frac{3}{10}$

10. $18\frac{1}{6} + 24\frac{2}{3}$

11. $3\frac{3}{4} + 4\frac{1}{3}$

12. $1\frac{1}{2} + 5\frac{5}{8}$

13. $12\frac{3}{5} + 16\frac{1}{3}$

14. $16\frac{7}{8} + 38\frac{2}{3}$

15. $20\frac{1}{5} + 19\frac{7}{10}$

16. $10\frac{1}{9} + 21\frac{2}{3}$

17. $5\frac{3}{5} + 12\frac{1}{3}$

18. $22\frac{1}{10} + 17\frac{2}{5}$

19. $31\frac{5}{8} + 24\frac{3}{4}$

20. $17\frac{1}{2} + 18\frac{4}{5}$

21. $\frac{3}{4} + \frac{3}{20} + \frac{2}{5} =$

22. $\frac{5}{8} + \frac{1}{6} + \frac{1}{4} =$

23. $1\frac{1}{4} + 3\frac{1}{10} + 5\frac{2}{3} =$

24. $3\frac{1}{2} + 1\frac{2}{8} + 2\frac{3}{4} =$

25. $1\frac{3}{4} + \frac{7}{8} + 3\frac{5}{6} =$

26. $\frac{5}{9} + 3\frac{2}{3} + 5\frac{1}{6} =$

27. $1\frac{7}{20} + 2\frac{1}{4} + 4\frac{2}{5} =$

28. $3\frac{9}{10} + 5\frac{1}{2} + 6\frac{1}{4} =$

29. $2\frac{1}{3} + 7\frac{2}{5} + 1\frac{7}{10} =$

30. $4\frac{1}{6} + 3\frac{2}{3} + 1\frac{5}{12} =$

Directions: Subtract. Reduce answers to lowest terms.

31. $\frac{3}{4} - \frac{1}{2}$

36. $\frac{5}{9} - \frac{1}{12}$

41. $10\frac{1}{2} - 3\frac{3}{4}$

46. $20\frac{3}{4} - 16\frac{2}{3}$

32. $\frac{2}{3} - \frac{1}{6}$

37. $14 - 3\frac{1}{2}$

42. $15\frac{2}{3} - 7\frac{1}{4}$

47. $18\frac{3}{10} - 10\frac{4}{5}$

33. $\frac{5}{6} - \frac{1}{2}$

38. $27\frac{1}{4} - 15\frac{3}{8}$

43. $20 - 12\frac{5}{6}$

48. $16\frac{1}{8} - 3\frac{1}{4}$

34. $\frac{3}{4} - \frac{1}{3}$

39. $3\frac{1}{5} - 1\frac{4}{5}$

44. $40\frac{7}{8} - 15\frac{3}{4}$

49. $12 - 10\frac{7}{8}$

35. $\frac{9}{10} - \frac{1}{2}$

40. $26\frac{1}{4} - 18\frac{3}{8}$

45. $35\frac{1}{3} - 27\frac{4}{5}$

50. $25\frac{1}{4} - 13\frac{3}{5}$

51. $\frac{7}{10} - \frac{1}{4} =$

56. $6\frac{1}{5} - 3\frac{1}{3} =$

61. $20 - \frac{2}{3} =$

52. $3\frac{2}{3} - 1\frac{1}{4} =$

57. $8\frac{5}{6} - 2\frac{1}{4} =$

62. $18\frac{3}{4} - 9\frac{7}{8} =$

53. $4 - 2\frac{4}{5} =$

58. $10\frac{9}{10} - \frac{3}{5} =$

63. $11 - 3\frac{2}{5} =$

54. $9\frac{1}{2} - 2\frac{5}{8} =$

59. $12 - \frac{7}{8} =$

64. $16\frac{1}{4} - 12\frac{1}{6} =$

55. $15\frac{1}{9} - 11\frac{4}{9} =$

60. $13\frac{3}{4} - \frac{15}{16} =$

65. $10 - 5\frac{2}{7} =$

Answers are on page 347.

Strategies for Solving Word Problems

Use the 4-step approach to solve word problems containing fractions.

Step 1: **Read** the problem carefully. Make sure you understand what you are asked to find.

Step 2: **Plan.** Think about the operations you need to solve the problem.

Step 3: **Do** the work; perform all calculations.

Step 4: **Check** your work. Use estimation to make sure your answer seems reasonable.

Example: A recipe calls for $1\frac{5}{8}$ cups of milk and $3\frac{3}{4}$ cups of water. What is the total amount of liquid called for in the recipe?

(1) $4\frac{1}{2}$ cups

(2) $5\frac{1}{4}$ cups

(3) $5\frac{3}{8}$ cups

(4) $5\frac{7}{8}$ cups

(5) 6 cups

Step 1: Read the problem carefully. You are asked to find the total of the two amounts.

Step 2: Plan to find a total. You need to add $1\frac{5}{8}$ and $3\frac{3}{4}$.

Step 3: Do the work. Remember to rewrite the fractions using a common denominator if necessary. Reduce to lowest terms.

$$\begin{array}{rcl} 1\frac{5}{8} & = & 1\frac{5}{8} \\ +\,3\frac{3}{4} & = & 3\frac{6}{8} \\ \hline & & 4\frac{11}{8} = 5\frac{3}{8} \end{array}$$

Step 4: Check your work. Estimate. Is your answer reasonable?

$$\begin{array}{rlr} 1\frac{5}{8} & \text{rounds to} & 2 \\ +\,3\frac{3}{4} & \text{rounds to} & +\,4 \\ \hline & & 6 \end{array}$$

The answer is reasonable.

The correct answer is **(3) $5\frac{3}{8}$ cups.**

Solving Word Problems

Directions: Choose the best answer to each item.

1. Belinda kept records of her gasoline purchases for one month. She bought $8\frac{5}{10}$ gallons, $9\frac{3}{10}$ gallons, 8 gallons, and $7\frac{7}{10}$ gallons. How many gallons did she buy that month?

   (1) $32\frac{1}{2}$
   (2) $32\frac{7}{10}$
   (3) $33\frac{1}{10}$
   (4) $33\frac{2}{5}$
   (5) $33\frac{1}{2}$

2. Paul earns 10 vacation days per year. He has used $4\frac{1}{2}$ days this year. How many vacation days does he have left?

   (1) 5
   (2) $5\frac{1}{2}$
   (3) 6
   (4) $6\frac{1}{2}$
   (5) 7

3. Mary planned to spend $3\frac{1}{2}$ hours organizing the stock room. She has been working for $1\frac{3}{4}$ hours. How many more hours does she plan to work?

   (1) $1\frac{3}{4}$
   (2) $2\frac{1}{4}$
   (3) $2\frac{3}{4}$
   (4) $3\frac{1}{4}$
   (5) $5\frac{1}{4}$

4. A tailor bought $12\frac{1}{2}$ yards of brown wool, $8\frac{7}{8}$ yards of blue tweed, and $6\frac{3}{4}$ yards of brown plaid. How many yards did he buy in all?

   (1) $26\frac{7}{8}$
   (2) $27\frac{1}{2}$
   (3) $28\frac{1}{8}$
   (4) $28\frac{1}{2}$
   (5) $28\frac{3}{4}$

5. Carmine worked $5\frac{2}{5}$ hours on Monday, $6\frac{1}{2}$ hours on Tuesday, and $8\frac{4}{5}$ hours on Wednesday. How many hours did he work for the three days?

   (1) $19\frac{7}{10}$
   (2) $19\frac{4}{5}$
   (3) $20\frac{1}{5}$
   (4) $20\frac{7}{10}$
   (5) $21\frac{1}{2}$

6. Ed needs to shorten an $8\frac{1}{3}$-foot pole to a length of $5\frac{3}{4}$ feet. How much should he cut off?

   (1) $2\frac{7}{12}$ feet
   (2) $2\frac{3}{4}$ feet
   (3) $3\frac{1}{8}$ feet
   (4) $3\frac{1}{3}$ feet
   (5) $3\frac{1}{2}$ feet

Answers are on page 350.

Multiplying and Dividing Fractions

Multiplying Fractions

The Easton Hotel gives an employment test that lasts $1\frac{3}{4}$ hours. Two-thirds of the time is spent on math and reading questions. How long are the math and reading parts?

To solve this problem, you need to answer this question:

$$\text{What is } \frac{2}{3} \text{ of } 1\frac{3}{4}\text{?}$$

The word "of" means multiply. Use these steps to multiply fractions:

Step 1: Change a mixed number to an improper fraction. Write a whole number as a fraction with a denominator of 1. $\left(3 = \frac{3}{1}\right)$

$$\frac{2}{3} \times 1\frac{3}{4} = \frac{2}{3} \times \frac{7}{4}$$

Step 2: Multiply the numerators. This is the numerator of the answer.

$$\frac{\mathbf{2}}{3} \times \frac{\mathbf{7}}{4} = \frac{\mathbf{14}}{} \quad (2 \times 7 = 14)$$

Step 3: Multiply the denominators. This is the denominator of the answer.

$$\frac{2}{\mathbf{3}} \times \frac{7}{\mathbf{4}} = \frac{14}{\mathbf{12}} \quad (3 \times 4 = 12)$$

Step 4: Reduce the fraction to lowest terms. Write improper fractions as mixed numbers.

$$\frac{14 \div 2}{12 \div 2} = \frac{7}{6} = 1\frac{1}{6}$$

Remember, reducing a fraction means to divide the numerator and the denominator by the same number. Using this principle, you can reduce as you work the problem.

$$\frac{1}{3} \times \frac{3}{2} = \frac{1 \times \overset{1}{\cancel{3}}}{\underset{1}{\cancel{3}} \times 2} = \frac{1}{2}$$

The numerator and the denominator are both divided by 3.

Drawing a slash through the 3 and writing 1 shows that 3 divided by 3 equals 1. This reducing process is called **cancellation.**

As a shortcut, you can cancel without rewriting the problem. Make sure you divide a number in both the numerator and the denominator by the same number. Step 4 becomes Step 2!

Correct:

$$\frac{1}{\underset{3}{\cancel{6}}} \times \frac{\overset{1}{\cancel{2}}}{3}$$

The numerator and the denominator are both divided by 2. The answer is $\frac{1}{9}$.

Incorrect:

$$\frac{1}{\underset{2}{\cancel{6}}} \times \frac{2}{\underset{1}{\cancel{3}}}$$

Although 6 and 3 can both be divided by 3, both numbers are in the denominator.

Example: Multiply $1\frac{2}{3}$ by $7\frac{1}{2}$.

Use these steps:

1. Change to improper fractions.
2. Divide by the same number.
3. Multiply.
4. Write as a mixed number.

$$1\frac{2}{3} \times 7\frac{1}{2} = \frac{5}{3} \times \frac{15}{2} = \frac{5}{\cancel{3}_1} \times \frac{\cancel{15}^5}{2} = \frac{25}{2} = 12\frac{1}{2}$$

(15 and 3 are divided by 3.)

Example: A coffee shop had 125 customers during the lunch rush hour on Monday. One-fifth of the customers ordered the lunch special. How many customers ordered the special?

This problem asks you to find a fractional part of a whole number. Use the same steps.

$$125 \times \frac{1}{5} = \frac{125}{1} \times \frac{1}{5} = \frac{\cancel{125}^{25}}{1} \times \frac{1}{\cancel{5}_1} = \frac{25}{1} = 25$$

Dividing Fractions

A candy company receives a shipment of 12 pounds of lemon drops. If the company sells the lemon drops in $\frac{1}{4}$-pound bags, how many bags can they make from the shipment?

Divide 12 by $\frac{1}{4}$ to solve the problem. The number being divided, 12, is written first: $12 \div \frac{1}{4}$ means "how many $\frac{1}{4}$s are there in 12?"

Follow these steps to divide fractions:

Step 1: Change any mixed numbers to improper fractions. Write a whole number as a fraction with a denominator of 1.

$$12 \div \frac{1}{4} = \frac{12}{1} \div \frac{1}{4}$$

Step 2: Invert the divisor (the fraction you are dividing by) and change the operation to multiplication. For example, when you invert the fraction $\frac{1}{4}$, it becomes $\frac{4}{1}$.

$$\frac{12}{1} \div \frac{1}{4} = \frac{12}{1} \times \frac{4}{1}$$

Step 3: Complete the problem as you would any multiplication problem and reduce to lowest terms.

$$\frac{12}{1} \times \frac{4}{1} = \frac{48}{1} = 48$$

There are 48 one-fourths in 12. The candy company can make 48, $\frac{1}{4}$-pound bags from 12 pounds of lemon drops.

ged Practice **Lesson 8**

Directions: Multiply. Reduce answers to lowest terms.

1. $\frac{1}{2} \times \frac{2}{3} =$
2. $\frac{7}{8} \times \frac{4}{5} =$
3. $\frac{3}{4} \times \frac{3}{7} =$
4. Find $\frac{2}{15}$ of $\frac{3}{8}$.
5. What is $\frac{8}{9}$ of $\frac{5}{6}$?
6. What is $\frac{5}{12}$ of $\frac{9}{10}$?
7. Find $\frac{1}{3}$ of 6.
8. Find $\frac{3}{5}$ of 15.
9. Find $\frac{2}{3}$ of 18.
10. What is $\frac{7}{8}$ of 24?
11. Find $\frac{1}{2}$ of 11.
12. What is $\frac{2}{9}$ of 24?
13. $\frac{4}{5} \times 1\frac{2}{3} =$
14. $1\frac{1}{2} \times 1\frac{1}{2} =$
15. $9 \times 5\frac{1}{3} =$
16. $\frac{3}{8} \times 1\frac{1}{6} =$
17. $2\frac{2}{5} \times 2\frac{1}{2} =$
18. $3\frac{1}{3} \times 4\frac{1}{8} =$
19. $\frac{5}{6} \times \frac{9}{2} =$
20. $2\frac{3}{4} \times 6 =$
21. $3 \times 5\frac{1}{4} =$
22. $2\frac{5}{6} \times 4 =$
23. $2\frac{1}{3} \times 2\frac{1}{3} =$
24. $\frac{4}{5} \times 1\frac{2}{5} =$
25. $1\frac{3}{4} \times 1\frac{1}{4} =$
26. $2\frac{3}{8} \times 4 =$
27. $3\frac{1}{3} \times 3\frac{1}{5} =$
28. $4 \times 1\frac{7}{8} =$
29. $2\frac{2}{3} \times 1\frac{1}{2} =$
30. $3\frac{1}{9} \times 2\frac{1}{4} =$

Directions: Divide. Reduce answers to lowest terms.

31. $\frac{1}{3} \div \frac{5}{6} =$

32. $\frac{2}{3} \div \frac{2}{5} =$

33. $\frac{7}{10} \div 2 =$

34. $\frac{5}{8} \div \frac{5}{24} =$

35. $\frac{6}{7} \div 3 =$

36. $\frac{4}{9} \div \frac{2}{3} =$

37. $\frac{7}{8} \div \frac{1}{4} =$

38. $4\frac{1}{2} \div \frac{1}{8} =$

39. $15 \div \frac{1}{3} =$

40. $\frac{4}{5} \div \frac{1}{2} =$

41. $12 \div 1\frac{1}{2} =$

42. $3\frac{3}{4} \div 1\frac{2}{3} =$

43. $6\frac{1}{2} \div \frac{1}{4} =$

44. $2\frac{1}{4} \div 1\frac{1}{2} =$

45. $18 \div \frac{2}{3} =$

46. $2\frac{2}{5} \div \frac{6}{25} =$

47. $4\frac{9}{10} \div 1\frac{1}{6} =$

48. $6\frac{1}{9} \div 1\frac{5}{6} =$

49. $6 \div 2\frac{1}{4} =$

50. $3\frac{5}{8} \div 4 =$

51. $\frac{3}{4} \div 1\frac{1}{2} =$

52. $2\frac{2}{3} \div \frac{1}{3} =$

53. $4 \div 1\frac{1}{4} =$

54. $5\frac{1}{2} \div 1\frac{1}{2} =$

55. $10 \div 1\frac{1}{5} =$

56. $8\frac{3}{4} \div \frac{1}{4} =$

57. $12 \div \frac{4}{9} =$

58. $16 \div \frac{4}{5} =$

59. $9\frac{1}{8} \div 1\frac{2}{3} =$

60. $4 \div 2\frac{1}{5} =$

Answers are on page 350.

Strategies for Solving Word Problems

Identifying Necessary Information

Some fraction problems may not include all the information you need to solve them. Other problems may have information that is not necessary. When you solve problems, think carefully about the situation. Be sure you have the information you need and are considering only the necessary information.

Example: Lyle is making a bookcase that is to stand 4 feet tall. He needs to cut a board $5\frac{1}{4}$ feet long into equal pieces. How long will each piece be?

(1) $1\frac{1}{20}$ feet
(2) $1\frac{5}{16}$ feet
(3) $1\frac{3}{4}$ feet
(4) $2\frac{5}{8}$ feet
(5) Not enough information is given.

The correct answer is **(5) Not enough information is given.** To find out how long each of the equal pieces will be, you would divide the length of the board by the number of equal pieces. You have the length of the board, but you do not know the number of equal pieces. The height of the bookcase is not needed to solve the problem.

Example: In March, a flower shop made $\frac{3}{10}$ of their deliveries to homes, $\frac{2}{5}$ to hospitals, $\frac{1}{5}$ to restaurants, and $\frac{1}{10}$ to other places. What fraction of the deliveries were made to homes and hospitals?

(1) $\frac{1}{10}$
(2) $\frac{3}{25}$
(3) $\frac{1}{5}$
(4) $\frac{7}{10}$
(5) Not enough information is given.

You need only two facts: the fraction of deliveries made to homes and the fraction made to hospitals. Add the fractions to find the answer.

$$\frac{3}{10} = \frac{3}{10}$$
$$+\frac{2}{5} = \frac{4}{10}$$
$$\frac{7}{10}$$

The answer is **(4) $\frac{7}{10}$.** The fraction of deliveries made to restaurants and other places is not needed.

Solving Word Problems

Directions: Choose the best answer to each item.

1. A restaurant ordered $15\frac{3}{4}$ pounds of almonds. The nut supplier sent only $\frac{1}{2}$ the order. How many pounds of almonds did the supplier send?
   (1) $3\frac{11}{16}$
   (2) $5\frac{1}{4}$
   (3) $7\frac{7}{8}$
   (4) $16\frac{1}{4}$
   (5) Not enough information is given.

2. A short-order cook uses $\frac{1}{3}$ pound of hamburger to make the lunch special. How many specials can he make from 15 pounds of hamburger?
   (1) 5
   (2) 15
   (3) 30
   (4) 45
   (5) Not enough information is given.

3. Of the 60 animals at an animal shelter, $\frac{4}{5}$ are dogs. How many of the animals at the shelter are dogs?
   (1) 28
   (2) 32
   (3) 36
   (4) 48
   (5) Not enough information is given.

4. Sherril exercised for $\frac{1}{2}$ hour on Monday, $\frac{3}{4}$ hour on Wednesday, and $1\frac{1}{4}$ hours on Friday. How many hours did she exercise in those days?
   (1) $1\frac{3}{4}$
   (2) 2
   (3) $2\frac{1}{2}$
   (4) $2\frac{3}{4}$
   (5) 3

5. Elio works for a construction company. His boss recently gave him a raise. If the raise is $\frac{1}{8}$ of his present yearly salary, what is the dollar amount of the raise?
   (1) $2,750
   (2) $3,125
   (3) $3,500
   (4) $3,875
   (5) Not enough information is given.

6. Each board in a stack is $\frac{3}{4}$ inch thick and 8 feet long. If the stack is 24 inches high, how many boards are in the stack?
   (1) 3
   (2) 18
   (3) 32
   (4) 44
   (5) Not enough information is given.

7. A hiking trail in a national park is $6\frac{1}{4}$ miles long. If Steve averages $2\frac{1}{2}$ miles per hour, how many hours will it take him to reach the end of the trail?
   (1) 1
   (2) $2\frac{1}{2}$
   (3) $5\frac{1}{2}$
   (4) $6\frac{1}{4}$
   (5) $12\frac{3}{4}$

8. Sylvia has 3 yards of cotton. If she uses $1\frac{7}{8}$ yards for a blouse, how many yards will she have left?
   (1) 1
   (2) $1\frac{1}{8}$
   (3) $1\frac{1}{2}$
   (4) 2
   (5) $2\frac{1}{8}$

Answers are on page 351.

ged Mini-Test Lessons 5–8

Directions: Choose the best answer to each item.

1. A recent study shows that approximately $\frac{3}{5}$ of the cars driving south on Canyon Road exceed the speed limit. Of 250 cars driving on the road, which expression shows approximately how many exceed the speed limit?

   (1) $\frac{3}{5} + 250$
   (2) $250 - \frac{3}{5}$
   (3) $250 \times \frac{3}{5}$
   (4) $250 \div \frac{3}{5}$
   (5) Not enough information is given.

2. Henry is carpeting three rooms. He needs $12\frac{1}{2}$ square yards of carpet for one room, $20\frac{3}{4}$ square yards for the next room, and $13\frac{1}{3}$ square yards for a third room. If he has 50 square yards of carpet on hand, how many square yards will he have left after he finishes the job?

   (1) $2\frac{7}{12}$
   (2) $3\frac{1}{12}$
   (3) $3\frac{5}{12}$
   (4) $3\frac{3}{4}$
   (5) $4\frac{1}{6}$

3. A piece of art paper is $18\frac{1}{4}$ inches long. If you cut off $3\frac{7}{16}$ inches, how many inches are left?

   (1) $12\frac{9}{16}$
   (2) $13\frac{1}{4}$
   (3) $13\frac{3}{4}$
   (4) $14\frac{13}{16}$
   (5) $14\frac{15}{16}$

Items 4 to 6 refer to the following information.
Colson builds kitchen cabinets. He offers five different wood finishes.

Customer Selection Information
$\frac{1}{8}$ of his customers choose dark walnut.
$\frac{3}{8}$ of his customers choose golden oak.
$\frac{1}{4}$ of his customers choose natural grain.
$\frac{3}{16}$ of his customers choose fruitwood.
$\frac{1}{16}$ of his customers choose cherry.

4. Which finish do most customers choose?

   (1) dark walnut
   (2) golden oak
   (3) natural grain
   (4) fruitwood
   (5) cherry

5. What fraction of his customers choose golden oak, natural grain, and fruitwood finishes?

   (1) $\frac{7}{16}$
   (2) $\frac{1}{2}$
   (3) $\frac{13}{16}$
   (4) $\frac{3}{4}$
   (5) Not enough information is given.

6. Of 200 customers, how many would Colson expect to choose dark walnut?

   (1) 25
   (2) 30
   (3) 35
   (4) 40
   (5) Not enough information is given.

Items 7 to 9 refer to the following information.
Liza and Bill Saenz have a combined income of \$2,080 per month.

Liza and Bill's Monthly Budget	
Rent	$\frac{1}{4}$ of income
Food	$\frac{1}{5}$ of income
Clothes	$\frac{1}{8}$ of income
Bills	$\frac{3}{8}$ of income
Savings and Miscellaneous	$\frac{1}{20}$ of income

7. Which expression shows what fraction of their income Liza and Bill spend on food and clothes?

(1) $\frac{1}{5} + \frac{1}{8}$
(2) $\frac{1}{5} - \frac{1}{8}$
(3) $\frac{1}{8} - \frac{1}{5}$
(4) $\frac{1}{8} \times \frac{1}{5}$
(5) $\frac{1}{8} \div \frac{1}{5}$

8. How much do they spend on rent per month?

(1) \$416
(2) \$520
(3) \$658
(4) \$700
(5) Not enough information is given.

9. How much of their monthly earnings do they put in savings?

(1) \$ 20
(2) \$ 68
(3) \$104
(4) \$416
(5) Not enough information is given.

10. An insurance company estimates that 4 out of every 100 renters do not have any insurance on their personal belongings. What fraction is this?

(1) $\frac{1}{25}$
(2) $\frac{3}{50}$
(3) $\frac{1}{10}$
(4) $\frac{1}{4}$
(5) $\frac{2}{5}$

11. A store owner wants to have continuous music during store hours. The store is open for 10 hours daily. If each music tape lasts $1\frac{1}{4}$ hours, how many tapes will she need for 1 day?

(1) $6\frac{2}{3}$
(2) 8
(3) $9\frac{1}{2}$
(4) 40
(5) Not enough information is given.

12. Each Friday, Emily walks $1\frac{2}{3}$ miles from her home to work and $1\frac{2}{3}$ back. She also walks an additional $3\frac{3}{4}$ miles in the evening. How many miles does Emily walk on Friday?

(1) $5\frac{5}{12}$
(2) $6\frac{1}{4}$
(3) $6\frac{2}{3}$
(4) $7\frac{1}{12}$
(5) Not enough information is given.

13. A plant grows at a rate of $5\frac{1}{2}$ inches per week. At this rate, which expression shows how many inches the plant will grow in $2\frac{1}{2}$ weeks?

(1) $5\frac{1}{2} + 2\frac{1}{2}$
(2) $5\frac{1}{2} - 2\frac{1}{2}$
(3) $2\frac{1}{2} \times 5\frac{1}{2}$
(4) $2\frac{1}{2} \div 5\frac{1}{2}$
(5) $5\frac{1}{2} \div 2\frac{1}{2}$

Answers are on page 352.

Lesson 9 Decimals

Place Value, Comparing, and Rounding

In lessons 5–8, you learned about fractions. Another way to express fractions is with **decimal numbers.** A decimal is a fraction that uses a place value system. Look at the following fractions and their decimal equivalents.

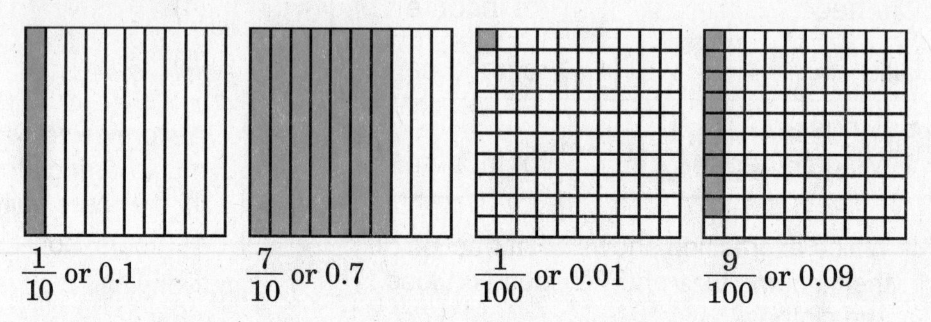

$\frac{1}{10}$ or 0.1 $\frac{7}{10}$ or 0.7 $\frac{1}{100}$ or 0.01 $\frac{9}{100}$ or 0.09

The whole is divided into an equal number of parts. The number of these equal parts is shown by the denominator, 10 or 100.

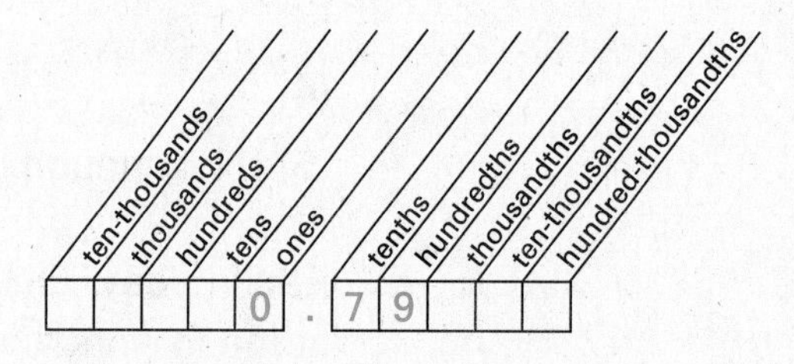

The chart shows the names of the first ten decimal place values. Notice that the whole numbers are to the left of the decimal point and the decimals are to the right. As you move from left to right along the place value chart, the value of the numbers becomes smaller.

As with whole numbers, the sum of the values of the digits equals the total value of the number.

Example: What is the value of each digit in 13.4782?

1 is in the tens place. $1 \times 10 = 10$
3 is in the ones place. $3 \times 1 = 3$
4 is in the tenths place. $4 \times 0.1 = 0.4$
7 is in the hundredths place. $7 \times 0.01 = 0.07$
8 is in the thousandths place. $8 \times 0.001 = 0.008$
2 is in the ten-thousandths place. $+\ 2 \times 0.0001 = 0.0002$

13.4782

Reading Decimals

Follow these steps to read a number with part as a decimal:

Step 1: Read the whole number part.

Step 2: Say the word "and" to indicate the decimal point.

Step 3: Read the digits to the right of the decimal point as a whole number.

Step 4: Say the place name of the last digit on the right.

Examples: What is the word form of 65.034?
Sixty-five and thirty-four thousandths
What is the word form of 1.82?
One and eighty-two hundredths

Comparing and Ordering Decimals

To compare decimals with the same number of decimal places, compare them as though they were whole numbers.

Example: Which is greater, 0.364 or 0.329?
364 is greater than 329, so 0.364 is greater than 0.329.

Use these steps to compare decimals with a different number of digits.

Step 1: Write zeros to the right of the decimal with fewer digits so that the numbers have the same number of decimal places. This does not change the value of the decimal.

Step 2: Compare the decimals as though they were whole numbers.

Example: Which decimal is greater: 0.518 or 0.52?
Add a zero: 0.518 0.520

Compare: 0.518 0.520 0.52 is greater than 0.518

Notice that the decimal with more decimal places may not be the greater decimal. When you compare numbers that have a whole number and a decimal, always compare the whole numbers first.

Example: Compare 32.001 and 31.999
Compare the whole numbers: 32 31
32.001 is greater than 31.999 because 32 is greater than 31. It does not matter that 0.999 is greater than 0.001. The whole number parts 31 and 32 determine which number is greater.

Rounding

Decimals are rounded according to the same rules as whole numbers.

Example: Round 16.959 to the nearest tenth.

Circle the digit in the tenths place.	16.(9)59
Look at the digit to the right.	16.(9)59
The digit is 5, so add 1 to the tenths place.	+ 1
Drop the remaining digits. Since 9 + 1 is 10, put a zero in the tenths place and add one to the ones place.	16.9 17.0

ged Practice Lesson 9

Directions: Write the value of the underlined digit in words.

Example: 1.5$\underline{4}$09 four hundredths

1. 10.92$\underline{5}$1 ______
2. 7.$\underline{8}$5 ______
3. 255.0$\underline{7}$ ______
4. 36.002$\underline{9}$ ______
5. 4.$\underline{1}$62 ______
6. 17.2$\underline{9}$6 ______

7. What is the value of each digit to the right of the decimal point in 0.18362?

1 ______

8 ______

3 ______

6 ______

2 ______

Directions: Match each number in Column A with its word form in Column B.

Column A		Column B
8. ______	1.35	a. one and three hundred five thousandths
9. ______	1.035	b. one and thirty-five thousandths
10. ______	1.305	c. one and thirty-five hundredths
11. ______	1.0305	d. one and three hundred five ten-thousandths

Column A		Column B
12. ______	1.2	a. twelve hundredths
13. ______	0.12	b. twelve thousandths
14. ______	0.0012	c. twelve ten-thousandths
15. ______	0.012	d. one and two tenths

Directions: Write each number in word form.

16. 5.25 ______
17. 6.008 ______
18. 0.37 ______
19. 1.01 ______
20. 2.005 ______
21. 4.05 ______
22. 3.9 ______
23. 0.08 ______
24. 3.004 ______
25. 12.6 ______

Directions: Compare the following numbers. Write >, <, or = between the numbers.

26. 0.32 0.3109
27. 0.98 1.9
28. 0.5 0.50
29. 0.006 0.06
30. 1.075 1.57
31. 0.18 0.108
32. 2.38 2.83
33. 1.09 1.009
34. 3.60 3.600

Directions: List these numbers in order from least to greatest.

35. 3.4 3.09 3.9 3.901 ____________________

36. 0.08 0.8 0.89 ____________________

37. 0.95 0.954 0.9054 ____________________

38. 12.608 12.001 12.8 12.04 ____________________

Directions: Round these numbers.

39. Round **3.5719** to the tenths place. ____________________

40. Round **125.0699** to the thousandths place. ____________________

41. Round **5.132** to the hundredths place. ____________________

42. Round **17.0813** to the tenths place. ____________________

43. Round **0.6415** to the hundredths place. ____________________

44. Round **3.8301** to the ones place. ____________________

45. Round **0.543** to the ones place. ____________________

46. Round **4.617** to the hundredths place. ____________________

47. Round **1.2491** to the tenths place. ____________________

48. Round **0.5104** to the thousandths place. ____________________

49. Round **3.389** to the ones place. ____________________

50. Russ inspects transistors at an electronics firm. In one test, he weighs the transistors. He reports the weight of the transistors in grams (g) rounded to the nearest tenth. For each of the following, what weight should he report?

A. 0.3619 g ____________________ g

B. 0.7082 g ____________________ g

C. 0.0561 g ____________________ g

D. 0.9357 g ____________________ g

Answers are on page 352.

Estimation and Money

Use Estimation with Money

To estimate means to find an approximate amount. Many money situations do not require exact amounts. In such cases, you can use amounts rounded to the nearest whole dollar (the ones place).

Auto Parts Price List	
Outside Wide-Angle Mirror	$13.45
Steering Wheel Cover	$15.95
Oil Drip Pan	$8.73
Windshield Washer Fluid	$2.85
Brake Fluid	$6.35

Example: Using the price list, about how much would Pat pay for a steering wheel cover, a wide-angle mirror, and an oil drip pan?

(1) between $31 and $33
(2) between $33 and $35
(3) between $35 and $37
(4) between $37 and $39
(5) between $39 and $41

Round the cost of each item to the nearest whole dollar and find the total of the estimates.

Item	**Cost**	**Estimate**
Steering wheel cover	$15.95	$16
Wide-angle mirror	13.45	13
Oil drip pan	8.73	+ 9
Total:		$38

The best answer is **(4) between $37 and $39.**

You can also use estimation in multiplication and division problems.

Example: Use the price list to estimate the number of bottles of brake fluid a customer could buy for $20.

(1) 1
(2) 2
(3) 3
(4) 4
(5) 5

Round the price of one bottle of brake fluid to the nearest whole dollar: $6.35 rounds to $6.

Divide $20 by $6: $20 ÷ $6 = 3 with a remainder of 2.
The customer could expect to buy 3 bottles and have some change left over. The correct answer is **(3) 3.**

Estimation and Money

Directions: Choose the best answer to each item.

Items 1 to 3 refer to the following information.

Computer Game Sales		
Game	Regular Price	Sale Price
Fast Pitch	$11.79	$8.99
Par 4	8.85	6.29
Dugout Derby	17.25	12.78
Crown of Power	13.72	10.09
Thunderclap Mine	12.99	9.25
Batwing	10.77	7.98

1. Lee wants to buy Fast Pitch, Crown of Power, and Dugout Derby. About how much will the three games cost on sale?

   (1) between $21 and $24
   (2) between $24 and $27
   (3) between $27 and $30
   (4) between $30 and $33
   (5) between $33 and $36

2. Ana can spend $20 on computer games. She plans to buy Par 4 for $6.29 and Batwing for $7.98. At the sale price, which other game can she also afford?

   (1) Fast Pitch
   (2) Dugout Derby
   (3) Crown of Power
   (4) Thunderclap Mine
   (5) She cannot afford any other game.

3. Mark wants to buy Dugout Derby and Crown of Power. About how much will he save by buying the games on sale?

   (1) $ 7
   (2) $ 8
   (3) $ 9
   (4) $10
   (5) $11

Items 4 and 5 refer to the following information.

Bosco's Burgers
******************** **Menu** ********************

No-Frills Burger		$1.19
Bacon Burger		1.89
Bosco Special		2.95
Hot Dog		1.04
Chili Dog		1.76

	Small	Large
French Fries	$1.05	$1.95
Onion Rings	1.20	2.11

Drinks		
	Small	$0.89
	Medium	1.19
	Large	1.99
	Jumbo	2.25

****

4. Lilia orders 4 Bosco Specials, 4 small orders of fries, and 4 medium drinks for her family. Approximately how much will her order cost?

   (1) between $15 and $17
   (2) between $17 and $19
   (3) between $19 and $21
   (4) between $21 and $23
   (5) between $23 and $25

5. Mr. Lau has six children. He wants to buy each child a No-Frills Burger, a small order of fries, and a large drink. About how much will this cost?

   (1) $20
   (2) $21
   (3) $22
   (4) $23
   (5) $24

6. Len is buying 4 grocery items: lettuce for \$1.29, cereal for \$2.99, milk for \$1.59, and carrots for \$1.19. Which is the nearest estimate of how much the groceries will cost?

   (1) \$5
   (2) \$6
   (3) \$7
   (4) \$8
   (5) \$9

7. At the hardware store, Lori is buying: 2 packages of light bulbs at \$2.97 each, 1 quart of paint for \$6.47, and a wrench for \$14.99. Which is the nearest estimate of how much Lori spent?

   (1) \$24
   (2) \$25
   (3) \$26
   (4) \$27
   (5) \$28

8. Oscar pays for these things at the drug store: soap for \$1.99, toothpaste for \$2.19, tissues for \$1.14, and a razor for \$.85. He pays with a 10-dollar bill. Which is the nearest estimate of how much change he should get back?

   (1) \$2
   (2) \$3
   (3) \$4
   (4) \$5
   (5) \$6

9. Harry is buying eight blank tapes for his tape recorder. The tapes cost \$3.98 each. What is the nearest estimate of how much change he should get back from \$40?

   (1) \$4
   (2) \$5
   (3) \$6
   (4) \$7
   (5) \$8

Items 10 to 12 refer to the following information.

THE CHICKEN ROASTER

Menu Chicken			
$\frac{1}{4}$-Chicken		\$3.99	
$\frac{1}{2}$-Chicken		\$5.99	
Sides	Small	Large	
Vegetables	\$1.59	\$3.19	
Stuffing	\$1.39	\$2.79	
Potatoes	\$1.29	\$2.69	
Drinks	Small	Medium	Large
Sodas	\$0.79	\$1.09	\$1.89

10. Maria orders a $\frac{1}{2}$-chicken, 2 large orders of stuffing, and 3 medium drinks. About how much will her order cost?

    (1) between \$8 and \$10
    (2) between \$10 and \$12
    (3) between \$12 and \$14
    (4) between \$14 and \$16
    (5) between \$16 and \$18

11. Charles orders a $\frac{1}{4}$-chicken, a small order of vegetables, a small order of potatoes, and a small soda. About how much will this cost?

    (1) between \$5 and \$7
    (2) between \$7 and \$9
    (3) between \$9 and \$11
    (4) between \$11 and \$13
    (5) between \$13 and \$15

12. Mr. Turner orders three $\frac{1}{2}$-chickens, 3 large orders of stuffing, and 3 large orders of vegetables. Approximately how much will his order cost?

    (1) between \$15 and \$18
    (2) between \$18 and \$21
    (3) between \$21 and \$27
    (4) between \$27 and \$37
    (5) between \$37 and \$40

Items 13 to 15 refer to the following information.

PLANTS PLUS Price List	
3-inch Potted Plant	$1.79
4-inch Potted Plant	$2.89
5-inch Potted Plant	$3.69
Potting Soil	$3.19
Watering Can	$1.89

13. How many 3-inch potted plants can Sylvia buy with $10?

(1) 2
(2) 3
(3) 5
(4) 8
(5) 10

14. Ernesto buys 2 5-inch potted plants and 1 watering can. What is the greatest number of 4-inch potted plants he can buy so that the total cost of his entire purchase is about $20?

(1) 1
(2) 2
(3) 3
(4) 4
(5) 5

15. Julia buys 3 bags of potting soil and 1 watering can. About how much change should she get from $15?

(1) $1
(2) $2
(3) $3
(4) $4
(5) $5

16. Hank's Hardware sells a carbon monoxide detector for $38.82 and a smoke detector for $12.39. Which is the nearest estimate of the total cost of the 2 items?

(1) $26
(2) $27
(3) $50
(4) $51
(5) $52

17. Hank's Hardware sells 6 packages of nails that cost $5.75 for each package. Which is the nearest estimate of the total cost of the nails?

(1) $ 5
(2) $ 6
(3) $25
(4) $30
(5) $36

18. Pete pays $185.60 each month on the loan for his delivery van and $46.36 each month for the insurance. Which is the nearest estimate of the combined total Pete pays for his van loan and insurance?

(1) $139
(2) $140
(3) $226
(4) $232
(5) $240

19. Mark wants to buy a sofa for the recreation center that costs $685. He has saved $110. About how much more money does he need to buy the sofa?

(1) $200
(2) $300
(3) $400
(4) $500
(5) $600

Answers are on page 354.

Decimal Operations

Adding and Subtracting Decimals

Examples: Anna assembles machine parts. One part comes in two sections with lengths of 4.875 and 3.25 centimeters. What is the total length of the two sections?

Cesar has $213 in a checking account. If he writes a check for $36.68, how much will be left in the account?

Follow these steps to add and subtract decimals:

Step 1: Write the numbers so that the decimal points are in line. If necessary, write zeros to the right of the last digit so that all the numbers have the same number of decimal places. A number without a decimal point is understood to have one to the right of the ones place.

$$\begin{array}{r} 4.875 \\ +\ 3.25\mathbf{0} \\ \hline \end{array}$$

$$\begin{array}{r} \$213.\mathbf{00} \\ -\ \ 36.68 \\ \hline \end{array}$$

($213 is written as $213.00)

Step 2: Add or subtract as you would with whole numbers. Carry or borrow as normal.

$$\begin{array}{r} \scriptstyle 1\ 1 \\ 4.875 \\ +\ 3.250 \\ \hline 8.125 \end{array} \qquad \begin{array}{r} \scriptstyle 1\,1\ \ 9 \\ \scriptstyle 10\,2\ \ 11 \\ \$213.00 \\ -\ \ 36.68 \\ \hline \$176.32 \end{array}$$

Step 3: Line up the decimal point in the answer with the decimal points in the problem.

$$\begin{array}{r} 4.875 \\ +\ 3.250 \\ \hline 8.125 \end{array} \qquad \begin{array}{r} \$213.00 \\ -\ \ 36.68 \\ \hline \$176.32 \end{array}$$

The total length of Anna's machine part is **8.125** centimeters. Cesar will have **$176.32** left in his checking account.

Multiplying Decimals

Example: In the meat department at a grocery store, a cut of meat weighs 1.6 pounds and costs $1.79 per pound. To the nearest whole cent, what is the cost of the meat?

This problem can be solved by multiplying the weight of the meat by the cost per pound. The problem tells you to round your answer.

To multiply decimals, follow these steps:

Step 1: Multiply as you would with whole numbers. The decimal points do not need to line up. Ignore the decimal points until you are finished.

$$\begin{array}{r} \$1.79 \\ \times\ \ \ 1.6 \\ \hline 1074 \\ 179\ \ \\ \hline 2864 \end{array}$$

Step 2: Count the decimal places in the original problem to find how many places are needed in the answer.

$$\begin{array}{rl} \$1.\mathbf{79} & \text{2 decimal places} \\ \times\ \ 1.\mathbf{6} & \text{1 decimal place} \\ \hline \end{array}$$

There are three decimal places in the problem.

Step 3: Place the decimal point in the answer. Starting on the right, count three decimal places.

```
  $ 1.79
×   1.6
  1 074
  1 79
  $2.864
```

$2.864 rounded to the nearest hundredth is $2.86

The cut of meat costs **$2.86.**

Dividing Decimals

Use these steps to divide a decimal by a whole number:

Step 1: Place the decimal point in the answer directly above the decimal point in the problem.

Step 2: Divide as you would with whole numbers.

Step 3: When a problem shows a remainder, write a zero to the right of the number you are dividing and continue. Keep dividing and writing zeros until either there is no remainder or you reach the place value you need. Carry out the division to one place to the right of the desired place value, then round.

Example: Marvin bought a dual cassette player for $64.55, which he will pay for in 12 equal payments. How much will each payment be? Round your answer to the nearest cent.

```
     $5.379
12) $64.550
    60
     4 5
     3 6
       95
       84
       110
       108
         2
```

Write 0 to the right of the 5 and continue dividing. Round to the nearest cent.

$5.379 rounds to $5.38. Each payment is **$5.38.**

To divide by a decimal, you must make the divisor (the number you are dividing by) a whole number. Move the decimal point in the divisor all the way to the right. Move the decimal point in the number you are dividing the same number of places to the right. Then divide as you would by a whole number.

Example: A druggist is preparing a medication. Each capsule requires 0.007 grams of aspirin. He has 14 grams of aspirin. How many capsules can he prepare?

```
        2,000.
0.007) 14.000
       14
```

Move the decimal point 3 places to the right in both numbers. Write zeros to the right of the number you are dividing.

$14 \div 0.007 = 2{,}000$

Multiplying and Dividing by Powers of Ten

The powers of 10 are 10, 100, 1,000 and so on. When you are multiplying or dividing by a power of ten, a special rule will make your work easier.

Use this rule to multiply and divide by powers of ten:

Count the number of zeros in the power of ten.
To **multiply,** move the decimal point the same number of places to the **right.**
To **divide,** move the decimal point the same number of places to the **left.**

Examples:

$$\begin{array}{r} 1.35 \\ \times\ 10 \\ \hline 13.50 \end{array}$$

Notice that multiplying by 10 has shifted the decimal point one place to the **right.**
$1.35 \times 10 = 13.5$

$$\begin{array}{r} 0.125 \\ 100 \overline{)\,12.500} \\ \underline{10\ 0} \\ 2\ 50 \\ \underline{2\ 00} \\ 500 \\ \underline{500} \end{array}$$

Notice that dividing by 100 has shifted the decimal point two places to the **left.**
$12.5 \div 100 = 0.125$

Example: Multiply 78.5 by 1,000.
There are three zeros in 1,000. Move the decimal point three places to the right. You will need to write zeros as placeholders.
$78.5 \times 1{,}000 = 78.500 = 78{,}500$

Example: Divide 0.5 by 1,000.
There are three zeros in 1,000. Move the decimal point three places to the left. You will need to write zeros in front of the 5.
$0.5 \div 1{,}000 = 0\,000.5 = 0.0005$

Lesson 10

Directions: Solve.

1. $\begin{array}{r} 0.03 \\ +\ 2.6 \\ \hline \end{array}$

3. $\begin{array}{r} 7.1 \\ +\ 8.003 \\ \hline \end{array}$

5. $\begin{array}{r} 5.075 \\ -\ 2.15 \\ \hline \end{array}$

7. $\begin{array}{r} 3.61 \\ +\ 1.2 \\ \hline \end{array}$

2. $\begin{array}{r} 1.35 \\ +\ 4.05 \\ \hline \end{array}$

4. $\begin{array}{r} 6.90 \\ -\ 1.353 \\ \hline \end{array}$

6. $\begin{array}{r} 10.3 \\ -\ 6.125 \\ \hline \end{array}$

8. $\begin{array}{r} 16.05 \\ -\ 4.27 \\ \hline \end{array}$

9. $\begin{array}{r} 1.85 \\ 0.03 \\ 19.007 \\ +\ 62 \\ \hline \end{array}$

10. $\begin{array}{r} 12.4 \\ 11.08 \\ 16.1 \\ +\ 4.575 \\ \hline \end{array}$

11. $\begin{array}{r} 16{,}004.1 \\ -\ 6{,}972.1 \\ \hline \end{array}$

12. $\begin{array}{r} 3.8 \\ -\ 1.006 \\ \hline \end{array}$

13. $\begin{array}{r} 12.87 \\ -\ 9.923 \\ \hline \end{array}$

14. $\begin{array}{r} 23.07 \\ -\ 5.965 \\ \hline \end{array}$

15. $14.01 + 8.6 + 0.058 =$

16. $56.8 - 24.95 =$

17. $0.95 + 1.843 + 3.008 + 0.9 =$

18. $0.6 - 0.3407 =$

19. $3.15 + 2.816 + 4.05 + 0.3 =$

20. $39.05 - 15.7 =$

21. $0.125 + 1.4 + 3.76 + 0.01 =$

22. $25.6 - 12.85 =$

Directions: Place the decimal point in each answer. You may need to add zeros.

Example: 0.5 × 0.1 5 = 7 5 There are 3 decimal places in the problem. Count three places from the right, writing zeros as placeholders. The correct answer is 0.075.

23. 8.5 × 0.4 = 3 4 0

24. 0.04 × 0.6 = 2 4

25. 5.6 × 0.002 = 0 0 1 1 2

26. 3.4 × 0.3 = 1 0 2

27. 12 × 3.06 = 3 6 7 2

28. 21.1 × 14.7 = 3 1 0 1 7

29. 0.008 × 12 = 9 6

30. 0.04 × 15 = 0 6 0

31. 1.2 × 0.07 = 0 0 8 4

32. 25 × 2.4 = 6 0 0

Directions: Multiply.

33. $\begin{array}{r} 1.07 \\ \times \quad 12 \\ \hline \end{array}$

34. $\begin{array}{r} 0.09 \\ \times \ 6.1 \\ \hline \end{array}$

35. $\begin{array}{r} 2.27 \\ \times \ 1.8 \\ \hline \end{array}$

36. $\begin{array}{r} 5.04 \\ \times \quad 15 \\ \hline \end{array}$

37. $\begin{array}{r} 0.008 \\ \times \quad 2.5 \\ \hline \end{array}$

38. $\begin{array}{r} 1.05 \\ \times 0.11 \\ \hline \end{array}$

39. $\begin{array}{r} 0.012 \\ \times \quad 12 \\ \hline \end{array}$

40. $\begin{array}{r} 7.15 \\ \times 0.03 \\ \hline \end{array}$

41. $\begin{array}{r} 12.25 \\ \times \quad 1.5 \\ \hline \end{array}$

Directions: Divide.

42. $8\overline{)20.48}$

43. $3\overline{)3.2916}$

44. $7\overline{)0.952}$

45. $5\overline{)2.09}$

46. $6\overline{)0.021}$

47. $0.07\overline{)4.34}$

48. $1.6\overline{)0.04}$

49. $1.05\overline{)6.3987}$

50. $3.6\overline{)7.704}$

51. $0.9\overline{)0.036}$

52. $1.5\overline{)0.45}$

53. $2.25\overline{)0.27}$

54. $1.2\overline{)0.0864}$

55. $0.03\overline{)4.38}$

56. $0.15\overline{)0.0372}$

Directions: Divide. Round to the nearest hundredth.

57. $7\overline{)2}$

58. $11\overline{)3}$

59. $13\overline{)6}$

60. $12\overline{)5}$

61. $6\overline{)5}$

62. $9\overline{)20}$

Answers are on page 356.

Strategies for Solving Word Problems

Solving Multi-step Problems

A multi-step problem may require more than one operation (addition, subtraction, multiplication, division) to find the solution. Read the problem carefully. Decide which operations you will need before you do any calculations. Then do the work. Finally, check your answer by asking if your answer makes sense. Use rounding and estimation to make sure your answer is reasonable.

The order in which you do the operations may affect your answer. Look at this problem: $5 \times 3 + 6 = ?$ This problem uses both multiplication and addition. Let's solve the problem two ways:

Multiply first ($5 \times 3 = 15$); then add ($15 + 6 = 21$).
Add first ($3 + 6 = 9$); then multiply ($5 \times 9 = 45$).

The answer can be 21 or 45 depending on the order of operations. Because the order of operations can affect the solution, follow these rules:

Rule 1: First do the multiplication and division operations in order from left to right. $5 \times 3 = 15$

Rule 2: Then do the addition and subtraction operations in order from left to right. $15 + 6 = 21$

Example: $7 \times 4 + 15 \div 3 - 12 \times 2 = ?$

Multiply and divide: $7 \times 4 + 15 \div 3 - 12 \times 2 =$
$28 + 5 - 24 =$

Add and subtract: $33 - 24 = 9$

The order of operations can be changed using parentheses. Always do an operation in parentheses first. Then follow the rules of order.

Example: $(15 + 5) \div 4 - 2 = ?$

Do the operation in parentheses: $(15 + 5) \div 4 - 2 =$
Do the division: $20 \div 4 - 2 =$
Then do the subtraction: $5 - 2 = 3$

A number next to a number in parentheses means to multiply.

- 3(9) means $3 \times 9 = 27$
- 4(2 + 5) means $4 \times (2 + 5)$
  $4 \times 7 = 28$
- (2 + 1)(3 − 1) means $(2 + 1) \times (3 - 1)$
  $3 \times 2 = 6$

Example: $2(4 + 5) \div (4 - 1) = ?$

Do the operations in parentheses, then multiply and divide from left to right.

$$2(4 + 5) \div (4 - 1) =$$
$$2(\ 9\ ) \div \ 3 =$$
$$18 \div 3 = 6$$

These rules allow you to answer the questions on the GED test that ask how you would solve a problem that has more than one step. The right answer shows the order of operations needed to solve the problem.

Example: Vickie is a clerk in a minimarket. A customer buys an item that costs $5.24. The customer pays $0.31 sales tax on the purchase. The customer hands Vickie a $20 bill. Which expression shows how much change Vickie should give the customer?

(1) $20 + $5.24 + $0.31

(2) $5.24 + $0.31 − $20

(3) $20 − $5.24 + $0.31

(4) $5.24 − ($0.31 + $5.24)

(5) $20 − ($5.24 + $0.31)

You are not asked to solve the problem; you are asked to show how to solve it. The situation calls for two operations. Vickie must first add the cost of the item ($5.24) and the sales tax ($0.31). Then she subtracts the total from $20. Let's look at the answer options.

- Only options (2) and (5) show the addition of $5.24 and $0.31 as the first operation.
- Option (1) is incorrect because it adds the three amounts together. There is no subtraction step.
- In option (2), $20 is subtracted from the total of $5.24 and $0.31. Subtracting $20 from the total is not the same as subtracting the total from $20. Option (2) is incorrect.
- Option (4) is incorrect because the sum is subtracted from $5.24, not from $20.
- In option (5), the addition comes first because it is in parentheses. The total is subtracted from $20. Option **(5)** is the correct way to solve the problem. Vickie would give the customer $14.45 change.

Look at option (3). The numbers are in the same positions as in option (5), but the parentheses are missing. Although option (3) is similar to option (5), it does not yield the right answer. When solving this expression, only $5.24 is subtracted from $20. The parentheses make a difference.

In the following practice exercises, you will be asked to show how to solve the problems. Then you will solve the problems.

Solving Word Problems

Directions: Choose the best answer to each item.

Items 1 and 2 refer to the following information.

Wilma has \$35. She buys a blouse for \$12.98, a belt for \$10.67, and a poster for \$5.98.

1. Which expression can be used to find how much Wilma has left?

   (1) \$35 − \$12.98 + (\$10.67 + \$5.98)
   (2) \$35 − (\$12.98 + \$10.67 + \$5.98)
   (3) (\$35 + \$12.98) − (\$10.67 + \$5.98)
   (4) (\$35 − \$10.67) + (\$12.98 − \$5.98)
   (5) \$35 − \$12.98 + \$10.67 − \$5.98

2. How much does Wilma have left?

   (1) \$8.67
   (2) \$7.85
   (3) \$6.33
   (4) \$5.37
   (5) \$5.14

Items 3 and 4 refer to the following information.

The Computer Center sells 3.5-inch disks for \$0.89 each. Computer Warehouse sells them for \$1.05 each. A customer wants to buy 25 disks.

3. Which expression can be used to find how much the customer will save by shopping at the Computer Center?

   (1) 25(\$1.05 − \$0.89)
   (2) 25(\$1.05 + \$0.89)
   (3) 25 − (\$1.05 + \$0.89)
   (4) 25 + (\$1.05 − \$0.89)
   (5) (\$1.05 − \$0.89) − 25

4. How much will the customer save by shopping at the Computer Center?

   (1) \$ 0.41
   (2) \$ 2.80
   (3) \$ 3.48
   (4) \$ 4.00
   (5) \$10.25

Items 5 and 6 refer to the following information.

Getting ready for a trip, Mr. Valdez bought two tires for \$45.79 each. He also paid \$18.25 for an oil change.

5. Which expression can be used to find how much Mr. Valdez spent?

   (1) 2(\$45.79) + \$18.25
   (2) 2(\$45.79) − \$18.25
   (3) 2(\$45.79 + \$18.25)
   (4) 2(\$45.79 − \$18.25)
   (5) 2(\$45.79) + 2(\$18.25)

6. How much did Mr. Valdez spend?

   (1) \$ 73.33
   (2) \$ 91.58
   (3) \$109.83
   (4) \$119.53
   (5) \$128.08

7. Angelo's salary was \$18,575 a year. It was raised to \$21,000. What was the monthly increase to the nearest dollar?

   (1) \$ 202
   (2) \$1,548
   (3) \$1,750
   (4) \$2,021
   (5) \$2,425

Answers are on page 358.

Lesson 11

Decimals and Fractions

Both decimals and fractions can be used to show part of a whole. Sometimes it is easier to calculate using fractions. At other times, decimals are more useful. The ability to change a number from one form to the other is an important skill.

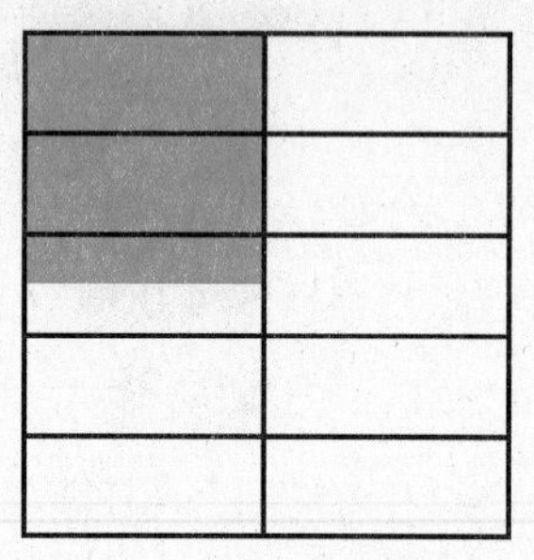
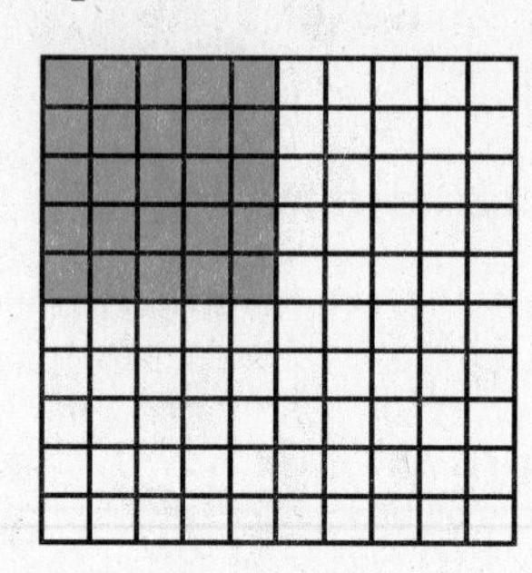
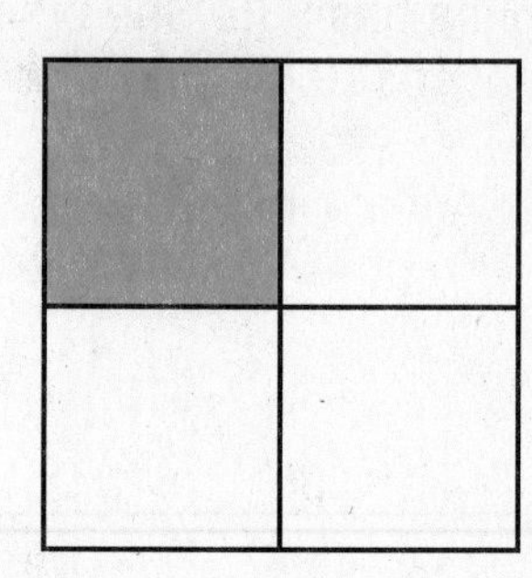

$.2\frac{1}{2}$ ($2\frac{1}{2}$ tenths) $=$ $\frac{25}{100}$ or 0.25 $=$ $\frac{1}{4}$

Changing Decimals to Fractions

Example: Change 0.375 to a fraction.
To change a decimal to a fraction, follow these steps:

Step 1: Write the number without the decimal point as the numerator of a fraction. $0.375 = \frac{375}{?}$

Step 2: The denominator of the fraction is the place value of the decimal digit on the right. Hint: The number of decimal places will be the number of zeros in the denominator. $0.375 = \frac{375}{1{,}000}$

Step 3: Reduce the fraction to lowest terms. $0.375 = \frac{375 \div 125}{1{,}000 \div 125} = \frac{3}{8}$

Example: Change $0.33\frac{1}{3}$ to a fraction.
This decimal has a fraction part. To change this type of decimal to a fraction, follow these steps:

Step 1: Write the fraction as you did in steps 1 and 2 above.

$0.33\frac{1}{3} = \frac{33\frac{1}{3}}{100}$ This means $33\frac{1}{3} \div 100$.

Step 2: Follow the steps to divide mixed numbers.

1. Change to improper number.
2. Invert the divisor.
3. Reduce and multiply.

$$33\frac{1}{3} \div \frac{100}{1} = \frac{100}{3} \div \frac{100}{1} = \frac{100}{3} \times \frac{1}{100} = \frac{\overset{1}{\cancel{100}}}{3} \times \frac{1}{\underset{1}{\cancel{100}}} = \frac{1}{3}$$

Look again at the drawing on page 104. Can you show that $0.2\frac{1}{2} = \frac{1}{4}$?

$$0.2\frac{1}{2} = \frac{2\frac{1}{2}}{10} = 2\frac{1}{2} \div \frac{10}{1} = \frac{5}{2} \div \frac{10}{1} = \frac{\overset{1}{\cancel{5}}}{2} \times \frac{1}{\underset{2}{\cancel{10}}} = \frac{1}{4}$$

Changing Fractions to Decimals

Example: Change $\frac{3}{5}$ to a decimal.

To change a fraction to a decimal, divide the numerator by the denominator: $5\overline{)3}$.

Step 1: Set the decimal point in the problem.

Step 2: Set the decimal point in the answer directly above the decimal point in the problem.

Step 3: Add zeros to make extra decimal places.

Step 4: Divide.

$5\overline{)3.0}$ ← 2. ← 3. 1.↑

$$\begin{array}{r} .6 \\ 5\overline{)3.0} \\ \underline{3\,0} \end{array}$$

Some problems may require you to divide to two, three, or four decimal places. Some problems will always have a remainder. When this happens, you can write the remainder as a fraction (for an exact answer), or you can round the answer to a chosen decimal place.

Example: Change $\frac{2}{9}$ to a decimal. Divide to three decimal places. Write the remainder as a fraction by writing the remainder, 2, over the divisor, 9.

$$\begin{array}{r} 0.222\frac{2}{9} \\ 9\overline{)2.000} \\ \underline{1\,8} \\ 20 \\ \underline{18} \\ 20 \\ \underline{18} \\ 2 = \text{r} \end{array}$$

$$\frac{2}{9} = 0.222\,\frac{2}{9}$$

To change $\frac{2}{9}$ to a decimal, round to a chosen decimal place:

0.222 rounds to 0.22

Working with Money

The unit price of an item is often stated as a decimal with a fraction. The fraction expresses a part of one cent.

Example: The unit price of orange juice is $4\frac{1}{2}$ cents per ounce. What is the cost of 32 ounces of juice?

Multiply 32 by $4\frac{1}{2}$ cents or $\$0.04\frac{1}{2}$ to solve the problem. Change the fraction part of the decimal to a decimal digit. The fraction $\frac{1}{2}$ converts to 0.5 ($1 \div 2 = 0.5$). So $4\frac{1}{2}$ cents could also be written 4.5 cents or \$0.045. Multiply 32 by \$0.045.

$$\begin{array}{r} 32 \\ \underline{\times\ \$0.045} \\ 160 \\ \underline{128} \\ \$1.440 \end{array}$$

At $4\frac{1}{2}$ cents per ounce, the cost of 32 ounces is \$1.44.

Example: What is the cost per ounce of a 40-ounce bottle of dishwashing liquid that sells for \$1.50?

Divide \$1.50 by 40 to two decimal places. Write the remainder as a fraction. Reduce to lowest terms.

$$\begin{array}{r} \$0.03\frac{30}{40} = \$0.03\frac{3}{4} \\ 40\overline{)\$1.50} \\ \underline{1\,20} \\ 30 \end{array}$$

The unit price of one ounce is $3\frac{3}{4}$ cents.

ged Practice Lesson 11

Directions: Change these decimals to fractions. Reduce to lowest terms.

1. $0.25 =$
2. $0.4 =$
3. $0.35 =$
4. $0.875 =$
5. $0.375 =$
6. $0.76 =$
7. $0.128 =$
8. $0.05 =$
9. $0.31\frac{1}{4} =$
10. $0.08\frac{1}{3} =$
11. $0.46\frac{2}{3} =$
12. $0.93\frac{3}{4} =$
13. $0.91\frac{2}{3} =$
14. $0.16\frac{2}{3} =$
15. $0.06\frac{2}{3} =$
16. $0.23\frac{3}{4} =$
17. $0.675 =$
18. $0.08\frac{1}{6} =$
19. $0.125 =$
20. $0.10\frac{5}{6} =$
21. $0.55 =$
22. $0.05\frac{1}{3} =$
23. $0.28 =$
24. $0.12\frac{3}{4} =$

Directions: Change fractions to decimals. Round to three decimal places.

25. $\frac{4}{5} =$

26. $\frac{3}{8} =$

27. $\frac{2}{3} =$

28. $\frac{5}{12} =$

Directions: Change fractions to decimals. Divide to two decimal places and write the remainder as a fraction. Reduce.

29. $\frac{5}{6} =$

30. $\frac{8}{9} =$

31. $\frac{7}{15} =$

32. $\frac{1}{16} =$

Directions: Choose the best answer to each item.

33. A breakfast cereal costing $2.82 contains 12 servings per box. What is the cost of 1 serving?

(1) $22\frac{7}{8}$ cents
(2) $23\frac{1}{4}$ cents
(3) $23\frac{1}{2}$ cents
(4) $23\frac{3}{4}$ cents
(5) $24\frac{1}{8}$ cents

34. A brand of peanut butter costs $13\frac{3}{4}$ cents per ounce. What is the cost of a 16-ounce jar?

(1) $2.09
(2) $2.12
(3) $2.15
(4) $2.16
(5) $2.20

35. A can of spaghetti sauce contains 7 servings. If each serving costs $15\frac{1}{2}$ cents, what is the cost of the can to the nearest whole cent?

(1) $1.08
(2) $1.09
(3) $1.10
(4) $1.11
(5) $1.12

36. A can of chili beans costs $0.59 and contains 4 servings. What is the cost of each serving?

(1) 14 cents
(2) $14\frac{1}{2}$ cents
(3) $14\frac{3}{4}$ cents
(4) 15 cents
(5) $15\frac{1}{2}$ cents

Answers are on page 358.

Strategies for Solving Word Problems

Use the Same System

Either decimals or fractions can be used to show part of a whole. Some word problems contain both.

To solve such problems, all measurements need to be written using the same system. But which measurement should you change? Look at the possible answers. Solve the problem using the same system as that of the answer options. If the options are written as fractions, solve the problem using fractions. Or use the system that will take less time. Sometimes decimal operations take longer than fraction operations. Multiplying by $0.33\frac{1}{3}$ generally takes longer than multiplying by $\frac{1}{3}$.

Memorize this table to convert fractions and decimals more quickly.

Halves	Thirds	Fourths	Fifths	Eighths
$0.5 = \frac{1}{2}$	$0.33\frac{1}{3} = \frac{1}{3}$	$0.25 = \frac{1}{4}$	$0.2 = \frac{1}{5}$	$0.12\frac{1}{2}$ or $0.125 = \frac{1}{8}$
	$0.66\frac{2}{3} = \frac{2}{3}$	$0.75 = \frac{3}{4}$	$0.4 = \frac{2}{5}$	$0.37\frac{1}{2}$ or $0.375 = \frac{3}{8}$
			$0.6 = \frac{3}{5}$	$0.62\frac{1}{2}$ or $0.625 = \frac{5}{8}$
			$0.8 = \frac{4}{5}$	$0.87\frac{1}{2}$ or $0.875 = \frac{7}{8}$

Example: One sheet of cardboard is 0.045 inch thick. How many sheets of cardboard are in a stack that is $29\frac{1}{4}$ inches high?

(1) 575
(2) 600
(3) 625
(4) 650
(5) 675

The answer options do not tell you what to do in this problem. From the table, you know that $\frac{1}{4}$ equals 0.25 so $29\frac{1}{4}$ can be changed easily to 29.25.

Solve:

$$\begin{array}{r} 650. \\ 0.045\overline{)29.250} \\ \underline{27\,0} \\ 2\,25 \\ \underline{2\,25} \\ 0 \end{array}$$

The correct answer is **(4) 650.**

Solving Word Problems

Directions: Choose the best answer to each item.

1. One of the sewing machines at Dreyer's Dress Company broke down. The replacement part for the broken part should be 0.007 inch thick. The part on hand is $\frac{1}{100}$ inch thick. What part of an inch too thick is the part that is on hand?

   (1) 0.0093
   (2) 0.009
   (3) 0.003
   (4) 0.09
   (5) 0.3

2. At Wherry Bat Company, $\frac{7}{8}$ of the bats passed inspection on Monday. On Tuesday, only 0.75 of the bats passed inspection. What fraction more passed on Monday than on Tuesday?

   (1) $\frac{1}{16}$
   (2) $\frac{1}{8}$
   (3) $\frac{3}{16}$
   (4) $\frac{1}{4}$
   (5) $\frac{5}{8}$

3. Unleaded gasoline sells for $1.36 per gallon at U-Save Gas. How much would $12\frac{1}{2}$ gallons cost?

   (1) $12.62
   (2) $15.65
   (3) $16.32
   (4) $17.00
   (5) $17.75

4. In a union poll, $\frac{3}{8}$ of those surveyed agreed with Mr. Samuels. Of those surveyed, 0.4 agreed with Ms. Havel. If 400 people were surveyed, how many more people agreed with Ms. Havel than with Mr. Samuels?

   (1) 10
   (2) 35
   (3) 80
   (4) 150
   (5) 160

5. Alicia needs to know how many cans of corn to open for dinner. First, she figures how many servings are in one can. A can of corn contains $35\frac{3}{4}$ ounces. If one serving is 3.25 ounces, how many servings are in the can?

   (1) 13
   (2) 11
   (3) 9
   (4) 7
   (5) 5

6. Frank jogs each evening to stay fit. He jogged 2.75 miles on Monday, $3\frac{1}{2}$ miles Tuesday, $3\frac{7}{8}$ miles Wednesday, $2\frac{3}{4}$ miles on Thursday, and 3.375 miles on Friday. How many miles did he jog in the five days?

   (1) $15\frac{7}{8}$
   (2) $16\frac{1}{4}$
   (3) $16\frac{3}{4}$
   (4) $17\frac{3}{8}$
   (5) $17\frac{1}{2}$

Answers are on page 360.

Strategies for Solving Word Problems

Using Estimation to Narrow Options

Earlier lessons showed how estimation strategies, such as rounding, might be used to help check your actual solution or to solve problems that do not require an exact solution. **Estimation,** in some cases, can also help you narrow down, or find the likely answer from, a set of options. This strategy can be beneficial if you are unsure of a solution or have limited time. **Note:** this strategy may not be helpful if the answer options are close to one another in value, or if you are unclear about a problem's solution process. Before using this strategy, make sure that the options are far apart from one another.

Example: A movie complex is planning to show a movie. Each day, there will be 6 showings of the movie in 2 of its theaters. Each theater can seat 218 people. How many people will the movie complex be able to seat for the movie if it shows the movie for 31 days?

(1) 257
(2) 6,758
(3) 13,516
(4) 40,548
(5) 81,096

Follow these steps. Use estimation to eliminate options.

Step 1: Decide what steps you need to find the actual solution.
Multiply (6)(2)(218)(31).

Step 2: Round each factor that is greater than 10 to its greatest place. The greatest place is the place farthest to the left.

(6)(2)(218)(31). 218 rounds to the nearest hundred. 31 rounds to the nearest ten.

(6)(2)(200)(30)

Step 3: Work through the problem with the rounded numbers.
$6 \times 2 \times 200 \times 30 = 72{,}000$

Step 4: Compare your estimates to the options. Options (1), (2), (3), and (4) are all much lower than the estimate so they can be eliminated. Option **(5)** is the most likely answer.

Lesson 11

Directions: Choose the best answer to each item. Use rounding when appropriate.

1. Ana worked $7\frac{1}{2}$ hours on Thursday and $6\frac{3}{4}$ hours on Friday. What fraction of an hour more did she work on Thursday than on Friday?

   (1) $\frac{1}{4}$
   (2) $\frac{1}{2}$
   (3) $\frac{3}{4}$
   (4) $\frac{5}{8}$
   (5) $\frac{7}{8}$

2. Gas Station A charges $1.35 per gallon for gas and Gas Station B charges $1.39 per gallon. How much more would 18 gallons of gas cost at Station B than at Station A?

   (1) $ 0.04
   (2) $ 0.72
   (3) $ 1.17
   (4) $ 2.74
   (5) $25.02

3. The CitiPark parking garage charges $6.50 for the first hour of parking and $1.25 for each $\frac{1}{2}$ hour after the first hour. Paul parks his car for $3\frac{1}{2}$ hours. How much will he owe?

   (1) $ 7.75
   (2) $ 9.63
   (3) $10.25
   (4) $12.75
   (5) $20.75

4. How many lengths of fabric, each $\frac{1}{3}$-yard long, can be cut from a 12-yard-long piece?

   (1) $\frac{1}{36}$
   (2) 4
   (3) 12
   (4) $12\frac{1}{3}$
   (5) 36

5. Cari worked 35 hours at $6 per hour and 4 hours at $9 per hour. Which of the following expressions represents her earnings?

   (1) 39×15
   (2) $15(35 + 4)$
   (3) $(35 \times \$6) + (4 \times \$9)$
   (4) $35 \times \$6 \times 4 \times \$9$
   (5) $\frac{35 + 4}{\$6 + \$9}$

6. A carton of 50 tiles covers $12\frac{1}{2}$ square feet. How many square feet will 8 cartons cover?

   (1) 100
   (2) 400
   (3) 625
   (4) 5,000
   (5) Not enough information is given.

7. On Saturday, 5,136 people attended a basketball game. On Sunday, 2,852 people attended. How many people attended the games on the 2 days?

   (1) 2,284
   (2) 5,704
   (3) 7,988
   (4) 7,990
   (5) 10,272

8. Tran bought 120 shirts for the spring sale at his store for $9 each. He sold all the shirts for $18 each. How much was Tran's profit?

   (1) $ 147
   (2) $ 1,080
   (3) $ 2,160
   (4) $19,440
   (5) Not enough information is given.

Answers are on page 360.

Lessons 9–11

Directions: Choose the best answer to each item.

1. A delivery truck logged the following trips: to Hoptown, 12.3 miles; to Mendon, 15.8 miles; to Canton, 21.7 miles; return to office, 8.9 miles. What was the total mileage?

   (1) 30.6 miles
   (2) 40.9 miles
   (3) 49.8 miles
   (4) 58.7 miles
   (5) 71.3 miles

2. A brand of hair conditioner contains 0.52 grams of special ingredients. There are 0.06 grams of ingredient A and 0.1 gram of ingredient B in the conditioner. Which expression can be used to find how many grams of ingredient C are in the conditioner?

   (1) $0.52 + 0.06 + 0.1$
   (2) $0.52 - (0.06 + 0.1)$
   (3) $(0.52 - 0.06) + 0.1$
   (4) $0.52 - (0.06)(0.1)$
   (5) $0.52 - (0.1 - 0.06)$

3. A package of ground beef costs $6.98. The price per pound is $1.86. To the nearest hundredth, how many pounds of ground beef are in the package?

   (1) 3.55
   (2) 3.64
   (3) 3.75
   (4) 3.86
   (5) 3.92

Items 4 and 5 refer to the following information.

It costs the Quality Company $0.14 per mile to run its van. The van's mileage for the week was as follows: Monday, 87.6 miles; Tuesday, 17.3 miles; Wednesday, 14 miles; Thursday, 102.2 miles; Friday, 42.7 miles.

4. What was the van's total mileage for the week?

   (1) 263.8 miles
   (2) 268.9 miles
   (3) 277.8 miles
   (4) 281.6 miles
   (5) 284.8 miles

5. To the nearest whole cent, what was the cost to run the van on Monday?

   (1) $ 1.23
   (2) $ 6.57
   (3) $12.26
   (4) $13.14
   (5) Not enough information is given.

6. The Cheese Works sells cheddar cheese for $3.85 per pound. How much will a package cost that weighs 1.6 pounds?

   (1) $ 5.45
   (2) $ 6.16
   (3) $ 8.78
   (4) $10.01
   (5) Not enough information is given.

7. The gold filling on a finished ring is to be 0.03 millimeters thick. The gold already on the ring has a thickness of 0.021 millimeters. How many more millimeters of gold should be added to the ring?

   (1) 0.9
   (2) 0.18
   (3) 0.09
   (4) 0.018
   (5) 0.009

Items 8 and 9 refer to the following information.

Smith's Berry Barn Weekend Sales	
Saturday Morning	$ 82.37
Saturday Afternoon	136.29
Sunday Morning	51.04
Sunday Afternoon	127.76

8. Approximately how much more did the Berry Barn earn on Sunday afternoon than on Sunday morning?

   (1) between $60 and $65
   (2) between $65 and $70
   (3) between $70 and $75
   (4) between $75 and $80
   (5) between $80 and $85

9. Which expression can be used to find how much more the Berry Barn earned on Saturday than on Sunday?

   (1) ($82.37 + $136.29) − ($51.04 + $127.76)
   (2) ($136.29 − $82.37) + ($127.76 − $51.04)
   (3) ($136.29 + $127.76) − ($82.37 + $51.04)
   (4) ($136.29 − $127.76) − ($82.37 − $51.04)
   (5) 2($136.29 − $127.76)

10. A metal plating process covers metal parts with 0.005 inch of aluminum chrome. A second step covers each part with 0.01 inch more of aluminum. What is the total thickness of the aluminum after the second step?

   (1) 0.0051 inch
   (2) 0.006 inch
   (3) 0.015 inch
   (4) 0.06 inch
   (5) 0.15 inch

11. Kim works as a salesperson for a photo studio. To find his earnings for the week, he multiplies his total sales by 0.175. His sales for the week of October 10 total $2,507.47. To the nearest whole cent, what did he earn for the week?

   (1) $388.06
   (2) $438.81
   (3) $449.83
   (4) $880.60
   (5) Not enough information is given.

12. Andre's weekly pay is $226.40. His boss plans to give him a raise that is 0.1 of his present salary. Which of the following expressions can be used to find Andre's new weekly pay?

   (1) $226.40 − (0.1) ($226.40)
   (2) $226.40 + ($226.40 − 0.1)
   (3) 0.1($226.40 + 0.1)
   (4) 0.1($226.40) + 0.1
   (5) 0.1($226.40) + $226.40

Answers are on page 361.

The Meaning of Percent

Percent is another way to show part of a whole. You have already learned about fractions and decimals. With fractions, the whole can be divided into any number of equal parts—this number is shown as the denominator. With decimals, the number of parts is 10, 100, 1,000, 10,000, or another power of 10. With percents, the whole is always divided into **100** equal parts.

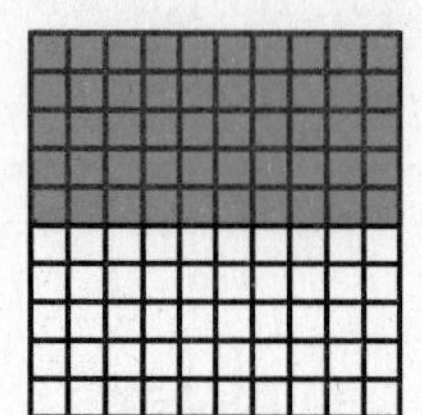

This square is divided into 100 equal parts. The entire square represents 100%. Fifty parts, or one-half of the whole square, are shaded. The shaded part is 50% of the whole. The percent sign, %, means "out of 100." Fifty out of 100 parts are shaded.

Percents greater than 100% are like mixed numbers. A percent greater than 100% is greater than one whole.

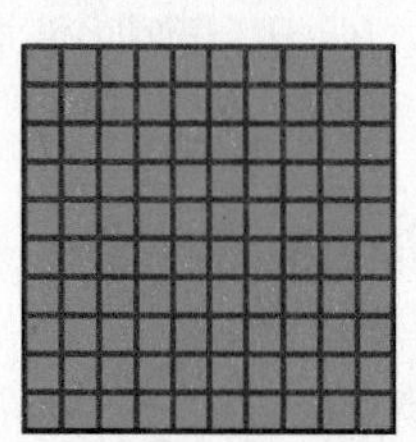

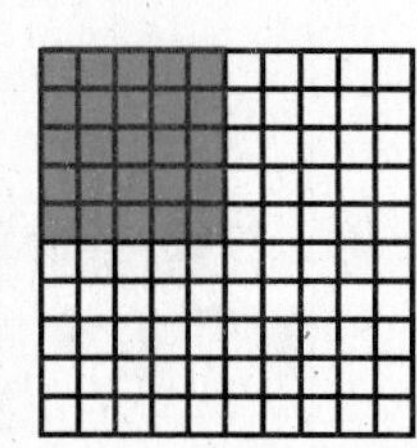

This drawing represents 125%. One hundred twenty-five parts are shaded. Since 100 parts equals 1 and 25 out of 100 parts is $\frac{1}{4}$, 125% equals the mixed number $1\frac{1}{4}$ and the decimal 1.25.

This drawing shows $\frac{1}{2}$%. Only one-half of one part is shaded. A percent that is less than 1% is less than $\frac{1}{100}$.

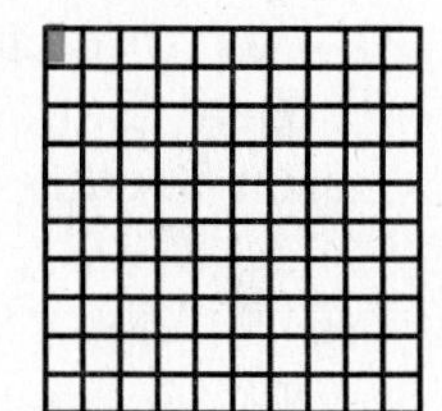

To solve percent problems, you need to change the percent to a decimal, a fraction, a mixed number, or a whole number.

Steps for changing percents are given below.

Changing a Percent to a Decimal

Follow these steps to change a percent, like 45%, to a decimal.

Step 1: Drop the percent sign. — 45

Step 2: Move the decimal point two places to the left. Remember, a number without a decimal point is understood to have a decimal point to the right of the ones place. Write the zero in the ones place (to the left of the decimal point) to serve as a placeholder. — 0.45

Example: Change 7.5% to a decimal.

Step 1: Drop the percent sign. — 7.5

Step 2: Move the decimal point two places to the left. Write zeros as placeholders as needed. — 0.075

A percent without a decimal point is understood to have one to the right of the ones place.

Example: Change $16\frac{2}{3}\%$ to a decimal.
The 6 is in the ones place, not the fraction $\frac{2}{3}$. Fractions do not occupy a place value. $16\frac{2}{3}\% = 0.16\frac{2}{3}$

Lesson 12

Directions: Change each percent to a decimal or whole number.

1. 60%
2. 4.8%
3. 4%
4. $5\frac{1}{2}\%$
5. 200%
6. 130%
7. $9\frac{1}{4}\%$
8. 5.6%
9. 225%

Changing a Decimal to a Percent

Follow these steps to change a decimal, like 0.15, to a percent.

Step 1: Move the decimal point two places to the right. $0.15. = 15$
It is not necessary to show the decimal point when it is at the end of a whole number. The zero placeholder may also be dropped.

Step 2: Write the percent sign. 15%

Examples: Change 2.5 to a percent. Change $0.33\frac{1}{3}$ to a percent.
Move the decimal point 2 places.
Write the zero as a placeholder. $2.50 = 250\%$
Do not show the decimal point after a whole number or between a whole number and a fraction. $.33\frac{1}{3} = 33\frac{1}{3}\%$

Directions: Change each decimal or whole number to a percent.

10. 0.85
11. 0.36
12. 0.144
13. 0.4
14. 4.5
15. $0.16\frac{2}{3}$
16. 8.75
17. 0.375
18. $0.07\frac{1}{3}$
19. 1.5
20. 0.175
21. 0.09

Changing a Fraction or Mixed Number to a Percent

You know how to make fraction and decimal conversions. You can convert a fraction to a percent by changing it first to a decimal, then to a percent. Or you can use the following steps to change a fraction directly into a percent:

Step 1: Multiply the fraction by $\frac{100}{1}$. Reduce and divide as needed.
Step 2: Write the percent sign.

Example: Change $\frac{1}{4}$ to a percent.

$$\frac{1}{\cancel{4}_1} \times \frac{\cancel{100}^{25}}{1} = \frac{25}{1} = 25 = \mathbf{25\%}$$

Example: Change $\frac{2}{3}$ to a percent.

$$\frac{2}{3} \times \frac{100}{1} = \frac{200}{3} = 66\frac{2}{3} = 66\frac{2}{3}\%$$

Follow these steps to change a mixed number, like $3\frac{1}{5}$, to a percent:

Step 1: Multiply the whole number by 100 and write the percent sign. $3 \times 100 = 300\%$
Step 2: Change the fraction to a percent. $\frac{1}{\cancel{5}_1} \times \frac{\cancel{100}^{20}}{1} = \frac{20}{1} = 20\%$
Step 3: Add the two percents. $300\% + 20\% = 320\%$

Directions: Change each fraction to a percent.

22. $\frac{2}{5}$
23. $\frac{3}{4}$
24. $\frac{5}{8}$
25. $\frac{1}{3}$
26. $2\frac{7}{20}$
27. $3\frac{19}{40}$
28. $1\frac{5}{8}$
29. $4\frac{2}{3}$
30. $2\frac{11}{30}$
31. $1\frac{7}{8}$
32. $3\frac{1}{10}$
33. $4\frac{1}{5}$

Changing a Percent to a Fraction or Mixed Number

Follow these steps to change a percent to a fraction or mixed number:

Step 1: Drop the percent sign and write the number with a denominator of 100.
Step 2: Reduce the fraction, if necessary.
Example: Change 35% to a fraction.

$$\frac{35}{100} = \frac{35 \div 5}{100 \div 5} = \frac{7}{20}$$

Example: Change 150% to a mixed number.

$$\frac{150}{100} = \frac{150 \div 50}{100 \div 50} = \frac{3}{2} = 1\frac{1}{2}$$

Directions: Change each percent to a fraction.

34. 20%

35. 65%

36. 84%

37. 12%

38. 140%

39. 275%

40. 39%

41. 120%

42. 325%

Percents with a fraction part are harder to convert to fractions. Remember, writing the number with a denominator of 100 is the same as dividing by 100.

Example: Change $41\frac{2}{3}$ % to a fraction.

Drop the % sign and divide by 100.

$$\frac{41\frac{2}{3}}{100} = 41\frac{2}{3} \div 100 = \frac{\cancel{125}^{5}}{3} \times \frac{1}{\cancel{100}_{4}} = \frac{5}{12}$$

Percents with a decimal part must be converted to a decimal and then changed to a fraction.

Example: Change 37.5% to a fraction.

Change the percent to a decimal by moving the decimal point two places to the left. $37.5\% = 375 = 0.375$

Write as a fraction and reduce. Notice the denominator is 1,000 because 0.375 has three decimal places. $\frac{375 \div 125}{1{,}000 \div 125} = \frac{3}{8}$

43. Fill in the missing numbers on the chart.

Decimal	Fraction	Percent
0.1		10%
.2		20%
.25	$\frac{1}{4}$	25%
0.3		
	$\frac{1}{3}$	
.4		40%
0.5		50%
		60%
$0.66\frac{2}{3}$		$66\frac{2}{3}$%
	$\frac{7}{10}$	
.75	$\frac{3}{4}$	75%
		80%
0.9		90%

Answers are on page 362.

Strategies for Solving Word Problems

Use Mental Math

"Mental math" means to do as much work in your head as possible. Since the GED Math Test is timed, you want to choose problem-solving strategies that are accurate, but take as little time as possible.

When you are solving a percent problem, you can use a fraction or a decimal to find the answer. You will want to choose the one that allows you to do more mental math.

Here are some hints to help you decide which problem-solving strategy to use.

Hint 1: Memorize these common percent/fraction equivalents.

Halves	Thirds	Fourths	Fifths
$50\% = \frac{1}{2}$	$33\frac{1}{3}\% = \frac{1}{3}$ $66\frac{2}{3}\% = \frac{2}{3}$	$25\% = \frac{1}{4}$ $75\% = \frac{3}{4}$	$20\% = \frac{1}{5}$ $40\% = \frac{2}{5}$ $60\% = \frac{3}{5}$ $80\% = \frac{4}{5}$

Sixths	Eighths	Tenths
$16\frac{2}{3}\% = \frac{1}{6}$ $83\frac{1}{3}\% = \frac{5}{6}$	$12\frac{1}{2}\%$ or $12.5\% = \frac{1}{8}$ $37\frac{1}{2}\%$ or $37.5\% = \frac{3}{8}$ $62\frac{1}{2}\%$ or $62.5\% = \frac{5}{8}$ $87\frac{1}{2}\%$ or $87.5\% = \frac{7}{8}$	$10\% = \frac{1}{10}$ $30\% = \frac{3}{10}$ $70\% = \frac{7}{10}$ $90\% = \frac{9}{10}$

Hint 2: Fractions are easier and faster to use than decimals when the fraction has a 1 in the numerator (such as $\frac{1}{2}, \frac{1}{4}, \frac{1}{5}$). Remember, when there is a 1 in the numerator, the shortcut is to divide by the number in the denominator.

Example: Find 20% of 45.
Instead of multiplying 45 by 20%, use the fraction $\frac{1}{5}$.

$$45 \times \frac{1}{5} = \frac{\overset{9}{\cancel{45}}}{1} \times \frac{1}{\cancel{5}_1} = \frac{9}{1} = 9$$

To multiply by $\frac{1}{5}$, think "divide by 5." $45 \div 5 = 9$.

Hint 3: Fractions are easier and faster to use when the percent has a fraction part.

Example: Find $33\frac{1}{3}\%$ of 36.
Use $\frac{1}{3}$ instead of $33\frac{1}{3}\%$. Think "divide by 3." 36 divided by 3 is 12. $33\frac{1}{3}\%$ of 36 is 12.

Hint 4: When the numerator of the fraction is greater than 1 (such as $\frac{3}{4}$), use whichever method seems easier to you.

Solving Word Problems

Directions: Choose the best answer to each item.

1. George Hoffman is remodeling his living room. His total cost for carpeting, draperies, and a new sofa is $1,800. If the carpeting is $33\frac{1}{3}$% of the total cost, how much does the carpeting cost?

   (1) $185
   (2) $250
   (3) $450
   (4) $600
   (5) $900

2. Estelle earns $5.50 per hour. She spends 20% of her hourly wage on child care. How much does she spend per hour on child care?

   (1) $1.10
   (2) $2.50
   (3) $3.00
   (4) $4.50
   (5) $5.00

3. Berry is selling baseball tickets for $66\frac{2}{3}$% of the original $9.00 price. What is the sale price of the ticket?

   (1) $3.00
   (2) $6.00
   (3) $7.50
   (4) $8.00
   (5) $8.50

4. At Piedmont Nursing Home, $83\frac{1}{3}$% of the 120 employees will be working on Thanksgiving. How many employees will be working?

   (1) 100
   (2) 90
   (3) 80
   (4) 70
   (5) 60

5. During elections for union steward, $62\frac{1}{2}$% of the 280 workers voted for Sarah. How many voted for Sarah?

   (1) 128
   (2) 162
   (3) 175
   (4) 200
   (5) 212

6. Paul is going to night school. The cost of his course is $1,600. Paul has a student loan that will cover most of the cost, but he must pay 12.5% from his own funds. How much money will Paul have to pay from his own funds?

   (1) $200
   (2) $240
   (3) $300
   (4) $400
   (5) $480

Answers are on page 363.

Solving Percent Problems (Part 1)

There are three basic elements in a percent problem—the base, the part, and the rate. Think about this statement:

15% or 30 of the 200 job applicants are not available to work on weekends.

- The **base** is the whole amount. The other numbers in the problem are compared to the base. In this statement, 200 is the base amount.
- The **part** is a piece of the whole or base. In this statement the number 30 tells what part of the 200 applicants (the base) are not available on weekends.
- The **rate** is always followed by the percent (%) sign. The rate tells the relationship of the part to the base.

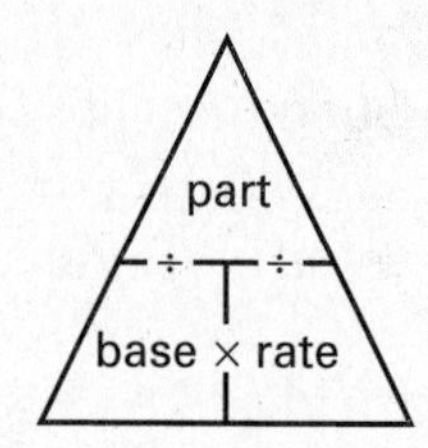

In a percent problem, one of these three elements is missing. You can find the missing element using the percent formula: **base × rate = part.**
The triangle to the left shows the relationship of these three elements.

Finding the Part

Example: Aretha puts 5% of her weekly paycheck in her savings account. If her weekly paycheck is $286.00, how much will she put in savings?

The base or whole amount is $286.00. The rate is 5%. You need to solve for the part. Use the triangle to find what operation to perform. Cover the word part. The remaining elements are connected by the "times" sign for multiplication. The problem can be solved by multiplying:

Base × Rate = Part

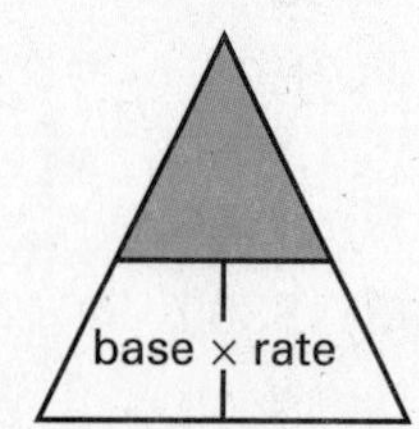

Convert the rate to a decimal: 5% = 0.05

Multiply:
$$\begin{array}{r} \$286 \\ \times\ 0.05 \\ \hline \$14.30 \end{array}$$

Aretha will put $14.30 of the $286.00 in her savings account.

In some problems, the rate may be greater than 100%. In such a case, the part will be greater than the base.

Lesson 13

Directions: Solve.

1. Find 3% of 500.
2. Find 15% of $950.
3. Find 90% of 72.
4. Find 85% of 140.
5. Find 125% of $220.
6. Find 150% of 184.
7. Find 75% of 80.
8. Find $5\frac{1}{2}$% of $200.
9. Find $6\frac{3}{4}$% of $500.
10. Find $2\frac{1}{2}$% of $50.
11. What is 4% of $1,200?
12. What is 70% of 490?
13. What is 55% of $20?
14. What is 40% of $96?
15. What is 210% of 150?
16. What is $1\frac{1}{2}$% of $50?
17. What is $33\frac{1}{3}$% of 120?
18. What is $66\frac{2}{3}$% of $360?

Directions: Choose the best answer to each item.

19. During the winter, Green's Lawn Care Service contacted 325 homeowners. Of these contacts, 36% hired the lawn care service. How many homeowners hired Green's Lawn Care Service?

(1) 72
(2) 91
(3) 117
(4) 253
(5) 300

20. Angelo earns $2,080 each month. His total deductions are 30% of his pay. Which expression can be used to find how much is deducted from his pay each month?

(1) 0.03($2,080)
(2) 0.30($2,080)
(3) 30($2,080)
(4) $2,080 − 30
(5) $2,080 ÷ 30

21. In May an automobile manufacturer produced 1,280 cars. In June the manufacturer produced 125% more cars. How many cars were produced in June?

(1) 320
(2) 400
(3) 520
(4) 670
(5) 1,600

22. Lucia sells electronics items. She makes a commission of 2.5% of her sales. She sold a VCR for $650. How much was her commission?

(1) $ 1.50
(2) $ 1.63
(3) $ 16.25
(4) $150.00
(5) $162.50

23. The Blakes sold their home in Sylmar for $65,000. They paid the real estate agent a 5% commission. How much did the agent make on the sale?

(1) $ 325
(2) $ 500
(3) $ 650
(4) $1,300
(5) $3,250

24. Vern purchased a new table for $310. He paid 6% sales tax on the purchase. How much sales tax did he pay?

(1) $ 15.50
(2) $ 18.60
(3) $ 21.70
(4) $ 31.00
(5) $328.60

25. Waiting tables at a restaurant, Emil made a 20% tip on a $45 meal. How much was the tip?

(1) $ 0.90
(2) $ 4.50
(3) $ 6.00
(4) $ 9.00
(5) $12.00

26. The Jacksons found a new apartment with an annual rent of $9,600. They paid the real estate agent a 9% commission. How much was the commission?

(1) $ 96.00
(2) $ 192.00
(3) $ 864.00
(4) $1,620.00
(5) $4,800.00

Finding the Rate

Example: Joel earns \$1,500 per month. If he spends \$375 on rent each month, what percent of his income does he spend on rent?

In this problem, the base is \$1,500—Joel's "whole" salary for the month. Hint: The base or the words that describe the base are often preceded by the word of. The phrase "what percent of his income . . ." tells you that his income is the base.

The part is \$375—the part of his salary that is spent on rent.

You need to find the rate or percent. Use the triangle to find which operation to perform. Cover the word rate. The remaining elements are connected by the division symbol. The problem can be solved by dividing:

part
÷
base

$$\frac{\textbf{Part}}{\textbf{Base}} = \textbf{Rate}$$

Divide:

$$\begin{array}{r} 0.25 \\ \$1{,}500\overline{)\ \$375.00} \\ \underline{300\ 0} \\ 75\ 00 \\ \underline{75\ 00} \\ 0 \end{array}$$

Convert: $0.25 = 25\%$

Joel spends 25% of his salary on rent.

Note: If the part is greater than the base, the rate will be greater than 100%.

Finding Percent of Increase or Decrease

Sometimes when finding the rate, you will be asked to solve for the percent of increase or decrease. To solve, follow these steps:

Step 1: Subtract the original amount from the new amount.
Step 2: Divide the difference by the original amount.
Step 3: Convert the decimal to a percent.

Example: Last month, Marty's Discount Store sold \$1,375 in small appliances. This month, \$1,540 worth of small appliances were sold. What is the rate of increase in the sale of small appliances?

Step 1: Subtract: $\$1{,}540 - \$1{,}375 = \$165$.
Step 2: Divide: $\$165 \div \$1{,}375 = 0.12 \left(\frac{\$165}{\$1{,}375} = 0.12\right)$.
Step 3: Convert the decimal to a percent: $0.12 = 12\%$.
Note: Use the same method to find rate of decrease. Subtract the new amount from the original amount.

Lesson 13

Directions: Solve.

27. 123 is what percent of 820?

28. 125 is what percent of 625?

29. $112.50 is what percent of $250.00?

30. $3.50 is what percent of $175.00?

31. What percent of $180 is $252?

32. What percent of 5,000 is 225?

33. What percent of $110.00 is $82.50?

34. What percent of $40.00 is $72.00?

Directions: Find the percent of increase or decrease.

	Original Amount	New Amount
35.	$1,500	$1,725
36.	$520.00	$582.40
37.	12	15
38.	350	525
39.	280	70
40.	$1,200	$1,140
41.	$11.00	$8.80
42.	6,000	4,920

Directions: Choose the best answer to each item.

43. The Coaches Corner sporting goods store had a sale on athletic shoes. Their top-of-the-line basketball shoe usually sells for \$90. During the sale, the same model sold for \$72. What percent did the store discount the shoe for the sale?

(1) 10%
(2) 15%
(3) 18%
(4) 20%
(5) 22%

44. Kent works in the housewares section of a department store. This year he set a record high for vacuum sales with 567 vacuums sold. The previous high was last year, when 540 vacuums were sold. Which expression can be used to find the percent of increase from last year to this year?

(1) $567 \div 540$
(2) $540 \div 567$
(3) $567 - (540 \div 567)$
(4) $(567 - 540) \div 567$
(5) $(567 - 540) \div 540$

45. At the warehouse where Rathena works, 140 cartons had to be loaded into a truck for delivery. By lunchtime, 119 cartons had been loaded. What percent of the cartons had been loaded?

(1) 15%
(2) 21%
(3) 79%
(4) 85%
(5) 119%

46. Mr. Cortez asked Jorge to stock the shelves of the minimart with 25 cases of soft drinks: 10 cola, 8 root beer, and 7 orange. Of the total cases of soft drinks, what percent were cola?

(1) 40%
(2) 35%
(3) 25%
(4) 15%
(5) 10%

47. Stacey, a postal clerk, just received her annual cost-of-living raise. She previously earned \$1,450 a month. Her new salary is \$1,508. What is the rate of increase?

(1) 1%
(2) 2%
(3) 3%
(4) 4%
(5) 5%

48. Ahmad's car insurance last year was \$630. This year he had two accidents, so the insurance company raised his yearly bill to \$1,008. What was the percent of increase in Ahmad's car insurance?

(1) 160%
(2) 60%
(3) 50%
(4) 40%
(5) 6%

Answers are on page 364.

Strategies for Solving Word Problems

Use Mental Math

Finding the amount to tip on your restaurant bill, estimating sales tax, and figuring out sale prices are calculations that can be done in your head. Look at the following example:

Evelyn bought a $40 dress at a 10% discount.
How much did she save?

Remember, 10% is equal to the decimal 0.10. To multiply by 0.10, move the decimal point one place to the left. $40.00
10% of $40.00 is $4.00. The store took $4 off the price of the dress.

Using the short method for finding 10%, we can develop strategies for quickly finding 5%, 15%, and 20% of an amount.

Example: Jan's monthly paycheck is $900. 5% is deducted from her check for medical and dental insurance. How much is deducted from her paycheck for medical and dental benefits?
To find 5% of $900, find 10% of $900 by moving the decimal point one place to the left. $900.00
Since 5% is half of 10%, divide $90.00 by 2. $90.00 ÷ 2 = $45.00
5% of $900.00 is **$45.00.**

To find 15% of a number, find 10%, divide that result by 2 to find 5%, and add the 10% and 5% amounts.

Example: Carl's restaurant bill is $20. He wants to leave a 15% tip. How much should he leave for the tip?
Find 10% of $20 by moving the decimal point one place to the left. $20.00
5% is half of 10%. Find 5% of $20 by finding half of $2.00. $2.00 ÷ 2 = $1.00
Since 10% + 5% = 15%, add $2.00 and $1.00 to find 15% of $20. $2.00 + $1.00 = $3.00
Carl should leave a **$3.00** tip.

To find 20% of a number, find 10% and multiply by 2.

Example: Sharon works as a truck driver for a furniture warehouse. One day she received $82 in tips. She gives 20% of the tips to her assistant. How much should Sharon give to her assistant?
Find 10% of $82.00 by moving the decimal point one place to the left. $82.00
20% is twice 10%. Multiply $8.20 by 2. $8.20 × 2 = $16.40
Sharon should give her assistant **$16.40.**

Solving Word Problems

Directions: Choose the best answer to each item.

1. The Gardner Public Library records the number of books borrowed each month. Of the 8,520 books borrowed in May, 10% were children's books. How many were children's books?

   (1) 85
   (2) 117
   (3) 852
   (4) 1,174
   (5) 2,500

2. Estelle works as a waitress. She received a 15% tip on a dinner check for $26.00. How much was her tip?

   (1) $1.30
   (2) $1.60
   (3) $2.75
   (4) $3.90
   (5) $5.00

3. Boyd bought some fishing equipment for $150 and paid 5% sales tax. How much sales tax did he pay?

   (1) $75.00
   (2) $35.00
   (3) $15.00
   (4) $12.00
   (5) $ 7.50

4. David's check for lunch was $8.80. He left a 15% tip. How much was the tip?

   (1) $0.88
   (2) $0.96
   (3) $1.22
   (4) $1.32
   (5) $1.76

5. Marvina sold a radio/cassette player originally priced at $70 and discounted 20%. How much was the discount?

   (1) $14.00
   (2) $18.00
   (3) $21.00
   (4) $34.50
   (5) $42.00

6. Of Zoila's $1,500 monthly earnings, 20% is paid for rent. How much does she pay each month for rent?

   (1) $240
   (2) $270
   (3) $300
   (4) $360
   (5) $412

7. A state charges 5% sales tax. If Vern makes a $24 purchase in that state, how much sales tax does he pay?

   (1) $0.90
   (2) $1.20
   (3) $1.50
   (4) $2.20
   (5) $2.50

8. The Johnson Community Center is selling tickets to their fundraising concert. They have sold 30% of the 1,800 tickets. How many tickets have they sold?

   (1) 600
   (2) 540
   (3) 60
   (4) 54
   (5) 6

Answers are on page 365.

Solving Percent Problems (Part 2)

Finding the Base

Example: Of the employees who work at Stalling Printing, 90% attended the safety procedures meeting. If 63 employees attended the meeting, how many employees work at Stalling Printing?

The base or whole amount is the number of employees at Stalling Printing. The base is not given in the problem. The number 63 is part of the whole group of employees. The 63 employees are 90% of the base. 90% is the rate.

Use the triangle to find what operation to perform. Cover the word base. The remaining elements are connected by the division sign. The problem can be solved by dividing:

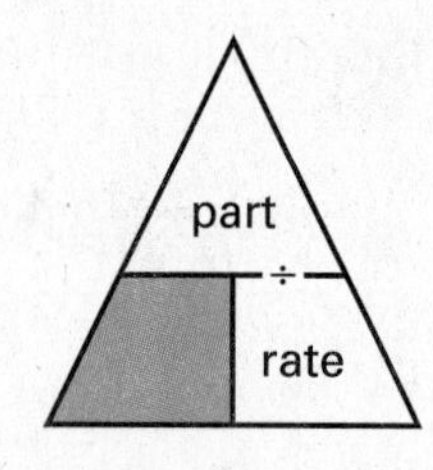

$$\frac{\textbf{Part}}{\textbf{Rate}} = \textbf{Base}$$

Convert the rate to a decimal: 90% = 0.9

Divide:

```
       70.
0.9) 63.0
     63
      00
```

There are **70** employees at the printing company.

The part may be greater than the base.

Example: Regina recently started a better job. Her new weekly salary is 120% of her weekly salary at her old job. If her new weekly salary is $326.40, what was her old weekly salary?

You know that the part is greater than the base because the rate is greater than 100%. The problem is solved as above:

Convert the rate to a decimal: 120% = 1.2

Divide the part by the rate:

```
       $27 2.0
1.2) $326.4 0
      24
       86
       84
        2 4
        2 4
          00
```

Regina's old weekly salary was **$272.00.**

Problems that require you to solve for the base frequently have more than one step. Make sure you know what you are to solve for. Decide what operations you will need to solve the problem. Then do the work.

ged Practice Lesson 14

Directions: Solve.

1. \$54 is 60% of what amount?

2. \$3.75 is 75% of what amount?

3. 9.6 is 15% of what number?

4. 720 is 9% of what number?

5. 35% of what amount is \$157.50?

6. 6% of what amount is \$1.92?

7. 17% of what number is 85?

8. \$26.88 is 24% of what amount?

9. 90% of what number is 495?

10. \$62.40 is 104% of what amount?

11. 270% of what number is 810?

12. 115% of what number is 207?

13. 3.8% of what amount is \$0.76?

14. 22.5% of what number is 36?

15. \$679.35 is $5\frac{1}{4}$% of what amount?

16. \$1.25 is $2\frac{1}{2}$% of what amount?

17. 6,000 is $66\frac{2}{3}$% of what number?

18. \$239.91 is $33\frac{1}{3}$% of what amount?

Directions: Choose the best answer to each item.

19. Jeff recently had a major operation. His insurance company paid $7,500 to the hospital. If the amount the insurance company paid was 80% of the total bill, what was the amount of the total bill?

(1) $ 8,500
(2) $ 9,000
(3) $ 9,375
(4) $10,500
(5) $12,000

20. Last year Florence paid 15% of her yearly salary in federal income tax. If her tax bill was $3,555, what was her yearly salary?

(1) $22,300
(2) $22,948
(3) $23,700
(4) $24,312
(5) $27,255

21. Carmen bought a new winter coat for $98. The sale price was 70% of the original price. What was the original price?

(1) $108
(2) $117
(3) $128
(4) $140
(5) $161

22. One hundred seventy-eight members of The Community Action Committee voted to elect Elena as president. If 89% of the members voted for Elena, how many club members did not vote for her?

(1) 43
(2) 39
(3) 31
(4) 28
(5) 22

23. A store owner found that $33\frac{1}{3}\%$ of customers who entered the store in the month of January made a purchase. If 207 customers made a purchase, how many customers entered the store in January?

(1) 330
(2) 427
(3) 546
(4) 621
(5) 715

24. Of the appliances from Assembly Line 13, 7% were defective. If there were 42 defective appliances, how many appliances were not defective?

(1) 600
(2) 558
(3) 532
(4) 509
(5) 498

25. A cereal coupon saved Amir $0.75. If $0.75 is 20% of the cereal's original price, what was the original price?

(1) $0.95
(2) $1.50
(3) $2.75
(4) $3.75
(5) $6.00

26. A serving of soup supplies 9 grams of carbohydrates. If this is 3% of the carbohydrates Keisha needs daily, how many grams of carbohydrates does Keisha need daily?

(1) 0.27
(2) 27
(3) 300
(4) 900
(5) 2,700

Solving Interest Problems

Interest is a fee charged for using someone else's money. **Interest rates** are familiar to anyone who borrows money for a purchase or puts money into a savings account. The borrower needs to pay back the interest in addition to the **principal** (the amount borrowed).

The formula for finding the amount of **simple interest** due after a certain number of years is ***i = prt.***

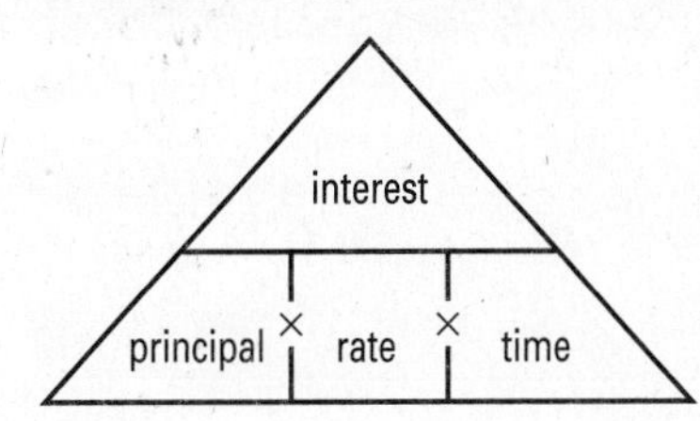

i = interest
p = principal, the amount borrowed or invested
r = interest rate, written as a percent
t = time of the loan, written in years

The formula is similar to the percent formula you have been using: part = base × rate. The interest formula can be written in a triangle.

To find simple interest, follow these steps:

Step 1: Write the rate as a decimal or a fraction.

Step 2: Write the time in terms of years. If the problem states the time in months, write the months as the numerator and the number 12 as the denominator. Reduce the fraction to lowest terms.

Step 3: Multiply: principal × rate × time. The result is the interest.

Example: Lynn Alvarez borrows \$2,500 for 3 months at an 8% rate of interest. How much interest will Lynn pay on the loan?

The principal is \$2,500. The rate is 8%, which can be converted to 0.08. The time is 3 months, or $\frac{3}{12}$ of a year, which reduces to $\frac{1}{4}$. You may choose to change $\frac{1}{4}$ to 0.25.

Multiply $p \times r \times t$: $\$2{,}500 \times 0.08 \times \frac{1}{4} =$

$$\$200 \times \frac{1}{4} =$$

$$\frac{\overset{50}{\cancel{200}}}{1} \times \frac{1}{\underset{1}{\cancel{4}}} = \$50$$

Lynn will owe **\$50** in interest.

Sometimes you will need to find the **amount paid back.** This is the total amount of money the borrower must pay back to the lender. It includes the principal and any interest owed.

In the last example, Lynn borrowed \$2,500 and paid \$50 interest. To find the pay back amount, add the principal and the interest: \$2,500 + \$50 = \$2,550.

Lynn must pay back **\$2,550.**

ged Practice Lesson 14

Directions: Find the interest on the following loans.

27. A loan of $1,250 at 12% for 2 years.

28. A loan of $2,400 at 9.5% for 1 year.

29. A loan of $900 at 12% for 6 months.

30. A loan of $4,000 at 5% for 3 months.

Directions: Find the interest earned on the following investments.

31. An investment of $3,200 at $5\frac{1}{2}$% interest for 1 year.

32. An investment of $500 at 8% interest for $6\frac{1}{2}$ years.

33. An investment of $2,300 at 9% interest for 6 months.

34. An investment of $800 at 7.5% interest for 3 months.

Directions: Find the interest earned on the following investments.

35. A loan of $600 at 10% for 6 months.

36. A loan of $5,400 at 8% for 9 months.

37. A loan of $8,000 at 12% for 5 years.

38. A loan of $300 at 6% for 4 months.

Directions: Choose the best answer to each item.

39. Mr. Bradford bought a delivery van. He borrowed $10,000 at $10\frac{1}{4}$% interest for 3 years. How much will he pay in interest on the loan?

(1) $2,925
(2) $3,000
(3) $3,075
(4) $3,150
(5) $3,200

40. Bart took out a student loan of $750 for 6 months at 3% interest. Which expression can be used to find how much Bart will pay back?

(1) ($750)(0.03)(0.5) + $750
(2) ($750)(0.3)$\left(\frac{1}{2}\right)$ + $750
(3) ($750)(0.03)(6) + $750
(4) ($750)(0.03)$\left(\frac{1}{2}\right)$ + 6
(5) ($750)(0.03)(0.5)$\left(\frac{1}{2}\right)$

41. To buy a used car, Rene borrowed $600 from the bank at 14% interest for 12 months. How much interest will she pay?

(1) $ 78
(2) $ 82
(3) $ 84
(4) $ 98
(5) $108

42. Dom borrows $3,450 to install a loading dock at his warehouse. He borrows the money at 12% for 2 years. What is the total amount he will have to pay back in 2 years?

(1) $3,741
(2) $3,866
(3) $4,008
(4) $4,278
(5) $4,416

43. Opal lent her sister $1,000 for $1\frac{1}{2}$ years at $7\frac{1}{2}$% interest. How much interest will her sister owe in $1\frac{1}{2}$ years?

(1) $112.50
(2) $112.75
(3) $114.50
(4) $115.25
(5) $116.00

44. The Yamatos borrow $900 for 3 months to pay the hospital bill for their daughter's birth. The interest rate is 18%. What is the total amount they will pay back in 3 months?

(1) $ 904.15
(2) $ 940.50
(3) $ 996.00
(4) $1,018.75
(5) $1,020.00

45. The Sandovals borrowed $2,500 to remodel their kitchen. They will pay 7.5% interest for 3 years. How much interest will they pay?

(1) $ 5.63
(2) $ 187.50
(3) $ 562.50
(4) $3,062.50
(5) $5,625.00

46. A credit card company charges 16.5% interest per year. How much interest will be charged on a balance of $1,235 for 6 months?

(1) $ 101.89
(2) $ 122.27
(3) $1,018.88
(4) $1,222.65
(5) $1,336.85

Answers are on page 365.

Strategies for Solving Word Problems

Solving Multi-step Problems

Sometimes you will use percents to solve problems that have more than one step. These problems are not difficult to solve if you organize the information you are given, decide what information you need to solve the problem, and break the problem down into steps. Remember to look for key words that tell you which operation to use.

Example: Wendy bought an $80 dress on sale for 40% off. How much did she pay for the dress?

Step 1: Find the discount amount.
Multiply the base by the rate. $\$80 \times 0.4 = \$32$

Step 2: Find the sale price.
Subtract the discount amount from the original price:
$\$80 - \$32 = \$48$

The sale price is **$48.**

Example: John made a small bookshelf using $8 worth of materials. John now wants to sell the bookshelf to make a 15% profit. How much should he sell the bookshelf for?

Step 1: Find the amount of profit.
Multiply base by rate:
$\$8.00 \times 15\% = \$8.00 \times 0.15 = 1.2 = \$1.20$

Step 2: Find the price of the bookshelf by adding the original cost to the profit: $\$8.00 + \$1.20 = \$9.20$

John should sell the bookshelf for **$9.20.**

Example: The discount price of a sofa at a 40%-off sale is $348. What is the original price of the sofa?

Step 1: Find the rate that represents the sale price of the sofa:
$100\% - 40\% = 60\%$

Step 2: Find the base or original price. Divide the part by the rate:
$\$348 \div 0.6 = \$580$

The original price of the sofa is **$580.**

Solving Word Problems

Directions: Choose the best answer to each item.

1. Murray's Bargain Basement is having a sale on household goods. A $20 king-size blanket is on sale for 15% off. What is the sale price of the blanket?

   (1) $ 3.00
   (2) $ 5.00
   (3) $ 7.50
   (4) $17.00
   (5) $18.67

2. Floyd received a 6% raise on his hourly wage of $9.00. One month later he was awarded a 2% increase on the new wage for being Employee of the Year. To the nearest whole cent, what is his current hourly wage?

   (1) $9.54
   (2) $9.72
   (3) $9.73
   (4) $9.80
   (5) $9.98

3. Alfredo bought a used car for $1,500. He replaced the engine and repaired and painted the body. He sold the car for $4,500. What was the rate of increase in the price of the car?

   (1) 67%
   (2) 133%
   (3) 200%
   (4) 300%
   (5) 400%

4. At a factory, Ms. Jackson supervises 30 employees. Next month, her staff will be increased by 40%. Which expression can you use to find the number of employees Ms. Jackson will supervise next month?

   (1) 0.4(30) + 0.4
   (2) 0.4(30) + 30
   (3) 0.4(30) − 0.4
   (4) 0.04(30) + 0.4
   (5) 0.4(30) − 30

5. Lavina wants to buy a rocking chair for $160. She will pay 10% down and pay the rest in 6 monthly installments. What will be the amount of each monthly payment?

   (1) $24
   (2) $26
   (3) $27
   (4) $30
   (5) $32

6. Grantsville Copper Mines employs 1,400 workers. The company must lay off 5% of its work force immediately and another 20% of the remaining work force by the end of the year. How many workers will be laid off?

   (1) 350
   (2) 336
   (3) 280
   (4) 210
   (5) 70

Answers are on page 366.

Lessons 12–14

Directions: Choose the best answer to each item.

1. On an employment test, George scored 80%. He had 16 correct answers. Which expression can be used to find the number of questions on the test?

   (1) $16(0.8) + 16$
   (2) $16 \div 0.8$
   (3) $16 \div 0.08$
   (4) $16(0.8)$
   (5) $16(0.08)$

2. At a playoff game, paid attendance was 39,900. The total attendance was 42,000. What percent of those attending had free tickets?

   (1) 2%
   (2) 3%
   (3) 4%
   (4) 5%
   (5) 6%

3. Of the people attending a concert, 40% received a 50%-off coupon for next month's concert. How many received the coupon if there were 1,840 people at the concert?

   (1) 368
   (2) 460
   (3) 613
   (4) 736
   (5) 920

4. The personnel department reported total employment is 120% of last year's total. If the total number employed last year was 1,735, what is this year's employment figure?

   (1) 1,388
   (2) 1,446
   (3) 1,615
   (4) 2,082
   (5) 20,820

5. Louise makes a purchase of $19.49 and pays 6% sales tax. To the nearest whole cent, what is the amount of the sales tax on her purchase?

   (1) $0.79
   (2) $0.84
   (3) $0.92
   (4) $1.03
   (5) $1.17

6. Brigham loans his brother-in-law $690 at 8% interest for 8 months. How much will the brother-in-law pay back when the loan is due?

   (1) $718.08
   (2) $726.80
   (3) $750.00
   (4) $764.25
   (5) $775.50

7. Merrill receives a fixed salary of $165 per week and a $4\frac{1}{2}$% commission on the merchandise he sells. In a week when his sales are $5,320, how much does he earn?

(1) $239.40
(2) $359.12
(3) $404.40
(4) $473.80
(5) Not enough information is given.

8. Jason pays $6\frac{1}{2}$% sales tax on a purchase of $15.60. Which expression can be used to find the total cost of his purchase?

(1) 0.065($15.60 + $6.50)
(2) $0.6\left(\frac{1}{2}\right)$($15.60) + $15.60
(3) $0.06\left(\frac{1}{2}\right)$($15.60) + $15.60
(4) 0.065($15.60) + $6.50
(5) 0.065($15.60) + $15.60

9. Ruth and Barry Souther borrow $3,500 at 18% for 2 years. How much interest will they pay on the loan?

(1) $1,242
(2) $1,260
(3) $1,340
(4) $1,356
(5) $1,470

10. The sales tax in the state where Alonzo lives is $5\frac{1}{4}$%. What is the total amount he pays for a radio with the price $40?

(1) $42.20
(2) $42.16
(3) $42.10
(4) $42.04
(5) $42.00

11. The discounted price of a new TV set at a 25%-off sale is $375.00. What was the original price of the TV?

(1) $400
(2) $450
(3) $500
(4) $550
(5) $600

12. Ana Tew invests $3,500 for one year in a credit union account where her investment will earn $9\frac{1}{2}$% annual interest. How much will she have in the account at the end of 1 year?

(1) $3,832.50
(2) $3,804.17
(3) $3,782.40
(4) $3,697.50
(5) $3,650.00

13. Jay buys a microwave regularly priced at $394.00. The microwave is on sale for 40% off the regular price. How much does Jay pay for the microwave?

(1) $226.95
(2) $232.60
(3) $236.40
(4) $240.25
(5) $250.00

14. Alina saved $12 on a pair of jeans that had been marked 60% off. What was the original price of the jeans?

(1) $ 4.80
(2) $ 7.20
(3) $12.00
(4) $20.00
(5) $36.00

Answers are on page 367.

ged Cumulative Review **Arithmetic**

Directions: Choose the best answer to each item.

1. Marco saves \$30 each week toward the \$600 cost of a vacation for his family. If he has already saved \$420, how many more weeks will it take him to save the whole amount?

   (1) 5
   (2) 6
   (3) 7
   (4) 8
   (5) 9

2. A fence will completely enclose a square field that is 25.4 yards on each of its four sides. How many yards of fencing will be needed?

   (1) 29.4
   (2) 97.6
   (3) 101.6
   (4) 104.8
   (5) 110.4

3. Arnold works in the office stockroom. One day he finds that the company has 26 boxes of ballpoint pens. Each box contains 12 pens. Two months later Arnold finds the company has only 59 pens left in stock. Which expression can be used to find how many pens the company used in the two-month period?

   (1) $(12 + 59) + 26$
   (2) $(12 + 59) - 26$
   (3) $12(59) - 26$
   (4) $12(26) - 59$
   (5) $12(59 - 26)$

4. Nick has a part-time job. One week, he worked $2\frac{3}{4}$ hours on Monday, $4\frac{1}{2}$ on Tuesday, $3\frac{1}{2}$ on Wednesday, $5\frac{1}{4}$ on Thursday, and $2\frac{3}{4}$ on Friday. How many hours did Nick work during the week?

   (1) $16\frac{3}{4}$
   (2) $17\frac{1}{2}$
   (3) $18\frac{3}{4}$
   (4) $19\frac{1}{4}$
   (5) $20\frac{1}{4}$

5. Jocelyn works as a salesperson at a computer store. She earns a 7% commission on each sale. Last week she sold 15 computers to a school district for a total of \$13,436. How much did Jocelyn make on the sale?

   (1) \$ 870.00
   (2) \$ 940.52
   (3) \$ 967.13
   (4) \$1,005.25
   (5) \$9,405.20

6. In a recent election, only 17% of the registered voters actually voted. The winning candidate won by a slim margin of 580 votes. How many votes did the winning candidate receive?

   (1) 580
   (2) 698
   (3) 3,411
   (4) 9,860
   (5) Not enough information is given.

7. In a catalog, the price of a lamp is \$45. A coupon reduces the price of the lamp 15%. With the coupon, what is the price of the lamp?

   (1) \$ 6.75
   (2) \$25.50
   (3) \$32.25
   (4) \$36.75
   (5) \$38.25

8. Desert covers 0.125 of the land on earth. Of the land on earth, $\frac{1}{10}$ can be used for farmland. How much greater is the desert area than the farmland?

   (1) 0.1
   (2) 0.2
   (3) 0.25
   (4) 0.01
   (5) 0.025

9. Hiro works at the reservations desk at a hotel. A company wants to rent a conference room for a 2-hour meeting that will meet 3 times a week for 3 weeks. For how many hours will the company need the conference room?

   (1) 6
   (2) 9
   (3) 11
   (4) 18
   (5) 36

Items 10 to 12 refer to the following information.

Estela works for a local newspaper. It is her job to figure the cost to run a want ad in the paper. The newspaper charges \$18 for the first 25 words and \$1.10 for each additional word in the ad. A 10% discount is given off the total price for ads 50 words or over.

10. Zoe wants to run a want ad containing 30 words. Which expression can be used to find out how much Estela should charge her?

   (1) \$1.10(30 − 25) + \$18
   (2) \$1.10(50 − 25) + \$18
   (3) \$1.10(30 − 18) + 25
   (4) \$1.10(30) − (25 + \$18)
   (5) \$1.10(25) − (30 − \$18)

11. Sally wants to spend no more than \$25 for an ad. What is the maximum number of words she can have in her ad?

   (1) 29
   (2) 30
   (3) 31
   (4) 32
   (5) 33

12. Jesse runs a want ad with 55 words. How much should Estela charge him?

   (1) \$33.00
   (2) \$45.90
   (3) \$48.40
   (4) \$51.00
   (5) \$56.10

13. The O'Rourkes want to borrow \$5,000 from the bank to fix the roof on their home. They borrow the money for 3 years at 14% interest. What is the total amount they will pay back to the bank?

   (1) \$2,100
   (2) \$5,042
   (3) \$6,400
   (4) \$7,100
   (5) \$8,000

14. A person walking uses about 3.5 calories each minute. How many calories will a person use in an 11-minute walk?

    (1) 30.5
    (2) 35
    (3) 38.5
    (4) 40.5
    (5) 45

15. Four families had a garage sale and took in $2,118.84. If they divide the money equally, how much will each family receive after equally sharing advertising expenses of $128.96?

    (1) $395.71
    (2) $419.07
    (3) $458.16
    (4) $497.47
    (5) $561.95

16. A local civic club has 300 members. At a recent meeting, 138 of the members attended. What percent of the club members attended the meeting?

    (1) 54%
    (2) 46%
    (3) 41%
    (4) 38%
    (5) 31%

17. An automobile odometer reading is 8,867.3 miles. In how many miles will the odometer read 10,000?

    (1) 832.7
    (2) 982.7
    (3) 995.4
    (4) 1,094.7
    (5) 1,132.7

18. A uniform shop has contracted to make 120 uniforms that require $4\frac{1}{2}$ yards of fabric each. The shop buys the fabric in 60-yard bolts. How many bolts should they buy to make this order?

    (1) 8
    (2) 9
    (3) 10
    (4) 11
    (5) 12

19. How many servings can be cut from 6 cheesecakes if each serving is $\frac{1}{8}$ of a cheesecake?

    (1) 24
    (2) 36
    (3) 48
    (4) 75
    (5) Not enough information is given.

20. On Valentine's Day, a florist shop had 246 orders for a dozen (12) roses. How many roses will the florist need to fill the orders?

    (1) 3,198
    (2) 2,952
    (3) 2,706
    (4) 2,460
    (5) Not enough information is given.

21. A recipe calls for $\frac{2}{3}$ cup of butter. How much butter should be used in order to make $\frac{1}{2}$ the recipe?

    (1) $\frac{1}{4}$ cup
    (2) $\frac{3}{10}$ cup
    (3) $\frac{1}{3}$ cup
    (4) $\frac{3}{8}$ cup
    (5) $\frac{1}{2}$ cup

22. Marta needs three boards to create a border for her flower bed. She needs these lengths: $3\frac{1}{2}$ feet, $5\frac{1}{4}$ feet, and $3\frac{3}{8}$ feet. What is the total length in feet of the pieces?

(1) $11\frac{7}{8}$

(2) $12\frac{1}{8}$

(3) $12\frac{3}{4}$

(4) $13\frac{1}{16}$

(5) $14\frac{1}{8}$

23. Maria has $\frac{1}{2}$ can of paint. If she uses $\frac{3}{4}$ of it, what fraction of the can is left?

(1) $\frac{3}{8}$

(2) $\frac{1}{3}$

(3) $\frac{3}{10}$

(4) $\frac{1}{4}$

(5) $\frac{1}{8}$

24. Rod rides his motorcycle to work and back 5 days each week. If his job is 3.2 miles from his home, how many miles does he travel to and from work each week?

(1) 12

(2) 16

(3) 18

(4) 24

(5) 32

25. Ben and four others tied for the big football pool. Among them, they split $735. How much did Ben win?

(1) $128

(2) $135

(3) $147

(4) $151

(5) $167

26. Phil is given $\frac{1}{3}$ of the company's display space for his department's products. He uses $\frac{3}{4}$ of his space for posters and the rest for samples. What fraction of the entire display space is used for Phil's samples?

(1) $\frac{3}{16}$

(2) $\frac{1}{12}$

(3) $\frac{1}{6}$

(4) $\frac{1}{4}$

(5) $\frac{3}{8}$

27. At a discount store, boxes of cereal are $1.59 each. A case holds 8 boxes. What is the price for $3\frac{1}{2}$ cases?

(1) $ 5.57

(2) $12.72

(3) $28.00

(4) $38.92

(5) $44.52

28. A giant-size pizza has 16 slices. Greta ordered four pizzas to serve 32 people at a party. How many slices will there be for each person?

(1) 2

(2) 6

(3) 8

(4) 64

(5) 128

29. Tony can pick $14\frac{1}{2}$ baskets of berries per hour. How many baskets of berries can he pick in $5\frac{1}{2}$ hours?

(1) 70

(2) $72\frac{1}{2}$

(3) $79\frac{3}{4}$

(4) $82\frac{1}{2}$

(5) 87

30. Vladimir borrowed $7,400 at $9\frac{1}{4}$% interest for 3 years. How much interest will he pay?

(1) $2,029.25
(2) $2,053.50
(3) $2,664.00
(4) $3,173.00
(5) $6,845.75

Items 31 and 32 refer to the following table.

La Habra Trucking Mileage	
Hector	756 miles
Katalina	236 miles
Jamie	524 miles
Freddie	483 miles
Vu	701 miles

31. Vu has driven more miles than Katalina. How many more?

(1) 375
(2) 445
(3) 465
(4) 495
(5) 505

32. What percent of the company's miles has Hector driven?

(1) $22\frac{1}{2}$%
(2) 25%
(3) 26%
(4) 28%
(5) $33\frac{1}{3}$%

33. Grace purchased 250 file folders at $0.12 each. If she received a 5% discount, how much did she pay for the file folders?

(1) $28.50
(2) $29.25
(3) $30.00
(4) $31.75
(5) $32.00

34. The county fair committee expects approximately 7,500 people to attend each day of the fair. If the fair lasts 6 days, what is the total attendance expected at the fair?

(1) 42,000
(2) 45,000
(3) 46,000
(4) 48,000
(5) 50,000

35. Geoff is a potter. For every 35 jugs he makes, 7 are damaged during glazing and cannot be sold. What fraction of the jugs can be sold?

(1) $\frac{3}{8}$
(2) $\frac{3}{7}$
(3) $\frac{1}{2}$
(4) $\frac{5}{7}$
(5) $\frac{4}{5}$

36. Fred sold a customer five videotapes totaling $149.45. Later, the customer returned one of the videotapes and asked for his money back. Fred refunded $24.99. What was the final amount of the sale?

(1) $115.99
(2) $118.84
(3) $120.96
(4) $124.46
(5) Not enough information is given.

37. Meg borrowed $2,500 from her employer for 2 years at $2\frac{1}{2}$% interest. What is the total amount she will pay back?

(1) $2,625
(2) $3,137
(3) $3,426
(4) $5,000
(5) $6,250

38. Last year the Quarles family paid $350 per month for rent. This year their landlord raised their rent to $378. What is the rate of increase?

(1) 2%
(2) 4%
(3) 5%
(4) 7%
(5) 8%

39. A 1.5% holiday bonus was given to 75% of a company's employees. If Kyle was one of the employees who received a bonus, how much did he receive?

(1) $ 50
(2) $ 85
(3) $115
(4) $130
(5) Not enough information is given.

40. Minh set a two-week company sales record by selling 140 car stereos. How many stereos did he sell the first day of the two-week period?

(1) 10
(2) 12
(3) 13
(4) 15
(5) Not enough information is given.

41. Sue needs to park for 4 hours. Lot A charges 75¢ per hour with a $5.00 minimum charge. Lot B charges $1.25 per hour with no minimum charge. Which statement is true for Sue?

(1) The charge will be the same.
(2) Lot A will charge $2 more than Lot B.
(3) Lot A will charge $3 more than Lot B.
(4) Lot B will charge $2 more than Lot A.
(5) Lot B will charge $3 more than Lot A.

42. Dan's health insurance requires him to go to the Eastside Health Clinic for treatment. He pays $33 per month for the insurance and has to pay $5 for each visit to the doctor. He goes to the clinic six times in two months. Which expression can be used to find the total amount Dan has paid for insurance and treatment for the two months?

(1) 6($33 − $5) + 2
(2) (2 + 6)($5 + $33)
(3) 2($5) + 6($33)
(4) 2($33) + 6($5)
(5) 12($33 + $5)

43. Alfredo purchased a new lawn mower for $249 on sale for 25% off the original price. Which expression can be used to find the original price of the lawn mower?

(1) $249 ÷ 0.25
(2) $249 ÷ (100% − 25%)
(3) 0.25($249)
(4) $249 + 25
(5) 0.25($249) + $249

44. Sadie paid $729 in state income taxes last year, which represents 4% of her yearly income. Which expression can be used to find her income last year?

(1) 0.04 ÷ $729
(2) $729 ÷ 0.04
(3) $729 ÷ 0.4
(4) $729(0.04)
(5) $729(0.04) + 729

45. Marvin received his credit card bill showing a prior balance of \$50. New purchases for the month were \$140. Marvin plans to send a check for $\frac{3}{4}$ of the bill. For how much should his check be written?

(1) \$105.00
(2) \$133.75
(3) \$142.50
(4) \$165.00
(5) \$253.50

Items 46 to 48 refer to the following information.

Recipe for Rice

2 servings ($\frac{2}{3}$ cup per serving)			
Rice uncooked	Water	Salt	Butter
$\frac{1}{2}$ cup	$1\frac{3}{4}$ cup	$\frac{1}{2}$ tsp.	$\frac{1}{2}$ tbsp.

46. Roberto plans to double the recipe. How many cups of water will he need?

(1) $3\frac{1}{2}$
(2) $2\frac{3}{4}$
(3) $2\frac{1}{2}$
(4) $2\frac{1}{4}$
(5) 2

47. Teri plans to make 30 two-third-cup servings of rice. After cooking, how many cups of rice will she have?

(1) 12
(2) 15
(3) 18
(4) 20
(5) 24

48. Francisco plans to make $1\frac{1}{2}$ times the recipe. Which expression can be used to find how many cups of uncooked rice Francisco will need?

(1) $1\frac{1}{2} \div \frac{1}{2}$
(2) $1\frac{1}{2}\left(\frac{1}{2}\right)$
(3) $1\frac{1}{2}\left(1\frac{1}{2}\right)$
(4) $\frac{1}{2} \div 1\frac{1}{2}$
(5) $1\frac{1}{2} \div \frac{1}{2}$

49. Lucinda bought a suit for \$69.50, shoes for \$39.99 and a blouse for \$24.90. How much did she spend?

(1) \$134.39
(2) \$144.40
(3) \$159.49
(4) \$162.29
(5) \$190.75

50. Kumiko made a fruit salad for the cafeteria. She used 3.32 pounds of apples, 2.45 pounds of oranges, 4.67 pounds of bananas, 2.25 pounds of grapes, and 3.08 pounds of peaches. To the nearest pound, how many pounds of fruit did she use?

(1) 14
(2) 15
(3) 16
(4) 17
(5) Not enough information is given.

51. Maria plans to make 4 dresses. Each dress requires $4\frac{7}{8}$ yards of material. How many yards of material will Maria use altogether?

(1) $16\frac{1}{2}$
(2) $16\frac{7}{8}$
(3) $18\frac{7}{8}$
(4) $19\frac{1}{2}$
(5) $20\frac{1}{8}$

Answers are on page 367.

Performance Analysis
Unit 1 Cumulative Review: Arithmetic

Name:_________________________________ **Class:**_______________ **Date:**_______________

Use the Answer Key on pages 367–370 to check your answers to the GED Unit 1 Cumulative Review: Arithmetic. Then use the chart to figure out the skill areas in which you need additional review. Circle on the chart the numbers of the test items you answered correctly. Then go back and review the lessons for the skill areas that are difficult for you. For additional review, see the *Steck-Vaughn GED Mathematics Exercise Book*, Unit 1: Arithmetic.

Use the chart below to find your strengths and weaknesses in application skills.

Concept	Item Numbers
Whole Numbers pages 26–57	1, 3, 9, 20, 25, 28, **31**, 34, 40, 42
Fractions pages 58–87	4, 18, 19, 21, 22, 23, 26, 29, 35, **46, 47, 48,** 51
Decimals pages 88–113	2, 8, 10, 11, 12, 14, 15, 17, 24, 27, 36, 38, 41, 45, 49, 50
Percent pages 114–137	5, 6, 7, 13, 16, 30, **32,** 33, 37, 38, 39, 43, 44

Boldfaced numbers indicate items based on charts, graphs, illustrations, and diagrams.

Lesson 15

Ratio and Proportion • Mean • Median • Probability

Ratio and Proportion

A **ratio** is a comparison of two numbers. When two ratios are written as equal ratios, the equation is called a **proportion.**

Example: In a mixture of 4 quarts of red paint and 2 quarts of blue paint, the ratio of red paint to blue paint is 4 to 2, 4:2 or $\frac{4}{2}$. The fraction $\frac{4}{2}$ equals $\frac{2}{1}$ when reduced to lowest terms, so for every 2 quarts of red paint, there is 1 quart of blue paint. The ratios $\frac{4}{2}$ and $\frac{2}{1}$ are **equal ratios.**

Ratios are often used to express **rates.**

Example: If Paul earns \$60 in 8 hours, what is the rate of earnings to hours?

Divide to find the unit rate: $\frac{\text{dollars earned}}{\text{hours}} = \frac{60}{8} = \frac{60 \div 8}{8 \div 8} = \frac{7.5}{1}$

When the denominator of a ratio is 1, the ratio is called a **unit rate.** The unit rate $\frac{7.5}{1}$ tells us that Paul earns **\$7.50** each (per) hour.

If cross-products are equal, then the ratios are equal. If two ratios are equal, they form a **proportion.**

Examples:

Are $\frac{8}{12}$ and $\frac{12}{18}$ equal ratios?

$\frac{8}{12} \overset{?}{\times} \frac{12}{18}$

$8 \times 18 \overset{?}{=} 12 \times 12$

$144 = 144$

The cross-products are equal, so $\frac{8}{12} = \frac{12}{18}$ are equal ratios. Thus, $\frac{8}{12} = \frac{12}{18}$ is a proportion.

Are $\frac{15}{25}$ and $\frac{7}{10}$ equal ratios?

$\frac{15}{25} \overset{?}{\times} \frac{7}{10}$

$15 \times 10 \overset{?}{=} 25 \times 7$

$150 \neq 175$

The cross-products are <u>not</u> equal, so $\frac{15}{25} = \frac{7}{10}$ are not equal ratios. Thus, $\frac{15}{25} = \frac{7}{10}$ is not a proportion.

In a proportion problem, one of the four values is missing. The proportion can be solved using this rule:

$\frac{N_1}{D_1} \times \frac{N_2}{D_2}$

$N_1 \times D_2 = N_2 \times D_1$

Cross-product Rule: To find the missing number in a proportion, cross multiply and divide the product by the third number.

Example: Paula can drive her car 350 miles on 14 gallons of gasoline. How many gallons of gasoline will Paula need to drive 875 miles?

Solve this problem by writing a proportion. Use the Cross-product Rule:

Step 1: Write a proportion. $\frac{350 \text{ miles}}{14 \text{ gallons}} = \frac{875 \text{ miles}}{? \text{ gallons}}$

Step 2: Cross multiply. $14 \times 875 = 12{,}250$

Step 3: Divide by the third number. $12{,}250 \div 350 = 35$

Paula will need to buy **35 gallons** of gasoline.

Mean and Median

A list of numbers is sometimes called **data.** The average, or **mean,** is the sum of the data divided by the number of items on that list. When the data is arranged in order, the middle number is called the **median.**

Follow these steps to find an average:

Step 1: Add the data you need to average.

Step 2: Divide by the number of data items.

Example: For five days Paula recorded the time it took to drive to work. Her data is shown on the chart. What is her average driving time? Add the five values: $28 + 40 + 30 + 25 + 32 = 155$. Divide by 5, the number of data items: $155 \div 5 = 31$

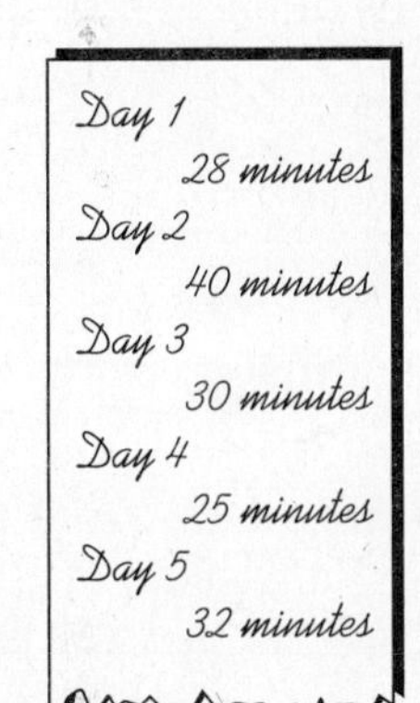

Paula's average or mean driving time is **31 minutes.**

Other information can be found by listing the data in order from lowest to highest: 25, 28, 30, 32, 40. The lowest number is 25 and the highest is 40. So we say that the **range** of the data is from 25 to 40. When the data is listed in order this way, the middle number is called the **median.**

Paula's median driving time is **30 minutes.**

If there is an even number of data items, the median is the average of the two middle numbers.

Example: George bowled four games and had scores of 128, 157, 155, and 160. He computed his average correctly at 150. What is his median score?

List the scores from lowest to highest: 128, 155, 157, 160. Because there is an even number of scores, average the two middle numbers to find the median score.

$155 + 157 = 312 \qquad 312 \div 2 = 156$

George's median score is **156.**

Probability tells how likely it is that an event will happen. In a probability situation, there are always two numbers to consider:

1. The total number of possible results.
2. The number of favorable results.

To find the probability that an event will happen, follow these steps:

Step 1: Write this ratio: $P = \frac{\text{number of favorable results}}{\text{total number of possible results}}$

Step 2: Reduce the ratio to lowest terms.

Example: If you buy 4 tickets in a drawing and 1,000 total tickets are sold, what is the probability that you will win?

$$P = \frac{\text{number of favorable results}}{\text{total number of possible results}} = \frac{4}{1{,}000} = \frac{1}{250} = 0.004$$

Thus, probability can be expressed three ways:
- as a ratio—1 out of 250
- as a fraction—$\frac{1}{250}$
- as a decimal—0.004

Lesson 15

Directions: Write each ratio as a fraction in lowest terms.

1. 16 to 12
2. 15 to 45
3. 6 to 36
4. 14 to 35
5. 9 to 36
6. 25 to 10
7. 24 to 36
8. 12 to 9
9. 85 to 100
10. 18 to 54
11. 18 to 6
12. 35 to 7
13. 80 to 100
14. 14 to 50
15. 26 to 50

Directions: Find each unit rate.

16. 400 miles in 5 hours

17. 5 pounds for 85 cents

18. $225 in 12 hours

19. 112 calories in 28 grams of cheese

20. 512 people for 8 teams

21. 5,400 oranges in 24 bags

22. 1,032 feet in 16 seconds

23. 16 ounces for $2.40

Directions: Solve each proportion problem. (Hint: Cross multiply.)

24. $\frac{2}{3} = \frac{\quad}{15}$

25. $\frac{14}{\quad} = \frac{28}{12}$

26. $\frac{\quad}{20} = \frac{3}{10}$

27. $\frac{\quad}{18} = \frac{3.5}{9}$

28. $\frac{4.2}{3} = \frac{\quad}{10}$

29. $\frac{5}{\quad} = \frac{15}{24}$

30. $\frac{12}{15} = \frac{24}{\quad}$

31. $\frac{14}{6} = \frac{7}{\quad}$

32. $\frac{\quad}{6} = \frac{11.5}{23}$

33. $\frac{7}{\quad} = \frac{3.5}{16}$

34. $\frac{4.9}{7} = \frac{\quad}{10}$

35. $\frac{\quad}{15} = \frac{3.2}{8}$

36. $\frac{3}{\quad} = \frac{1.8}{6}$

37. $\frac{6}{1.2} = \frac{5}{\quad}$

38. $\frac{6}{2.1} = \frac{\quad}{7}$

Directions: Find the range, median and mean for the following data sets.

39. Test scores: 85, 100, 65, 100, 94, 80, 85

40. Weekly pay: $215.35, $219.82, $245.82, $227.83, $199.48

41. Average temperatures: 61.5°F, 64.8°F, 69.0°F, 67.3°F, 65.6°F, 60.8°F, 61.1°F

Directions: Find the range, median and mean for the following sets.

42. Grams of vitamin C in a group of cereal samples: 0.06, 0.04, 0.055, 0.052, 0.048

43. Attendance figures: 135, 174, 128, 215

44. Number of video tapes rented each day for a week: 27, 24, 12, 35, 120, 150, 87

45. Running times for 5 days: 41 minutes, 65 minutes, 67 minutes, 52 minutes, 35 minutes

---

Directions: Choose the best answer to each item.

46. Roger worked 9 hours on Friday and 6 hours on Saturday. What is the ratio of the work done on Friday to the total hours worked on both days?

(1) $\frac{2}{5}$

(2) $\frac{3}{5}$

(3) $\frac{2}{3}$

(4) $\frac{3}{2}$

(5) $\frac{5}{3}$

47. An architect is planning a city parking lot. For every 12 commuters, the parking lot will need 5 parking spaces. How many parking spaces will be needed by 132 commuters?

(1) 7

(2) 12

(3) 17

(4) 55

(5) 60

48. A recipe that serves 8 people calls for 2 cups of milk. How many cups of milk will be needed for 36 servings?

(1) 8
(2) 9
(3) 10
(4) 12
(5) 18

49. A person uses about 315 calories to jog 3 miles. How many calories will be used in a 10-mile jog?

(1) 945
(2) 1,005
(3) 1,050
(4) 3,150
(5) 9,450

Items 50 and 51 refer to the following dart board.

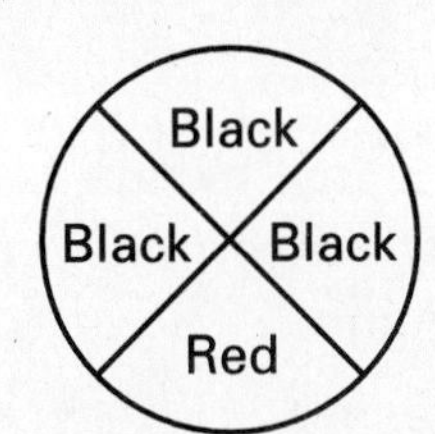

50. What is the probability of hitting red?

(1) 0.1
(2) 0.2
(3) 0.25
(4) 0.4
(5) Not enough information is given.

51. What is the probability of hitting black?

(1) $\frac{1}{4}$
(2) $\frac{1}{3}$
(3) $\frac{1}{2}$
(4) $\frac{2}{3}$
(5) $\frac{3}{4}$

Items 52 to 55 refer to the following information.

An appliance store has identical boxes containing electric mixers on a stockroom shelf. There are 5 of Model A, 8 of Model B, 10 of Model C, and 2 of Model D. A stock person takes one box from the shelf to fill an order.

52. What is the probability that the box chosen contains the Model A mixer?

(1) $\frac{1}{10}$
(2) $\frac{1}{5}$
(3) $\frac{2}{5}$
(4) $\frac{1}{2}$
(5) $\frac{3}{4}$

53. Which model has a probability of 0.08 of being chosen?

(1) Model A
(2) Model B
(3) Model C
(4) Model D

54. Which model has the greatest probability of being chosen?

(1) Model A
(2) Model B
(3) Model C
(4) Model D

55. What is the probability that the box chosen contains the Model C mixer?

(1) 0.04
(2) 0.2
(3) 0.25
(4) 0.35
(5) 0.4

Answers are on page 370.

Strategies for Solving Word Problems

Use Proportion to Solve Percent Problems

Percents are ratios: 15% means 15 out of 100. Every percent problem can be set up as a proportion by writing two ratios that are equal.

$\frac{\text{Part}}{\text{Base}} = \frac{\text{Rate}}{100}$ To find the missing number in the proportion, cross multiply and divide by the third number.

Example: Find 40% of 2,500.
Set up the proportion: $\frac{?}{2{,}500} = \frac{40}{100}$

Cross multiply: $2{,}500 \times 40 = 100{,}000$
Divide by 100: $100{,}000 \div 100 = 1{,}000$
40% of 2,500 is **1,000.**

Example: 12 is 75% of what number?
Set up the proportion: $\frac{12}{?} = \frac{75}{100}$

Cross multiply: $12 \times 100 = 1{,}200$
Divide by 75: $1{,}200 \div 75 = 16$
12 is **75%** of 16.

Example: 16 is what percent of 80?
Set up the proportion: $\frac{16}{80} = \frac{?}{100}$

Cross multiply: $16 \times 100 = 1{,}600$
Divide by 80: $1{,}600 \div 80 = 20$
16 is **20%** of 80.

To solve percent of increase or decrease problems, set up the proportion as follows:

$$\frac{\text{difference between the amounts}}{\text{original amount}} = \frac{\text{rate}}{100}$$

Example: Last year Sheila earned $6.40 per hour. This year she earns $7.36 per hour. What is the percent of increase?

Find the difference: $\$7.36 - \$6.40 = \$0.96$

Set up the proportion: $\frac{\$0.96}{\$6.40} = \frac{?}{100}$

Cross multiply: $\$0.96 \times 100 = 96.$

Divide: $96 \div 6.4 = 15$ so $\frac{\$0.96}{\$6.40} = \frac{15}{100} = .15 = 15\%.$

There was a **15% increase** from last year to this year.

Solving Word Problems

Directions: Choose the best answer to each item.

1. Masoud bought a used car for $1,200. Later he sold the car for $1,800. What was the percent of increase in the price of the car?

   (1) 5%
   (2) 25%
   (3) $33\frac{1}{3}$%
   (4) 50%
   (5) 75%

2. A sweater was originally priced $35. If the sweater is on sale for 20% off, how much would you save by buying it on sale?

   (1) $ 7
   (2) $ 9
   (3) $12
   (4) $14
   (5) $15

3. Sidney's insurance paid 90% of the cost of getting his car fixed. If the repair bill was $625, how much did the insurance pay?

   (1) $437.50
   (2) $468.75
   (3) $500.00
   (4) $562.50
   (5) $605.15

4. Aldora earns $1,344 per month. If 2.5% of her earnings goes to state income tax, how much does she pay per month in state income tax?

   (1) $ 20.16
   (2) $ 26.88
   (3) $ 33.60
   (4) $ 42.20
   (5) $336.00

5. On a test, a student got 80% of the items correct. If the student got 56 items correct, how many items were on the test?

   (1) 64
   (2) 70
   (3) 72
   (4) 84
   (5) 90

6. The Bulldogs won 18 games out of 45. What percent of their games did the Bulldogs win?

   (1) 40%
   (2) 45%
   (3) 50%
   (4) 55%
   (5) 60%

7. A telephone was originally priced $112.80. Now it is on sale for $84.60. What is the percent of decrease from the original price?

   (1) 15%
   (2) 18%
   (3) 25%
   (4) 28%
   (5) $33\frac{1}{3}$%

8. 80% of the Usagi Express Company's employees are drivers. If there are 300 drivers in the company, how many employees work for Usagi Express?

   (1) 320
   (2) 335
   (3) 342
   (4) 365
   (5) 375

Answers are on page 371.

Lesson 15

Directions: Choose the best answer to each item.

1. Juanita used 2 quarts of blue paint, 3 quarts of green and 5 quarts of white to make a paint mixture for her home. If it takes 25 quarts of paint to cover her home, how many quarts of green paint must Juanita use?

   (1) 5
   (2) 7
   (3) 7.5
   (4) 10
   (5) 15

2. A baseball team has won 16 games and lost only 4 games. If the team plays 30 games, how many more games must they win to keep the same winning ratio?

   (1) 2
   (2) 4
   (3) 6
   (4) 8
   (5) 10

3. Gail, a hairdresser, kept a record of her sales for 5 days: Monday, \$120; Tuesday, \$120; Wednesday, \$170; Thursday, \$140; Friday, \$160. She reported that her average sales were \$142. Which of the following is true?

   (1) Only the median is \$142.
   (2) Only the mean is \$142.
   (3) Neither the median nor mean is \$142.
   (4) Both the median and mean are \$142.
   (5) More information is needed.

Items 4 and 5 refer to the following information.

A used car lot has 12 cars in 5 different colors:

Blue	Black	Red	Red	Yellow	Green
Black	Blue	Yellow	Green	Red	Red

4. One car is picked at random. What is the probability that the car is red?

   (1) $\frac{1}{3}$
   (2) $\frac{1}{4}$
   (3) $\frac{1}{5}$
   (4) $\frac{1}{6}$
   (5) $\frac{1}{12}$

5. Three more yellow cars are brought to the lot. One car is picked at random. What is the probability that the car is yellow?

   (1) $\frac{1}{2}$
   (2) $\frac{1}{3}$
   (3) $\frac{1}{4}$
   (4) $\frac{1}{5}$
   (5) $\frac{1}{6}$

6. Nancy received 150 votes for union president. Stanley received 56 votes and Marc received 34. What is the ratio of Nancy's votes to the total number of votes?

   (1) $\frac{3}{5}$
   (2) $\frac{5}{8}$
   (3) $1\frac{3}{5}$
   (4) $1\frac{5}{8}$
   (5) $1\frac{2}{3}$

7. In Northville, 2 out of 5 adults have part-time jobs. If the city of Northville has an adult population of 4,500, how many adults work part-time?

(1) 900
(2) 1,250
(3) 1,800
(4) 2,250
(5) 4,500

8. Mario runs each morning before going to work. His recorded times for 7 days are 45 minutes, 35 minutes, 30 minutes, 42 minutes and 55 minutes on the other 3 days. What is the mean (average) of his running times?

(1) about 30 minutes
(2) between 30 minutes and 42 minutes
(3) between 42 minutes and 46 minutes
(4) between 46 minutes and 50 minutes
(5) more than 50 minutes

Items 9 to 11 refer to the following information.

Gem City Volleyball League

	Games Won	Games Lost	Total Games Played
Team A	3	11	14
Team B	5	9	14
Team C	12	2	14

9. What is the ratio of Team A wins to Team C wins?

(1) $\frac{3}{11}$
(2) $\frac{1}{4}$
(3) $\frac{5}{12}$
(4) $\frac{11}{14}$
(5) $\frac{11}{15}$

10. Which team has the highest ratio of games won to games played?

(1) Team A with $\frac{3}{11}$
(2) Team A with $\frac{3}{14}$
(3) Team B with $\frac{5}{14}$
(4) Team C with $\frac{12}{2}$
(5) Team C with $\frac{12}{14}$

11. Team B will play a total of 42 games this season. How many of these must they win to keep the same ratio of games won to games played as they now have?

(1) 8
(2) 9
(3) 12
(4) 15
(5) 35

12. A vocational school admits 4 out of every 10 people who apply for admission. One year, 640 students were admitted. How many people applied for admission?

(1) 256
(2) 1,600
(3) 2,560
(4) 6,400
(5) 12,800

13. A computer printer prints 1,000 characters every 3 seconds. If there are approximately 3,300 characters on a page, how many seconds will it take to print 5 pages?

(1) 4.95
(2) 9.9
(3) 30
(4) 49.5
(5) 495

Items 14 to 16 refer to the following information.

A number cube has 6 sides, labeled with the numbers 1 to 6.

14. What is the probability that you will roll an even number?

(1) $\frac{5}{6}$
(2) $\frac{4}{5}$
(3) $\frac{3}{4}$
(4) $\frac{1}{2}$
(5) $\frac{1}{3}$

15. What is the probability that you will roll a number less than 5?

(1) $\frac{1}{3}$
(2) $\frac{2}{3}$
(3) $\frac{1}{2}$
(4) $\frac{4}{5}$
(5) $\frac{5}{6}$

16. What is the probability that you will roll a 4 or a 5?

(1) $\frac{1}{5}$
(2) $\frac{1}{4}$
(3) $\frac{1}{3}$
(4) $\frac{2}{3}$
(5) $\frac{5}{6}$

17. Alicia drove 220 miles in 4 hours. How far can she drive in 7 hours at the same speed?

(1) 227 miles
(2) 385 miles
(3) 660 miles
(4) 880 miles
(5) 1,080 miles

18. Jamal earned the following amounts in the last 3 years:
$21,200 $24,500 $25,100
What is the mean (average) amount he earned?

(1) $21,000
(2) $22,000
(3) $23,600
(4) $24,500
(5) $25,000

19. The scale on a map is $\frac{1}{2}$ inch = 15 miles. What is the distance between two towns that are 3 inches apart on the map?

(1) 30 miles
(2) 45 miles
(3) 60 miles
(4) 90 miles
(5) 180 miles

20. Alison earned $35 in tips in 7 hours of work at the diner. How much can she expect to earn in 10 hours if she continues to earn at the same rate?

(1) $17
(2) $42
(3) $45
(4) $50
(5) $52

21. Fred spends $15 out of every $20 he earns, and he saves the rest. His weekly take-home pay is $560. How much money does he save each week?

(1) $ 5
(2) $ 28
(3) $112
(4) $140
(5) $160

Answers are on page 371.

Measurement Systems

In this lesson you will learn about the two most common measurement systems: the standard system and the metric system. You will need to understand both these systems to solve problems about length, weight, volume, and time.

The Standard System

You are probably most familiar with the standard system. This is the system that is used in the United States. The basic units of measure used in this system are shown in the chart below. You should learn the values on this chart.

Standard Measurement System		
Length		
1 foot (ft)	=	12 inches (in.)
1 yard (yd)	=	3 ft
Weight		
1 pound (lb)	=	16 ounces (oz)
1 ton	=	2,000 lb
Volume		
1 cup (c)	=	8 fluid ounces (fl oz)
1 pint (pt)	=	2 c
1 quart (qt)	=	2 pt
1 gallon (gal)	=	4 qt
Time		
1 minute (min)	=	60 seconds (sec)
1 hour (hr)	=	60 min
1 day	=	24 hr
1 week	=	7 days
1 year	=	12 months or 365 days

Converting Measurements

When solving problems you must always keep in mind what units are being used. Converting measurements involves using ratios correctly to change from one unit to another.

Example: How many seconds are there in one year?

This is a problem that takes several steps to solve. You must know the number of days in a year (365), the number of hours in a day (24), the number of minutes in an hour (60) and the number of seconds in a minute (60).

Most of this information can be written as ratios or rates to solve the problem.

$$\frac{365 \text{ days}}{1 \text{ year}} \quad \frac{24 \text{ hours}}{1 \text{ day}} \quad \frac{60 \text{ minutes}}{1 \text{ hour}} \quad \frac{60 \text{ seconds}}{1 \text{ minute}}$$

Notice that in all these ratios, the numerator equals the denominator, so the ratios are all equal to the number 1. Multiplying by 1 does not change the value of an expression.

$$1 \cancel{\text{year}} \times \frac{365 \cancel{\text{days}}}{1 \cancel{\text{year}}} \times \frac{24 \cancel{\text{hours}}}{1 \cancel{\text{day}}} \times \frac{60 \cancel{\text{minutes}}}{1 \cancel{\text{hour}}} \times \frac{60 \text{ seconds}}{1 \cancel{\text{minute}}} = \underline{?} \text{ seconds}$$

We only need to calculate 365 × 24 × 60 × 60 to find the correct number of seconds. We have changed the quantity "1 year" into the equivalent quantity "31,536,000 seconds." Notice the technique here: We built a chain of numerators and denominators so that all units would cancel except "seconds," which is what we were looking for.

Example: How many inches are in 4 feet? [larger (feet) to smaller (inches)]

There are 12 inches in 1 foot. Make a conversion ratio and multiply: $4 \cancel{\text{feet}} \times \frac{12 \text{ inches}}{1 \cancel{\text{foot}}} = 48 \text{ inches}$

There are **48 inches** in 4 feet.

Example: Roger threw a rock 48 feet. How many yards was that?

There are 3 ft. in 1 yard. You are converting smaller (feet) to larger (yards) so your conversion ratio is $\frac{1 \text{ yard}}{3 \text{ feet}}$.

$$\overset{16}{\cancel{48}}\ \cancel{\text{feet}} \times \frac{1 \text{ yard}}{\cancel{3}\ \cancel{\text{feet}}} = 16 \text{ yards}$$

Notice that when you are converting from a larger unit to a smaller unit (feet to inches), the smaller unit is on the top of the conversion ratio. When converting from a smaller unit to a larger unit, the larger unit is on the top of the ratio. Just remember to set up your ratio so the units cancel leaving the unit you want on top.

Not all conversions come out even. Often you will have a remainder. The remainder can be written as a whole number and labeled with a smaller unit name, or it can be written as a fraction and labeled with the larger unit name.

Example: Convert 56 ounces to pounds.

There are 16 ounces in 1 pound. You want to convert smaller units (ounces) to larger units (pounds) so the pounds go on top of the conversion ratio: $\frac{1 \text{ pound}}{16 \text{ ounces}}$.

Multiply: $56\ \cancel{\text{ounces}} \times \frac{1 \text{ pound}}{16\ \cancel{\text{ounces}}}$; $\require{enclose} \textcircled{16}\enclose{longdiv}{56}\overset{3}{}$, 48, $r = \textcircled{8}$; $\frac{8}{16} = \frac{1}{2}$.

The remainder can be written as a whole number, 8 ounces,

or as a fraction: $\frac{\text{remainder}}{\text{divisor}} = \frac{8}{16} = \frac{1}{2}$ pound

$$\overset{1}{\cancel{8}}\ \cancel{\text{ounces}} \times \frac{1 \text{ pound}}{{}_2\cancel{16}\ \cancel{\text{ounces}}} = \frac{1}{2} \text{ pound.}$$

The answer is **3 lb. 8 oz.** or **$3\frac{1}{2}$ pounds.**

Operations with Measurements

You can add and subtract measurements. But **all** measurements **must** be in the **same** units. You cannot add feet and inches (unlike units). You must first convert one measurement into the other. Then you can perform the operation.

Remember that smaller units are fractions of the larger unit (1 inch = $\frac{1}{12}$ of a foot). Often you will have an improper fraction in your answer that you will need to reduce:

$$15 \text{ inches} = \frac{15}{12} \text{ foot} = 1\frac{3}{12} \text{ foot} = 1 \text{ foot } 3 \text{ inches.}$$

Example: Two pieces of wood are 3 ft. 8 in. and 4 ft. 9 in. What is the total length of the two pieces?

Add like units:

$$\begin{array}{r} 3 \text{ ft. } \ 8 \text{ in.} \\ + \ 4 \text{ ft. } \ 9 \text{ in.} \\ \hline 7 \text{ ft. } 17 \text{ in.} \end{array}$$

Reduce your answer:

$7 \text{ ft. } 17 \text{ in.} = 7\frac{17}{12} \text{ feet} = 7 \text{ feet} + 1\frac{5}{12} \text{ ft.} = 8 \text{ ft. } 5 \text{ in.}$

The total length is **8 ft. 5 in.**

Example: John's son weighed 11 lb. 11 oz. at 2 months of age. At 4 months he weighed 14 lb. 8 oz. How much weight has he gained?

Subtract:

$$\begin{array}{r} \overset{13}{\cancel{14}} \text{ lb. } \ \overset{24}{\cancel{8}} \text{ oz.} \\ - \ 11 \text{ lb. } 11 \text{ oz.} \\ \hline 2 \text{ lb. } 13 \text{ oz.} \end{array}$$

Borrow 1 lb. from the 14 lbs. and convert it to 16 oz. Add 16 oz. to 8 oz., and then subtract.

John's son has gained **2 lb. 13 oz.** in two months.

You can also multiply and divide measurements. As before, use like units and reduce when necessary.

Example: A plan calls for 4 boards, each 2 ft. 3 in. in length. What is the total length of the boards?

Multiply:

$$\begin{array}{r} 2 \text{ ft. } \ 3 \text{ in.} \\ \times \qquad \quad 4 \\ \hline \end{array}$$

Multiply each part of the measure (2 ft. × 4 = 8 ft.; 3 in. × 4 = 12 in.)

Reduce: $8 \text{ ft. } 12 \text{ in.} = 8 \text{ ft.} + 1 \text{ ft.} = 9 \text{ ft.}$ (12 in. = 1 ft.)

Example: Mel has a pipe that is 14 feet 2 inches long. If he cuts the pipe into five equal pieces, what will be the length of each piece?

Convert to the smaller unit (inches):

$$14 \text{ ft. } 2 \text{ in.} = 14 \cancel{\text{ft.}} \times \frac{12 \text{ in.}}{1 \cancel{\text{ft.}}} + 2 \text{ in.} = 168 \text{ in.} + 2 \text{ in.} = 170 \text{ in.}$$

Divide by 5: $170 \text{ in.} \div 5 = 34 \text{ in.}$

Simplify: $34 \text{ in.} \div 12 \text{ in./ft.} = 2 \text{ r}10$ or 2 ft. 10 in.

The Metric System

The metric system is used throughout most of the world. It uses these basic metric units:

Length: meter (m) A meter is a few inches longer than one yard.
Weight or Mass: gram (g) It takes 28 grams to make one ounce.
Volume: liter (L) A liter is slightly more than one quart.
Other units of measure are made by adding the prefixes shown below to the basic units listed above.

milli- means $\frac{1}{1,000}$	A milligram is $\frac{1}{1,000}$ of a gram.
centi- means $\frac{1}{100}$	A centimeter is $\frac{1}{100}$ of a meter.
deci- means $\frac{1}{10}$	A deciliter is $\frac{1}{10}$ of a liter.
deka- means 10	A dekagram is 10 grams.
hecto- means 100	A hectoliter is 100 liters.
kilo- means 1,000	A kilometer is 1,000 meters.

Metric Conversions

Because the metric system is based on the powers of ten, you can convert metric units by moving the decimal point. Use this chart to make metric conversions.

kilo	hecto	deka	meter	deci	centi	milli
1,000	100	10	gram liter	$\frac{1}{10}$	$\frac{1}{100}$	$\frac{1}{1,000}$

To convert to smaller units ——————————→
←—————————————— To convert to larger units

Follow these steps to convert from one metric unit to another.

Step 1: Count from the unit given to the unit to which you are converting.

Step 2: Move the decimal point that number of places in the direction shown on the chart.

Operations

Operations with metric units are performed as you would with decimals. Remember, you can add and subtract like units only.

Example: Troy walked 2 kilometers on Monday. On Tuesday, he walked 1.575 kilometers. How many total kilometers did he walk on the two days?

You can add 2 kilometers and 1.575 kilometers.

$$\begin{array}{r} 2 \quad\quad \text{kilometers} \\ +\ 1.575 \text{ kilometers} \\ \hline 3.575 \text{ kilometers} \end{array}$$

Troy walked **3.575 kilometers.**

Lesson 16

Directions: Solve.

1. Change $3\frac{1}{2}$ yards to feet and inches.
2. Change 4 minutes to seconds.
3. Change 18 cups to pints.
4. Change 7,500 pounds to tons.
5. Change $2\frac{1}{3}$ hours to minutes.
6. Change 72 fluid ounces to cups.
7. Change $5\frac{3}{4}$ gallons to quarts.
8. Change 6 feet 9 inches to yards.
9. Which is greater, 3 feet or 30 inches?
10. Which is greater, $1\frac{1}{2}$ gallons or 13 pints?
11. Which is less, $3\frac{1}{4}$ hours or 200 minutes?
12. Which is less, $4\frac{1}{4}$ pounds or 75 ounces?
13. Add 3 gal. 3 qt. and 2 gal. 2 qt.
14. Add 8 ft. 8 in., 3 ft. 5 in., and 4 ft. 9 in.
15. Subtract 20 min. 45 seconds from $\frac{1}{2}$ hour.
16. Subtract 1 ft. 7 in. from 2 ft. 2 in.
17. Multiply 3 lb. 4 oz. by 5.
18. Multiply 2 yds. 1 ft. by 8.
19. Divide 1 yd. 4 in. by 5. Write your answer in inches.
20. Divide 3 gallons by 4. Write your answer in quarts.

Directions: Choose the best answer to each item.

21. Dorie is canning raspberry jam. She makes 2 gallons of jam. If she plans to keep 6 pints for herself, how many pints will she have to give away?

(1) 8
(2) 9
(3) 10
(4) 11
(5) 12

22. Zola buys $3\frac{1}{4}$ yards of fabric to make a blouse. She actually uses 2 yards 2 feet 9 inches of fabric. How many inches of fabric are left?

(1) 12
(2) 13
(3) 14
(4) 15
(5) 16

23. Hank has three lengths of wire: 75 centimeters, 126 centimeters, and 460 centimeters. What is the total length in centimeters of the wire?

(1) 0.661
(2) 6.61
(3) 66.1
(4) 661
(5) 6,610

24. Meg needs 15 lengths of wood to make the railings for a stairway bannister. If each railing is 2 feet 9 inches, what is the total length of wood (in feet) that Meg will need?

(1) $11\frac{1}{4}$
(2) $36\frac{1}{4}$
(3) $41\frac{1}{4}$
(4) $48\frac{1}{2}$
(5) $53\frac{3}{4}$

25. Three shipments weigh 3 tons 750 pounds, 1 ton 150 pounds, and 4 tons 100 pounds. What is the total weight of the shipments in tons?

(1) $7\frac{3}{4}$
(2) 8
(3) $8\frac{1}{2}$
(4) $9\frac{1}{4}$
(5) $9\frac{1}{2}$

26. Marco can assemble one cabinet in 25 minutes. How many cabinets can he assemble in $3\frac{3}{4}$ hours?

(1) 7
(2) 8
(3) 9
(4) 10
(5) 11

27. A package of soup mix makes 24 ounces of soup. If $1\frac{1}{2}$ cups of soup equals one serving, how many servings will 5 packages of soup mix make?

(1) 9
(2) 10
(3) 11
(4) 12
(5) 13

28. Gregg Pharmaceuticals makes cough syrup. If 60 milliliters of the cough syrup contains 15 milliliters of the medication, how many milliliters of the medication are there in 1,000 ml of the syrup?

(1) 135
(2) 160
(3) 200
(4) 220
(5) 250

Answers are on page 373.

Strategies for Solving Word Problems

Use Mental Math

Estimation can help you narrow the choices in a multiple choice problem. Because the metric system is based on powers of ten, you can use rounding like you do with decimals.

Example: Over three days, Paula ran 1.9 kilometers, 2.8 kilometers, and 3.1 kilometers. What was the total distance in kilometers that she ran?

(1) 6.3
(2) 6.8
(3) 7.5
(4) 7.8
(5) 8.3

→ means "rounds to"

To estimate the answer:	1.9	rounds to	2
Round to the nearest kilometer:	2.8	→	3
	+ 3.1	→	3
Add the estimates:			8

Options (4) and (5) are closest to the estimate of 8.

Now solve: $1.9 + 2.8 + 3.1 = 7.8$

Option **(4) 7.8** is the correct answer.

You can also use estimation with the standard measurement system. To round to a standard unit, think "what is $\frac{1}{2}$ of this unit?" Since 1 pound equals 16 ounces, 8 ounces is $\frac{1}{2}$ of a pound; one-half of a foot is 6 inches.

Example: Jeradine worked for 3 hours 50 minutes on Monday and 5 hours 15 minutes on Tuesday. How long did she work over the two-day period?

(1) 7 hrs. 35 min.
(2) 8 hrs. 10 min.
(3) 8 hrs. 45 min.
(4) 9 hrs. 5 min.
(5) 9 hrs. 25 min.

To estimate the answer "$\frac{1}{2}$ of an hour is 30 min.":

3 hrs. 50 min. rounds to	4 hrs.
5 hrs. 15 min. rounds to	+ 5 hrs.
Add:	9 hrs.

From the estimate, choices 3 and 4 are possibilities.

Solve:

  3 hrs. 50 min.
+ 5 hrs. 15 min.
  8 hrs. 65 min. = 8 hrs. + 1 hr. 5 min. = 9 hrs. 5 min.

Option **(4) 9 hrs. 5 min.** is the correct answer.

Caution: Estimation with multiplication and division is less accurate.

Solving Word Problems

Directions: Estimate the best answer to each item, then solve.

1. Dennis needs 8 pieces of 4 ft. 10 in. molding. What is the total length of molding that he needs?

   (1) 35 ft. 6 in.
   (2) 38 ft. 8 in.
   (3) 44 ft. 10 in.
   (4) 49 ft. 3 in.
   (5) 55 ft. 11 in.

2. A plumber has a pipe 5 feet 8 inches long. If he cuts off a piece 1 foot 2 inches long, what is the length of the remaining piece?

   (1) 2 ft. 6 in.
   (2) 3 ft. 6 in.
   (3) 4 ft. 6 in.
   (4) 5 ft. 6 in.
   (5) 6 ft. 6 in.

3. Wanitta uses 1 gallon 3 quarts of pesticides three times per year on her garden. What is the total amount of pesticide that she puts in her garden during a year?

   (1) 2 gallons 2 quarts
   (2) 3 gallons 1 quart
   (3) 5 gallons 1 quart
   (4) 6 gallons 3 quarts
   (5) 7 gallons

4. Mark mailed 3 packages. One weighed 5.8 kilograms, one weighed 6.5 kilograms, and the third weighed 7.2 kilograms. What was the total weight in kilograms that he mailed?

   (1) 18
   (2) 18.5
   (3) 19
   (4) 19.5
   (5) 20

5. A tailor needs 4 yards 6 inches of material to make a pair of men's pants. How many pairs of pants could he make from 21 yards?

   (1) 4
   (2) 5
   (3) 6
   (4) 7
   (5) 8

6. A warehouse receives shipments of the following weights: 2 tons 1,500 pounds, 4 tons 300 pounds, and 6 tons 1,800 pounds. What is the total weight of these shipments?

   (1) 10 tons 200 pounds
   (2) 10 tons 1,900 pounds
   (3) 11 tons 800 pounds
   (4) 12 tons 1,500 pounds
   (5) 13 tons 1,600 pounds

7. Miles has a roll of wire that is 126 centimeters long. How many 4.2 centimeter lengths can he cut from the roll?

   (1) 18
   (2) 24
   (3) 30
   (4) 38
   (5) 45

8. At Benlow's Shirt Factory, it takes $2\frac{1}{4}$ yards of cotton fabric to make 1 shirt. How many yards will it take to make 3 shirts?

   (1) 4
   (2) $4\frac{1}{2}$
   (3) 5
   (4) 6
   (5) $6\frac{3}{4}$

Answers are on page 374.

Perimeter, Circumference, and Area

Perimeter measures the distance around the edge of any flat object. **Circumference** is the perimeter of a circle. **Area** measures the surface inside any flat object.

Perimeter

To find the perimeter of any figure, add the lengths of its sides.

Example: The county supervisor plans to fence in the area shown below for a playground. How much fencing is needed?

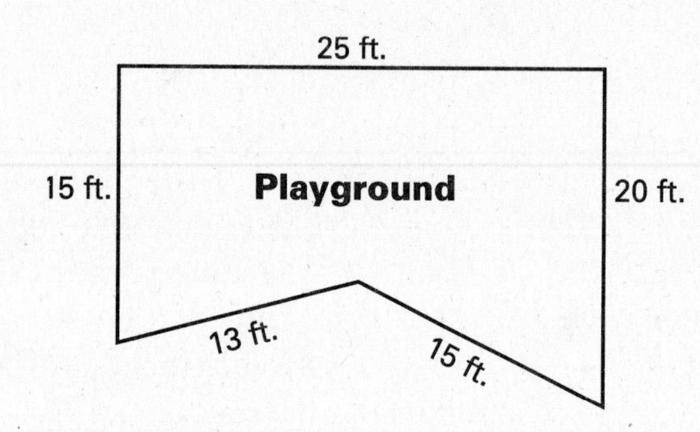

The fence will go around the perimeter of the playground. Find the perimeter by adding the lengths of the sides.
$15 + 13 + 15 + 20 + 25 = 88$

It will take **88 feet** of fencing to enclose the playground.

A **rectangle** has 4 sides, square corners, and opposite sides of equal length. Here is a special formula for finding the perimeter *(P)* of a rectangle:

$\boldsymbol{P = 2l + 2w}$, where l = length and w = width.

(Note: $P = 2l + 2w$ is the same as $P = l + l + w + w$ which is the sum of the 4 sides.)

Example: Use the formula to find the perimeter of this rectangle.

l = 12.25 inches

w = 7.5 inches

$$\begin{aligned} P &= (2 \times l) + (2 \times w) \\ &= (2 \times 12.25) + (2 \times 7.5) \\ &= 24.5 + 15 \\ &= 39.5 \text{ inches} \end{aligned}$$

The perimeter of the rectangle is **39.5 inches.**

The formula works for any unit: inches, feet or metric measurements. Note that all measurements must be the same units.

A **square** has four sides and square corners with all sides equal in length. This special formula is used to find the perimeter of a square:

$\boldsymbol{P = 4s}$, where s = side.

Example: What is the perimeter of this square?

5.25 feet (each side)

$$\begin{aligned} P &= 4(5.25) \\ &= 21 \text{ feet} \end{aligned}$$

The perimeter of the square is **21 feet.**

A **triangle** has three sides. The formula for finding the perimeter of a triangle is $\boldsymbol{P = a + b + c}$, where a, b, and c are the sides of the triangle. (For all other figures (with 5 or more sides) there is no special formula. Simply add the lengths of all the sides.)

Example: Find the perimeter of this triangle.

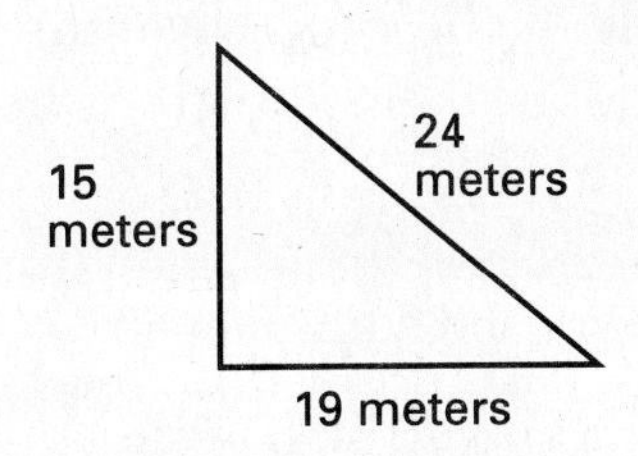

$P = a + b + c$
$= 15 + 19 + 24$
$= 58$ meters

The perimeter of the triangle is **58 meters.**

Circumference

The shapes you have already learned about all have straight sides. A **circle** has a curved edge. The perimeter of a circle, or the distance around the circle, is called the **circumference.** To find the circumference, you need to know either the diameter or the radius of the circle.

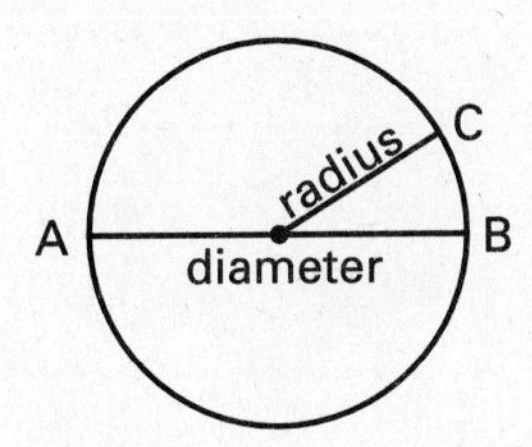

The **diameter** of a circle is a line segment drawn through the center of the circle.

The **radius** is a line segment connecting the center of the circle to a point on the circle. The length of the radius is $\frac{1}{2}$ the diameter. The length of the diameter is 2 times the radius.

The circumference of a circle is the distance around the circle. For all circles, the ratio of the circumference (C) to the diameter (d) is the same. The Greek letter π (pi) is used to represent this constant ratio whose value is $\frac{22}{7}$ which is approximately 3.14. $\left(\pi = \frac{C}{d}\right)$

Using this ratio, a formula can be written for finding circumference. The circumference of a circle can be found using this formula:

$\boldsymbol{C = \pi d}$, where $\pi = 3.14$ and d = diameter.

Example: Joe plans to put a low railing around a circular fish pond. The diameter of the pond is 20 meters. What will be the length of the railing?

$C = \pi d$
$= 3.14 \times 20$
$= 62.8$

20 m.

The railing will be about **62.8 meters.**

Area

Squares and Rectangles

Area measures the surface inside an object. Area is measured in square units, such as square inches, square yards or square meters. Imagine the rectangle at the right covered with 1-inch by 1-inch squares. Each is 1 square inch. Area tells how many square inches it takes to cover a rectangle.

Example: This rectangle is 6 inches along the bottom, so 6 squares can be lined up on the bottom row. There are 5 of these rows, so it will take 5×6, or 30 squares to cover the surface.

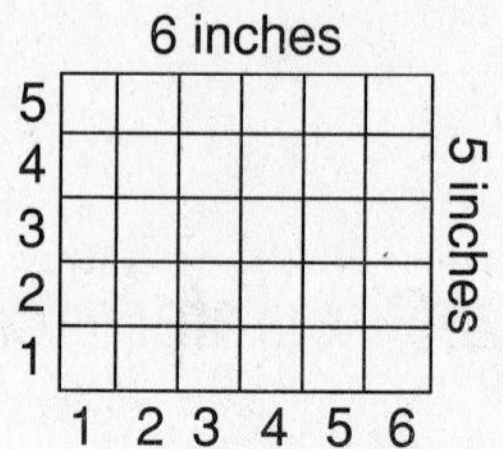

A formula that tells the number of square units in any rectangular or square figure is:

Area = $\boldsymbol{l \times w}$, where l = length and w = width.

In this figure, area = 6×5 = **30 square inches.**

Parallelograms

A **parallelogram** has 4 sides and the opposite sides are parallel. The area of a parallelogram is found by multiplying the length of the base by the height. **Height** is the distance straight down from a point on one non-slanting side to its opposite side, or the **base.**

The formula for the area of a parallelogram can be written:

$\boldsymbol{A = bh}$, where b = base and h = height.

Example: Find the area of the parallelogram shown below.

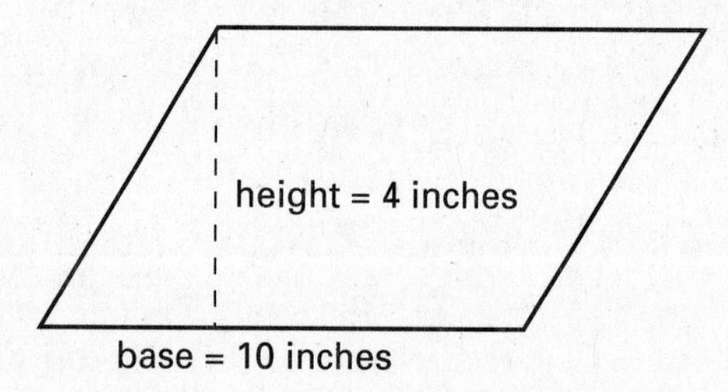

Use the formula for finding the area of a parallelogram:

$A = bh$
$= 10 \times 4$
$= 40$ square inches

The area of the parallelogram in the drawing is **40 square inches.**

Triangles

The formula used to find the area of a triangle is:

$\boldsymbol{A = \frac{1}{2}bh}$, where b = base and h = height.

The **base** of a triangle is the measurement along the bottom. The **height** is the distance from the top corner (called the **vertex**) straight down to the base line.

Example: Find the area of the triangle shown below.

Use the formula for finding the area of a triangle.

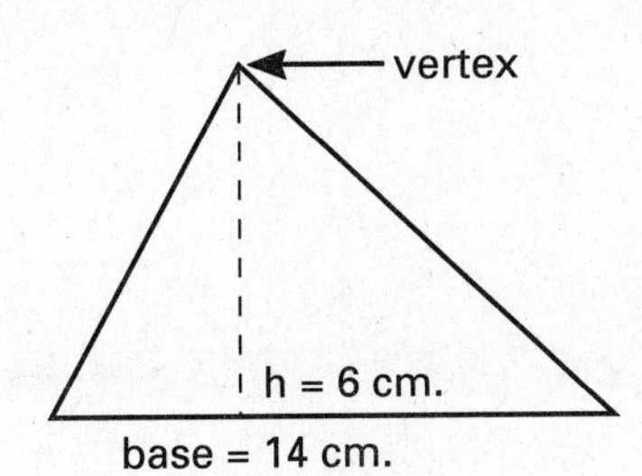

$A = \frac{1}{2}bh$

$= \frac{1}{2} \times 14 \times 6 = 7 \times 6$

$= 42$ square centimeters

The area of the triangle in the drawing is **42 square centimeters.**

Circles

The formula for finding the area of a circle is:

$\boldsymbol{A = \pi r^2}$, where $\pi = 3.14$ and r = radius.

In other words, the area of a circle is found by multiplying π (3.14) by the square of the radius. To square a number, multiply that number by itself.

Example: Javier plans to buy fertilizer for a circular flower bed. One bag of fertilizer will cover 25 square feet. How many bags will Javier need to buy?

First find the area of the garden.

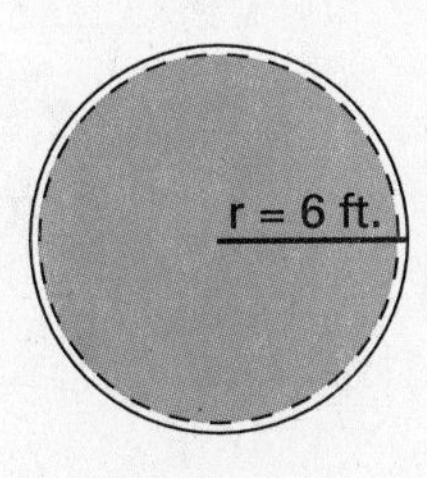

$A = \pi r^2$

$= 3.14 \times 6 \times 6$

$= 3.14 \times 36$

$= 113.04$ square feet

The area of the flower bed is about **113 square feet.**

To find the number of bags of fertilizer Javier must buy, divide 113 by 25.

```
     4.52
25)113.00
   100
    13 0
    12 5
       50
       50
```

Round to the nearest whole number: 4.52 rounds to 5. Javier needs to buy **5 bags** of fertilizer.

ged Practice Lesson 17

Items 1 to 6 refer to the following information.

Identify each figure as a square, rectangle, triangle, circle or parallelogram. Give the perimeter and area for each figure. All measurements are in inches.

1. 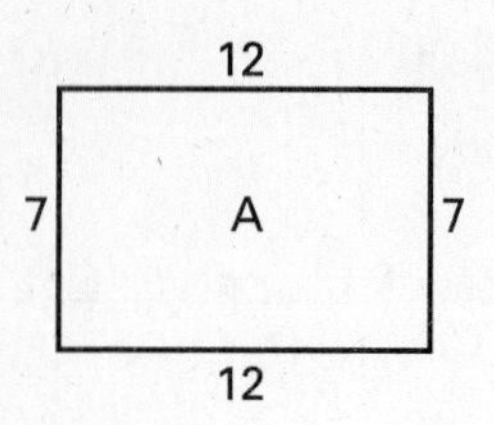

Figure name ______________________

Perimeter ______________________

Area ______________________

2. 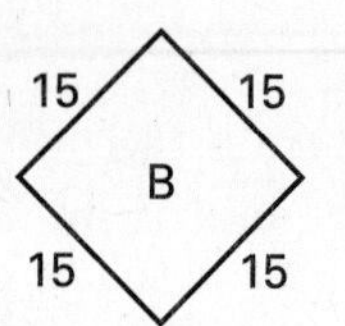

Figure name ______________________

Perimeter ______________________

Area ______________________

3. 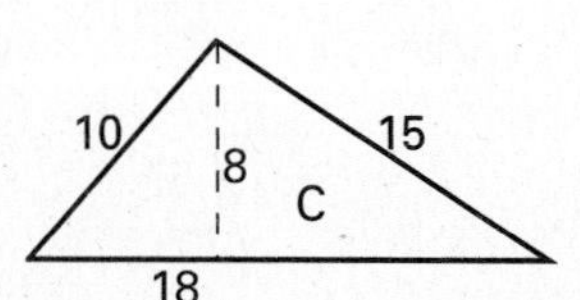

Figure name ______________________

Perimeter ______________________

Area ______________________

4. Figure name ______________________

Perimeter ______________________

Area ______________________

5. 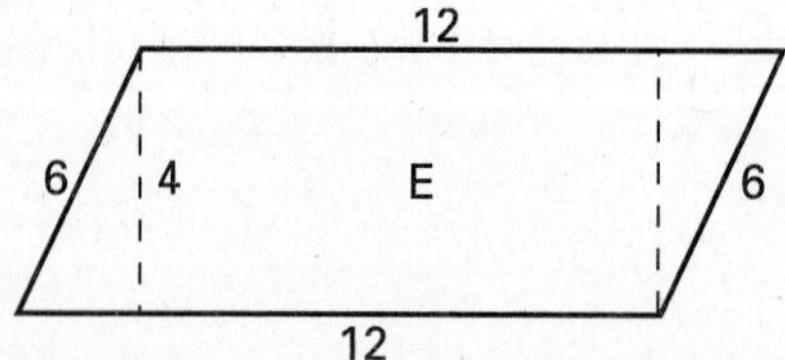

Figure name ______________________

Perimeter ______________________

Area ______________________

6. Figure name ______________________

Circumference ______________________

Area ______________________

Directions: Choose the best answer to each item.

Items 7 and 8 refer to the following figure.

A rectangular room is 8.5 feet by 25 feet.

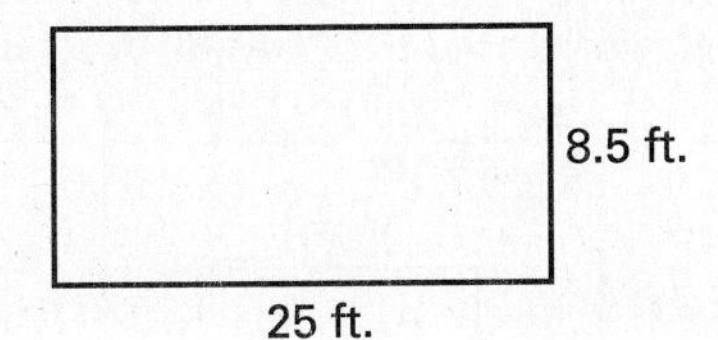

7. What is the perimeter in feet of the room in the drawing?

   (1) 59
   (2) 63
   (3) 67
   (4) 71
   (5) 74

8. What is the area in square feet of the room in the drawing?

   (1) 105.5
   (2) 144
   (3) 192.5
   (4) 212.5
   (5) Not enough information is given.

9. What is the area in square feet of the shaded portion of the parallelogram?

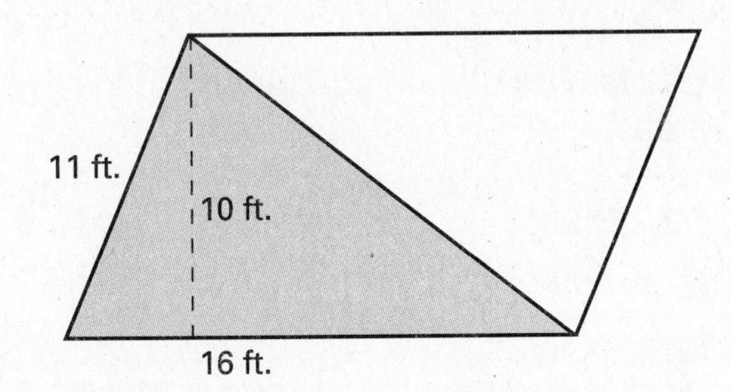

   (1) 60
   (2) 72
   (3) 75
   (4) 80
   (5) 96

10. What is the area in square inches of a triangle with a base of 18 inches and a height of 15 inches?

   (1) 135
   (2) 138
   (3) 140
   (4) 141
   (5) 144

11. The three sides of a triangle are 18 feet, 20 feet, and 24 feet. What is the area in square feet of the triangle?

   (1) 360
   (2) 400
   (3) 432
   (4) 480
   (5) Not enough information is given.

12. The design for a company logo display is shown below. If sides A and C are 14.3 centimeters long, sides D and E are 13.5 centimeters long, and side B is 12.4 centimeters long, what is the perimeter in centimeters of the logo?

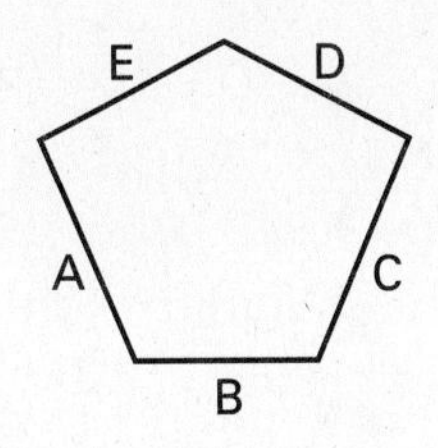

   (1) 40.2
   (2) 52.6
   (3) 58.6
   (4) 68
   (5) 70.6

13. One side of a square measures 4.6 meters. What is the area to the nearest square meter of the square?

(1) 16
(2) 17
(3) 19
(4) 20
(5) 21

14. The base of a parallelogram is 16 inches. Its height is 4 inches. What is the area of the parallelogram in square inches?

(1) 20
(2) 40
(3) 48
(4) 60
(5) 64

15. A carpenter is measuring a roof. There are two equal sections of the roof. Each section is 35 feet long and 18 feet wide. What is the total area of the roof in square feet?

(1) 648
(2) 1,096
(3) 1,225
(4) 1,260
(5) Not enough information is given.

16. To the nearest tenth of a square meter, what is the area of a circle with a radius of 7 meters?

(1) 146.4
(2) 149.7
(3) 150.2
(4) 153.9
(5) 155.3

Items 17 to 19 refer to the following diagram.

The city is planning to improve a section of Beverly Park. A cement area will be constructed in a grassy area for basketball and other sports.

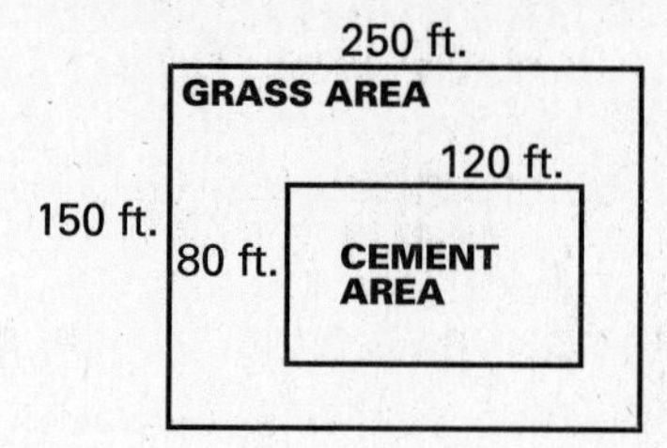

17. The city plans to build a fence around the grass area shown in the drawing. What is the perimeter of the grass area in feet?

(1) 500
(2) 600
(3) 650
(4) 700
(5) 800

18. What is the area in square feet of the cement area?

(1) 8,600
(2) 8,800
(3) 9,460
(4) 9,600
(5) 9,840

19. What is the area in square feet of the grass area?

(1) 25,900
(2) 26,400
(3) 27,900
(4) 30,100
(5) 33,500

20. Circle A has a diameter of 7 inches. Circle B has a radius of $3\frac{1}{2}$ inches. Circle C has a circumference of 22 inches. Using $\frac{22}{7}$ for pi, which of the following is a true statement?

(1) Circle A is the largest.
(2) Circle B is the largest.
(3) Circle C is the largest.
(4) Only circles A and B are the same size.
(5) All three circles are the same size.

21. What is the circumference (to the nearest whole centimeter) of a circle with a diameter of 10 centimeters?

(1) 29
(2) 30
(3) 31
(4) 32
(5) 33

Items 22 to 24 refer to the following drawing.

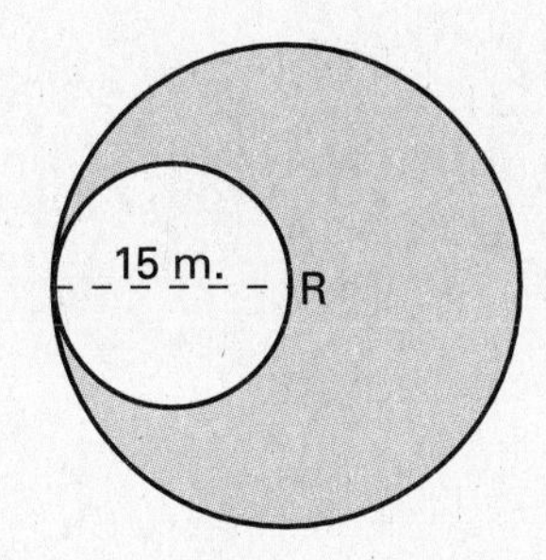

22. What is the diameter in meters of the larger circle?

(1) 7.5
(2) 12
(3) 15
(4) 30
(5) 33

23. What is the area of the inner circle to the nearest square meter?

(1) 165
(2) 177
(3) 184
(4) 189
(5) 195

24. What is the area of the shaded area of the larger circle to the nearest square meter?

(1) 507
(2) 526
(3) 530
(4) 577
(5) 589

Items 25 and 26 refer to the following drawing.

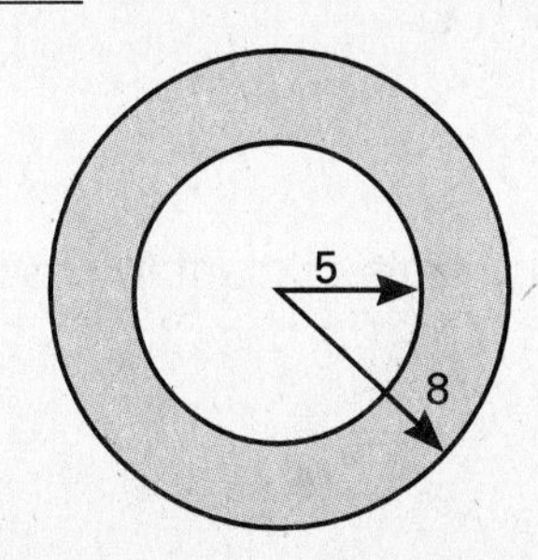

25. What is the area to the nearest square inch of the outer circle with the radius of 8?

(1) 187
(2) 192
(3) 196
(4) 198
(5) 201

26. What is the area to the nearest square inch of the shaded ring area?

(1) 121
(2) 122
(3) 126
(4) 138
(5) 142

Answers are on page 374.

ged Mini-Test Lessons 16 and 17

Directions: Choose the best answer to each item.

Items 1 to 3 refer to the following information.

A gardener has planted three square flower beds, each 6 feet on a side. In each square he plants a bush that needs an area of 5 square feet. The rest of each flower bed is planted with flowers. All three flower beds will be sprayed with liquid plant food. One pint of plant food will feed 5 square feet of planted area.

1. How many feet of rope will he need to fence off all three flower beds?

   (1) 18
   (2) 24
   (3) 36
   (4) 72
   (5) 108

2. How many pints of plant food will he need for all three flower beds?

   (1) 7
   (2) 8
   (3) 18
   (4) 19
   (5) 22

3. How many square feet are left for planting flowers in each flower bed after leaving space for the bush?

   (1) 25
   (2) 30
   (3) 31
   (4) 88
   (5) 120

4. A planting area is a half-circle with a radius of 63 feet. To the nearest whole foot, how many feet of fencing will be needed to enclose it?

   (1) 396
   (2) 324
   (3) 315
   (4) 261
   (5) 224

5. Which expression can be used to find the area in square feet of the parallelogram?

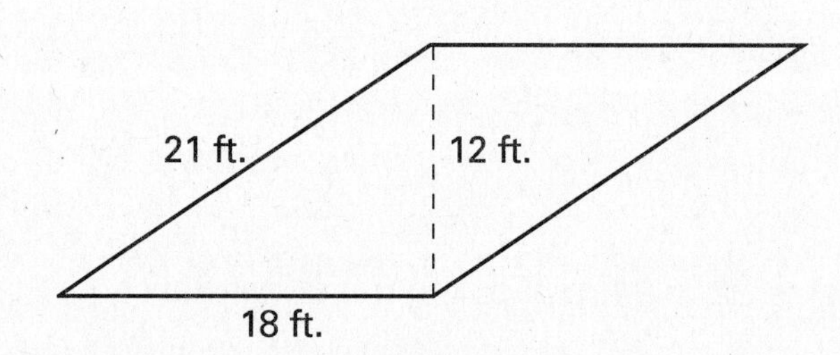

   (1) $21 + 21 + 12 + 12$
   (2) $21 + 21 + 18 + 18$
   (3) $2(12 + 18)$
   (4) 12×18
   (5) 21×18

6. What is the perimeter of a rectangular flower bed that is 25 feet by 3 feet?

   (1) 56
   (2) 60
   (3) 65
   (4) 66
   (5) 75

7. What is the area in square meters of this parallelogram?

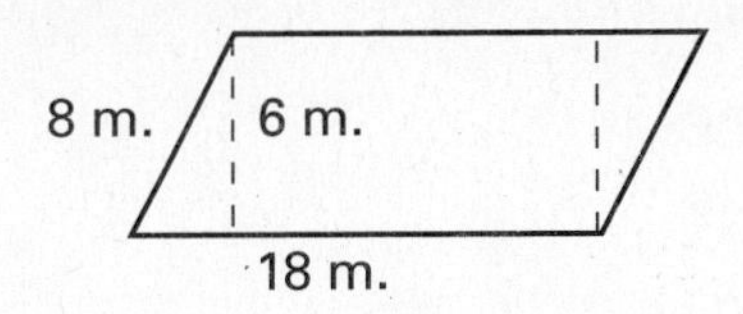

(1) 48
(2) 54
(3) 72
(4) 108
(5) 144

Items 8 and 9 refer to the following figure.

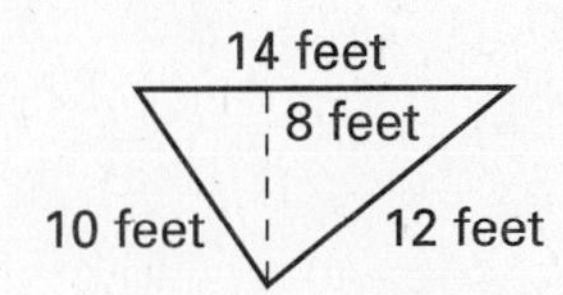

8. What is the perimeter of the triangle?

(1) 22
(2) 24
(3) 26
(4) 36
(5) Not enough information is given.

9. What is the area of the triangle?

(1) 56
(2) 63
(3) 72
(4) 112
(5) 126

10. The length of a rectangle is 36 centimeters. What is its area in square centimeters?

(1) 324
(2) 432
(3) 504
(4) 576
(5) Not enough information is given.

Items 11 and 12 refer to the following figure.

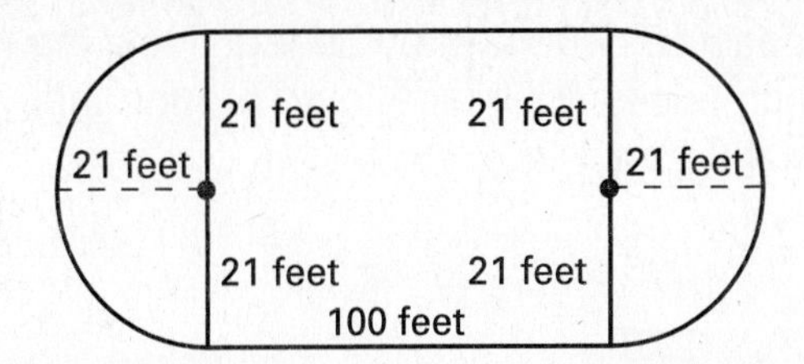

11. This aerial view of a running track shows that it is really a rectangle with two half circles at the ends. To the nearest whole foot, what is the distance around the track?

(1) 282 feet
(2) 316 feet
(3) 332 feet
(4) 364 feet
(5) 388 feet

12. The inside of the track is planted with grass. What is the area (to the nearest square foot) of grass?

(1) 4,735 square feet
(2) 4,925 square feet
(3) 5,295 square feet
(4) 5,585 square feet
(5) 5,750 square feet

13. What is the perimeter in centimeters of the figure in the drawing?

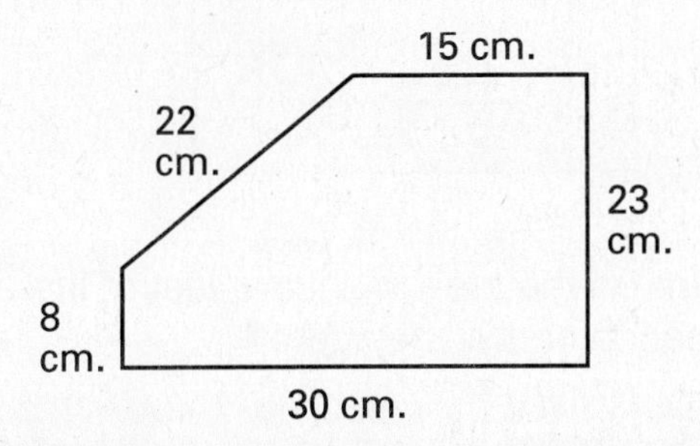

(1) 95
(2) 98
(3) 104
(4) 112
(5) 138

14. Arnold is putting brick trim around the edge of a new rectangular pool. The pool measures 100 feet by 35 feet. How many 9-inch bricks will it take to go around the edge of the pool?

(1) 200
(2) 270
(3) 360
(4) 2,430
(5) 3,240

15. Selina needs to buy paint for the cement basement of the Youth Center. The floor is a rectangle 40 feet wide by 60 feet long. One gallon of paint will cover about 200 square feet of concrete. How many gallons of paint will Selina need for one coat of paint on the floor?

(1) 12
(2) 24
(3) 40
(4) 60
(5) 120

Items 16 and 17 refer to the following drawing.

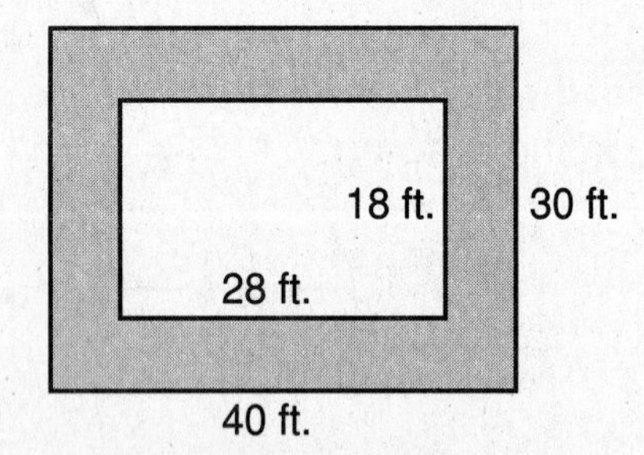

16. What is the area in square feet of the shaded part of the figure?

(1) 144
(2) 504
(3) 696
(4) 1,200
(5) 1,704

17. How many more feet is the perimeter of the larger rectangle than the perimeter of the smaller rectangle?

(1) 696
(2) 144
(3) 94
(4) 48
(5) 24

18. A circular pond in Franklin Park has a radius of 30 meters. What is the distance around the pond to the nearest meter?

(1) 85
(2) 90
(3) 94
(4) 188
(5) 848

Items 19 and 20 refer to the following figure.

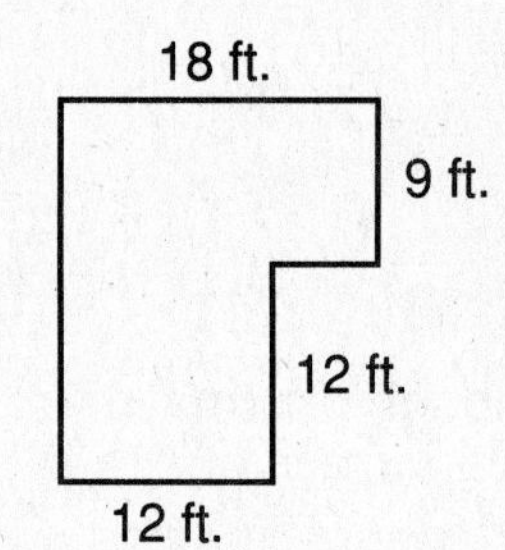

19. What is the perimeter of the figure in feet?

(1) 78
(2) 72
(3) 66
(4) 60
(5) 57

20. What is the area of the figure in square feet?

(1) 378
(2) 306
(3) 252
(4) 162
(5) 144

Answers are on page 376.

Graphs, Charts, and Tables

Graphs

Graphs are used to give data a visual meaning. They are used to compare data from different sources, to show change over a period of time, and to make projections about the future.

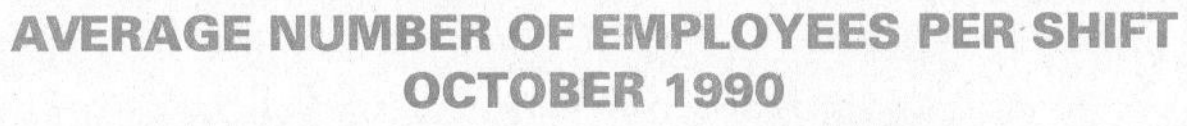

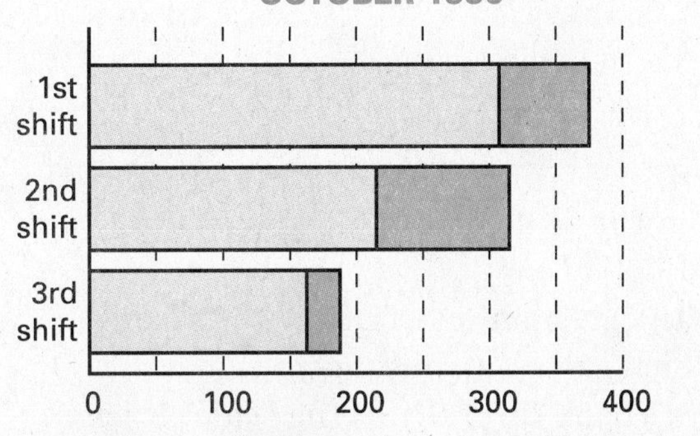

Full-time Part-time

To make a visual impact, the data in graphs is often condensed, rounded or incomplete. Yet the comparisons should stand out.

Example: The personnel manager at Morton Products wants to emphasize the difference in employment levels for each of the company's three shifts. She makes the **bar graph** at the left.

Notice these parts of the graph:

1. **Title**—labeled clearly to indicate the type of information found on the graph.
2. **Axis lines and scales**—graph has two axis lines.
   - **Horizontal axis** runs across the bottom of the graph. The sample graph shows the average number of employees.
   - **Vertical axis** runs up the side of the graph. The sample graph shows the number of shifts.
   - **Scale** is like a ruler. Each mark represents units. Some graphs have scale lines on only the vertical axis. Other graphs have a scale on both the horizontal and vertical axis.
3. **Key**—gives you any information you need to read the graph. The key is also called the legend.

To read the graph, follow across any of the three bars representing the three shifts from left to right. Examine where the endpoint of the bar falls with relation to the scale on the horizontal axis.

What information can you get from the graph?

- The first shift has the most employees.
- On the second shift, there are more part-time employees than on any other shift.
- The third shift has the smallest crew of full-time employees, with very few part-time employees.

What information is *not* available from the graph?

- The exact number of employees on any shift
- How these employment figures compare with a previous month
- The comparison between office, manufacturing, and shipping department employees

The bars in the graph below are vertical. By looking at the title and the horizontal axis at the bottom, we see that each bar represents sales from a different month from January through June. The sales amount for any month can be found from the numbers on the vertical scale at the left.

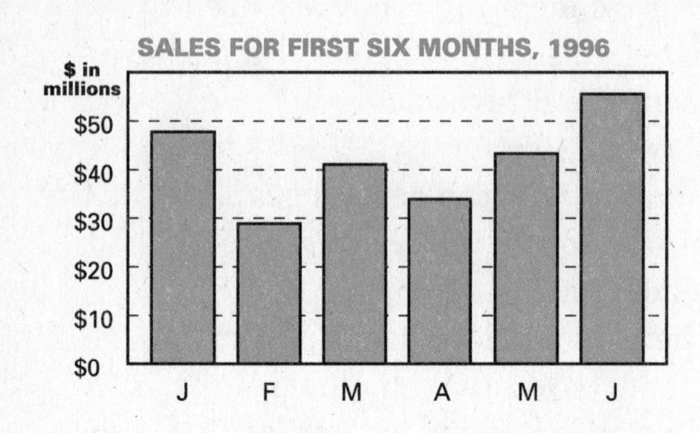

Notice the labels "$ in millions." This means that $10 on the vertical axis means $10 million. The scale runs from $0 to $50 million.

A graph can be misleading when the scale does not begin at zero.

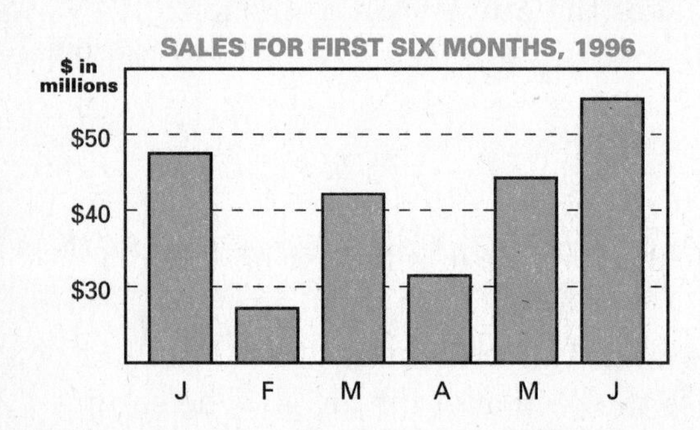

Compare the graph at the right with the one above. Which sales amounts seem the most impressive? Look carefully at the scale. It begins at $30 and ends at $50. The information is exactly the same as in the graph above.

Circle graphs show percentages. The whole circle is 100%.

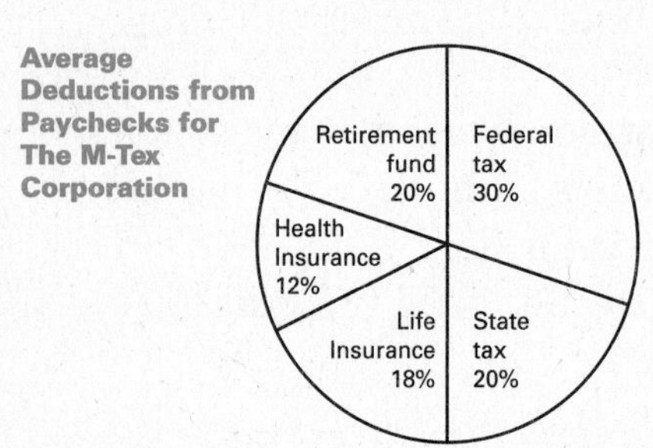

The various fractions are shown in pie-shaped pieces. A half-circle would be 50%. A quarter-circle is 25%. This graph shows how deductions from paychecks are used.

Most **line graphs** are made up of connected points. A point connects a horizontal value and a vertical value. The circled point below connects "10 A.M." with "55 degrees."

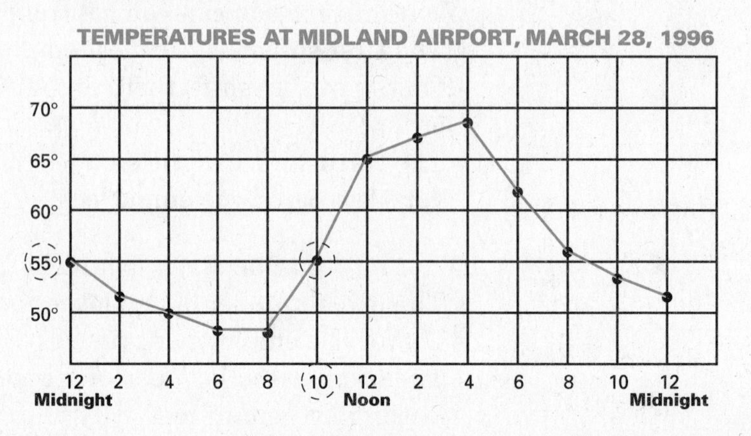

Tables and Charts

Another way to organize information is in a table or chart. Unlike a graph, tables show numbers instead of bar lines, pie-shaped pieces, or lines connecting points.

Tables group data into similar categories. Look at the following table.

Money Received by the Federal Government from Taxes

Tax Sources	1994	1995
Income Taxes		
Corporations	104.8*	117.2
Individuals	364.0	392.8
Social Security Taxes	(A) 273.2	307.4
Unemployment Taxes	23.8	(B) 22.2
Excise Taxes	32.6	33.4
Estate and Gift Taxes	6.0	5.8

*Dollars in billions

The table groups the amounts received by year. This grouping helps you to compare one year to the other for the same tax source (see A) or the amounts received from different sources in the same year (see B).

Remember, like graphs, tables and charts do not always give you all the information you may need. For example, this table does not tell you the amounts received in 1993. Nor does it tell you how much was received from Estate Taxes alone. You must remember to interpret data presented in tables carefully. Sometimes extra information is given in footnotes to the table. Notice the asterisk (*) which tells you to look at the bottom of the table for more information.

Lesson 18

Directions: Choose the best answer to each item.

Items 1 to 3 refer to the bar graph below.

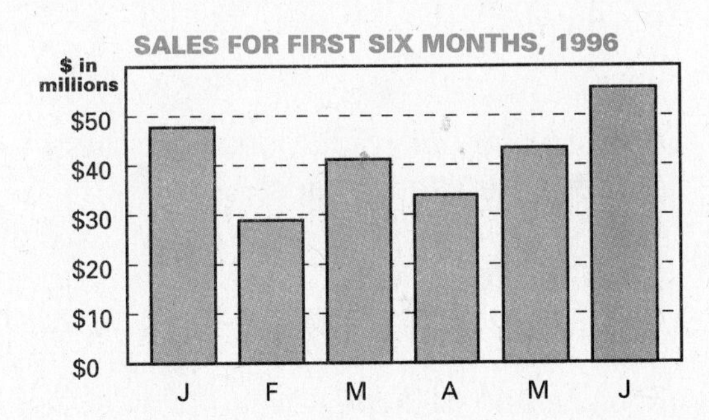

1. In which two months were sales the lowest?
   (1) January and April
   (2) February and March
   (3) February and April
   (4) February and May
   (5) March and May

2. Estimate the total sales for January.
   (1) between $30 and $35 million
   (2) between $35 and $40 million
   (3) between $40 and $45 million
   (4) between $45 and $50 million
   (5) between $50 and $60 million

3. The average monthly sales for the time period shown on the graph is $40 million. Which month's sales were closest to the average?
   (1) February
   (2) March
   (3) April
   (4) May
   (5) June

Items 4 to 7 refer to the circle graph below.

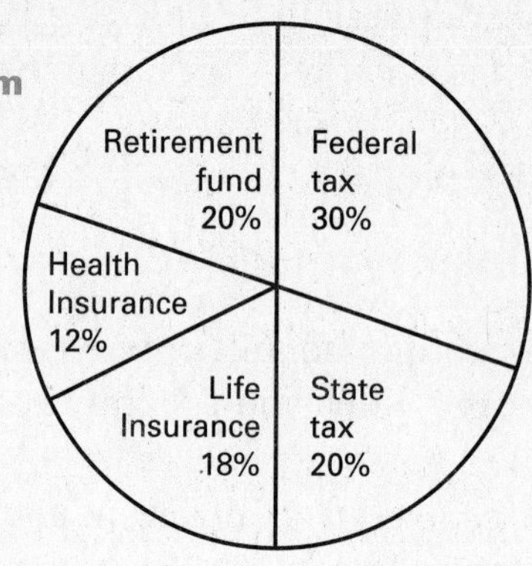

4. Which deduction takes the highest percent of the total paycheck deductions?

   (1) Federal tax
   (2) State tax
   (3) Life insurance
   (4) Health insurance
   (5) Retirement fund

5. What is the total percent deducted for health and life insurance?

   (1) 18%
   (2) 20%
   (3) 28%
   (4) 30%
   (5) 50%

6. If an employee's deductions for a pay period totaled $500, what amount would go to state and federal taxes?

   (1) $ 50
   (2) $100
   (3) $175
   (4) $225
   (5) $250

7. What percent of an employee's earnings is deducted for life insurance?

   (1) 30%
   (2) 20%
   (3) 18%
   (4) 12%
   (5) Not enough information is given.

Items 8 to 12 refer to the line graph below.

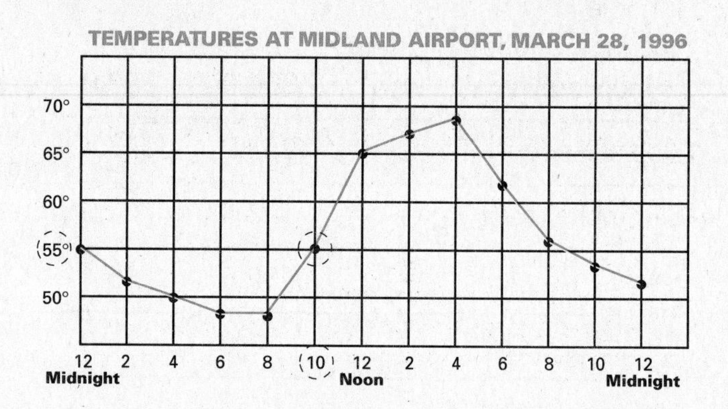

8. At what time was the highest temperature recorded for the day?

   (1) 12 noon
   (2) 1 P.M.
   (3) 2 P.M.
   (4) 3 P.M.
   (5) 4 P.M.

9. In which two-hour time period did the temperature rise the most?

   (1) 8 A.M. to 10 A.M.
   (2) 10 A.M. to 12 noon
   (3) 12 noon to 2 P.M.
   (4) 2 P.M. to 4 P.M.
   (5) 4 P.M. to 6 P.M.

10. About how many degrees did the temperature fall between 4 P.M. and 6 P.M.?

(1) 2
(2) 4
(3) 6
(4) 10
(5) 12

11. At approximately what time did the temperature first reach 60 degrees?

(1) 10 A.M.
(2) 11 A.M.
(3) 12 A.M.
(4) 1 P.M.
(5) 2 P.M.

12. About how many degrees difference was there between the lowest and the highest temperatures for the day?

(1) between 10 and 15 degrees
(2) between 15 and 20 degrees
(3) between 20 and 25 degrees
(4) between 25 and 30 degrees
(5) between 30 and 35 degrees

Items 13 to 16 refer to the table below.

Money Received by the Federal Government from Taxes

Tax Sources	1994	1995
Income Taxes		
Corporations	104.8*	117.2
Individuals	364.0	392.8
Social Security Taxes	273.2	307.4
Unemployment Taxes	23.8	22.2
Excise Taxes	32.6	33.4
Estate and Gift Taxes	6.0	5.8

*Dollars in billions

13. What amount was received in excise taxes in 1995?

(1) $32.6 billion
(2) $33.4 billion
(3) $36 billion
(4) $38.6 billion
(5) $66 billion

14. What was the total amount received from income taxes in 1994?

(1) $222 billion
(2) $468.8 billion
(3) $510 billion
(4) $756.8 billion
(5) Not enough information is given.

15. To the nearest whole percent, what was the percent of increase in social security taxes from 1994 to 1995?

(1) 10%
(2) 12%
(3) 13%
(4) 15%
(5) 18%

16. How much more was received from unemployment taxes in 1994 than in 1995?

(1) $ 1,600,000
(2) $ 16,000,000
(3) $ 160,000,000
(4) $ 1,600,000,000
(5) $16,000,000,000

Answers are on page 377.

Strategies for Solving Word Problems

Organize Information

Information can be organized into lists or tables to help solve problems. Organizing information this way helps you find the data you need quickly.

Example: Gordon and Sue are looking for a new apartment. At the Alpenrose apartment complex, three of the available apartments face east and four face south. Five of the apartments at the Briarwood complex face north and two face west. Six apartments face east at the Cornish Arms Apartments, two face north and four face south. Gordon and Sue prefer an apartment that faces east or south. How many choices do they have?

(1) 12
(2) 15
(3) 17
(4) 20
(5) Not enough information is given.

One way to solve this problem is to make a table.

	N	S	E	W
Alpenrose		4	3	
Briarwood	5			2
Cornish Arms	2	4	6	

Using the table, it is easy to see that 8 apartments face south and 9 face east. Total facing south or east: $8 + 9 = 17$. The correct answer is option **(3) 17.**

Example: Joleen works in a clothing store. She is preparing a window display featuring summer sportswear. Tank tops come in three colors: red, black, and green. Shorts also come in three colors: tan, white, and yellow. How many possible combinations are there of tops and shorts?

To solve the problem, make a table of all the possible combinations. Start with one color top and pair it with each color of shorts before considering the next top. The possible combinations are:

Tops \ Shorts	Tan	White	Yellow
Red	R-T	R-W	R-Y
Black	B-T	B-W	B-Y
Green	G-T	G-W	G-Y

She can make **nine** possible combinations.

Solving Word Problems

Directions: Make a table. Choose the best answer to each item.

Items 1 to 4 refer to the following information.

Sheila and Paul Jackson are looking for an apartment.

- One apartment on Mulberry Lane rents for $565 a month. There is also a $30 monthly parking fee, a $580 deposit, and $60 lease fee.
- The second apartment on Parke Boulevard rents for $485 a month, but they have to pay a $970 deposit and $50 for a key to the laundry room. There is no lease fee.
- A third apartment on Meridan Street rents for $525 per month. They would have to pay a $1,000 deposit and a $50 lease fee. There is also a $25 parking fee per month.
- A fourth apartment at 680 W. Tenth Avenue is only $495 per month, but they have to pay two months' rent as a deposit. They also have to pay a lease fee of 10% of one month's rent.
- The Jacksons would have to buy rugs for either the Mulberry Lane or Parke Boulevard apartment. They would need to buy drapes for either the Mulberry Lane or Meridan Street apartment. The Jacksons estimate that rugs would cost $450 and drapes would cost $250.

1. How much rent would the Jacksons pay per year if they rent the Parke Boulevard apartment?

   (1) $4,950
   (2) $5,280
   (3) $5,550
   (4) $5,820
   (5) $6,300

2. Which apartment requires the highest deposit?

   (1) Mulberry Lane
   (2) Parke Boulevard
   (3) Meridan Street
   (4) Tenth Avenue
   (5) Not enough information is given.

3. Including the first month's rent, any deposits and fees, and costs for rugs and/or drapes, what is the total move-in cost for the Mulberry Lane apartment?

   (1) $1,695
   (2) $1,935
   (3) $2,015
   (4) $2,265
   (5) $2,400

4. Not counting the cost of rugs or drapes, how much more is the move-in cost for the Tenth Avenue apartment than the Parke Boulevard apartment?

   (1) $6.00
   (2) $29.50
   (3) $30.00
   (4) $34.50
   (5) $62.10

Answers are on page 378.

ged Mini-Test Lesson 18

Directions: Choose the best answer to each item.

Items 1 to 6 refer to the following graph.

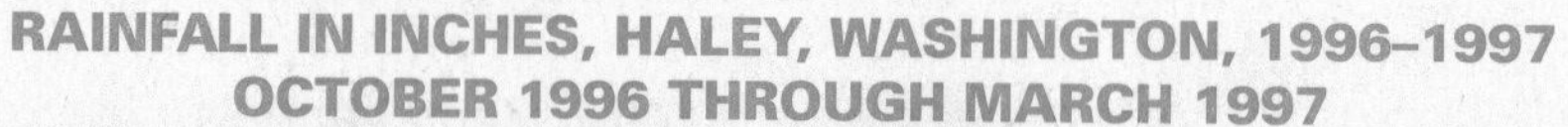

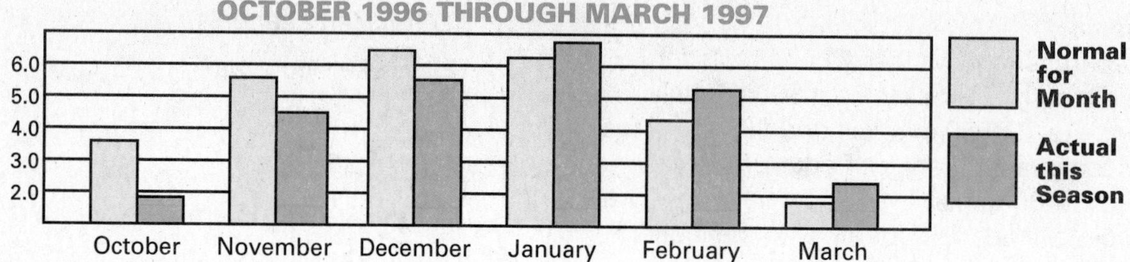

1. For which months was the rainfall less than normal?

   (1) October, November, and December
   (2) October, November, and January
   (3) October, December, and March
   (4) October, February, and March
   (5) January, February, and March

2. For how many months was the actual rainfall over 5.0 inches?

   (1) 2
   (2) 3
   (3) 4
   (4) 5
   (5) 6

3. For how many months was the actual rainfall less than 4.0 inches?

   (1) 2
   (2) 3
   (3) 4
   (4) 5
   (5) 6

4. About how many more inches of rain actually fell in January than in March?

   (1) 2.6
   (2) 3.5
   (3) 4.0
   (4) 4.4
   (5) 5.8

5. What is the approximate average normal rainfall for November, December, and January?

   (1) 4.5
   (2) 5.0
   (3) 5.5
   (4) 6.0
   (5) 6.5

6. About how many fewer inches of rain actually fell in October than in November?

   (1) 1.5
   (2) 1.7
   (3) 2.7
   (4) 3.7
   (5) 4.1

Items 7 to 10 refer to the following graph.

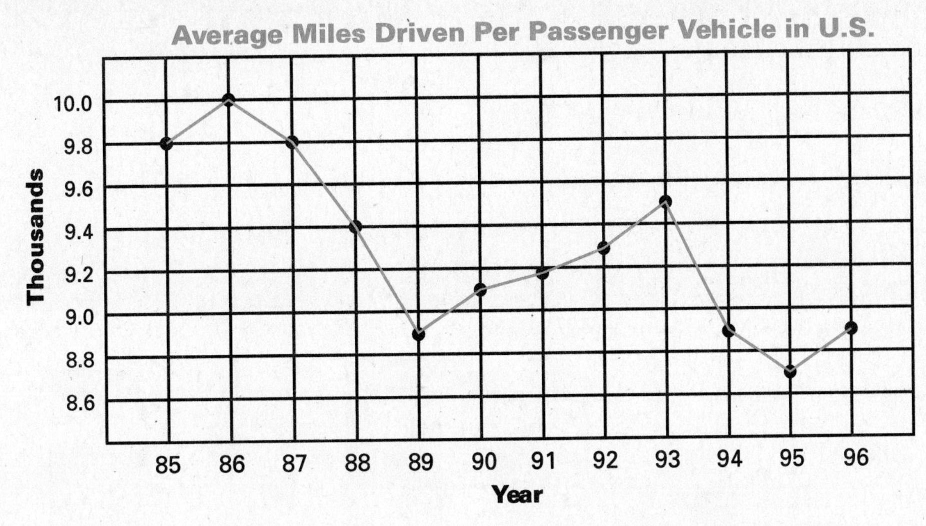

7. What is the range of miles driven, from the lowest to the highest?

   (1) from 8,400 to 10,200
   (2) from 8,700 to 9,800
   (3) from 8,700 to 10,000
   (4) from 8,800 to 10,000
   (5) from 9,800 to 10,000

8. A turn-around is a point on the graph where the direction changes. How many turn-arounds are shown on this graph?

   (1) two
   (2) three
   (3) four
   (4) five
   (5) six

9. Which year did *not* show a decline in miles driven per vehicle?

   (1) 1986 to 1987
   (2) 1987 to 1988
   (3) 1993 to 1994
   (4) 1994 to 1995
   (5) 1995 to 1996

10. What was the percent increase (to the nearest whole percent) in miles driven from 1989 to 1993?

    (1) 4%
    (2) 5%
    (3) 6%
    (4) 7%
    (5) 8%

Item 11 refers to the following graph.

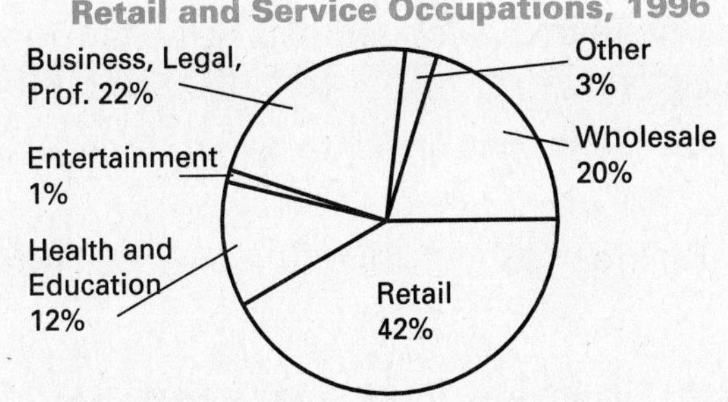

11. Which group has about $\frac{1}{2}$ of the total amount of workers as Retail occupations?

    (1) Entertainment occupations
    (2) Health and Education occupations
    (3) Other
    (4) Business, Legal, and Professional occupations
    (5) Not enough information.

Answers are on page 378.

ged Cumulative Review Measurement

Directions: Choose the best answer to each item.

Items 1 to 4 refer to the following information.

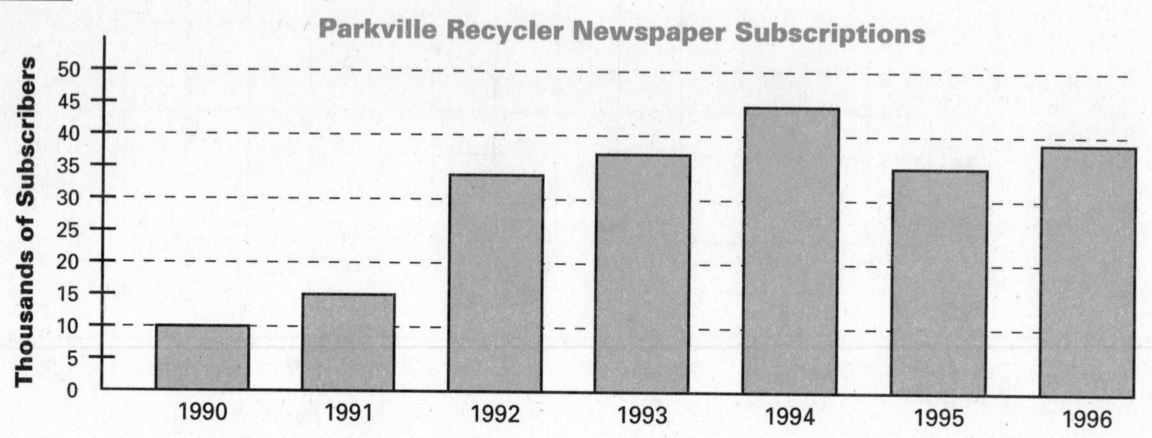

1. The number of subscriptions fell from 1994 to 1995. What was the percent of decrease to the nearest whole percent?

   (1) 10%
   (2) 13%
   (3) 16%
   (4) 22%
   (5) 34%

2. The mean number of subscriptions for the seven years was 30,500. Which year's subscriptions were closest to the mean?

   (1) 1996
   (2) 1995
   (3) 1994
   (4) 1993
   (5) 1992

3. About how many more subscriptions were there in 1994 than in 1993?

   (1) 2,500
   (2) 3,000
   (3) 5,000
   (4) 8,000
   (5) 13,000

4. If 45% of the subscriptions for 1995 were new subscriptions, how many of the subscriptions for that year were renewals by the previous year's subscribers?

   (1) 3,850
   (2) 14,530
   (3) 19,250
   (4) 24,680
   (5) 25,140

5. Five teams are canvassing the town before an election. With two weeks to go, they compare their results. Which statement is true?

Team Results

	A	B	C	D	E
Fraction of goal reached	$\frac{2}{3}$	$\frac{1}{3}$	$\frac{3}{10}$	$\frac{1}{2}$	$\frac{3}{8}$

   (1) Team E is closer to its goal than Team A.
   (2) Team E is closer to its goal than Team D.
   (3) Team B and Team C are exactly even.
   (4) Team C is farthest from its goal.
   (5) Team D is farthest from its goal.

Items 6 to 8 refer to the following information.

Hiking Trail Distances

	Crow's Point	Morning Peak	Rock Face
Eagle's Nest	3.8 mi.	5.6 mi.	2.1 mi.
Rock Face	7.2 mi.	3.3 mi.	
Morning Peak	3.4 mi.		3.3 mi.

6. The Weekend Hiking Club hiked from Eagle's Nest to Morning Peak and on to Rock Face. After lunch they hiked back the same trail. How many miles did they cover?

   (1) 7.8
   (2) 8.9
   (3) 15.6
   (4) 16.8
   (5) 17.8

7. In comparing the lengths of the shortest and longest trails, it is correct to say the longest is

   (1) about $3\frac{1}{2}$ times as long as the shortest.
   (2) $\frac{1}{3}$ the length of the shortest.
   (3) about twice the length of the shortest.
   (4) more than 4 times the length of the shortest.
   (5) less than twice as long as the shortest.

8. Charles hiked from Rock Face to Crow's Point at the rate of 1.8 miles an hour. Going on to Eagle's Nest he increased his rate to 2 miles per hour. How long did this hike take?

   (1) about 3 hr.
   (2) about 4 hr.
   (3) about 5 hr.
   (4) about 6 hr.
   (5) about 7 hr.

9. Janice took a speed reading course. In a 4-minute test at the beginning of the course she read 820 words. After two weeks she could read 4,100 words in 10 minutes. Which statement is true?

   (1) She reads almost 5 times as fast now.
   (2) Her reading rate has doubled.
   (3) Her new rate is 1,025 words per minute.
   (4) At the end of 4 weeks she will be able to read 8,200 words in 10 minutes.
   (5) Her beginning rate was 82 words per minute.

Items 10 and 11 refer to the following information.

310 yards

CITY GOLF COURSE DRIVING RANGE

120 yards

10. The city parks committee plans to fence in the driving range shown in the drawing. What is the perimeter in yards of the driving range?

    (1) 240
    (2) 430
    (3) 620
    (4) 860
    (5) 980

11. The city needs to buy grass seed to plant the driving range area. If one bag of seed plants 75 square yards of land, how many bags will the city need to buy to plant the driving range area?

    (1) 160
    (2) 357
    (3) 496
    (4) 653
    (5) 744

Items 12 and 13 refer to the following figure.

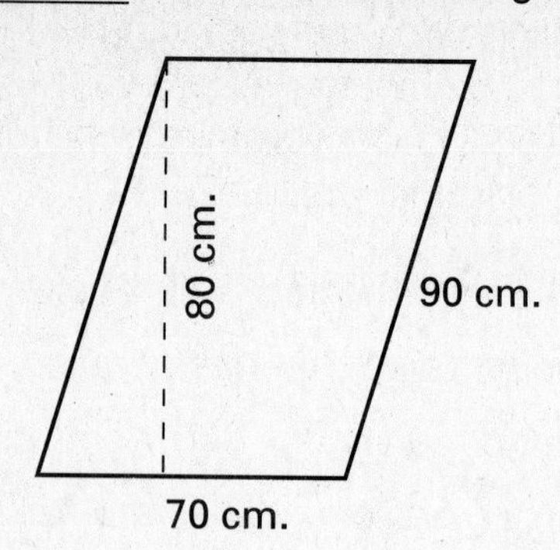

12. What is the area in square centimeters of the parallelogram shown in the figure?

(1) 5,500
(2) 5,600
(3) 5,800
(4) 6,300
(5) 6,400

13. What is the perimeter in centimeters of the parallelogram?

(1) 160
(2) 220
(3) 240
(4) 290
(5) 320

14. Jerry has three projects she must finish by noon. Project 1 will take 20 minutes. Project 2 will take three times as long as Project 1 and Project 3 will take 30 minutes. What time should she start work to finish all three projects?

(1) 10:10
(2) 10:50
(3) 10:53
(4) 11:07
(5) 11:10

15. A recent health study in a certain county showed that stroke victims had a survival rate of 60 per 100 cases. If there were 3,200 cases reported to the County Health Department that year, how many stroke victims survived their illness?

(1) 32
(2) 53
(3) 320
(4) 533
(5) 1,920

Items 16 and 17 refer to the following information.

Gross Earnings for Manuel Botta 1992 through 1996

1992	1993	1994	1995	1996
$16,907	$17,756	$19,174	$19,558	$20,730

16. Find the mean gross earnings for the five-year period.

(1) $17,824
(2) $18,205
(3) $18,750
(4) $18,825
(5) $19,174

17. Manuel Botta's earnings for 1991 were $17,204. Find the median gross earnings for the six-year period from 1991 to 1996.

(1) $16,907
(2) $17,756
(3) $18,465
(4) $18,825
(5) $19,174

Items 18 to 21 refer to the following information.

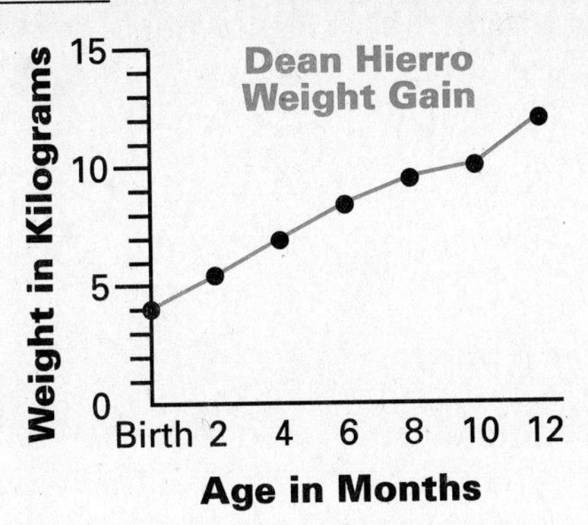

18. The Hierro infant showed the greatest weight gain over which two-month period?

    (1) birth to 2 months
    (2) 4 to 6 months
    (3) 6 to 8 months
    (4) 8 to 10 months
    (5) 10 to 12 months

19. From birth to 2 months, the Hierro infant gained 1.5 kg. During which other 2-month period did the infant show the same weight gain?

    (1) 2 to 4 months
    (2) 6 to 8 months
    (3) 8 to 10 months
    (4) 10 to 12 months
    (5) Not enough information is given.

20. The average weight for a male infant at one year of age is 10.5 kilograms. About how much more than the average does the Hierro baby weigh at one year?

    (1) 0.5 kg.
    (2) 0.8 kg.
    (3) 1.1 kg.
    (4) 1.3 kg.
    (5) 1.5 kg.

21. During which 2-month period did the infant experience the smallest weight gain?

    (1) birth to 2 months
    (2) 4 to 6 months
    (3) 8 to 10 months
    (4) 10 to 12 months
    (5) Not enough information is given.

22. The manager of the Ear-Ring Boutique recorded the number of sales during the week: Monday—21; Tuesday—42; Wednesday—34; Thursday—45; Friday—51; Saturday—47. What was the average number of sales per day for the week?

    (1) 39
    (2) 40
    (3) 41
    (4) 43
    (5) 45

23. Attendance for the first three days of a 15-day fair was 19,600. If attendance continues at this rate, what will be the total attendance for the fair?

    (1) 78,400
    (2) 98,000
    (3) 100,000 and 200,000
    (4) more than 200,000
    (5) Not enough information is given.

24. The length of one side of a square is 1 ft. 4 in. What is the perimeter of the square?

    (1) 4 ft. 6 in.
    (2) 5 ft. 4 in.
    (3) 5 ft. 8 in.
    (4) 6 ft. 0 in.
    (5) 6 ft. 2 in.

Items 25 and 26 refer to the following figure.

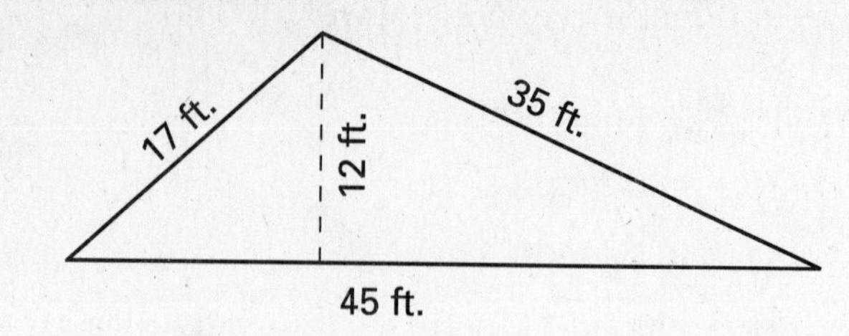

25. Find the area in square feet for the triangle shown in the figure.

    (1) 204
    (2) 270
    (3) 345
    (4) 540
    (5) 765

26. Which expression can be used to find the perimeter of the triangle shown in the figure?

    (1) 17×35
    (2) 17×45
    (3) $17 + 35 + 45$
    (4) $\frac{1}{2}(45)(17)$
    (5) $\frac{1}{2}(17 + 45)$

27. There are 50 marbles in a jar. They are all alike except for their color. There are 12 red marbles, 28 blue ones and 10 green ones. If they are mixed thoroughly and one is drawn at random, what is the probability that it is red?

    (1) 0.02
    (2) 0.08
    (3) 0.2
    (4) 0.24
    (5) 0.32

28. A circular patio has a diameter of 8 feet. To the nearest whole foot, what is the circumference of the patio?

    (1) 13
    (2) 16
    (3) 21
    (4) 25
    (5) 29

29. A wall in Joan's living room is 12 feet long. She buys four 35-inch bookcases to fit along this wall. Which statement is true?

    (1) They will be 8 inches too wide.
    (2) They will be 4 inches too wide.
    (3) They will fit exactly.
    (4) They will fit with 4 inches left over.
    (5) They will fit with 8 inches left over.

30. Last year the Hawks soccer team won 2 out of every 3 games it played. By the end of the season it had lost only 6 games. How many games were won?

    (1) 4
    (2) 9
    (3) 12
    (4) 16
    (5) 18

31. A photograph is $3\frac{3}{4}$ inches by $5\frac{1}{4}$ inches. Nelle orders a "double size" enlargement. What size will the finished photo be?

    (1) $7\frac{1}{4}$ in. by $10\frac{1}{2}$ in.
    (2) $6\frac{3}{4}$ in. by $10\frac{1}{4}$ in.
    (3) $7\frac{1}{2}$ in. by $10\frac{1}{2}$ in.
    (4) $6\frac{3}{4}$ in. by $10\frac{1}{2}$ in.
    (5) $7\frac{1}{2}$ in. by $10\frac{1}{4}$ in.

32. Cristina has two pieces of pipe. One is 2 yards in length. The other is 4 feet 10 inches. How much longer than the second is the first piece?

(1) 9 in.
(2) 1 ft.
(3) 1 ft. 2 in.
(4) 1 ft. 6 in.
(5) 2 ft. 10 in.

33. Luke buys 4 raffle tickets at the County Fair. If a total of 200 raffle tickets are sold, what is the probability that one of Luke's tickets will win?

(1) $\frac{1}{20}$
(2) $\frac{1}{25}$
(3) $\frac{1}{50}$
(4) $\frac{1}{100}$
(5) $\frac{1}{200}$

Items 34 to 36 refer to the following table.

Melendez Family Utilities	
Period	**Amount**
Jan–Feb	$89.36
Mar–Apr	$90.12
May–Jun	$74.47
Jul–Aug	$63.15
Sep–Oct	$59.76
Nov–Dec	$84.31

34. What was the total amount spent on utilities for the first six months of the year?

(1) $179.48
(2) $193.53
(3) $224.08
(4) $253.95
(5) $461.17

35. What is the median of the amounts on the table?

(1) $84.31
(2) $79.39
(3) $74.47
(4) $63.15
(5) Not enough information is given.

36. What is the average amount spent on utilities for a 2-month period?

(1) $ 76.86
(2) $ 79.28
(3) $ 79.39
(4) $230.58
(5) Not enough information is given.

37. There are 36 10-year employees, 152 5-year employees and 260 2-year employees honored at a company meeting. What is the ratio of 10-year employees to total employees working for the company?

(1) $\frac{410}{36}$
(2) $\frac{36}{260}$
(3) $\frac{36}{448}$
(4) $\frac{36}{152}$
(5) Not enough information is given.

38. Patricia's Nut Shoppe charges $5.75 for a 1.25-pound gift box of mixed nuts. At this rate, how much should a 5-pound box cost?

(1) less than $5
(2) between $5 and $10
(3) between $10 and $20
(4) more than $20
(5) Not enough information is given.

39. What is the area in square inches of the shaded portion of the figure?

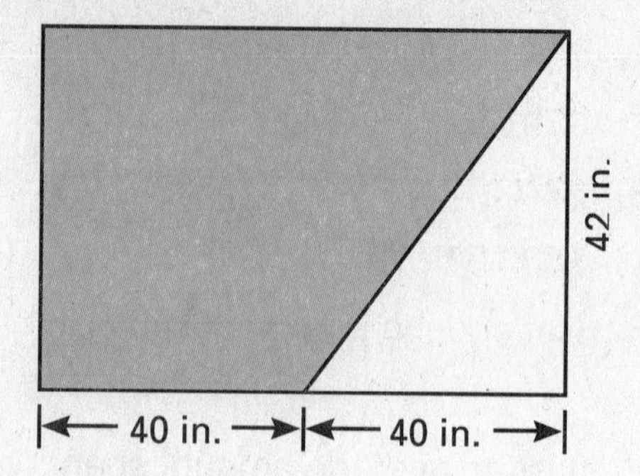

(1) 840
(2) 1,680
(3) 2,520
(4) 3,360
(5) Not enough information is given.

Items 40 and 41 refer to the following information.

Fifteen students took a writing class. Their final exam grades were as follows: 100, 100, 98, 95, 92, 87, 84, 84, 84, 80, 76, 72, 68, 67, 60.

40. To the nearest whole number, what was the mean score for the final exam?

(1) 98
(2) 87
(3) 84
(4) 83
(5) 76

41. What was the median score for the final exam?

(1) 98
(2) 87
(3) 84
(4) 83
(5) 76

42. At Garden City Electronics School, each class is $\frac{3}{4}$ of an hour. How many classes can they schedule from 8 A.M. to 5:45 P.M. each day?

(1) 9
(2) 10
(3) 11
(4) 12
(5) 13

43. A tailor has a piece of fabric that is $1\frac{1}{2}$ yards in length. He trims off 2 feet 3 inches. How long is the remaining piece?

(1) 2 feet 3 inches
(2) 2 feet 6 inches
(3) 2 feet 8 inches
(4) 2 feet 9 inches
(5) 2 feet 10 inches

44. The height of a triangle is 15 inches. If the base is 20 inches in length, what is the area of the triangle (in square inches)?

(1) 150
(2) 170
(3) 175
(4) 180
(5) 225

45. Brad bowled games of 220, 186, and 212. What was Brad's average for the three-game series?

(1) 192
(2) 201
(3) 206
(4) 210
(5) 212

Answers are on page 379.

Performance Analysis
Unit 1 Cumulative Review: Measurement

Name: ____________________ **Class:** __________ **Date:** __________

Use the Answer Key on pages 379–381 to check your answers to the GED Unit 1 Cumulative Review: Measurement. Then use the chart to figure out the skill areas in which you need additional review. Circle on the chart the numbers of the test items you answered correctly. Then go back and review the lessons for the skill areas that are difficult for you. For additional review, see the *Steck-Vaughn GED Mathematics Exercise Book,* Unit 1: Measurement.

Use the chart below to find your strengths and weaknesses in application skills.

Concept	Item Numbers
Ratio and Proportion pages 146–156	9, 15, 23, 30, 37, 38
Mean and Median pages 146–156	**16, 17,** 22, **35, 36,** 40, 41, 45
Probability pages 146–156	27, 33
Measurement pages 157–165	14, 29, 31, 32, 42, 43
Perimeter, Circumference, and Area pages 166–176	**10, 11, 12, 13,** 24, **25, 26,** 28, **39,** 44
Graphs, Charts, and Tables pages 177–185	**1, 2, 3, 4, 5, 6, 7, 8, 18, 19, 20, 21, 34**

Boldfaced numbers indicate items based on charts, graphs, illustrations, and diagrams.

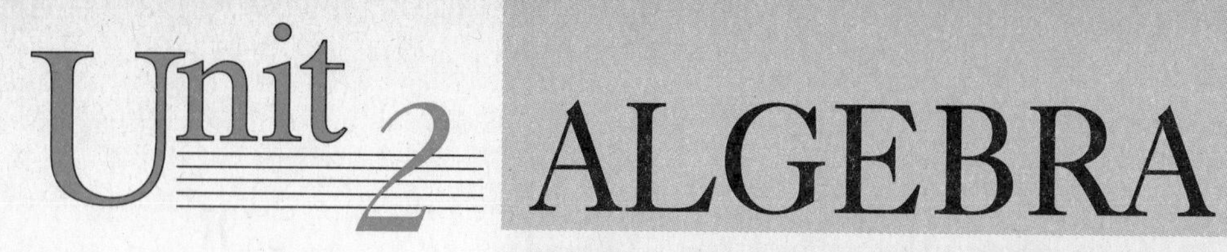

Unit 2 ALGEBRA

Many people use algebra as part of their jobs.

variable
any letter used to stand for a number

algebraic expression
a mathematical expression that uses letters and numbers instead of words

equation
a statement that says two expressions are equal

In this section, you will learn the basic concepts of algebra. The most famous characteristic of algebra is the use of letters called **variables** to stand for numbers. The letter x is often used to stand for an unknown number.

Variables and numbers are used together to make **algebraic expressions.** An algebraic expression uses letters and numbers instead of words to express mathematical information. In lesson 19 you will learn about algebraic expressions. You will learn how to add, subtract, multiply, and divide with variables and numbers. You will also learn how to write an algebraic expression.

Algebraic expressions can be used to solve problems. When an algebraic expression equals some number, you can write an equation. An **equation** is a statement that says two expressions are equal.

Think about this example:
A certain number added to 5 equals 9. What is the number?

solving an equation
finding a number that makes the equation true when the number is put in the place of the variable

If you answered 4 because $5 + 4 = 9$, you have just solved an algebra problem. We can write an equation to solve the problem by using x to represent the unknown number.

A certain number	x
added to 5	$5 + x$
equals 9.	$5 + x = 9$

$5 + x = 9$ is an equation. You will learn how to solve equations in lesson 20. You will also learn how to translate word problems into algebraic expressions and equations.

formula
an equation in which the letter stands for specific kinds of quantities

Actually, any letter of the alphabet can be used as a variable. When the letters in an equation stand for specific kinds of quantities, the equation is called a **formula.** You have already used several formulas.

For example, the formula for finding simple interest $i = prt$ uses letters to stand for specific quantities. The variable i stands for interest, p stands for principal, r stands for rate, and t stands for time. Formulas can be used to solve many algebra problems.

In lesson 21 you will learn about some special topics in algebra. An **inequality** is a statement that tells how two expressions compare. Inequalities use the signs $<$ less than, $>$ greater than, $\leq$ less than or equal to, and $\geq$ greater than or equal to.

Lesson 21 also covers factoring algebraic expressions and graphing equations.

In lesson 22 you will learn about powers and roots. **Powers** are used to show that a number is multiplied by itself a number of times. A small number called an **exponent**, written slightly above and to the right of a number, tells how many times the number should be multiplied.

5^2 means 5×5.

The number that is multiplied by itself is called the **root.**

In lesson 22 you will learn how to simplify expressions with powers and roots. You will also learn to write very large or very small numbers using scientific notation.

SEE ALSO: Steck-Vaughn GED Mathematics Exercise Book, Unit 2: Algebra

Signed Numbers • Algebraic Expressions

Integers

Signed numbers can be used to show quantity, distance, and direction. **Positive numbers** show an increase, a gain or upward motion. **Negative numbers** show the opposite: a decrease, a loss or downward direction. All whole numbers, both the positive numbers and the negative numbers, and zero are called **integers.**

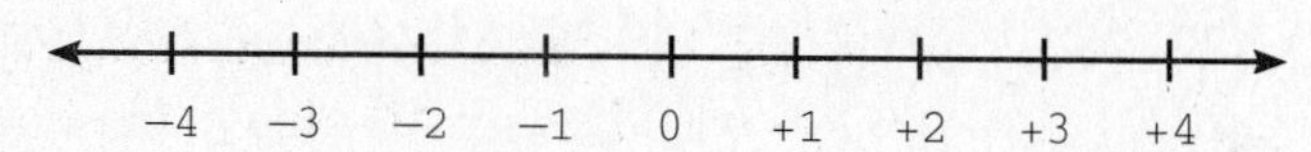

The distance an integer is from zero is called its **absolute value,** which is shown by enclosing the number between two vertical bars. The absolute value is the number value without the sign.

Examples: Absolute value of +3 is 3. We write $|+3| = 3$.
Absolute value of −7 is 7. We write $|-7| = 7$.
Absolute value of 0 is 0. We write $|0| = 0$.

We can show the addition of integers on a **number line.** Move to the right for positive numbers and to the left for negative numbers.

Example: Add: $(+6) + (-4)$

1. Start at 0 and move 6 units to the right.
2. From +6, move 4 units to the left to add −4.

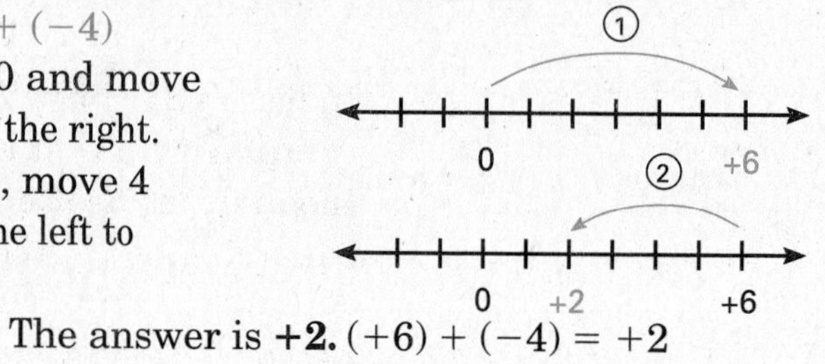

The answer is **+2.** $(+6) + (-4) = +2$

Addition

Use these rules to add signed numbers:

Rule 1: To add two integers with like signs:
Add the absolute value $(+3) + (+5) = |+3| + |+5| = +8$
and keep the same sign. $(-2) + (-6) = |-2| + |-6| = -8$

Rule 2: To add two integers with unlike signs:
Subtract the absolute $(-9) + (+3) = |-9| - |+3| = -6$
value of the lesser $(-4) + (+7) = |+7| - |-4| = +3$
integer from the absolute value of the greater. Keep the sign of the integer with the greater absolute value.

Subtraction

To subtract integers, use this rule:

Subtraction rule:

1. Change the subtraction operation to addition.
2. Change the sign of the number to be subtracted.
3. Complete the problem as addition.

Example: Subtract $(+5) - (-2)$
Change the operation to addition, and change -2 to $+2$.
$(+5) - (-2) = (+5) + (+2) = +7$

Study these examples to see how the subtraction rule is applied.
$(-5) - (-2)$ becomes $(-5) + (+2)$, which equals -3.
$(-5) - (+2)$ becomes $(-5) + (-2)$, which equals -7.
$(+5) - (+2)$ becomes $(+5) + (-2)$, which equals $+3$.
Notice that the sign of the first number in the expression does not change.

Multiplication and Division

The rule for multiplying or dividing integers is:

Multiplication/Division rule:
If the signs are the same, the answer is positive.
If the signs are different, the answer is negative.

Examples: $3 \times 5 = 15$ $\quad -3 \times 5 = -15$ $\quad 3 \times -5 = -15$ $\quad -3 \times -5 = +15$
$24 \div 6 = 4$ $\quad -24 \div 6 = -4$ $\quad 24 \div -6 = -4$ $\quad -24 \div -6 = +4$

Note: Positive numbers can be written without the plus sign. Negative numbers are always written with the minus sign.

Order of Operations

When an algebraic expression contains several operations, you must follow the order of operations rule.

Order of operations rule:
1. If there are parentheses, do all work in them first in order: multiply, divide, add, or subtract.
2. Then, multiply and divide in order from left to right through the expression.
3. Last, add and subtract through the expression from left to right.

Example: $3 + (4 + 6 \times 3)$

1. Do the work in the parentheses first. Multiply, then add.
   $3 + (4 + 6 \times 3)$
   $3 + (4 + 18)$
   $3 + (22)$
2. Then do the operations in order (multiply, divide, add, subtract). So add $3 + 22$.
   25

If there is a fraction bar, work on the numerator and denominator separately, applying the order of operations rule to each. Then do the final division.

Example: $\dfrac{3 + 5 \div 5}{(6 - 4) \div 2} = \dfrac{3 + 1}{2 \div 2} = \dfrac{4}{1} = 4$

GED Practice Lesson 19

Directions: Write the absolute value for each.

1. +6
2. −7
3. −14.6
4. $+\frac{2}{3}$
5. 0

Directions: Write a number for each expression.

6. an increase of 45 points
7. a change of $\frac{1}{4}$ inch downward
8. below sea level by 10 feet
9. a loss of $200
10. an increase of 10 seconds
11. a gain of 30 points

Directions: Add or subtract.

12. $(+7) + (+5)$
13. $(-10) + (-6)$
14. $(-6) + (+5)$
15. $(+10) - (+7)$
16. $(-3) - (+7)$
17. $(-1) - (+5)$
18. $(+6) + (-8)$
19. $\left(+3\frac{1}{2}\right) + (-2)$
20. $(+10) - (+17)$
21. $(+2) - (-7)$
22. $(+8) - (-4)$
23. $(-4) + (+9)$
24. $(-4.5) + (-1.5)$
25. $(+4) - (-1)$
26. $(-6.1) - (+2.6)$

Directions: Multiply.

27. $(-2)(+3)$
28. $(-2.5)(+2)$
29. $(-4)(-7)$
30. $(+7)(0)$
31. $(+5)(-3)$
32. $(-4)(-1)(+2)$
33. $(0)(-2)$
34. $(+6)(-2)\left(-\frac{1}{3}\right)$

Directions: Divide.

35. $(-64) \div (+4)$
36. $(+15) \div (-3)$
37. $(+20) \div (+5)$
38. $(-42) \div (-3)$
39. $(+36) \div (-12)$
40. $(-55) \div (+11)$
41. $(-49) \div (-7)$
42. $(+126) \div (-9)$

Directions: Solve each expression.

43. $6 + 8 \times 4$

44. $\frac{-2 - (+8)}{(6) \div (-6)}$

45. $(4 \times -9) - (2 \times -3)$

46. $10 + (-3 + 4 \times (-2))$

47. $(-25) - 4 \times 3$

48. $\frac{(-4) + (-6)}{(+4) - (-1)}$

49. $6 - (4 \times 8 + (-1))$

50. $(10 \times -2) + (6 \times -3)$

Answers are on page 381.

Algebraic Expressions

An **algebraic expression** is a group of numbers, operation signs, and variables.

Examples: $2x + 3$ $\frac{4x + 1}{3}$ $3x - 4$ $3(6x)$

Expressions like these are formed by translating number relationships into symbols. Analyze the following expressions carefully.

4 times a number $4x$
6 more than a number $x + 6$
2 less than a number $x - 2$
one-half a number increased by 7
$\frac{x}{2} + 7$ or $\frac{1}{2}x + 7$

the product of 6 and x $6x$
the quotient of x and 5 $\frac{x}{5}$ or $x \div 5$
the product of x and 8 added to the sum of 2 and x
$8x + (2 + x)$

An algebraic expression always contains **variables.** Variables are letters that are used to represent numbers. Whenever the value of x changes, the expression also changes its value.

x	$4 + x$	becomes		value
3	$4 + x$	$4 + 3$	=	7
−8	$4 + x$	$4 + (-8)$	=	−4
8	$4 + x$	$4 + 8$	=	12
−3	$4 + x$	$4 + (-3)$	=	1

This table shows how the value of the expression $4 + x$ changes as the value of x changes. An algebraic expression has a value only when all the variables are replaced by numbers.

Example: Find the value of $3 - x + y$, when $x = -1$ and $y = +5$.

Substitute −1 for x and +5 for y. $3 - (-1) + (+5)$
Apply the subtraction rule. $3 + (+1) + (+5)$
Complete by adding. $+9$

In algebra, multiplication is indicated in two ways. A raised dot between two numbers means multiplication. Also putting the variable and numbers next to each other without any sign means multiplication. Parentheses may be used to clarify the meaning.

Examples: $2x$ means 2 times x. $5 \cdot 4$ means 5 times 4.
$4(-3)$ means 4 times -3. $(+5)(+4)$ means $+5$ times $+4$.

The division sign can be used in an algebraic expression, but it is more common to use a fraction bar.

Example: $\frac{x}{5}$ means x divided by 5.

Example: Find the value of $\frac{2x - y}{3}$ when $x = +5$ and $y = -2$.

Substitute: $\frac{2(+5) - (-2)}{3}$

Multiply: $\frac{+10 - (-2)}{3}$

Use the subtraction rule: $\frac{+10 + (+2)}{3}$

Add: $\frac{12}{3}$

Divide: 4

Simplifying an expression means performing all the operations you can within the expression containing a variable. Without knowing the values of the variables, you can add and subtract **like terms. Terms** are the number and variable combinations separated by addition and subtraction signs. Like terms contain exactly the same variables.

Examples of like terms: $3x, 5x, x, -4x$
Examples of unlike terms: $2x, 3y, z, -2xy, 4$

Example: Simplify the following expression: $4x + y - 2x + 2y$
Combine like terms: $4x - 2x = 2x;\ y + 2y = 3y$
(The number part of the variable by itself is understood to be one. So, $y + 2y$ is the same as $1y + 2y$.) $2x + 3y$

ged Practice **Lesson 19**

Directions: Simplify each expression.

1. $5x + y - 3x + 2y$
2. $8x + 3 + 2x - 5$
3. $a + 6b + a - 8b + 2a$
4. $14 - 7x - 4x$
5. $4z + z + 5y - 3$
6. $3x + 12y - 5x - 10y$

Directions: Find the value of each expression.

7. $x - (+5) + (+3)$, when $x = +2$
8. $(+3) + x + (+5) - x$, when $x = +5$
9. $(-1) + x + (+8)$, when $x = -4$
10. $(+2) - x - x - (-3)$, when $x = -0.5$
11. $5x + y - 4 + 2y$, when $x = 3$ and $y = 2$
12. $3xy + 2x + y$, when $x = -2$ and $y = 4$
13. $\frac{6xyz}{x}$, when $x = -3$, $y = 2$, $z = -1$
14. $\frac{a + 5b}{2a - 1}$, when $a = 4$ and $b = 2$

Directions: Translate the following into mathematical expressions.

15. a number minus 2
16. a number divided by 7
17. twice a number, then increased by -4
18. twice a number subtracted from three times that same number
19. three times a number, then minus 9
20. four times a number, then divided by 3
21. Darcia has \$11. She gets a certain amount more, then spends \$2.
22. Richard has 19 points in a game. He loses 5 points, gains a certain number more, and then loses 2 points.

Answers are on page 382.

Lesson 20 Equations

Equations

An **equation** is a mathematical statement which shows that two quantities are equal. When an equation contains a variable, we use algebra to find the value of the variable. To solve an equation means to find the number that makes the statement true.

Look at this equation:	$2x - 1 = 9$
When x equals 5,	$2(5) - 1 = 9$
the statement is true.	$10 - 1 = 9$

Any other value of x makes the equation false. For example,

$2(6) - 1 =$	$2(3) - 1 =$	$2(-5) - 1 =$
$12 - 1 = 11$	$6 - 1 = 5$	$-10 - 1 = -11$

Only 5 makes the equation $2x - 1 = 9$ true. So 5 is the **solution** of the equation. The variable x must equal 5.

One-step Equations

To solve an equation, you must keep the sides of the equation equal. Whatever you do to one side of the equation you must do to the other side.

Example: Solve: $x - 13 = 25$

When 13 is subtracted from some number, the result is 25. Think what this means: If 13 is subtracted from some number and the result is 25, then by adding 13 to 25 we should be able to find the missing number.

$$x - 13 = 25$$

Add 13 to both sides: $x - 13 + 13 = 25 + 13$

$$x = 38$$

Check: $38 - 13 \stackrel{?}{=} 25$ $\quad 25 = 25$

From this example, we can draw two conclusions:

- The basic strategy in solving an equation is to isolate the variable. **Isolating the variable** means to have the variable alone on one side of the equation.
- In an equation with only one operation, perform the **inverse** (opposite) operation on both sides of the equation.

Addition and subtraction are inverse operations.

Multiplication and division are inverse operations.

In the equation $x - 13 = 25$, the operation is subtraction. By performing its inverse, adding 13 to both sides, we isolated the variable and solved the equation.

Example: Solve: $x + 40 = 75$

What number added to 40 equals 75?

The operation is addition: $x \oplus 40 = 75$

Perform the inverse operation:

subtract 40 from both sides: $x + 40 - 40 = 75 - 40$

Solve the equation: $x = 35$

Check: $35 + 40 \stackrel{?}{=} 75$ $\quad 75 = 75$

Example: Solve: $5x = 35$

What number multiplied by 5 equals 35?

The operation is multiplication: $5x = 35$

To isolate the variable,

divide both sides by 5: $\frac{5x}{5} = \frac{35}{5}$

Solve: $x = 7$

Check: $5(7) \stackrel{?}{=} 35$ $\quad 35 = 35$

Example: Solve: $\frac{x}{9} = 4$

What number divided by 9 equals 4?

The operation is division: $\frac{x}{9} = 4$

To isolate the variable,

multiply both sides by 9: $(9)\frac{x}{9} = 4(9)$

Solve: $x = 36$

Check: $\frac{36}{9} \stackrel{?}{=} 4$ $\quad 4 = 4$

Two-step Equations

Some equations require more than one step in order to find the solution. In the following example, you need to perform the inverse operations for subtraction <u>and</u> multiplication to isolate the variable.

When solving equations:

- *Reverse* the normal order of operations,
- perform the inverse operations for addition and subtraction first,
- perform multiplication and division operations second.

Example: $5x - 10 = 35$

Add 10 to both sides:	$5x - 10 + 10 = 35 + 10$
Add and subtract first.	$5x = 45$
Divide both sides by 5: Multiply and divide second.	$\frac{5x}{5} = \frac{45}{5}$
Solve the equation:	$x = 9$
Check by substituting 9 for x in the original equation.	$5(9) - 10 \stackrel{?}{=} 35$ $45 - 10 \stackrel{?}{=} 35$ $35 = 35$

This is a true statement, so **9** is the solution of the equation.

Some equations have variable terms on both sides. The basic strategy is to isolate the variable on one side of the equation.

Example: $12x - 9 = 10x - 1$

Subtract $10x$ from both sides:

$$12x - 10x - 9 = 10x - 10x - 1$$

Now the variable is only on one side of the equation.	$2x - 9 = -1$
Add 9 to both sides:	$2x - 9 + 9 = -1 + 9$ $2x = 8$
Divide both sides by 2:	$\frac{2x}{2} = \frac{8}{2}$
Solve the equation:	$x = 4$
Check by substitution:	$12(4) - 9 \stackrel{?}{=} 10(4) - 1$ $48 - 9 \stackrel{?}{=} 40 - 1$ $39 = 39$

Some equations contain parentheses. To solve these equations, you need to remove the parentheses by multiplying each number within the parentheses by the multiplier.

Example: $2(x + 1) = -8$

Multiply both numbers inside the parentheses by 2:	$2(x + 1) = -8$ $2x + 2 = -8$
Subtract 2 from both sides:	$2x + 2 - 2 = -8 - 2$ $2x = -10$
Divide both sides by 2:	$\frac{2x}{2} = \frac{-10}{2}$
Solve the equation:	$x = -5$
Check by substitution:	$2(-5 + 1) = -8$ $2(-4) = -8$ $-8 = -8$

Equations help us solve problems. We translate problems into algebraic equations, using x (or any other letter) to represent the unknown number. We can then solve to find the value of the variable.

Notice how the words are translated into symbols in these examples:

Word Problem	Solution
The **sum** of a number and 9 is 14. What is the number? "The sum of" means two numbers are to be added.	$x + 9 = 14$ $x + 9 - 9 = 14 - 9$ $x = 5$
6 less than some number is 10. Find the number. "Less than" means 6 is subtracted from some number.	$x - 6 = 10$ $x - 6 + 6 = 10 + 6$ $x = 16$
9 more than twice a number is 15. What is the number? "More than" means addition. Twice a number gives the **product** 6. "Product" is the answer in multiplication.	$2x + 9 = 15$ $2x + 9 - 9 = 15 - 9$ $2x = 6$ $\frac{2x}{2} = \frac{6}{2}$ $x = 3$
The **quotient** of 12 and x is 3. What is the value of x? "Quotient" is the answer in division.	$\frac{12}{x} = 3$ $\cancel{x}\left(\frac{12}{\cancel{x}}\right) = 3x$ $12 = 3x$ $\frac{12}{3} = \frac{3x}{3}$ $4 = x$

To simplify an equation, we often "combine like terms." This means that we add together variable terms separately and number terms separately. Study the following examples carefully. The variable x always means 1 times x, or $1x$.

Examples: $2x + 3x = 5x$ $\quad x + 3x = 4x$ $\quad x + x = 2x$ $\quad 4x - 2x = 2x$

$5x - x = 4x$ $\quad 6x - 6x = 0$ $\quad x - x = 0$ $\quad 4x - 7x = -3x$

Solve: $8x + 4 - 5x - 4 = 6x + 9 - 6x$

$3x + 0 = 0 + 9$

Simplify each side: $(4 - 4 = 0 \text{ and } 6x - 6x = 0)$

$3x = 9$

Divide both sides by 3: $\frac{3x}{3} = \frac{9}{3}$

Simplify: $1x = 3$

Solution: $x = 3$

ged Practice Lesson 20

Directions: Solve each equation.

1. $x - 15 = 4$
2. $x - 7 = 3$
3. $2x = 12$
4. $-6x = -42$
5. $x + 12 = 22$
6. $x - 8 = -10$
7. $-3x = 18$
8. $5x = -45$
9. $6x + 7 = 37$
10. $4x + 5x - 10 = 35$
11. $3x - 6x + 2 = -4x$
12. $6 - x + 12 = 10x + 7$
13. $5x + 7 - 4x = 6$
14. $9x + 6x - 12x = -7x + 2x - 12 + 5x$
15. $7x + 3 = 31$
16. $3x - 8 = 28$
17. $8x + 6 = 5x + 9$
18. $11x - 10 = 8x + 5$
19. $-2x - 4 = 4x - 10$
20. $5x + 8 = x - 8$
21. $11x - 12 = 9x + 2$
22. $5(x + 1) = 75$
23. $5(x - 7) = 5$
24. $6(2 + x) = 5x + 15$

Directions: Translate each statement into an algebraic equation. Then solve.

25. The sum of an unknown number and 12 is 25.

26. 4 less than a certain number is 11.

27. The product of 12 and a certain number is 60.

28. The quotient of a certain number and 10 is −2.

29. Multiply a number by 7, then subtract 8. The result is 5 times the number.

30. Add a number and −5. Multiply by 8. The result is 2 more than the number.

31. Six less than twice a certain number is 24.

32. Eight times a number, added to 10, is equal to that number multiplied by 3.

33. A certain number increased by 20 equals 5 times the number.

34. 8 more than a certain number is 19.

35. A certain number subtracted from 21 is equal to 10.

36. The product of an unknown number and −8 is −32.

37. The quotient of a certain number and −6 is 4.

38. Add 7 to twice a number. The result is 10 more than the number.

39. Multiply a number by 2 and add 2. The result is 20.

40. The sum of 8 and a certain number is increased by 12. The result is the same as that number multiplied by 3.

41. The product of a certain number and 5 is 12 less than the number.

42. The product of a certain number and 4 is 12 less than that number.

Answers are on page 383.

Strategies for Solving Word Problems

Writing and Solving an Equation

Now you can use the skills you developed to solve word problems.

Example: A busboy in a sandwich shop noticed that there were 8 more women than men in the shop. If there were 32 people in the shop, how many men were there?

Using this example we can develop steps for placing the words into the equation to be solved:

Step 1: **Identify the unknown amount, x.** Let x equal the number of men.

Step 2: **Label the other quantities in terms of x.** Since there were 8 more women than men, let $x + 8$ equal the number of women.

Step 3: **Write the equation.** The number of men added to the number of women equals 32, so $x + (x + 8) = 32$.

Step 4: **Solve for x.**

$$x + (x + 8) = 32$$
$$2x + 8 = 32$$
$$2x + 8 - 8 = 32 - 8$$
$$2x = 24$$
$$\frac{2x}{2} = \frac{24}{2}$$
$$x = 12$$

Step 5: **Solve the problem.** Since x = the number of men, 12 is the answer to the problem. However, sometimes the problem asks a different question. For this example, the question might have been "how many women were there?" The answer to the problem would be $x + 8$, or 20.

Step 6: **Check your answer.** Read the problem again, substituting your answer, to make sure the answer makes sense. There were 12 men and 20 women (12 + 8). The total is 32. The answer makes sense.

In many problems, a chart is useful to clarify the information and to show necessary relationships.

Example: Ralph is three times as old as his daughter Suellen. In ten years, he will be only two times as old. How old are Ralph and Suellen now?

There are four unknowns: two present ages and two future ages.

Box 1: Let x equal the smallest unknown quantity (Suellen's age now).

Box 2: Ralph is now three times as old as Suellen, so his age is $3x$.

Box 3: In ten years Suellen's age will be ten years more: $x + 10$.

Box 4: In ten years Ralph's age will be ten years more: $3x + 10$.

	Suellen's Age	Ralph's Age
Now	1 x	2 $3x$
In 10 years	3 $x + 10$	4 $3x + 10$

We also know that Ralph's age in ten years will be two times Suellen's age in 10 years. Use this fact to write the equation: the expression in box 4 is equal to two times the expression in box 3.

$$3x + 10 = 2(x + 10)$$
$$3x + 10 = 2x + 20$$
$$x = 10$$

Now return to the chart to find the answer to the problem: Suellen's age now is 10; Ralph's age now is 30. As a check, note that in 10 years, Suellen will be 20 years old and Ralph will be 40 years old ($20 \times 2 = 40$).

Formulas are another kind of equation. A formula tells you how to use information to solve a certain kind of problem. For example, the formula for finding distance (d) is $d = rt$; where r = rate and t = time. To use formulas, just substitute the known quantities and solve.

Note: The three variables must use the <u>same</u> units of measure. If the rate is in <u>miles per hour</u>, the distance will be in <u>miles</u> and the time will be in <u>hours</u>.

Example: Kiel's model rocket traveled 250 feet in 10 seconds. What was the rate of travel?

$$d = rt$$

Substitute the known quantities: $250 = r(10)$

Solve for r. Divide both sides of the equation by 10.

$$\frac{250}{10} = \frac{r(10)}{10}$$

$$25 = r$$

The rocket traveled at a rate of 25 **feet per second.**

The formula for finding total cost (c) is $c = nr$; where n = number of units and r = cost per unit.

Example: The total cost of a shipment of chairs is $2,250. If each chair costs $75, how many chairs are in the shipment?

$$c = nr$$

Substitute the known quantities: $\$2{,}250 = n(\$75)$

Solve for n. Divide both sides of the equation by $75.

$$\frac{\$2{,}250}{\$75} = \frac{n(\$75)}{\$75}$$

$$30 = n$$

There are **30** chairs in the shipment.

We have already used formulas for interest ($i = prt$), perimeter, area and circumference. We will use more formulas in geometry.

ged Practice | Solving Word Problems

Directions: Choose the best answer to each item.

1. A certain number is added to 9. This sum is multiplied by 5, and then 15 is subtracted from the result. The answer is 40. What is the number?

   (1) 2
   (2) 8
   (3) $9\frac{1}{5}$
   (4) 11
   (5) 17

2. The sum of two consecutive numbers is 49. What is the lesser number? (Note: If x is one number, the next greater consecutive number is just one more than x, namely $x + 1$.)

   (1) 16
   (2) 18
   (3) 24
   (4) 27
   (5) 33

3. The formula for the perimeter of a triangle is $P = a + b + c$. The perimeter P is 53. Which equation should you use to find the value of b?

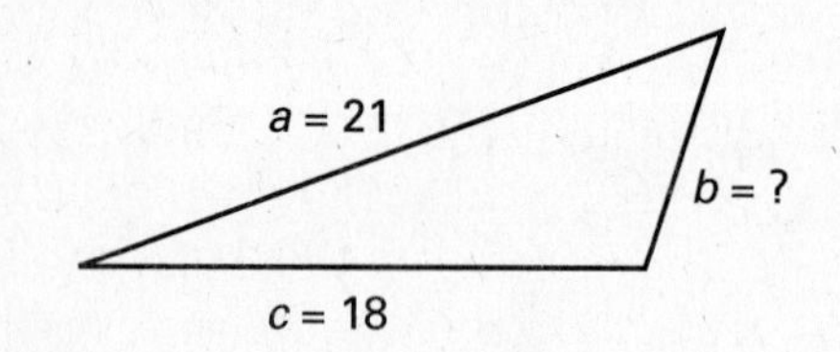

   (1) $53^2 = a + b + c$
   (2) $53 + 21 + b$
   (3) $53 - 21 - 18 = b$
   (4) $53 + 21 + 18 = b$
   (5) $53 - 18 + 21 = b$

4. Birnham Mills has 360 employees. Twelve more than 3 times the number of employees who work in management work in production. How many employees work in management?

   (1) 68
   (2) 87
   (3) 103
   (4) 112
   (5) 124

5. The formula for finding the perimeter of a rectangle (P) is $P = 2l + 2w$, where l = length and w = width. Which equation could be used to find the length of a rectangle with a perimeter of 30 centimeters and a width of 6 centimeters?

   (1) $\frac{30 - 12}{2} = l$
   (2) $30 + 12 = 2l$
   (3) $2(30) + 2(6) = l$
   (4) $30 - (2)(6) = l$
   (5) $30 - 12 - l$

6. Bonnie's weekly income is $150 less than twice her husband's weekly earnings. Together the couple earns $750. How much does her husband earn per week?

   (1) $280
   (2) $300
   (3) $315
   (4) $320
   (5) $345

7. Eight times a number, decreased by five, equals seven times that number. What is the number?

(1) 2
(2) 3
(3) 4
(4) 5
(5) 6

8. At a gym Frank did a certain number of pushups. Tom did 12 more than Frank. The total number both men did was 66. How many did Tom do?

(1) 27
(2) 29
(3) 34
(4) 39
(5) 41

9. Al got two parking tickets. The fine for the second ticket was twice as much as the fine for the first ticket. If the total fines were $54, what was the fine for the first ticket?

(1) $ 9
(2) $15
(3) $18
(4) $20
(5) $22

10. George is 5 times as old as his son. In 19 years he will be only twice as old as his son. Let x represent the son's present age. Which equation shows the correct age relationship?

(1) $x = 5x + 19$
(2) $5x = 2(5x + 19)$
(3) $2(x + 19) = 5x + 19$
(4) $x + 19 = 2(5x + 19)$
(5) $x + 19 = 2(5x)$

11. Nora is 4 years older than Diana. Two years from now Nora will be twice as old as Diana. How old is Diana now?

(1) 2
(2) 4
(3) 5
(4) 8
(5) 12

12. A plane travels 2,125 miles in 5 hours. Using the formula $d = rt$, find the plane's rate of travel.

(1) 315 miles per hour
(2) 385 miles per hour
(3) 425 miles per hour
(4) 455 miles per hour
(5) 500 miles per hour

13. A hardware store purchased 6 dozen hammers for a total cost of $104.40. Using the formula $c = nr$, what was the cost of one dozen hammers?

(1) $ 1.45
(2) $ 8.30
(3) $12.55
(4) $17.40
(5) $18.90

14. The formula $F = \frac{9}{5}C + 32$ shows the relationship between the Fahrenheit (F) and Celsius (C) temperature scales. Using the formula, what is the Celsius temperature that equals 212 degrees Fahrenheit?

(1) 82 degrees
(2) 85 degrees
(3) 90 degrees
(4) 95 degrees
(5) 100 degrees

Answers are on page 384.

Special Topics

Solving Inequalities

An **inequality** says that two algebraic expressions are *not* equal. The inequality symbols are:

$>$ "greater than" $\geq$ "greater than or equal to"
$<$ "less than" $\leq$ "less than or equal to"

Note: The arrow points to the lesser quantity.

Examples: A < B says A is less than B.
A > B says A is greater than B.
A ≤ B says A is less than or equal to B.
A ≥ B says A is greater than or equal to B.

For two numbers on a number line, the one to the left is "less than" the other. All of these are true statements.

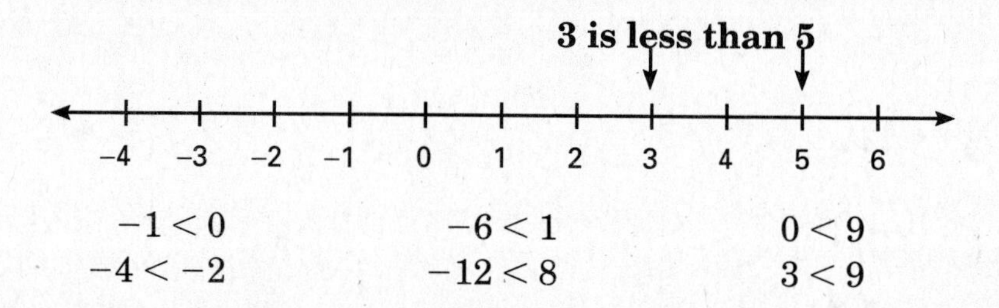

$-1 < 0$ $\quad -6 < 1$ $\quad 0 < 9$
$-4 < -2$ $\quad -12 < 8$ $\quad 3 < 9$

When a variable appears in an inequality, many numbers make the statement true. Consider this inequality: $x < 5$.

	$x < 5$	
Replace x by 2:	$2 < 5$	**True**
Replace x by 0:	$0 < 5$	**True**
Replace x by −12:	$-12 < 5$	**True**

Every number to the left of 5 on the number line is a solution. We show the solution by drawing a solid line on the number line.

Graph the solution:

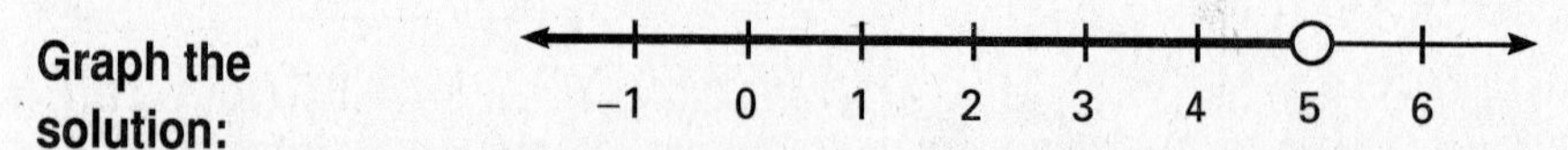

Notice that 5 itself is represented by an empty circle. This shows that the number 5 is not included. Five is not "less than" 5.

The solution of the inequality $x \geq -3$ is the number -3 and all numbers to the right of -3 on the number line.

Graph the solution:

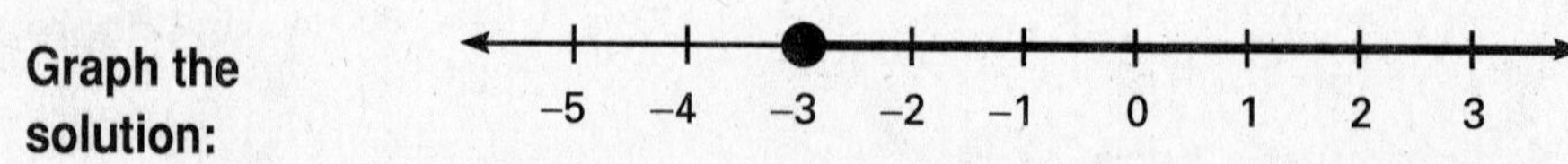

The circle at -3 is filled in because -3 is included in the solution.

An inequality can be solved much like an equation. The same number can be added to or subtracted from both sides of an inequality.

Example: Solve $2x + 7 < x + 10$

Subtract x: $x + 7 < 10$

Subtract 7: $x < 3$

Graph the solution:

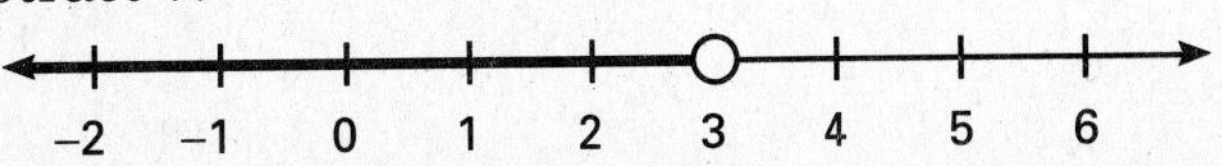

Check: Pick a number in the solution part of the graph, like 1, and substitute it in the inequality.

$2x + 7 < x + 10$

$2(1) + 7 < (1) + 10$

$2 + 7 < 11$

$9 < 11$ **True**

Any number to the left of 3 should make the statement true, and any number to the right of 3 should make the statement false.

Both sides of an inequality can be multiplied or divided by the same number. **But** multiplication and division have an important difference: if the number used is negative, the inequality sign must be **reversed.**

Example: Solve $3x - 4 < 5x$

Subtract $5x$: $-2x - 4 < 0$

Add 4: $-2x < 4$

Divide by -2: $x > -2$

Graph the solution:

−4 −3 −2 −1 0 1 2

Note: "Less than" changes to "greater than."

Check: Pick a number in the solution part of the graph, like -1, and substitute it for x.

$3(-1) - 4 < 5(-1)$

$-3 - 4 < -5$

$-7 < -5$ **True**

Inequality signs are also used to show "between." If a number is between -1 and 3, it is greater than -1 and it is also less than 3:

$$-1 < x < 3$$

Graph the solution:

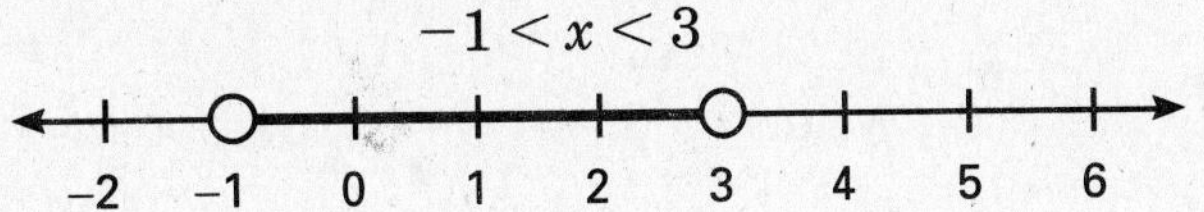

Word problems are sometimes solved by using an inequality.

Example: Sandy has three tasks to complete. The second will take three times as long as the first, and the third will take 30 minutes. If Sandy wants to be finished with all three tasks in less than 90 minutes, what is the longest amount of time the first task can take?

Let x stand for the time of the first task. Then $3x$ stands for the time of the second task. Since all three tasks must take less than 90 minutes, the inequality can be written:

$$x + 3x + 30 < 90$$

Solve:

$$4x + 30 < 90$$
$$4x < 60$$
$$x < 15$$

As long as the first task takes **less than 15 minutes,** Sandy will finish the three tasks in less than 90 minutes.

Lesson 21

Directions: Which of these numbers are solutions for these inequalities?

$-3 \quad -2 \quad -1 \quad 0 \quad 1 \quad 2 \quad 3$

1. $2x < 5$
2. $4x - 2 < 3x$
3. $5x \leq 3x - 4$
4. $8x < 7x$
5. $x > 0$
6. $3x - 1 \geq 2$

Directions: Graph the following on separate number lines.

7. Numbers that are greater than 2 but are less than 6.

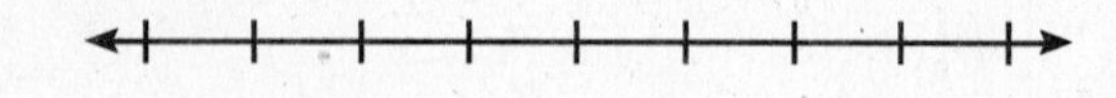

8. Numbers that are greater than or equal to 7.

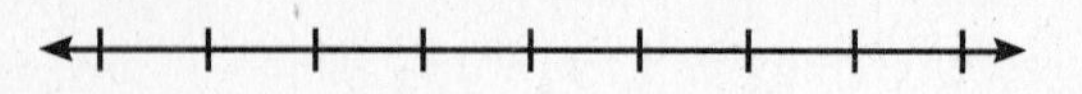

9. Numbers that are greater than -3 but are less than 1.

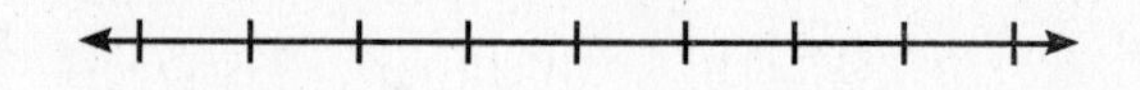

10. Numbers that are less than or equal to -1.

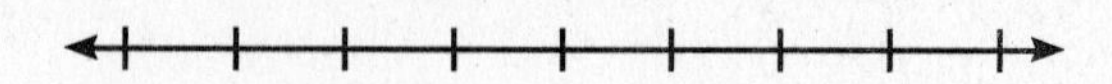

Directions: Write the inequality for each number line.

11. 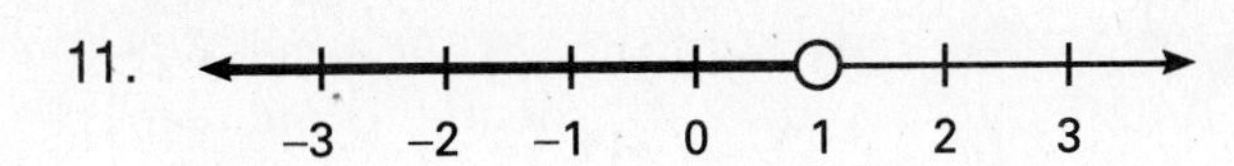

12. 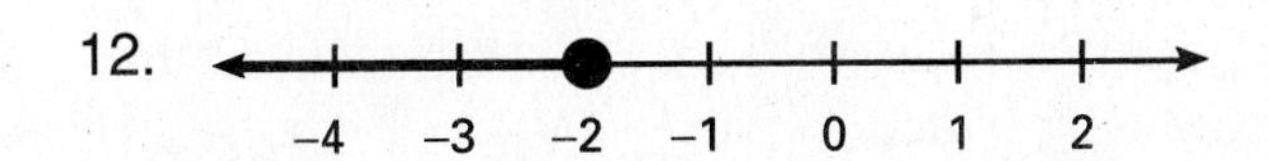

13. 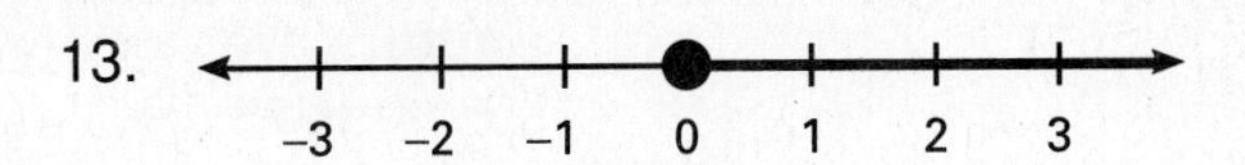

14. 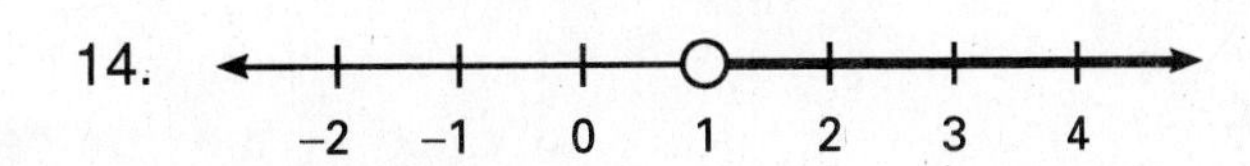

Directions: Solve the following inequalities.

15. $3x - 7 < 2x + 1$

16. $5x + 2 > 4x + 1$

17. $6x - 4 < 3x + 2$

18. $3(x + 1) > x + 4x - 5$

19. $5 + 8(x - 2) < x + 3$

20. $x + 12 < 5(x + 8)$

21. $2x + (4 - 3x) < 21$

22. $7x - 3x - x < 3x + 2x + 10$

Directions: Choose the best answer to each item.

23. If five times a number is added to 6, the result is less than 4 times that same number added to 10. Which of the following is the solution?

(1) $x < 3$
(2) $x < 4$
(3) $x > 3$
(4) $x > 4$
(5) $x < -4$

24. The perimeter of a triangle must be less than or equal to 65 inches. One side is 21 inches. The second side is 18 inches. What is the longest the third side can be (in inches)?

(1) 22
(2) 23
(3) 25
(4) 26
(5) 28

Answers are on page 385.

Factors are numbers that are multiplied together. For instance, the factors of 12 are 3 and 4, 2 and 6, and 1 and 12. In the algebraic term, $7x$, 7 and x are factors.

This algebraic expression, $4x + 10$, has two terms. Both terms can be divided by the number 2, so 2 is one factor of $4x + 10$. Divide both terms by 2 to find the other factor.

$\frac{4x + 10}{2} = 2x + 5$ Check: $2(2x + 5) = 4x + 10$

Follow these steps to factor an algebraic expression with two terms:

Step 1: Find a number and/or a variable that divides evenly into both terms.

Step 2: Divide to find the other factor.

Example: Factor $x^2 + 2x$.
The term x^2 means $x \cdot x$. The term $2x$ means $2 \cdot x$. The variable x divides evenly into both terms. Divide to find the other factor: $\frac{x^2 + 2x}{x} = x + 2$

Check: $x(x + 2) = x^2 + 2x$

A quadratic expression is one which contains the variable raised to the second power, or "squared," as in $x^2 + 2x$. Quadratic expressions will always have the variable *(x)* in both factors—in this case, x and $x + 2$ are the factors.

You may be asked to factor a quadratic expression with three terms: for example, $x^2 - 3x - 4$. Both factors for this kind of expression will have two terms: the variable and an integer.

To factor a quadratic expression with three terms, follow these steps:

Step 1: Find all possible factors of the whole number third term.

Step 2: Find the integer factors from Step 1 that can be combined to make the number part of the middle term.

Step 3: Write the two factors with the variable as the first term in each factor.

Step 4: Check your work. Multiply both terms of the second factor by each term in the first factor.

Example: Factor $x^2 - 3x - 4$.

Step 1: The possible factors for -4 are: $-4 \cdot 1$, $4 \cdot -1$, $-2 \cdot 2$, and $2 \cdot -2$.

Step 2: Combining -4 and 1 will give you the number part of the middle term: $-4 + 1 = -3$.

Step 3: Write the factors. Put the variable as the first term in both factors: $(x \quad)(x \quad)$. Then add the integers from Step 2, -4 and 1: $(x - 4)(x + 1)$.

Step 4: Check: $(x - 4)(x + 1) = x^2 + x - 4x - 4 = x^2 - 3x - 4$.

ged Practice **Lesson 21**

Directions: Factor each expression.

1. $5x + 30$
2. $6y + 15$
3. $8x - 2$
4. $4z - 14$
5. $b^2 + 9b$
6. $y^2 + 3y$
7. $2x^2 + 4x$
8. $3x^2 + 9x$
9. $7y^2 - y$
10. $4x^2 + 2x$
11. $x^2 + 9x + 20$
12. $x^2 - 5x + 6$
13. $x^2 + 5x - 6$
14. $x^2 - 3x - 28$
15. $x^2 - 2x - 3$
16. $x^2 + 8x + 12$
17. $x^2 - 7x + 12$
18. $x^2 + 7x - 8$
19. $x^2 + 3x - 10$
20. $x^2 + 10x + 21$
21. $x^2 - 13x + 40$
22. $x^2 - x - 12$
23. $x^2 - 8x - 20$
24. $x^2 - 11x + 18$
25. $x^2 - 6x - 55$
26. $x^2 + 16x + 48$
27. $x^2 + 7x - 18$
28. $x^2 - 10x + 24$
29. $x^2 + 10x + 25$
30. $x^2 - 6x - 7$

Answers are on page 386.

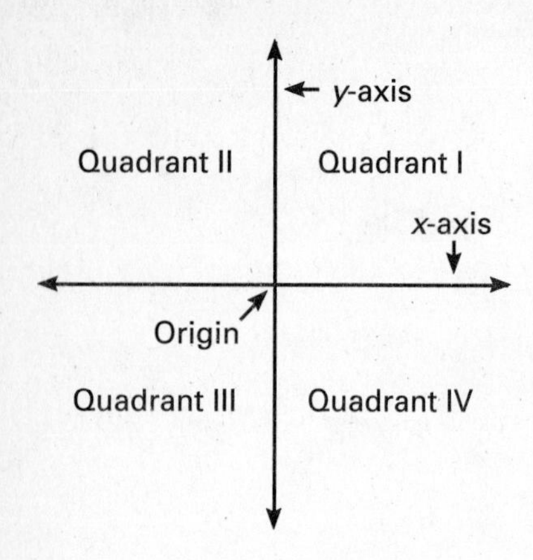

A set of points on a **graph** is called a **coordinate system.** By using a fixed starting point, called the **origin,** we can find any point by using two numbers, called the **coordinates** of that point.

On a **grid,** a horizontal line is called the ***x*-axis** and a vertical line is called the ***y*-axis.** The two axes cross at a point called the **origin.** The grid is divided into four quadrants, labeled as shown in the diagram.

The graphs used earlier were located only in Quadrants I and II. Now we will use all four quadrants.

Starting from the origin, we count positive numbers to the right on the x-axis and count negative numbers to the left. On the y-axis, we count up to locate positive numbers, and we count down to locate negative numbers.

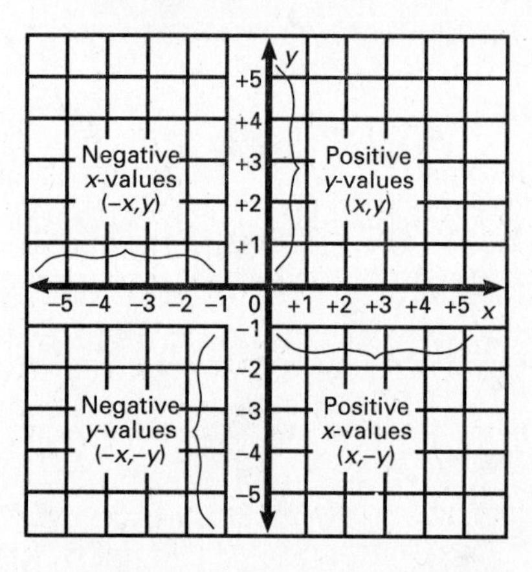

Each point on the grid is named by two numbers, an x-coordinate (always written first) and a y-coordinate (always written second). They are called an **ordered pair** and are enclosed in parentheses, separated by a comma: (x,y).

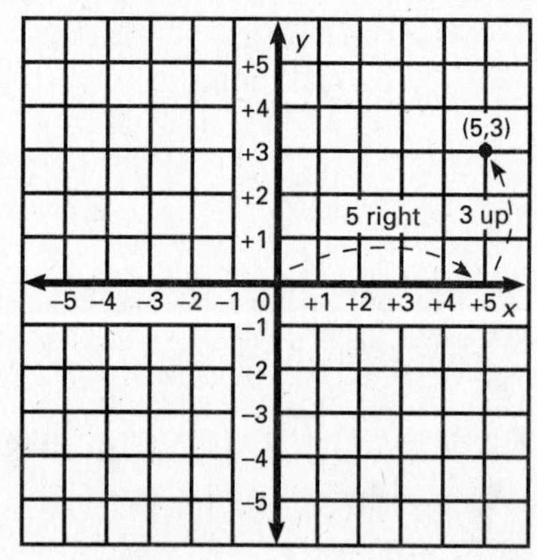

Example: Plot the point (5,3).

1. Start at the origin, which has coordinates (0,0). Move 5 units to the right, the positive x direction.

2. From that point, move 3 units upward, the positive y direction. This locates the point (5,3).

Note: The ordered pair (5,3) has a different location than the point (3,5). Remember, the point on the x-axis is *always* named first.

Note carefully the location of each point on this grid. Be sure you know how to plot each one.

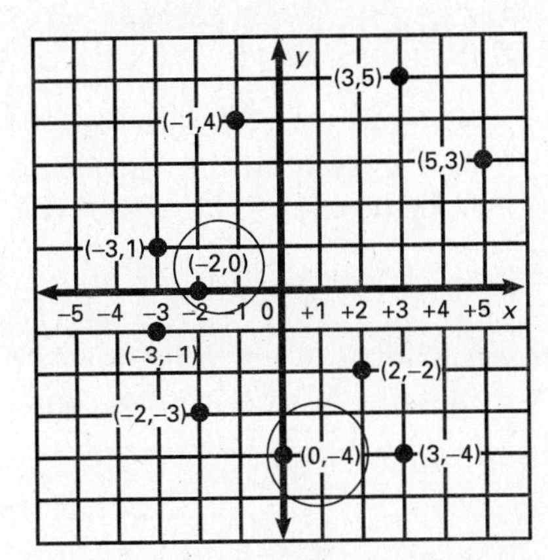

Plot a point with x-coordinate 0, like $(0,-4)$:
1. Start at the origin, move 0 on the x-axis (which is no move at all).
2. Then move -4 on the y-axis.

The point $(0,-4)$ is 4 units down on the y-axis.

Plot a point with y-coordinate 0, like $(-2,0)$:
1. Start at the origin, move 2 units to the left on the x-axis.
2. Then move 0 units on the y-axis (which is no move at all).

The point $(-2,0)$ is 2 units to the left of the origin on the x-axis.

Graphing Equations

An equation with two variables might look like this: $2x - y = 4$. Graphing techniques help us to understand and solve such equations.

To solve an equation with two variables, we need to find a value for x and a value for y that makes the equation true. Follow these steps:

Step 1: Choose any value for one variable and substitute it into the equation.

Step 2: Solve the equation for the other variable.

Step 3: Write the solution as an ordered pair (x,y)

There are many ordered pairs that are solutions to an equation with two variables.

Example: Solve $2x - y = 4$

Substitute –2 for x and solve for y:

$$\begin{aligned} 2x - y &= 4 \\ 2(-2) - y &= 4 \\ -4 - y &= 4 \\ -y &= 8 \\ y &= -8 \end{aligned}$$

So, **$(-2,-8)$** is one solution of the equation.

Now find a second solution to the same equation:

Substitute 0 for y and solve for x:

$$\begin{aligned} 2x - y &= 4 \\ 2x - 0 &= 4 \\ 2x &= 4 \\ x &= 2 \end{aligned}$$

So, **$(2,0)$** is another solution.

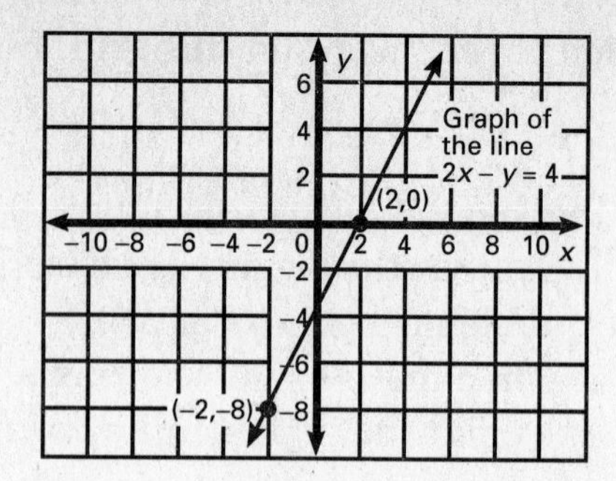

Locate both ordered pairs, $(-2,-8)$ and $(2,0)$ on a grid, and draw a line through them. The result is the graph of the equation $2x - y = 4$.

The graph of a **linear equation** is a straight line. Every point on the line satisfies ("solves") the equation.

A graph of the linear equation $x + y = 5$ is shown below. As we have just seen, every pair of coordinates on the line solves the equation, that is, the value for x and y add up to 5.

Now let's reconsider the inequalities that we studied earlier. We solved inequalities with one variable on a number line. An inequality with two variables can be solved by plotting points on the grid.

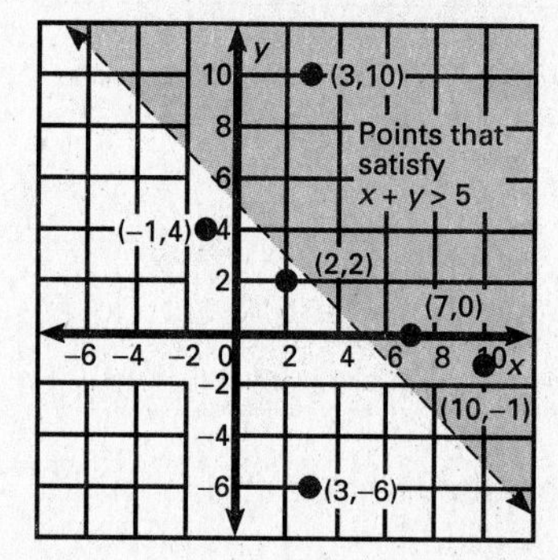

The shaded portion above the line shows the points that satisfy the inequality

$$x + y > 5.$$

(3,10):	$3 + 10 = 13$
(7,0):	$7 + 0 = 7$
(10,−1):	$10 + (-1) = 9$

All coordinate pairs result in numbers greater than 5.

The points below the line satisfy the inequality $x + y < 5$.

(2,2):	$2 + 2 = 4$
(3,−6):	$3 + (-6) = -3$
(−1,4):	$-1 + 4 = -3$

All result in numbers less than 5.

The graph of the solution of the inequality $x + y > 5$ is the shaded region above the line. The graph of the solution of the inequality $x + y < 5$ is the region below the line.

The use of rectangular coordinates gives us the chance to see math solutions. As we plot the lines on the grid, we can see how math computation can be displayed and that it is predictable.

Slope

Slope is a number that measures the steepness of a line. To find the slope of a line, choose any two points on the line. Then use this formula to find the slope (m):

$m = \frac{y_2 - y_1}{x_2 - x_1}$ Note: Either ordered pair can be thought of as (x_1, y_1), but you must be consistent.

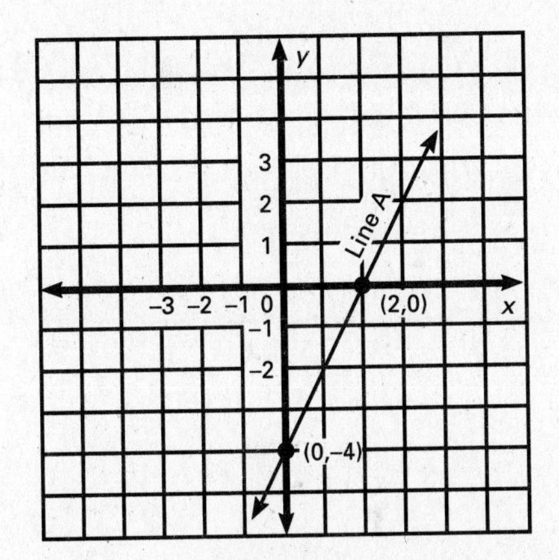

Example: Find the slope of line A.

Choose 2 points on the line. The coordinates of the two points are: (2,0) and (0,−4).

Use the formula: $m = \frac{y_2 - y_1}{x_2 - x_1}$

$m = \frac{-4 - 0}{0 - 2}$

$m = \frac{-4}{-2} = 2$

The slope of line A is positive 2. All lines that rise as they move from left to right have a positive slope. All lines that fall as they move from left to right have a negative slope.

Example: What is the slope of line B?

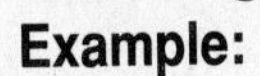

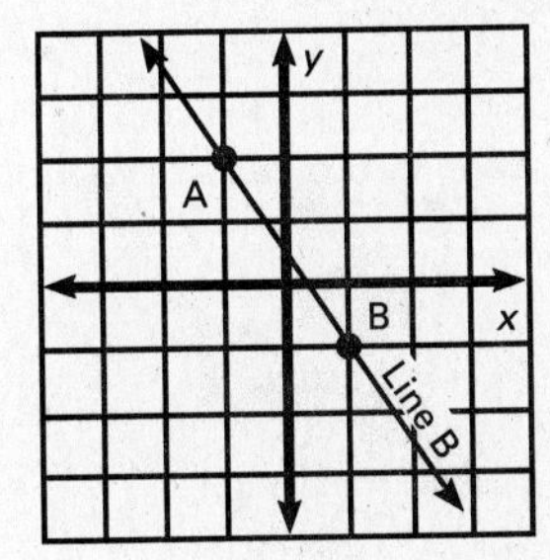

Choose two points on the line.
The coordinates of point A are (−1,2).
The coordinates of point B are (1,−1).

Use the slope formula: $m = \frac{-1 - 2}{1 - (-1)} = \frac{-3}{2}$

The slope of line B is $-\frac{3}{2}$.

Some points to remember about slope:

> The slope of the x-axis is 0.
> The y-axis has no slope.
> The slope of any 45-degree angle is 1.
> The slope of steeper lines is greater than 1.
> All lines with the same slope are parallel.

Lesson 21

Directions: Name the points and the ordered pairs for each point.

Items 1 to 10 refer to the following graph.

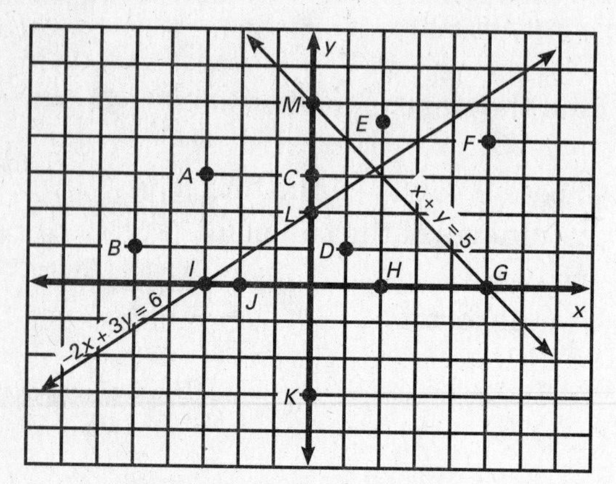

1. Which points have positive x-values and positive y-values?

2. Which points have negative x-values and positive y-values?

3. Which points are on the y-axis?

4. For which points are the x-values equal to the y-values?

5. For which points are the x-values greater than the y-values?

Directions: Name the points that satisfy each equation or inequality.

6. Points on the line $-2x + 3y = 6$.

7. Points in the region $-2x + 3y > 6$.

8. Points in the region $x + y < 5$.

9. Points in the region $x + y > 5$ and also in the region $-2x + 3y > 6$.

10. Points in the region $x + y < 5$ and also in the region $-2x + 3y > 6$.

Directions: Find the slope of each line.

11. 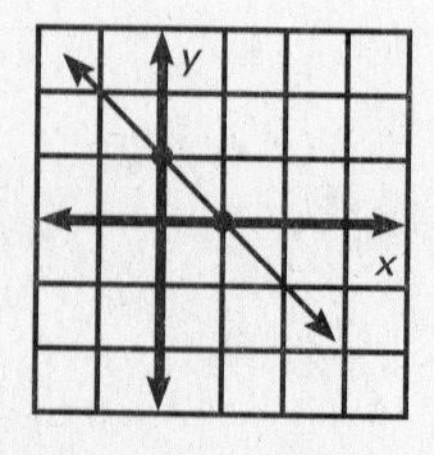

12. 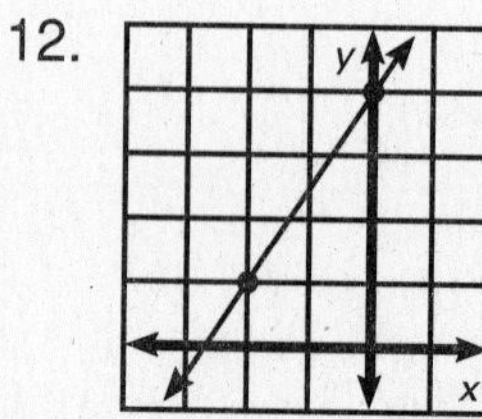

13. 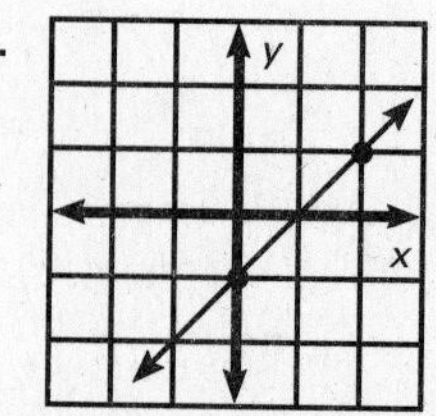

Directions: Choose the best answer to each item.

14. The points graphed satisfy which of the following equations?

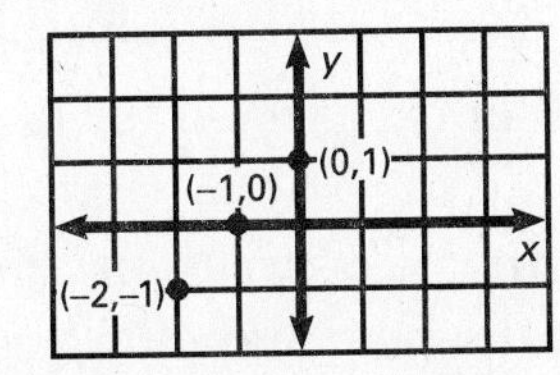

(1) $x - y = 1$
(2) $x - y = -1$
(3) $2y - x = 0$
(4) $x = 0$
(5) $y = 0$

15. What is the missing y-value if (2,?) is a solution of $3x - 4y = 10$?

(1) $y = -4$
(2) $y = -3$
(3) $y = -1$
(4) $y = 1$
(5) $y = 4$

16. The ordered pair (2,1) is a solution of which of these equations?
A: $x + y = 3$
B: $2x - y = 3$
C: $-10x + 15y = -5$

(1) A
(2) B
(3) C
(4) A and B
(5) A, B, and C

17. Which ordered pair is a solution of $x - y = 1$?

(1) (−3,−4)
(2) (−3,−2)
(3) (−1,0)
(4) (0,1)
(5) (1,−2)

18. The point (−4,2) belongs to the graph of which of the following?
A: $2x - 3y = 5$
B: $2x - 3y < 5$
C: $2x - 3y > 5$

(1) A
(2) B
(3) C
(4) both A and B
(5) (−4,2) is not part of any of them.

19. Which of these points, A, B or C, is not a solution of $2x - y = 1$?
A: (3,5) B: (1,−3) C: (0,1)

(1) A
(2) B
(3) C
(4) Neither A nor B is a solution.
(5) Neither B nor C is a solution.

20. What is the missing x-value if (?,1) is a solution of $-4x + 7y = 15$?

(1) $\frac{-11}{2}$
(2) −2
(3) $\frac{19}{7}$
(4) 2
(5) $\frac{11}{2}$

21. What is the equation of the line graphed here?

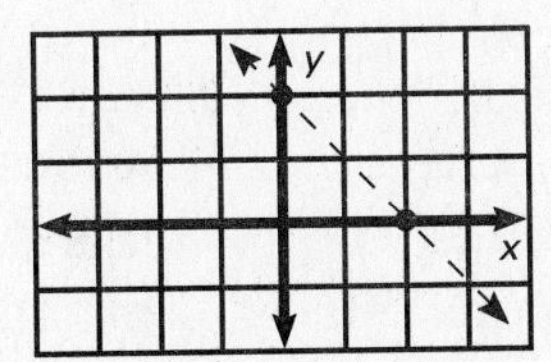

(1) $x + y = -4$
(2) $x + y = -2$
(3) $x + y = 0$
(4) $x + y = 2$
(5) $x + y = 4$

Answers are on page 387.

Exponents and Roots

Exponents

Multiplication is used to simplify problems that call for repeated addition. **Exponents** are used to simplify problems that call for repeated multiplication. The raised 3 in the expression 5^3 is an exponent. The exponent (or "power") tells how many 5's are to be multiplied.

Example: $5^4 = 5 \times 5 \times 5 \times 5 = 625$
We say, "the fourth power of 5 is 625," or "five to the fourth power is 625," or "625 is the fourth power of 5."

Examples: $12^2 = 12 \times 12 = 144$
$2^5 = 2 \times 2 \times 2 \times 2 \times 2 = 32$

Note: The exponent is the same as the number of zeros. For example:

$10^6 = 1{,}000{,}000$

Our number system is based on the idea of grouping things by tens. Powers of 10 are especially important.

$10^1 = 10$
$10^2 = 10 \times 10 = 100$ (10^2 is called 10 squared.)
$10^3 = 10 \times 10 \times 10 = 1{,}000$ (10^3 is called 10 cubed.)
$10^4 = 10 \times 10 \times 10 \times 10 = 10{,}000$
$10^5 = 10 \times 10 \times 10 \times 10 \times 10 = 100{,}000$
$10^6 = 10 \times 10 \times 10 \times 10 \times 10 \times 10 = 1{,}000{,}000$

An exponent can also be 1, 0 or a negative number.

Examples: Any number to the first power is itself.

$2^1 = 2$ $56^1 = 56$ $10^1 = 10$

Any number (except 0) to the zero power is 1.

$4^0 = 1$ $3^0 = 1$ $10^0 = 1$

Any number to a negative power represents a fraction.

Note: The "negative exponent" and the number of decimal places are the same.

$10^{-4} = 0.0001$

Notice this pattern:

$10^1 = 10$

$10^0 = 1$

$10^{-1} = \frac{1}{10} = 0.1$

$10^{-2} = \frac{1}{10 \times 10} = 0.01$

$10^{-3} = \frac{1}{10 \times 10 \times 10} = 0.001$

Exponents are a convenient way of writing very large numbers (such as the distance to the sun, which is 93,000,000, or 9.3×10^7, miles) and very small fractions (such as the weight of an atom, which could be 0.0000008, or 8×10^{-7}, grams). These numbers have so many zeros that they are difficult to read and take up a lot of space. **Scientific notation** is a way of writing these very large or very small numbers. A number written in scientific notation is expressed as a product of a number between one and ten and a power of ten.

Example: Write 5,260,000 in scientific notation.

To find a number between one and ten, move the decimal point 6 places to the left.

$$5{,}260{,}000 = 5.260000 \times 1{,}000{,}000$$

$1{,}000{,}000 = 10^6$

$$= 5.26 \times 10^6$$

Notice the exponent is the number of places the decimal point was moved.

Example: Write 3.1156×10^{-7} in standard notation.

Write the given number with a string of zeros in front of it. This does not change its value. 00000000003.1156

Move the decimal point to the left as many places as the exponent number. 00000000003.1156
Discard any extra zeros.

Therefore, $3.1156 \times 10^{-7} = 0.00000031156$

Roots

Example: What is the side of a square if the area is 25 square inches?

x = ?

A = 25 sq. in.

Remember that the area of a square is found by multiplying the length of one side by itself: $A = 5 \times 5$. If the side is 5 inches long, the area is $5^2 = 5 \times 5 = 25$, or 25 square inches.

Since $5^2 = 25$, the length of each side of the square in the example is 5 inches. The number 5 is called a square root of 25. A **square root** is a number that when multiplied times itself equals the given number. The symbol for square root is $\sqrt{\ }$.

$$5^2 = 25 \quad \sqrt{25} = 5$$

To find the square root of a number, ask yourself "What number times itself equals the number?"

Remember, if you know the "squares" you can do square roots!

$1^2 = 1$	$5^2 = 25$	$9^2 = 81$
$2^2 = 4$	$6^2 = 36$	$10^2 = 100$
$3^2 = 9$	$7^2 = 49$	$11^2 = 121$
$4^2 = 16$	$8^2 = 64$	$12^2 = 144$

Example: What is $\sqrt{144}$? (Say, "What is the square root of 144?")

Since $12^2 = 144$, $\sqrt{144} = 12$.

Sometimes a square root is not a whole number.

Example: What is $\sqrt{55}$?

What number times itself equals 55? You know that 7^2 is 49 and 8^2 is 64. So the square root of 55 must be **between 7 and 8** because 55 is between 49 and 64.

When a number comes before the square root symbol, that number is multiplied by the square root.

Example: What is $2\sqrt{81}$?

$2\sqrt{81}$ means "2 times the square root of 81."

Since the square root of 81 is 9, the expression equals 18.

$$2\sqrt{81} = 2(9) = \mathbf{18}$$

We will come back to square roots in the geometry unit.

Lesson 22

Directions: Find each value.

1. 2^4
2. 1^3
3. 3^3
4. 5^2
5. 4^5
6. 3^4
7. 5^3
8. 2^6
9. 3^2
10. 6^2
11. 5^4
12. 2^5
13. 10^4
14. 8^2
15. 4^3

Directions: Complete.

16. The third power of 2 is __?__.

17. 100 is the __?__ power of 10.

18. 36 is the second power of __?__.

19. In the expression 4^5, the exponent is the number __?__.

20. __?__ is the fifth power of 2.

21. One million is 10 to the __?__ power.

Directions: Find each value.

22. 6×10^3

23. 3×10^2

24. 8×10^2

25. 4×10^1

26. 56×10^4

27. $2^2 + 2^3$

28. $3^2 + 2^2$

29. $5^2 \times 4$

30. $7^3 + 3^5$

31. $5^2 - 3^2 - 1^2$

32. $2^5 - 5^2 - 2^2$

33. $4^2 + 3^2 + 2^2$

Directions: Find these values involving special exponents.

34. 10^0

35. 5^1

36. 10^{-4}

37. 10^{-2}

38. 2^0

Directions: Apply your knowledge of exponents to evaluate the following.

39. $(6 \times 10^3) + (3 \times 10^2)$

40. $(4 \times 10^2) - (5 \times 10^1)$

41. $(3 \times 10^4) - (2 \times 10^3)$

42. $(5 \times 10^5) + (4 \times 10^3)$

43. $(3.1 \times 10^3) - (1.3 \times 10^1)$

44. $(6.25 \times 10^3) + (5 \times 10^2)$

45. $(2.5 \times 10^4) - (6 \times 10^3)$

46. $(1.75 \times 10^2) + (4.5 \times 10^1)$

Directions: Evaluate these fractional and decimal values.

47. $\left(\frac{1}{2}\right)^2$

48. $\left(\frac{2}{3}\right)^2$

49. $\left(\frac{1}{3}\right)^2$

50. $\left(\frac{1}{2}\right)^3$

51. $\left(\frac{1}{4}\right)^3$

52. $\left(\frac{1}{5}\right)^3$

53. $(0.2)^2$

54. $(0.1)^3$

55. $(0.4)^3$

56. $(0.05)^2$

57. $(0.04)^3$

58. $(0.02)^3$

Directions: Complete this chart to evaluate each number.

	Scientific Notation		Decimal Form		Value
59.	4.1×10^{-1}	=	4.1×0.1	=	
60.	6.2×10^{-1}	=		=	0.62
61.	1.8×10^{-1}	=	1.8×0.1	=	
62.		=	6.1×0.01	=	0.061
63.	4.65×10^{-2}	=		=	0.0465

Directions: Find the side of each square.

64. $x = ?$

$A = 36$ sq. cm.

65. $x = ?$

$A = 100$ sq. ft.

66. $x = ?$

$A = 16$ sq. yd.

67. $x = ?$

$A = 49$ sq. in.

68. $x = ?$

$A = 81$ sq. m.

69. $x = ?$

$A = 64$ sq. cm.

Directions: Write the square roots.

70. $\sqrt{16}$

71. $\sqrt{0}$

72. $\sqrt{100}$

73. $\sqrt{81}$

74. $\sqrt{49}$

75. $\sqrt{121}$

76. $\sqrt{169}$

77. $\sqrt{1}$

78. $\sqrt{144}$

Directions: Choose the best answer to each item.

79. The distance across the sun, in scientific notation, is 8.65×10^5 miles. What is this distance in standard notation?

 (1) 86,500,000 mi.
 (2) 8,650,000 mi.
 (3) 865,000 mi.
 (4) 0.00000865 mi.
 (5) 0.00865 mi.

80. Decide which of these are true and which are false.
 $3^4 = 64$ $8^2 = 64$ $4^3 = 64$

 (1) All are true.
 (2) Only $8^2 = 64$ is true.
 (3) Only $4^3 = 64$ is true.
 (4) Only $3^4 = 64$ is false.
 (5) All are false.

81. The square root of 22 is between which of the following pairs of numbers?

 (1) 2.2 and 2.3
 (2) 3 and 4
 (3) 4 and 5
 (4) 5 and 6
 (5) 21 and 22

82. Which of the following has the numbers in order from least to greatest?

 (1) 4.7×10^{-1}, 2.34×10^2, 5.2×10^2
 (2) 4.7×10^{-1}, 5.2×10^2, 2.34×10^2
 (3) 2.34×10^2, 4.7×10^{-1}, 5.2×10^2
 (4) 2.34×10^2, 5.2×10^2, 4.7×10^{-1}
 (5) 5.2×10^2, 4.7×10^{-1}, 2.34×10^2

83. Since $10^2 = 100$ and $10^4 = 10{,}000$, what power of 10 represents $10^2 \times 10^4$?

 (1) 10^3
 (2) 10^4
 (3) 10^5
 (4) 10^6
 (5) 10^7

84. Arrange these numbers in order from least to greatest value: 10^2, 2^7, 3^4

 (1) 2^7 is the least and 10^2 is the greatest.
 (2) 3^4 is the least and 10^2 is the greatest.
 (3) 3^4 is the least and 2^7 is the greatest.
 (4) 10^2 is the least and 2^7 is the greatest.
 (5) All are false.

85. Mario and Lucia disagree on the meaning of the expression $3^2 - 2^3$. Mario says it means $9 - 6$, or 3. Lucia says it means $6 - 6$, or 0. Who is right?

 (1) Mario is right.
 (2) Lucia is right.
 (3) They are both right.
 (4) They are both wrong, since the expression equals 1.
 (5) They are both wrong, since the expression equals 2.

86. If you know that $2^{10} = 1{,}024$, how would you find 2^{12}?

 (1) Add 2.
 (2) Add 4.
 (3) Multiply by 12.
 (4) Multiply by 2.
 (5) Multiply by 2, and then multiply by 2 again.

Answers are on page 388.

Strategies for Solving Word Problems

Work Backwards

On the GED test each problem you will be asked to solve will have five answer options. Your task is to choose the best answer for each item.

For most items, it will be faster to solve the problem directly. Read the problem, decide which operation you need to solve the problem, do the work, and check your answer.

But in an algebra problem, you may be solving for a particular variable. Many of the facts are given. Perhaps the equation has already been written for you. In these cases, it may be faster to try each of the answer choices in the problem to see which choice is true.

Example: The sum of three consecutive numbers is 30. What are the numbers?

(1) 6, 7, and 8
(2) 8, 9, and 10
(3) 9, 10, and 11
(4) 11, 12, and 13
(5) 14, 15, and 16

Since you know the numbers add up to 30, why take time to write an equation? Add the numbers for each answer option. You can quickly eliminate options 4 and 5 since $10 + 10 + 10 = 30$. Clearly, options 4 and 5 total more than 30. Quickly add the numbers for the first three options.

Choice 1: $6 + 7 + 8 = 21$
Choice 2: $8 + 9 + 10 = 27$
Choice 3: $9 + 10 + 11 = 30$ Option (**3**) is correct.

By doing the calculations in your head, you save the time you would spend writing and solving this equation:

Let x equal the first number, $x + 1$ the second number, and $x + 2$ the third number.

$$x + x + 1 + x + 2 = 30$$
$$3x + 3 = 30$$
$$3x = 27$$
$$x = 9; x + 1 = 10; x + 2 = 11$$

Using the equation, the three numbers are **9, 10, and 11.**

Solving Word Problems

1. The sum of three consecutive numbers is 45. What are the three numbers?

   (1) 10, 11, and 12
   (2) 12, 13, and 14
   (3) 13, 14, and 15
   (4) 14, 15, and 16
   (5) 15, 16, and 17

2. Pam, Juan, and Eva took a 1,200-mile driving trip. Pam drove 100 more miles than Juan. Juan drove 100 more miles than Eva. How many miles did Eva drive?

   (1) 100
   (2) 150
   (3) 200
   (4) 250
   (5) 300

3. The sum of two consecutive numbers is 95. What are the two numbers?

   (1) 40 and 41
   (2) 42 and 43
   (3) 47 and 48
   (4) 52 and 53
   (5) 57 and 58

4. Four consecutive numbers total 38. What are the four numbers?

   (1) 7, 8, 9, and 10
   (2) 8, 9, 10, and 11
   (3) 9, 10, 11, and 12
   (4) 10, 11, 12, and 13
   (5) 11, 12, 13, and 14

5. Bill Jackson teaches at two schools. In the last 30 days, he taught in one of the schools for 2 more days than in the other school. How many days did he teach at each of the two schools in the last 30 days?

   (1) 9 and 11
   (2) 13 and 15
   (3) 14 and 16
   (4) 19 and 21
   (5) 24 and 26

6. Three consecutive odd numbers total 33. What are the three numbers?

   (1) 9, 11, and 13
   (2) 11, 13, and 15
   (3) 13, 15, and 17
   (4) 15, 17, and 19
   (5) 19, 21, and 23

7. Marta scored 93 points on her English test. She scored 5 points lower on the writing part of the test than on the reading part. What were her scores on each of the parts of the test?

   (1) 43 and 48
   (2) 44 and 49
   (3) 45 and 50
   (4) 47 and 52
   (5) 54 and 59

8. The sum of three consecutive odd numbers is 99. What are the three numbers?

   (1) 29, 31, and 33
   (2) 31, 33, and 35
   (3) 32, 33, and 34
   (4) 33, 33, and 33
   (5) 33, 35, and 37

Answers are on page 389.

ged Mini-Test Lessons 19–22

Directions: Choose the best answer to each question.

1. On a number line, which two answers are farthest apart?
   A: $(-2) + (-7)$ C: $(-3) - (-4)$
   B: $(-6) + (+8)$ D: $(+4) - (+10)$

   (1) A and B
   (2) A and C
   (3) B and C
   (4) B and D
   (5) C and D

2. Which of the following expressions shows the product of 9 and x, subtracted from the quotient of 2 and x?

   (1) $\frac{2}{x} - 9x$
   (2) $9x - 2 \div x$
   (3) $2x \times -9x$
   (4) $9 - x - 2 \div x$
   (5) $(2 + x) - 9x$

3. What is the slope of the line shown on the graph?

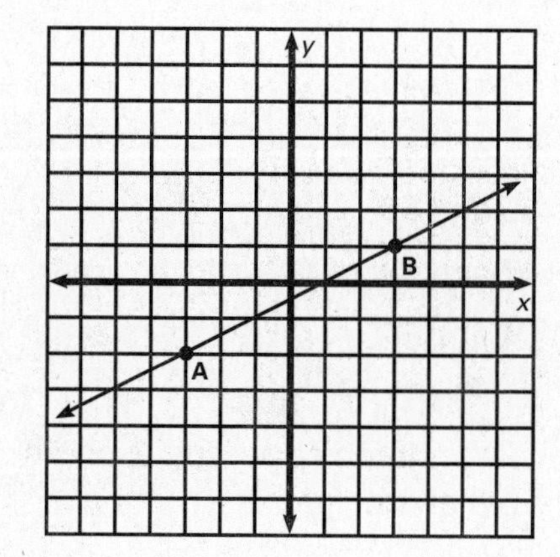

   (1) $-\frac{1}{2}$
   (2) $\frac{1}{3}$
   (3) $\frac{1}{2}$
   (4) 2
   (5) 3

4. Solve: $2(2x + 3) = -3(x + 5)$.

   (1) $x = -21$
   (2) $x = -3$
   (3) $x = \frac{2}{7}$
   (4) $x = -18$
   (5) $x = 14$

5. "Three more than the number I'm thinking of is 30," says Joe. "Then 10 less than your number," says Jill, "is the number __?__."

   (1) 17
   (2) 23
   (3) 33
   (4) 37
   (5) 43

6. Coop and Carni are two cats. Coop is 4 months older than Carni. Three months ago, 5 times Coop's age in months was the same as 7 times Carni's age in months. Complete the chart. What is the entry for box 4?

	Coop's Age in Months	**Carni's Age in Months**
Now	1 x	2 $x - 4$
3 months ago	3	4

   (1) $(x - 4) - 3$
   (2) $5(x - 4)$
   (3) $7(x - 4)$
   (4) $\frac{7}{5}(x - 4)$
   (5) $\frac{5}{7}(x - 4)$

7. What addition does this number line show?

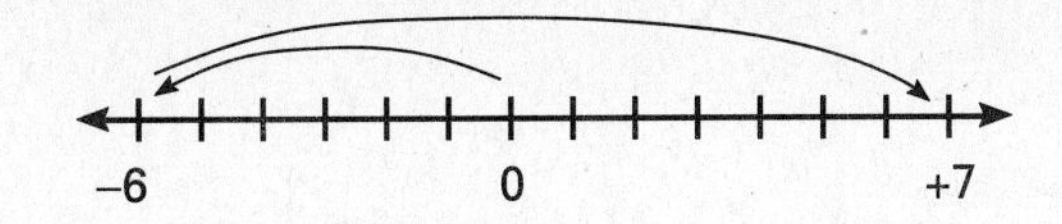

(1) $(-6) + (+7)$
(2) $(+7) + (-6)$
(3) $(-6) + (+13)$
(4) $(+7) + (-13)$
(5) $0 + (+6) - (+13)$

8. Becky's mother is twice as old as she is. Eleven years ago her mother was three times as old as Becky. Let x represent Becky's present age. Which equation shows their age relationship?

(1) $x - 11 = 3(2x - 11)$
(2) $x - 11 = 2x - 11$
(3) $2(x - 11) = 2x - 22$
(4) $3(x - 11) = 2x - 11$
(5) $3(x - 11) = 3x - 33$

9. When 13 is subtracted from the sum of two consecutive numbers, the answer is 18. What are the consecutive numbers?

(1) 6 and 7
(2) 9 and 10
(3) 10 and 11
(4) 15 and 16
(5) 31 and 32

10. Factor: $x^2 - 12x + 36$.

(1) $(x + 6)(x + 6)$
(2) $(x - 6)(x + 6)$
(3) $(x - 6)(x - 6)$
(4) $(x - 3)(x - 12)$
(5) $(x + 3)(x + 12)$

11. Solve: $-2(3 + x) - 4 = -3x - 1$.

(1) $x = -9$
(2) $x = \frac{4}{9}$
(3) $x = 4\frac{1}{2}$
(4) $x = 9$
(5) $x = 11$

12. Solve: $5x + 2 < 6x + 3x + 10$.

(1) $x > -2$
(2) $x < -2$
(3) $x < 3$
(4) $x > -3$
(5) $x < -3$

13. Which graph shows the solution of $x - 3 < -1$?

(1)
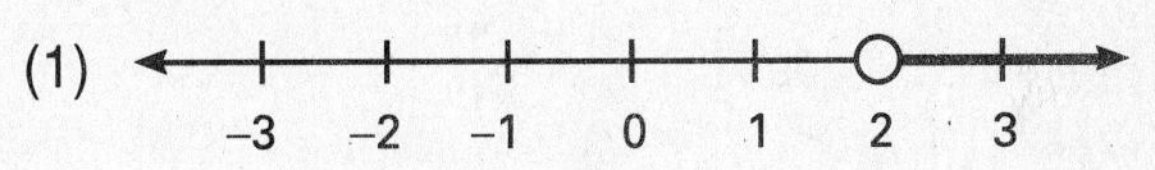

(2)
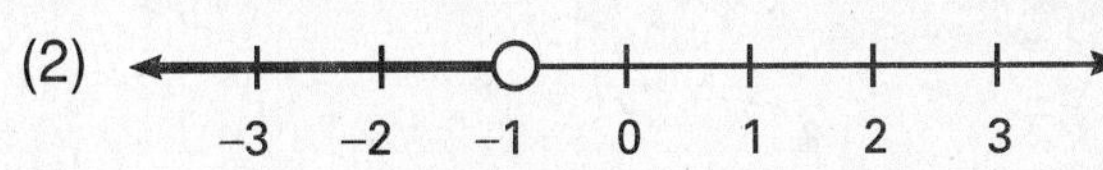

(3)
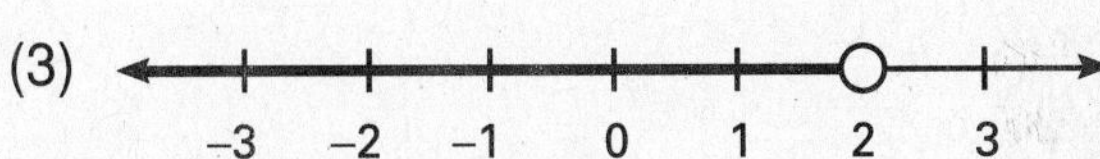

(4)
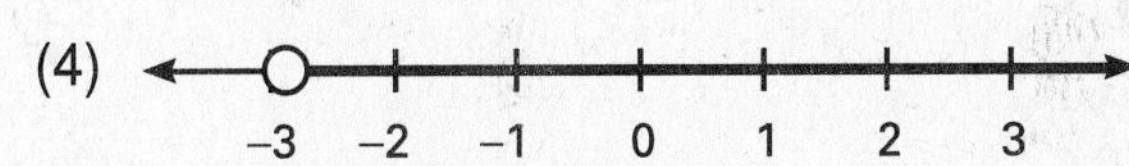

(5) −3 −2 −1 0 1 2 3

14. Which of the following represents 0.0024 written in scientific notation?

(1) 2.4×10^4
(2) 2.4×10^{-2}
(3) 2.4×10^2
(4) 2.4×10^{-4}
(5) 2.4×10^{-3}

Answers are on page 390.

ged Cumulative Review **Algebra**

Directions: Choose the best answer to each item.

1. What is the value of 10^{-3}?

   (1) 0.01
   (2) 0.03
   (3) 0.001
   (4) 0.003
   (5) 0.0001

2. What are the factors of $x^2 - 8x + 15$?

   (1) $(x - 3)(5 - x)$
   (2) $(x - 3)(x + 5)$
   (3) $(x + 3)(x + 5)$
   (4) $(x + 3)(x - 5)$
   (5) $(x - 3)(x - 5)$

3. The pair $(-1,-2)$ is a solution of which of these equations?
   A. $x + 2y = -3$
   B. $2x - y = -2$
   C. $x + y = -3$

   (1) A and B
   (2) A and C
   (3) B and C
   (4) B only
   (5) C only

4. What sum is shown in the number line?

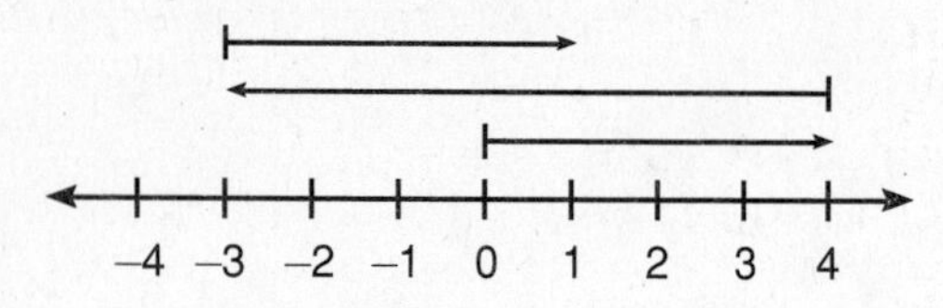

   (1) $4 + (-3) + 1 = 1$
   (2) $4 + 1 - 3 = 0$
   (3) $4 + (-7) + 4 = 1$
   (4) $4 - (-3) + 4 = 1$
   (5) $4 + (-7) + -4 = 1$

5. Simplify: $2 - (x + 7)$.

   (1) $x - 9$
   (2) $x - 9$
   (3) $-x - 5$
   (4) $-x - 9$
   (5) $-x + 9$

6. Find the value of this expression if $x = -1$ and $y = 5$. $4x - 2y + xy$

   (1) -26
   (2) -19
   (3) -9
   (4) -1
   (5) 1

7. What is the missing number?
   $7(x + 3) = 7x +$ ____

   (1) 7
   (2) 3
   (3) 21
   (4) $3x$
   (5) $10x$

8. Solve: $-6(x + 1) + 4 = 8x - 9$.

   (1) $\frac{1}{2}$
   (2) $-\frac{1}{2}$
   (3) -1
   (4) -3
   (5) -5

9. Solve the following inequality:
   $3x + 7 < -2x + 2$.

   (1) $x < -1$
   (2) $x < 1$
   (3) $x < -\frac{1}{5}$
   (4) $x < \frac{1}{5}$
   (5) $x < -5$

10. Which of the following expresses this equation?

$$x - 2 = \frac{x}{4} + 7$$

(1) Two less than x equals 7 more than 4 divided by x.

(2) Two less than x is the same as 7 more than the quotient of x and 4.

(3) Two subtracted from x equals 4 into 7 more than x.

(4) x less than 2 is the same as x divided by 4 plus 7.

(5) x subtracted from 2 equals 4 divided by x, increased by 7.

11. Which of the following shows the product of -6 and x, subtracted from the sum of -6 and y?

(1) $-6x - (-6 + y)$

(2) $(-6 + x) - (-6 + y)$

(3) $(-6 + x) - (-6y)$

(4) $(-6 + y) - (-6x)$

(5) $(-6y) - (-6x)$

12. Which expression is correct for the following: two less than the quotient of x and 6?

(1) $\frac{x-2}{6}$

(2) $\frac{2-x}{6}$

(3) $2 - \frac{x}{6}$

(4) $x - \frac{2}{6}$

(5) $\frac{x}{6} - 2$

13. Simplify this expression: $7x + 5 - 3x - 8$.

(1) $-4x - 13$

(2) $1x$ or x

(3) $4x - 3$

(4) $10x - 3$

(5) $7x - 2x - 8$

14. What is the formula for finding the perimeter P of the figure?

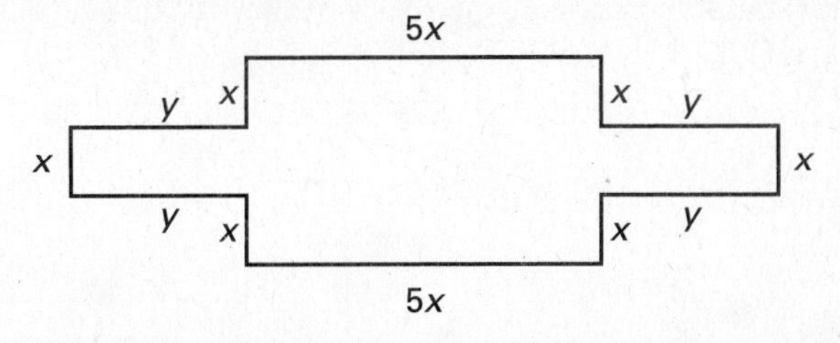

(1) $P = 11x + 4y$

(2) $P = 11x + 2y$

(3) $P = 16x + 2y$

(4) $P = 16x + 4y$

(5) $P = 10x + 4x + 4y$

15. Solve this inequality: $5x + 3 \leq 17 - 2x$.

(1) $x \leq -2$

(2) $x \leq 0$

(3) $x \leq 1$

(4) $x \leq 2$

(5) $x \leq 3$

16. The factors of $5x - 60$ are 5 and ____.

(1) $x - 6$

(2) $x - 12$

(3) $x + 12$

(4) $x - 15$

(5) $x + 20$

17. One number is 8 times another. When the lesser number is subtracted from the greater, the result is 2 more than 5 times the lesser number. What are the numbers?

(1) -2 and -16

(2) 1 and 8

(3) 3 and 24

(4) 4 and 32

(5) 6 and 48

18. Which statement is true?
A. $-5 < -6$
B. $3 < -2$
C. $0 < -7$

(1) A
(2) B
(3) C
(4) All are true.
(5) All are false.

19. Solve the following inequality:
$6 - 5x < 7x - 6$.

(1) $x < -1$
(2) $x < 0$
(3) $x < 1$
(4) $x > -1$
(5) $x > 1$

20. In a game, points you win are positive numbers, and points you lose are negative numbers. What is the value of this series of plays?
Win 7, Lose 9, Lose 2, Win 1, Win 10

(1) -7
(2) $+7$
(3) $+11$
(4) $+18$
(5) $+29$

21. Two of the following have the same value. Which two are they?
A. $4 + (1 - 4)$ C. $6 - (3 + 2)$
B. $(7 - 3) + 1$ D. $8 - (1 - 2)$

(1) A and B
(2) A and C
(3) B and C
(4) B and D
(5) C and D

22. Divide -12 by the sum of 4 and -2. The result is:

(1) -6
(2) -2
(3) 1
(4) 2
(5) 6

23. Which value for x makes the following inequality true: $x > 1{,}000$?

(1) 2^5
(2) 4^3
(3) 4^5
(4) 5^4
(5) 6^3

24. Express 9,800,000 in scientific notation.

(1) 9.8×10^3
(2) 9.8×10^4
(3) 9.8×10^5
(4) 9.8×10^6
(5) 9.8×10^7

25. The pair $(-1,1)$ is a solution of which of the following equations?

(1) $2x - 3y = 1$
(2) $3x - 2y = -5$
(3) $4x + 2y = 6$
(4) $-2x - 3y = 1$
(5) $-3x + 2y = -1$

26. Two points that belong to the line $2x - 5y = 10$ are:

(1) (0,2) and (5,0)
(2) (0,−2) and (5,2)
(3) (0,2) and (−5,0)
(4) (0,−2) and (5,0)
(5) (5,−2) and (−5,2)

27. Multiply: $(x + 9)(x - 2)$.

(1) $x^2 + 11x - 18$
(2) $x^2 - 11x - 18$
(3) $x^2 + 7x - 18$
(4) $x^2 + 7x + 18$
(5) $x^2 - 7x + 18$

28. Which inequality is graphed on the number line?

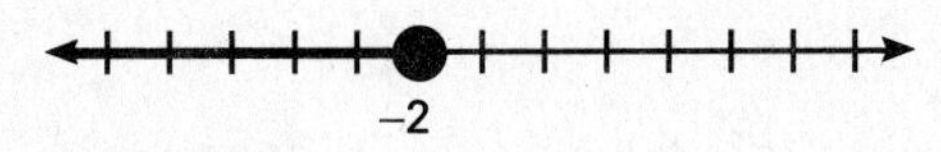

(1) $x \leq -2$
(2) $x < -2$
(3) $-2 \leq x$
(4) $-2 > x$
(5) $2 = x$

29. In solving this inequality, $-7x < 10$, what is the next step?

(1) $x < \frac{10}{7}$
(2) $x < \frac{-10}{7}$
(3) $-x < \frac{7}{10}$
(4) $x > \frac{10}{7}$
(5) $x > \frac{-10}{7}$

30. Ten less than a number is equal to the same number divided by 2. What is the number?

(1) 8
(2) 10
(3) 14
(4) 20
(5) 28

31. Factor: $8x^2 + 12x$.

(1) $x(4x + 3x)$
(2) $4(x + 3x)$
(3) $4x(2x + 3)$
(4) $4x(2 + 3x)$
(5) $4x(2x + 3x)$

Items 32 to 34 refer to the following graph.

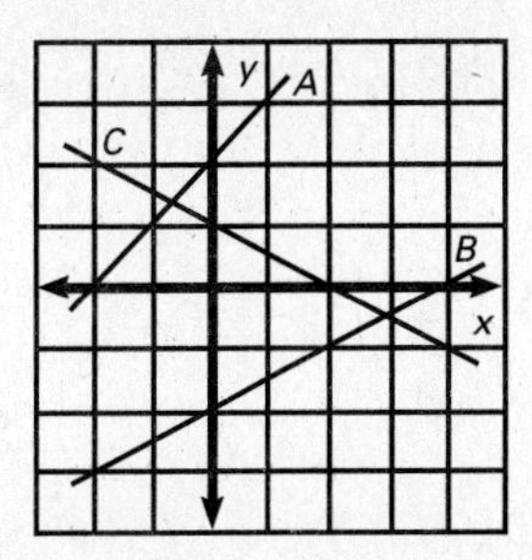

32. A point that belongs to both line *A* and line *C* would have a:

(1) positive *x*-value and a positive *y*-value
(2) positive *x*-value and a negative *y*-value
(3) negative *x*-value and a positive *y*-value
(4) negative *x*-value and a negative *y*-value
(5) There are no such points.

33. Which line has the greatest slope?

(1) *A*
(2) *B*
(3) *C*
(4) both *A* and *B*
(5) both *A* and *C*

34. Which line has a negative slope?

(1) *A*
(2) *B*
(3) *C*
(4) both *A* and *B*
(5) *A, B, C*

35. Factor: $x^2 - x - 2$.

(1) $(x - 1)(x - 1)$
(2) $(x - 1)(x + 2)$
(3) $(x - 1)(x - 2)$
(4) $(x + 1)(x + 2)$
(5) $(x + 1)(x - 2)$

36. Which inequality is graphed here?

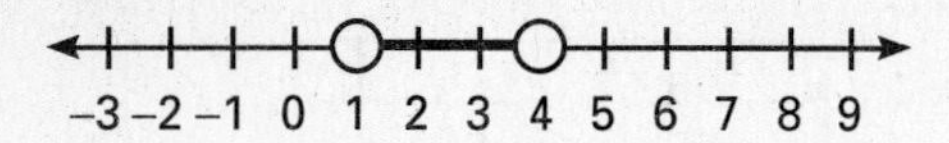

(1) $1 < x < 4$
(2) $1 > x > 4$
(3) $1 > x < 4$
(4) $1 < x > 4$
(5) None of the options is correct.

37. The square root of 45 is found between which pair of numbers?

(1) 4 and 5
(2) 5 and 6
(3) 6 and 7
(4) 7 and 8
(5) 8 and 9

38. Bill is 1 year less than twice as old as his sister Caroline. The total of their ages is 26. How old is Bill now?

(1) 12
(2) 17
(3) 19
(4) 24
(5) 27

39. Evaluate the following expression: $3^2 + 2^4 - 5^2$.

(1) -10
(2) 0
(3) 10
(4) 25
(5) 32

40. Solve this equation: $5x - 24 = 3x - x + 12$.

(1) $x = -4$
(2) $x = 4$
(3) $x = 8$
(4) $x = 12$
(5) $x = \frac{38}{5}$

41. Write an equation for the following: The product of a number and -4 is 8 more than 2 added to -5 times that number.

(1) $-4x = -5x + 2 + 8$
(2) $-4x - 8 = 2x + (-5x)$
(3) $-4x + 8 = -5x + 2$
(4) $-4x = 8 + 2 + (-5)$
(5) $-4x = 8 + (-5 + 2)x$

42. Find an equal expression for $6(x - 2)$.

(1) $6x - 2$
(2) $6x - 12$
(3) $6x + 6 - 2$
(4) $6x - 6 - 2$
(5) $6(x) - 2$

43. Add 12 to the product of 3 and a certain number, x. The result is 2 more than the number x. What is the number x?

(1) $x = -6$
(2) $x = -5$
(3) $x = 5$
(4) x is a fraction between 0 and 1.
(5) x is a fraction between -1 and 0.

44. Factor: $x^2 - 2x - 24$.

(1) $(x + 3)(x - 8)$
(2) $(x - 12)(x + 2)$
(3) $(x + 12)(x - 2)$
(4) $(x - 4)(x + 6)$
(5) $(x + 4)(x - 6)$

45. Which of the following statements is true?
A. $25 > 10$ B. $8^2 \leq 16$ C. $4^2 \geq 144$

(1) A only
(2) B only
(3) C only
(4) All of the statements are true.
(5) None of the statements is true.

46. Which of the following are equal?
A. $10 - (2)(2)$ B. $2 + \frac{8}{2}$ C. $3 - \frac{12}{-4}$

(1) A and B
(2) A and C
(3) B and C
(4) All are equal.
(5) All are different.

Items 47 to 52 refer to the following graph.

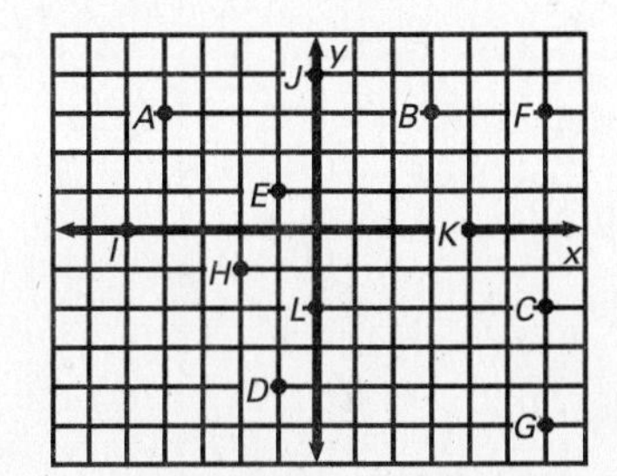

47. Which points have positive x-values and negative y-values?

(1) *A, E*
(2) *A, E, I*
(3) *C, G*
(4) *C, G, L*
(5) *C, G, L, K*

48. Which points have a y-value equal to 0?

(1) *K*
(2) *J* and *L*
(3) *I* and *K*
(4) *I, J, K, L*
(5) None have a y-value of 0.

49. If the line for $x - y = 8$ were drawn on the graph, which of the following points would be on the line?

(1) *A*
(2) *B*
(3) *C*
(4) *D*
(5) *E*

50. If the line for $x + 2y = -4$ were drawn on the graph, which of the following points would be on the line?

(1) *D*
(2) *I*
(3) *J*
(4) *K*
(5) *L*

51. What would be the slope of a line passing through points *I* and *E*?

(1) $\frac{1}{4}$
(2) $\frac{1}{2}$
(3) 1
(4) 2
(5) 3

52. What would be the slope of a line passing through points *J* and *K*?

(1) $\frac{1}{2}$
(2) 0
(3) $-\frac{1}{2}$
(4) -1
(5) -2

53. The length of a rectangle must be 6 times the width. The perimeter must be more than 110 feet. Which inequality can be solved to find the length and width?

(1) $7x > 110$
(2) $6x^2 > 110$
(3) $x + 6x > 110$
(4) $2(6x) + x > 110$
(5) $x + 6x + x + 6x > 110$

54. Solve this inequality: $2x + 5 \leq 4x - 1$.

(1) $x \geq 2$
(2) $x \geq 3$
(3) $x \geq 4$
(4) $x \geq 5$
(5) $x \geq 6$

55. Write an equation for the following: Twice a number, added to 3, is equal to the negative of the number.

(1) $3 + 2x = -x$
(2) $3 + 2x = -3$
(3) $2 + x + 3 = -x$
(4) $3 + 2 + x = -x$
(5) $2x + 3 = -(2x + 3)$

56. Solve: $-\frac{1}{3}x = 9$.

(1) $-\frac{1}{9}$
(2) $-\frac{1}{3}$
(3) -3
(4) -9
(5) -27

57. Multiply: $(x - 6)(x - 9)$.

(1) $x^2 - 3x + 54$
(2) $x^2 + 3x - 54$
(3) $x^2 - 3x - 54$
(4) $x^2 - 15x + 54$
(5) $x^2 + 15x - 54$

58. The square root of 70 is found between which pair of numbers?

(1) 5 and 6
(2) 6 and 7
(3) 7 and 8
(4) 8 and 9
(5) 9 and 10

59. Factor: $x^2 + x - 12$.

(1) $(x + 3)(x - 4)$
(2) $(x - 3)(x + 4)$
(3) $(x - 3)(x - 4)$
(4) $(x + 6)(x - 2)$
(5) $(x + 6)(x + 2)$

60. The pair $(0, -2)$ is a solution of which of the following equations?

(1) $2x - 3y = -6$
(2) $2x + 3y = 6$
(3) $-2x + 3y = -6$
(4) $-2x - 3y = 5$
(5) $x + 2y = 4$

61. Express 451,000 in scientific notation.

(1) 4.51×10^6
(2) 4.51×10^5
(3) 45.1×10^5
(4) 451×10^5
(5) $4{,}510 \times 10^5$

62. Which value for x makes the following inequality true: $x < 500$?

(1) 5^3
(2) 5^4
(3) 4^6
(4) 6^4
(5) 4^5

63. Solve this equation: $3x + 12 = 6x - x + 4$.

(1) $x = -4$
(2) $x = 3$
(3) $x = 4$
(4) $x = -3$
(5) $x = 2$

64. Find an equal expression for $5(x - 3)$.

(1) $5x - 3$
(2) $5x - 5$
(3) $5x - 15$
(4) $5x - 3 - 15$
(5) $5x + 15$

Answers are on page 391.

Performance Analysis
Unit 2 Cumulative Review: Algebra

Name: ______________________________ **Class:** ______________ **Date:** ______________

Use the Answer Key on pages 391–393 to check your answers to the GED Unit 2 Cumulative Review: Algebra. Then use the chart to figure out the skill areas in which you need additional review. Circle on the chart the numbers of the test items you answered correctly. Then go back and review the lessons for the skill areas that are difficult for you. For additional review, see the *Steck-Vaughn GED Mathematics Exercise Book,* Unit 2: Algebra.

Use the chart below to find your strengths and weaknesses in application skills.

Concept	Item Numbers
Signed Numbers and Algebraic Expressions pages 196–201	**4,** 5, 6, 11, 12, 13, 20, 21, 22, 46
Equations pages 202–211	7, 8, 10, **14,** 17, 30, 38, 40, 41, 43, 55, 56, 60, 63, 64
Inequalities pages 212–215	9, 15, 18, 19, **28,** 29, **36,** 53, 54
Factoring pages 216–217	2, 16, 27, 31, 35, 42, 44, 57, 59
Graphing, Slope, and Lines pages 218–223	3, 25, 26, **32, 33, 34, 47, 48, 49, 50, 51, 52**
Exponents and Roots pages 224–231	1, 23, 24, 37, 39, 45, 58, 61, 62

Boldfaced numbers indicate items based on charts, graphs, illustrations, and diagrams.

Unit 3 GEOMETRY

Geometric figures are everywhere.

line segment
part of a line

arc
part of a circle

angle
a pair of rays meeting at a point

right angle
an angle that makes a "square corner" that measures 90 degrees

dimension
a measurement of length, width, or thickness

plane
a set of points that forms a flat surface

In geometry you will study shapes that you see and use every day. These shapes are made of simple geometric ingredients—**line segments, arcs,** and **angles,** many of which are **right angles.**

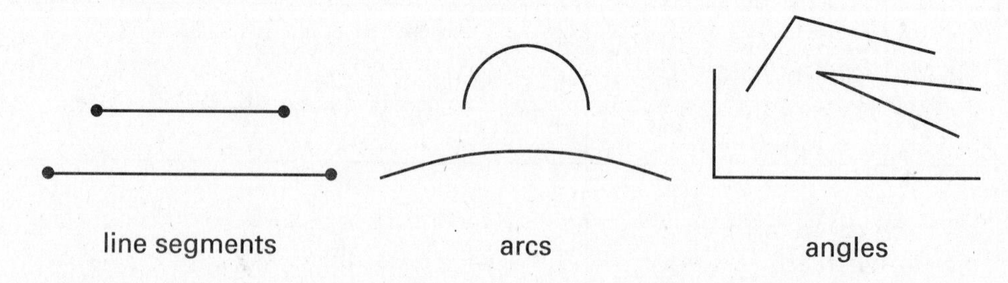

Look around your home at the rooms, furniture, clothing, and household objects, and outside at buildings, streets, and bridges. All around us are examples of these simple two-**dimensional** forms. We will learn about these figures in our study of **plane geometry.**

A **plane** is a set of points that forms a flat surface. A plane figure has only two dimensions. Some examples of plane figures are circles, squares, rectangles, and triangles.

The word **geometry** means "earth measure" (*geo:* earth, *metry:* measure). The three-dimensional, or solid, objects all around us are made up of plane figures: a cube is made up of squares; a sphere, or ball, is a circle that has been spun around its diameter. In lesson 23 you will review several kinds of measures of plane figures, such as perimeter and area, and you will learn how to measure the inside space, or volume of solid objects.

geometry
the study of points, lines, angles, and figures in space

In lesson 24 you will learn about angles. Many of the plane figures are made from angles. One of the most important figures is the **triangle.** It is formed from three lines and contains three angles. You will learn more about triangles and other plane figures in lesson 25.

triangle
a figure with 3 sides and 3 angles

Lesson 26 will show you how to tell when two figures have the same shape and size. Of course, you can always use a tape measure, but many things in our world are simply too large to measure in the mechanical way. In this lesson, you will learn how to use an object's shadow to find its height. You will also learn to interpret maps and **scale** drawings.

scale
a relationship between two sets of measurements

The early Egyptians are given credit for observing the first geometric relationships. In lesson 27 you will study one of their most famous discoveries, the **Pythagorean Theorem.** This theorem shows the relationship of the three sides of a right triangle. The discovery grew from the needs of the early Egyptians to survey their lands each year after the annual floods on the Nile River. While you may not need to survey land, an understanding of geometry is important and of practical application to modern life.

Pythagorean Theorem
in a right triangle, the square of the hypotenuse equals the sum of the squares of the other two sides

Many things in the world have irregular shapes. For example, the floors of some rooms are not square or rectangular but are combinations of those two shapes. In lesson 28, you will learn how to find the area and volume of **irregular figures.**

irregular figures
figures that are made up of several shapes

Sometimes you know how the area of two different shapes are related. In lesson 29, you will learn to use **area relationships,** area formulas, and what you know about solving equations to solve problems.

area relationships
relationships between areas of figures with different shapes

SEE ALSO: Steck-Vaughn GED Mathematics Exercise Book, Unit 3: Geometry

Volume

In lesson 17 you learned about three kinds of geometric measures: perimeter, circumference, and area. In this lesson we will review these and add a fourth—**volume.**

Perimeter

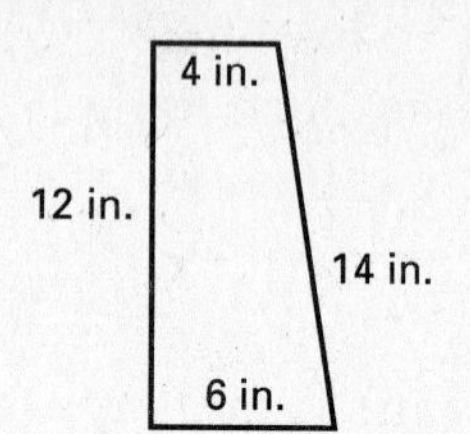

Perimeter is the measure of the distance around the edge of any flat object or figure. You can find the perimeter of an object by adding the measures of its sides.

Perimeter = 4 + 14 + 6 + 12 = 36 inches

These formulas are used to find the perimeter (P).

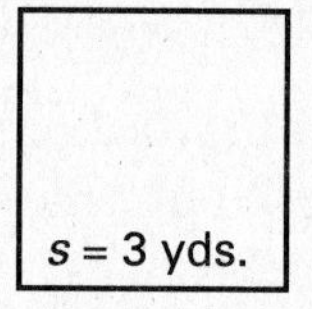

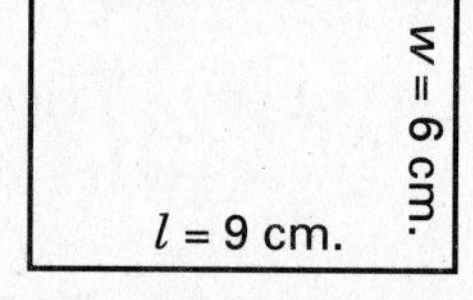

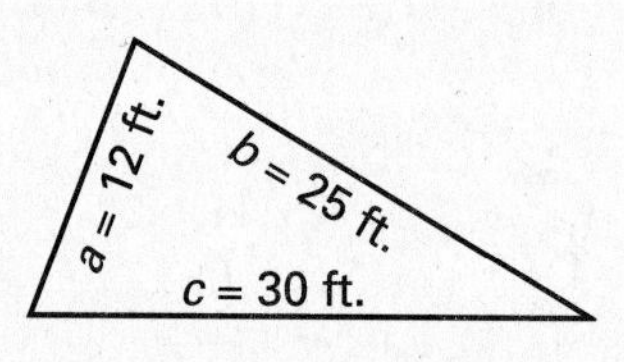

Square
$\boldsymbol{P = 4s}$
$= 4(3)$
$= 12$ yds.

Rectangle
$\boldsymbol{P = 2l + 2w}$
$= 2(9) + 2(6)$
$= 18 + 12$
$= 30$ cm.

Triangle
$\boldsymbol{P = a + b + c}$
$= 12 + 25 + 30$
$= 67$ ft.

Circumference

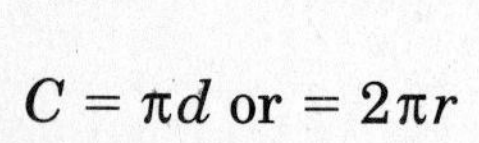

The perimeter of a circle is called the **circumference.**

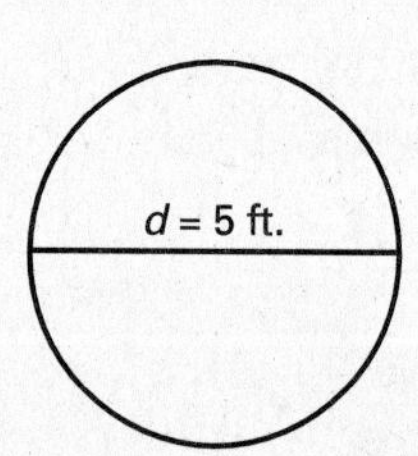

$\boldsymbol{C = \pi d}$
$= 3.14(5)$
$= 15.7$ ft.

(Remember that π, or pi, is a constant equal to approximately 3.14.)

Area

Area measures the surface inside a flat object or figure. Area is measured in square units. These formulas are used to find the area (A) of some figures.

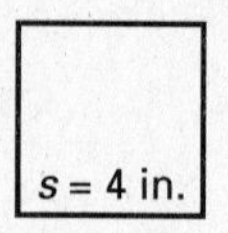

Square
$\boldsymbol{A = s^2}$
$= 4^2$
$= 16$ sq. in.

Rectangle
$\boldsymbol{A = lw}$
$= 22(11)$
$= 242$ sq. m.

Parallelogram
$\boldsymbol{A = bh}$
$= (10)7.5$
$= 75$ sq. ft.

Circle
$\boldsymbol{A = \pi r^2}$
$= 3.14(3^2)$
$= 28.26$ sq. in.

Triangle
$\boldsymbol{A = \frac{1}{2}bh}$
$= \frac{1}{2}(8)(10)$
$= 40$ sq. in.

Volume measures the amount of space inside a three-dimensional object. It is measured in cubic units—cubic inches, cubic feet, cubic yards. Each of these units is a cube made up of square sides, all the same size. To visualize volume, think of space filled with neat layers of ice cubes.

Rectangular Solid
all square corners
$V = A \times h$
$V = l \times w \times h$

Three of the most common solid shapes are the **rectangular solid** (cereal box), the **cube** (ice cube), and the **cylinder** (soup can). Formulas for finding their volumes are the same.

Volume = (area of the base) × height

The area of the base may be given as part of the data, or it may have to be calculated.

Example: What is the volume of a rectangular solid that is 14 inches long, 8 inches wide, and 4 inches high?

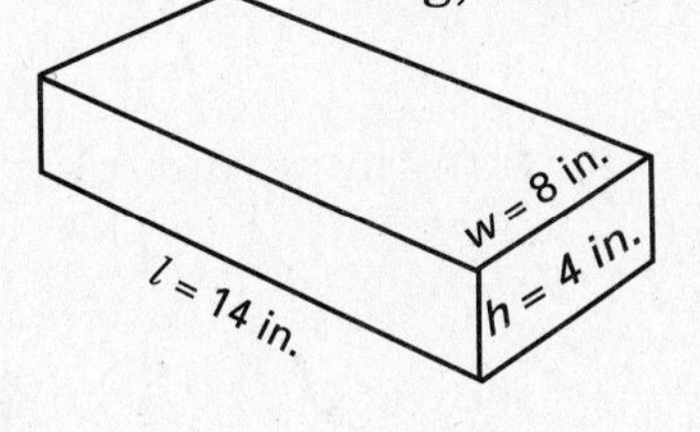

First find the area of the base:
$A = l \times w = 14 \times 8 = 112$ square inches

Now substitute this value into the volume formula:
V = (area of the base) × height
= 112 square inches × 4 inches
= 448 cubic inches

From this example you can see that for a rectangular solid:
Volume = length × width × height

Cube
rectangular solid
all sides square
$V = s \times s \times s$
or $V = s^3$

A **cube** is a special rectangular solid, so the same formula holds.

Example: Find the volume of a cube that is 3 feet along each side.

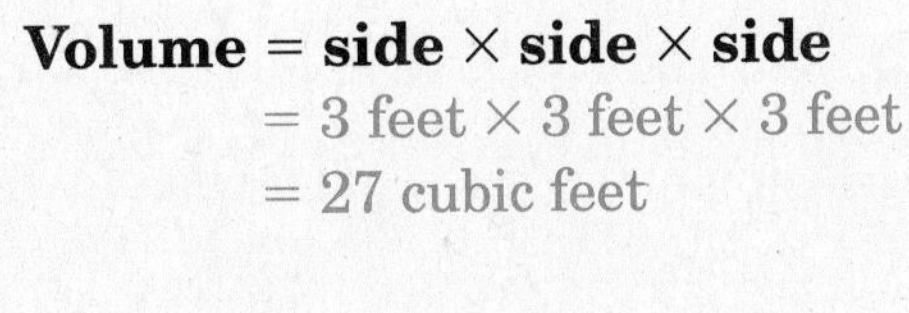

Volume = side × side × side
= 3 feet × 3 feet × 3 feet
= 27 cubic feet

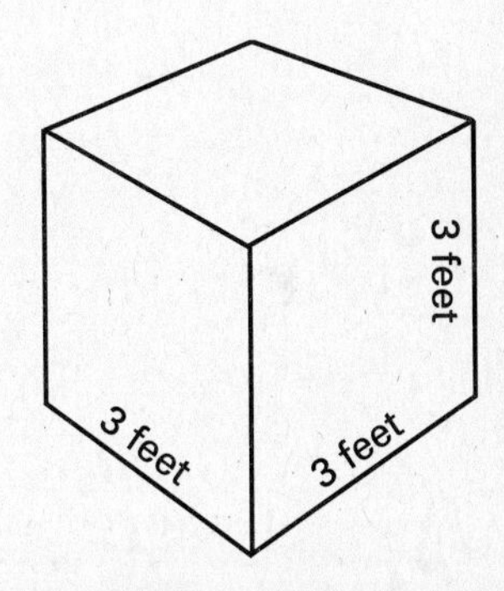

From this example, you can see that the volume of a cube is equal to the length of one side multiplied times itself three times, or "cubed."

Cylinder
circular base
straight sides
V = **Area of base** $\times$ h
or
$V = \pi \times r^2 \times h$

Example: Find the volume of a cylinder if the area of the base is 10 square inches and the cylinder is 14 inches high.

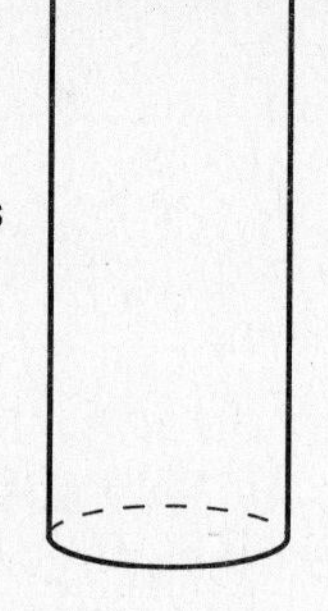

$V = A \times h$
$= 10 \times 14$
$= 140$ cubic inches

Example: Find the approximate volume of a cylinder if the radius of the base is 2 feet and the height of the cylinder is 10 feet.

First find the area of the base.

$A = \pi r^2$
Use 3.14 for π. $= 3.14 \times 4$
$= 12.56$ square feet

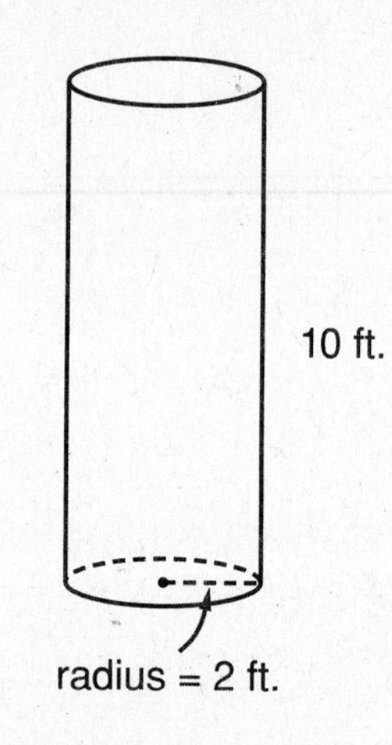

Use the area in the formula for volume.

$V = A \times h$
$= 12.56 \times 10$
$= 125.6$ cubic feet which rounds to 126 cubic feet

Two other shapes, **pyramids** and **cones,** have volume formulas that are the same. The volume of each solid is $\frac{1}{3}$ as great as its corresponding rectangular solid or cylinder.

Pyramid
base has equal sides

$$V = \frac{1}{3} \times A \text{ (Area of base)} \times h \text{ (height)}$$

Example: Find the volume of a pyramid if the Area of the base is 12 square feet and the height is 4 feet.

$V = \frac{1}{3} \times \text{Area of base} \times h$
$= \frac{1}{3} \times 12 \times 4$
$= 16$ cubic feet

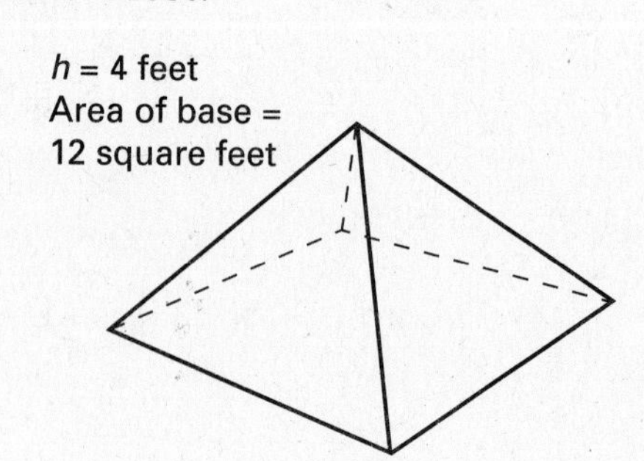

Example: The base of a pyramid is a triangle with height 12 inches and base 4 inches. The height of the pyramid is 16 inches. What is the volume of the pyramid?

First find the area of the base of the pyramid.

$A = \frac{1}{2}bh$

$= \frac{1}{2}(12)(4)$

$= 24$ square inches

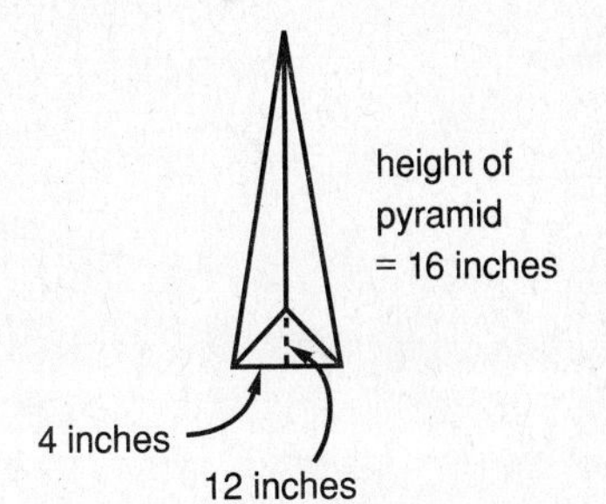

Use the area in the formula for volume.

$V = \frac{1}{3} \times \text{Area of base} \times h$

$= \frac{1}{3} \times 24 \times 16$

$= 128$ cubic inches

Cone
circular base
sides meet at a point
$\boldsymbol{V = \frac{1}{3} \times A \times h}$ **or**
$\boldsymbol{V = \frac{1}{3} \times \pi \times r^2 \times h}$

Example: Find the volume of a cone if the area of the base is 4 square inches and the height is 9 inches.

$V = \frac{1}{3} \times \text{Area of base} \times h$

$= \frac{1}{3} \times 4 \times 9$

$= 12$ cubic inches

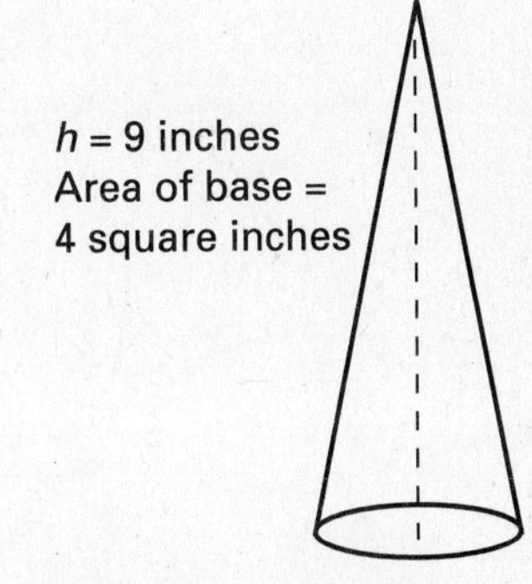

Note: The units of measurement you use in formulas must be the same. So *before* you calculate your answer, remember to *convert to like units.*

Lesson 23

Directions: For items 1 to 6, name the figure and find its volume.

1.

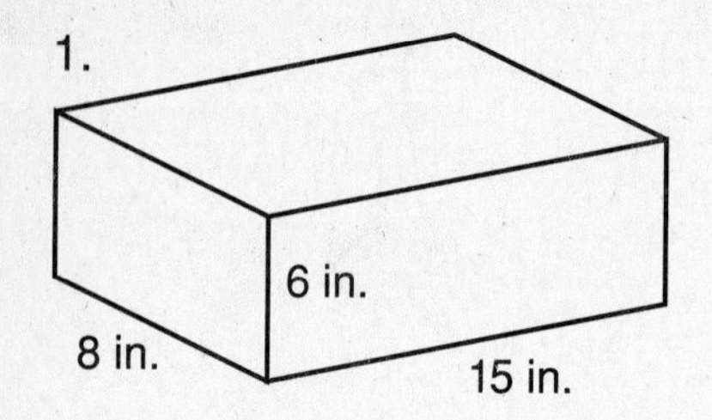

2.

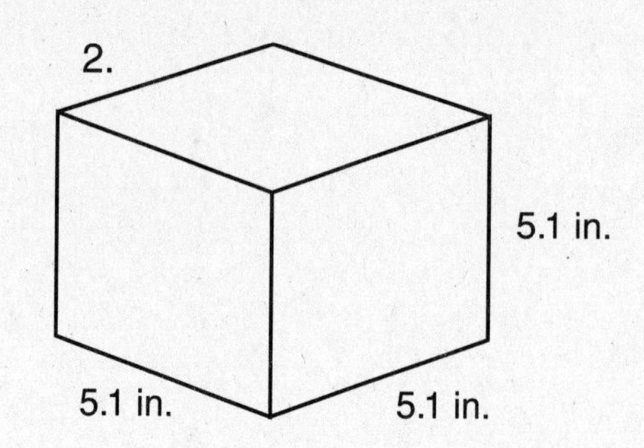

3.

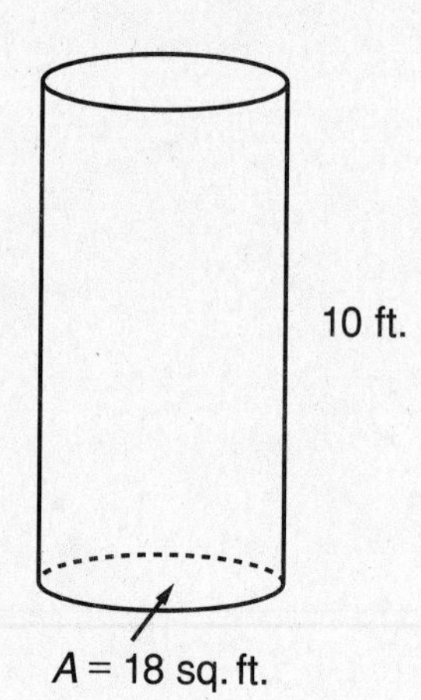

4.

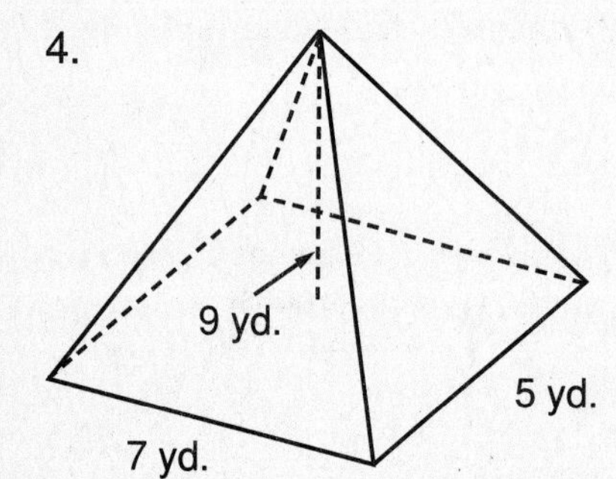

5.

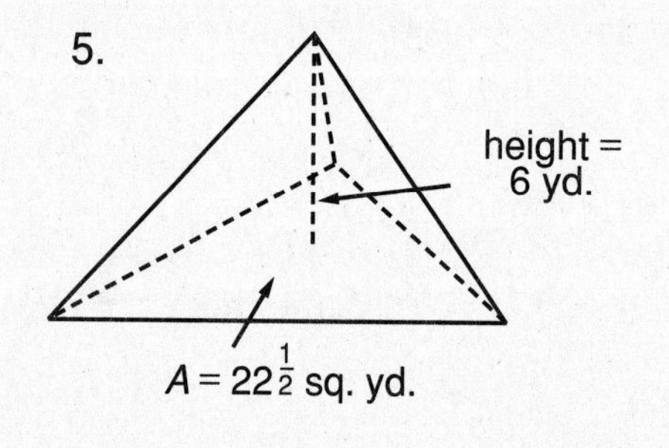

6.

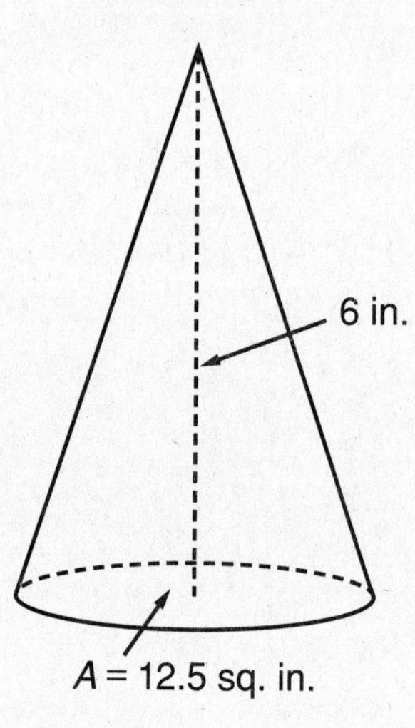

Directions: For items 7 to 9, find the volume of each figure. Round your answers to the nearest whole number.

7.

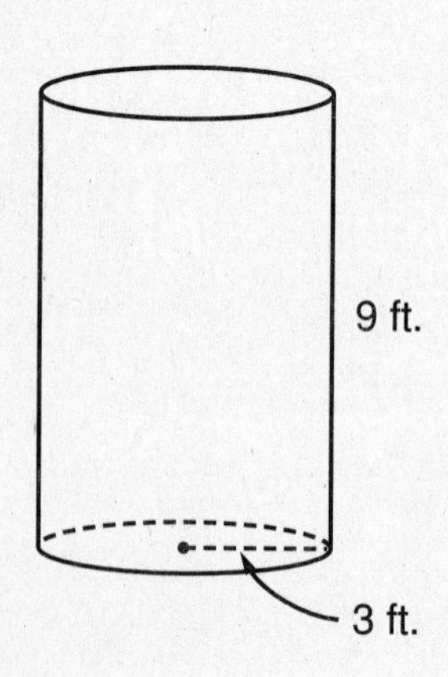

8.

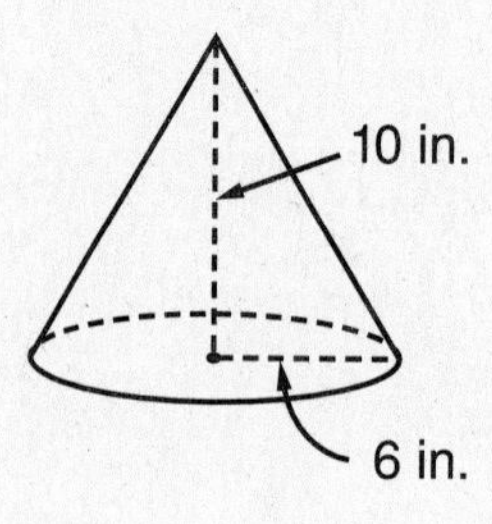

9.

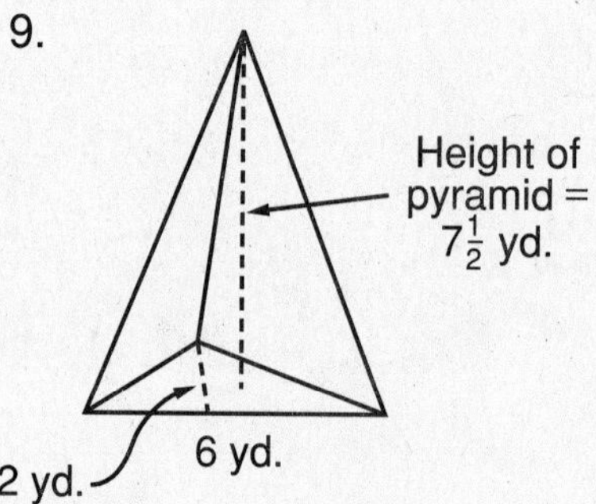

Directions: Choose the best answer to each item.

10. What is the volume in cubic inches of a rectangular solid with these measurements: length = 15 inches, width = 24 inches, and height = 4 inches?

(1) 1,280
(2) 1,410
(3) 1,440
(4) 1,520
(5) 1,640

11. What is the volume in cubic yards of a cube that measures 2 yards on one side?

(1) 6
(2) 7
(3) 7.5
(4) 8
(5) 9

12. The city is going to build a swimming pool. The hole needs to be 100 feet long by 25 feet wide. The depth needs to be 5 feet. How many cubic feet of dirt will have to be removed to dig the hole?

(1) 1,250
(2) 2,500
(3) 5,000
(4) 12,500
(5) 50,000

13. A contractor building a group of houses has a large amount of trash to be removed. The trash containers are 12 feet long, 8 feet wide, and 6 feet high. How many cubic feet of trash can each container hold?

(1) 48
(2) 72
(3) 96
(4) 192
(5) 576

14. A cylindrical tank has a base area of 9.1 square feet and a height of 6 feet. What is the volume of the tank in cubic feet?

(1) 5.46
(2) 27.3
(3) 54.6
(4) 273
(5) 546

15. What is the volume in cubic centimeters of a cone with height 8 centimeters and base area 12.6 square centimeters?

(1) 3.36
(2) 10.08
(3) 33.6
(4) 100.8
(5) Not enough information is given.

16. The base of a pyramid is a triangle with height 3 feet and base 6 feet. The height of the pyramid is $12\frac{1}{2}$ feet. What is the volume of the pyramid in cubic feet?

(1) 18
(2) $37\frac{1}{2}$
(3) 75
(4) $112\frac{1}{2}$
(5) 117

17. The base of a cone is a circle with a radius of 2 inches. The height of the cone is 6 inches. What is the approximate volume of the cone in cubic inches?

(1) 12
(2) 19
(3) 25
(4) 50
(5) 75

Answers are on page 394.

Strategies for Solving Word Problems

Converting Measurements

When solving problems you must always keep in mind what units are being used. Converting measurements involves using ratios and rates correctly to change from one unit to another.

Example: What is the volume of the rectangular solid?

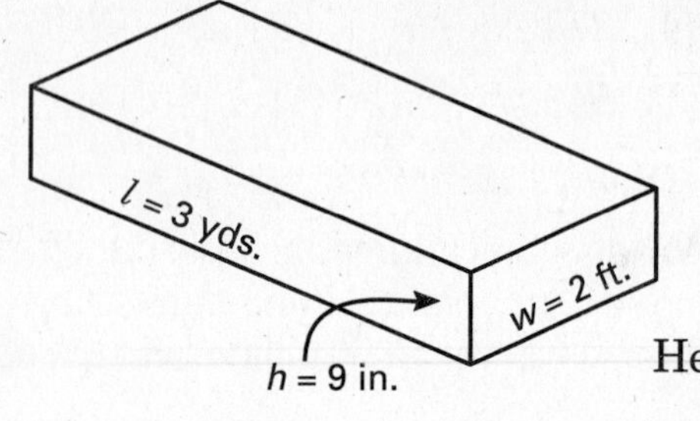

The formula you will need to solve the problem is $V = lwh$. But before you multiply you will need to convert the measurements given in the figure to the same unit of measurement.

Here the problem is solved by converting to inches.

Convert 3 yards to inches $\quad 3\,\cancel{\text{yds.}} \times \frac{36 \text{ in.}}{1\,\cancel{\text{yd.}}} = 108 \text{ in.}$

Convert 2 feet to inches $\quad 2\,\cancel{\text{ft.}} \times \frac{12 \text{ in.}}{1\,\cancel{\text{ft.}}} = 24 \text{ in.}$

$$\begin{aligned} \text{Volume} &= \text{length} \times \text{width} \times \text{height} \\ &= 108 \text{ in.} \times 24 \text{ in.} \times 9 \text{ in.} \\ &= \mathbf{23{,}328} \text{ cubic inches} \end{aligned}$$

This example can also be solved in terms of feet or yards.

Convert to feet:

$$\begin{aligned} \text{Volume} &= \text{length} \times \text{width} \times \text{height} \\ &= 3 \text{ yds.} \times 2 \text{ ft.} \times 9 \text{ in.} \\ &= 9 \text{ ft.} \times 2 \text{ ft.} \times \tfrac{3}{4} \text{ ft.} \\ &= 13\tfrac{1}{2} \text{ cubic feet} \end{aligned}$$

Convert to yards:

$$\begin{aligned} \text{Volume} &= \text{length} \times \text{width} \times \text{height} \\ &= 3 \text{ yds.} \times 2 \text{ ft.} \times 9 \text{ in.} \\ &= 3 \text{ yds.} \times \tfrac{2}{3} \text{ yd.} \times \tfrac{1}{4} \text{ yd.} \\ &= \tfrac{1}{2} \text{ cubic yard} \end{aligned}$$

Each answer represents the same quantity:

$$23{,}328 \text{ cubic inches} = 13\tfrac{1}{2} \text{ cubic feet} = \tfrac{1}{2} \text{ cubic yard}$$

How do you decide which unit to choose? Generally, it is easiest to convert to the smallest unit of measure so that you will not have to work with fractions. Another factor to consider is the answer options. Always convert to the unit of measure given in the answer options.

Solving Word Problems

Directions: Choose the best answer to each item.

Items 1 and 2 refer to the following figure.

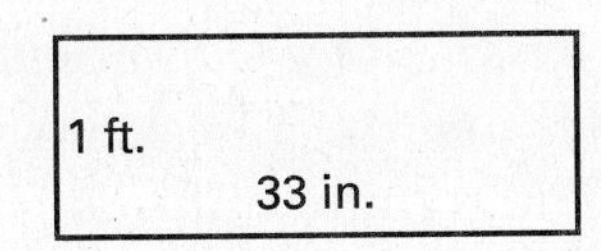

1. What is the perimeter in feet of the rectangle shown in the figure?

   (1) $6\frac{1}{2}$
   (2) $7\frac{1}{2}$
   (3) 8
   (4) $8\frac{1}{4}$
   (5) $8\frac{1}{2}$

2. What is the area of the rectangle in the figure in square inches?

   (1) 364
   (2) 386
   (3) 396
   (4) 412
   (5) 432

3. What is the volume of the cylinder shown to the nearest cubic inch?

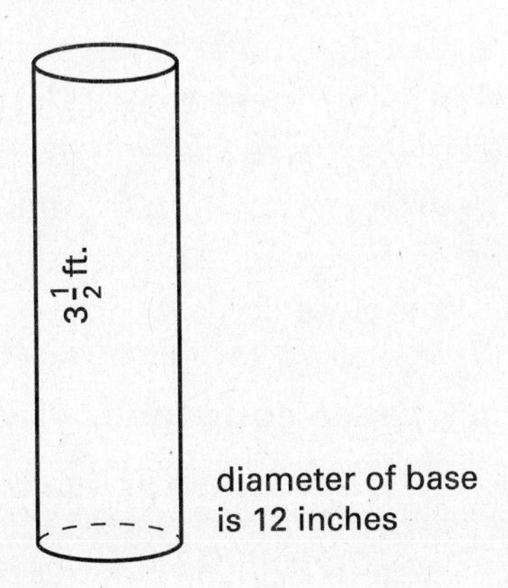

   (1) 132
   (2) 2,862
   (3) 4,224
   (4) 4,748
   (5) 4,918

4. What is the approximate volume in cubic feet of a box with a length of $2\frac{1}{2}$ ft., a width of 1 ft. 6 in., and a height of 1 ft. 9 in.?

   (1) between 3 and 4
   (2) between 4 and 5
   (3) between 5 and 6
   (4) between 6 and 7
   (5) between 7 and 8

5. What is the area in square centimeters of the triangle in the figure? (Hint: 1 meter = 100 centimeters.)

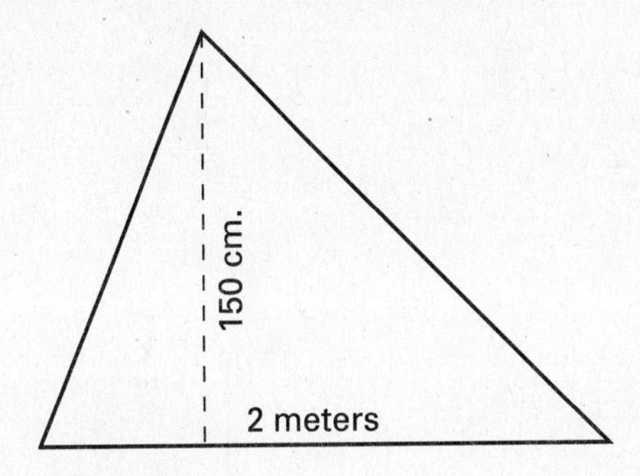

   (1) 12,000
   (2) 12,500
   (3) 15,000
   (4) 20,000
   (5) 22,500

6. A cone has a height of 18 in. and a base with an area of 1 ft. 4 in. What is the volume of the cone in cubic inches?

   (1) 48
   (2) 54
   (3) 96
   (4) 144
   (5) 288

Answers are on page 394.

Lesson 24 Angles

Angles

Ray

Vertex

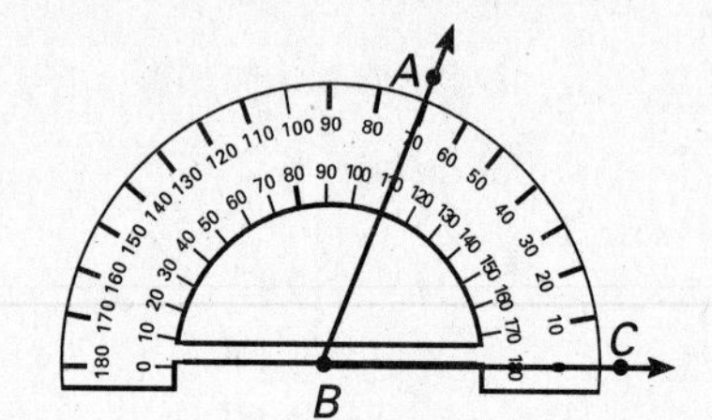

An **angle** is the space between a pair of lines, called **rays,** which extend away from a common point, the **vertex.** Angles are named by the three letters which label the rays and the vertex or the letter of the vertex alone. The sign for angle is ∠. Thus, at left, point B is the vertex of $\angle ABC$ (or $\angle B$.)

Degree

Angles are measured in **degrees** (1°) using a protractor. When ray *BC* is placed along the bottom of the protractor, the angle measure is read along the scale that starts at 0. The measure of the angle *ABC* is 70°, *not* 110°. Write $m\angle ABC = 70°$.

Right
Acute
Obtuse
Straight
Reflex

(1) Angles are identified by the measure of their angles.

A **right angle** measures 90°.

An **acute angle** measures less than 90°.

An **obtuse angle** measures more than 90° but less than 180°.

A **straight angle** measures 180°.

A **reflex angle** measures more than 180° but less than 360°.

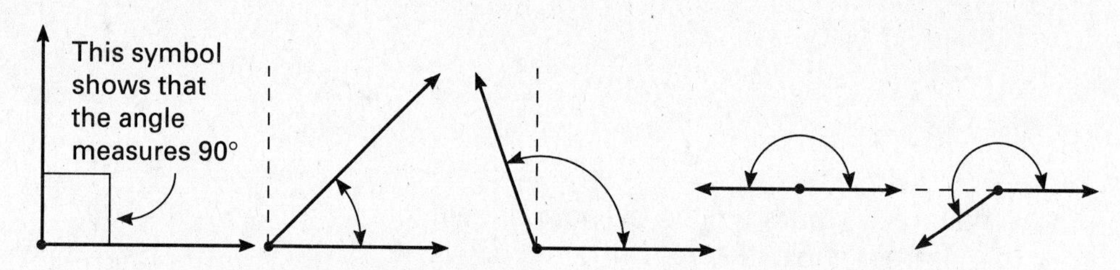

Notice that the right angle and the straight angle always have measures that are exact, 90° and 180°. The measures of the other angles are within a range. For example, an acute angle could be 1° or 89° because both of these measures are greater than 0° and less than 90°. A reflex angle could be 181° or 359° because both are greater than 180° and less than 360° (which is a complete circle).

Congruent

Angles that have equal measures are called **congruent ("equal") angles.**

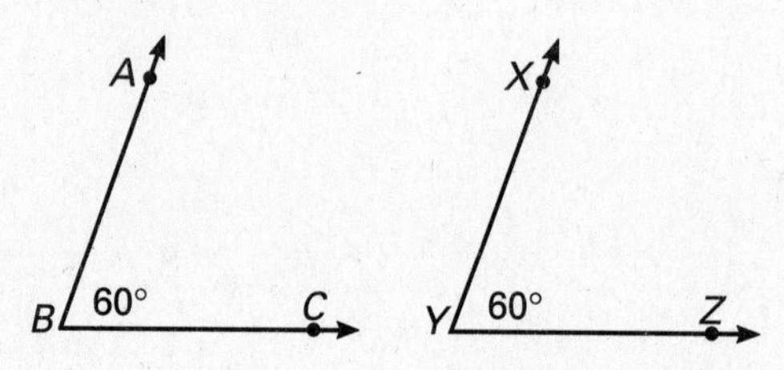

Angle *ABC* is congruent to angle *XYZ*.

$\angle ABC \cong \angle XYZ$

This symbol means "is congruent to".

Intersect
Vertical angles

(2) Angles have special relationships because of their location. When two lines intersect or cross, the angles that are across from each other are called **opposite angles** or **vertical angles**. Each pair of opposite angles is **congruent**.

$\angle 5 \cong \angle 6$ $\quad$ $\angle 7 \cong \angle 8$
$m\angle 5 = m\angle 6$ $\quad$ $m\angle 7 = m\angle 8$

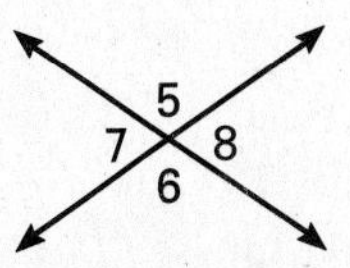

Perpendicular

If two lines form congruent adjacent angles, that is, 2 right angles, the lines are **perpendicular.** We write $l \perp m$.

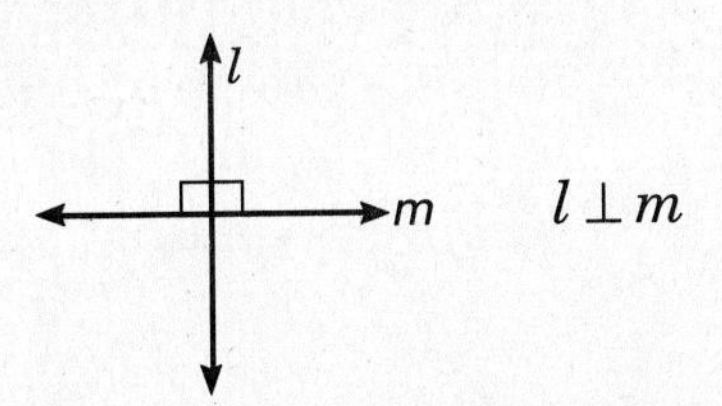

Adjacent
Non-adjacent

There are several other relationships between angles because of their location. **Adjacent angles** have a common vertex and a common ray. $\angle 1$ and $\angle 2$ are adjacent angles.

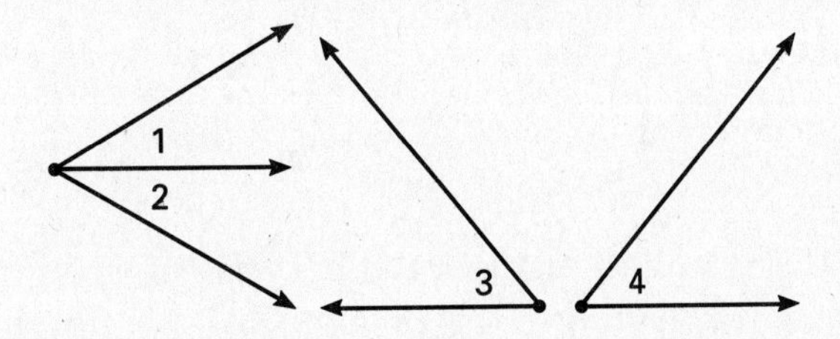

$\angle 3$ and $\angle 4$ are examples of **non-adjacent angles.**

Parallel
Transversal

Parallel lines are two lines on the same plane that do not intersect. The symbol used is $\|$. A line that crosses two or more parallel lines is called a **transversal.**

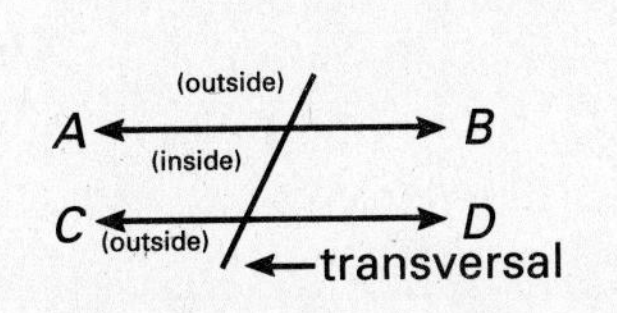

Line AB is parallel to line CD.
$\overleftrightarrow{AB} \parallel \overleftrightarrow{CD}$

Special pairs of angles are formed when two parallel lines are cut by a transversal.

Corresponding

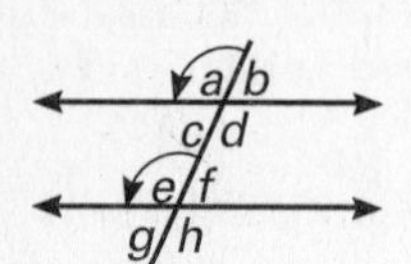

- **Corresponding angles** are those angles that are on the same side of the transversal and are either both above or both below the two parallel lines. They are always *equal* in measure.

$\angle a$ and $\angle e$ $\angle c$ and $\angle g$
$\angle b$ and $\angle f$ $\angle d$ and $\angle h$

Alternate exterior

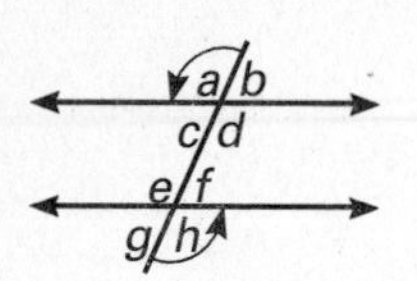

- **Alternate exterior angles** are always outside the parallel lines. They are on opposite sides of the transversal. One is above the top parallel line. The other is below the bottom parallel line. They are always *equal* in measure.

$\angle a$ and $\angle h$ $\angle b$ and $\angle g$

Alternate interior

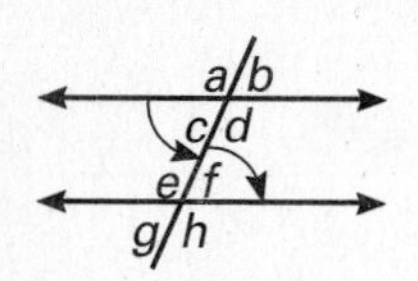

- **Alternate interior angles** are always inside the parallel lines. They are on opposite sides of the transversal. One is below the top parallel line. The other is above the bottom parallel line. They are always *equal* in measure.

$\angle c$ and $\angle f$ $\angle d$ and $\angle e$

Complementary

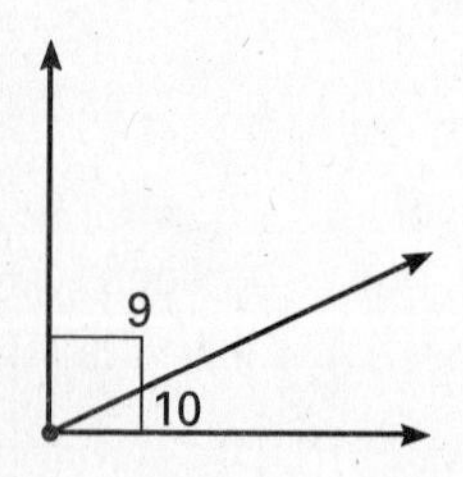

(3) Angles are also named because of the sum of their measures.

- If the sum of the measures of two angles is 90°, they are called **complementary angles.**

Example: If the measure of $\angle 10$ is 25°, what is the measure of $\angle 9$?

$m\angle 9 + m\angle 10 = 90°$
$m\angle 9 + 25° = 90°$
$m\angle 9 = 90° - 25° = 65°$
The measure of $\angle 9$ is **65°.**

Supplementary

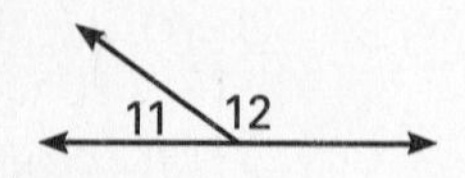

- If the sum of the measures of two angles is 180°, they are called **supplementary angles.**

Example: If the measure of $\angle 11$ is 35°, what is the measure of $\angle 12$?

$m\angle 11 + m\angle 12 = 180°$
$35° + m\angle 12 = 180°$
$m\angle 12 = 180° - 35° = 145°$
The measure of $\angle 12$ is **145°.**

Lesson 24

Directions: Label each angle as acute, obtuse, right or straight.

1. 150°
2. 35°
3. 90°
4. 180°
5. 7°
6. 95°
7. 175°
8. 45°

Directions: Answer each item.

Items 9 to 17 refer to the figure at the right.

9. Name an angle adjacent to $\angle 3$.
10. Name the angle that is the vertical angle to $\angle 2$.
11. Name the angle that is the vertical angle to $\angle 6$.
12. Name the angle that is the vertical angle to $\angle 1$.
13. Name two angles that are adjacent to $\angle 1$.
14. Name two angles that are adjacent to $\angle 5$.
15. Name the angle congruent to $\angle 5$.
16. Name the angle congruent to $\angle 4$.
17. Name the angle congruent to $\angle 3$.

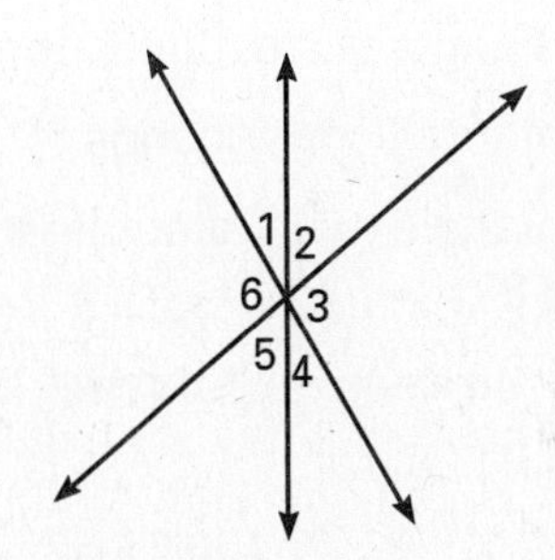

Items 18 to 21 refer to the figure at the right. $\angle AXB$ and $\angle BXC$ are complementary.

18. What is $m\angle BXC$?
19. What is $m\angle DXC$?
20. What is $m\angle BXD$?
21. Which angle forms a supplementary angle with $\angle DXC$?

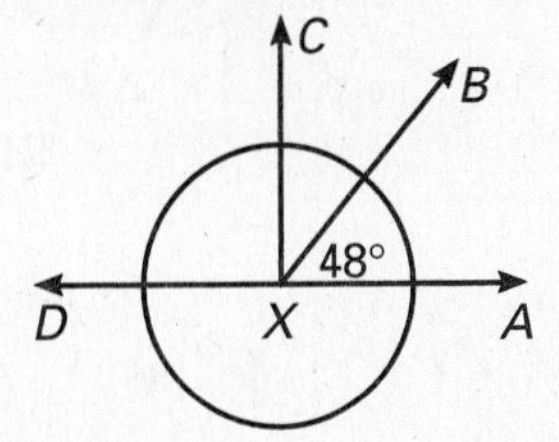

Items 22 to 25 refer to the figure at the right. $\angle ZXY$ and $\angle YXQ$ are complementary.

22. What is $m\angle ZXY$?
23. What is $m\angle ZXR$?
24. What is $m\angle QXR$?
25. What is the measure of the reflex angle marked by the curved arrow?

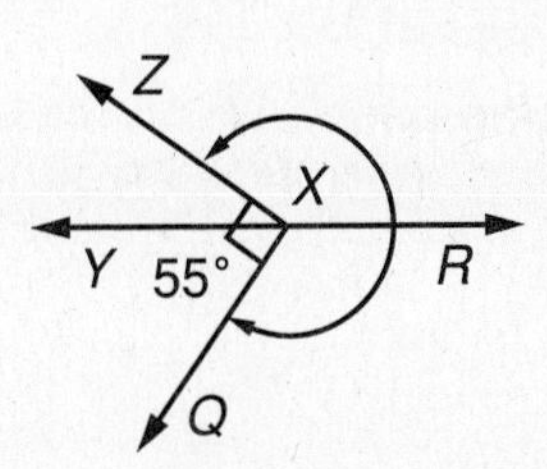

Items 26 to 35 refer to the following figure.

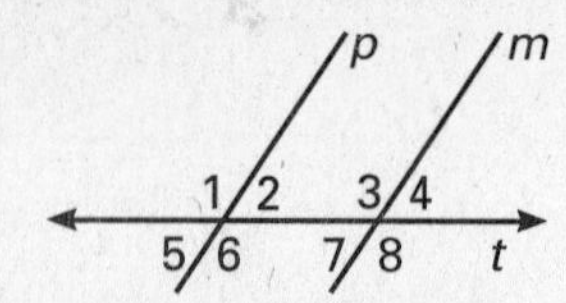

26. Name the interior angles.
27. Name the interior angle on the same side of the transversal as $\angle 6$.
28. Name the exterior angles.
29. Name the exterior angle on the same side of the transversal as $\angle 4$.
30. Which angle corresponds to $\angle 1$?
31. Which angle corresponds to $\angle 7$?
32. Which angle corresponds to $\angle 8$?
33. Which angle corresponds to $\angle 2$?
34. Which angle is an alternate interior angle with $\angle 2$?
35. Which angle is an alternate exterior angle with $\angle 5$?
36. How many acute angles are there in this figure?

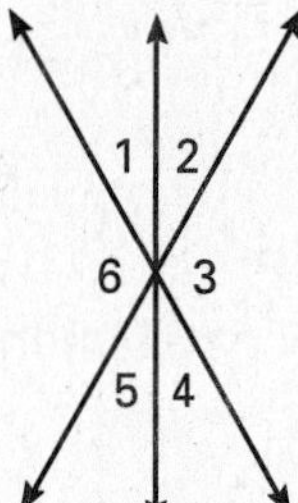

    (1) 2
    (2) 4
    (3) 5
    (4) 6
    (5) 8

37. The measure of $\angle A$ is 28°. The measure of $\angle B$ is 62°. Which of the following is true?

    (1) $\angle A$ is complementary to $\angle B$.
    (2) $\angle A$ and $\angle B$ are congruent.
    (3) $\angle A$ and $\angle B$ are obtuse angles.
    (4) $\angle A$ is adjacent to $\angle B$.
    (5) $\angle B$ is adjacent to $\angle A$.

38. $\angle M$ and $\angle R$ are complementary angles. The measure of $\angle M$ is 40°. What is the measure of $\angle R$?

    (1) 40°
    (2) 50°
    (3) 90°
    (4) 140°
    (5) 180°

Items 39 to 41 refer to the following figure.

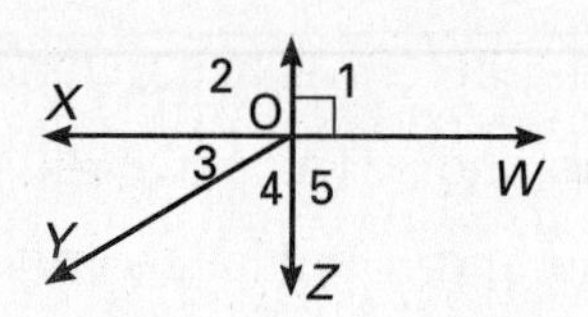

39. An angle that is supplementary, but not adjacent to, $\angle 2$ is

    (1) $\angle 1$
    (2) $\angle 5$
    (3) $\angle XOZ$
    (4) $\angle WOY$
    (5) $\angle WOX$

40. The measure of $\angle 3$ is 25°. What is the measure of $\angle WOY$?

    (1) 65°
    (2) 115°
    (3) 135°
    (4) 155°
    (5) 165°

41. An angle that is supplementary to $\angle XOZ$ must also be

    (1) an acute angle
    (2) a right angle
    (3) an obtuse angle
    (4) a vertical angle
    (5) congruent to $\angle 3$

42. Line A is parallel to line B. Which statement is true?

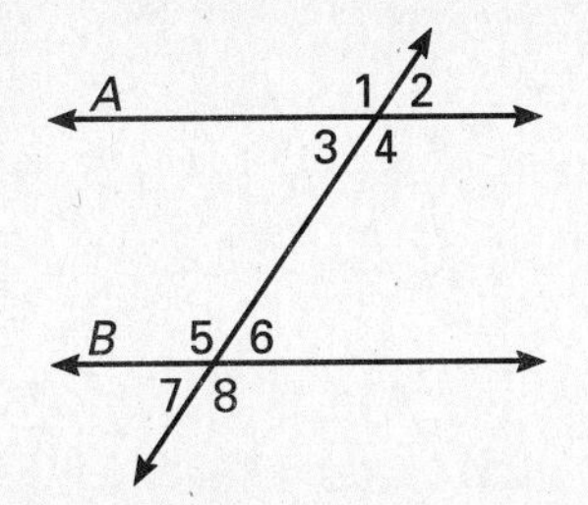

(1) $\angle 1 = \angle 7$

(2) $m\angle 2 + m\angle 6 = 90°$

(3) $\angle 5$ is complementary to $\angle B$.

(4) $m\angle 6 - m\angle 7 = 180°$

(5) $m\angle 1 + m\angle 7 = 180°$

43. $\angle Q$ is congruent to its vertical angle, $\angle R$. Which of the following must be true?

(1) $\angle Q$ and $\angle R$ are complementary angles.

(2) $\angle Q$ and $\angle R$ are both acute angles.

(3) $\angle Q$ and $\angle R$ are both obtuse angles.

(4) $\angle Q$ and $\angle R$ are supplementary angles.

(5) $\angle Q$ and $\angle R$ have the same degree measure.

44. The measure of $\angle 4 = 100°$. Which statement is true?

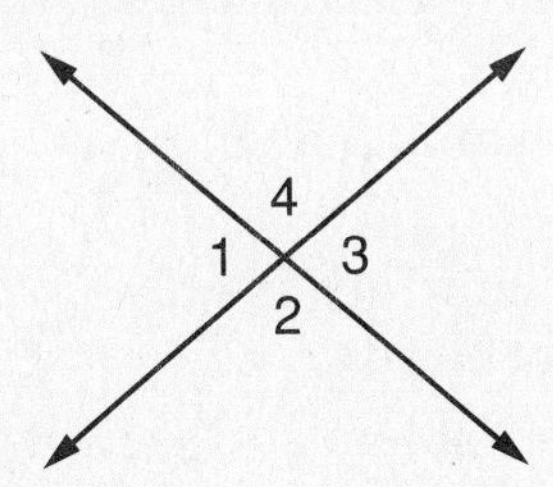

(1) $\angle 4$ and $\angle 2$ are complementary.

(2) $m\angle 2 = 100°$

(3) $m\angle 1 = 100°$

(4) $m\angle 1 + m\angle 3 = 100°$

(5) $m\angle 4 + m\angle 2 = 180°$

45. Line A is parallel to line B. Which statement is true?

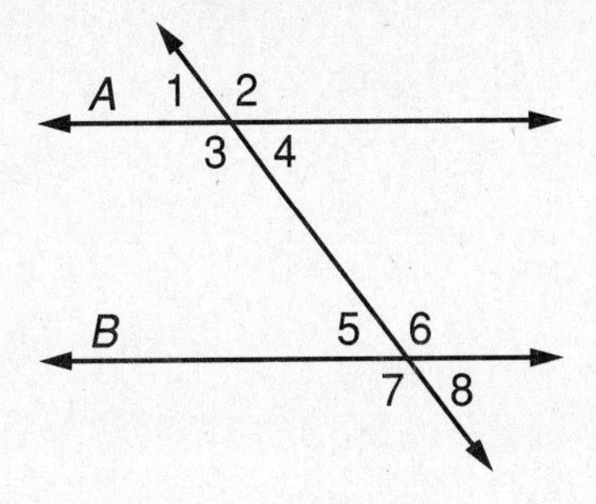

(1) $m\angle 2 + m\angle 5 = 180°$

(2) $\angle 1$ is complementary to $\angle 2$.

(3) $\angle 6$ is supplementary to $\angle 7$.

(4) $\angle 2$ and $\angle 6$ are vertical angles.

(5) $\angle 1$ and $\angle 6$ are corresponding angles.

Items 46 and 47 refer to the following figure.

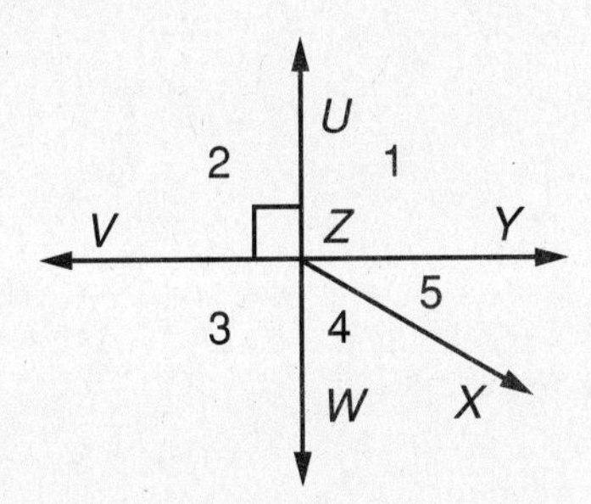

46. The measure of $\angle 4$ is 60°. What is the measure of $\angle UZX$?

(1) 30°

(2) 100°

(3) 120°

(4) 150°

(5) 180°

47. What is the measurement of $\angle XZY$?

(1) 30°

(2) 100°

(3) 120°

(4) 150°

(5) 180°

Answers are on page 395.

Strategies for Solving Word Problems

Draw a Diagram

Drawing a diagram is a strategy that helps you see how the information in a problem is related. Diagrams are especially useful in geometry.

Example: Kelly plans to put a low fence around a rectangular garden. The garden plot measures 4 feet in width. The length is 12 inches less than three times the width. How many feet of fencing will she need?

(1) 20
(2) 24
(3) 30
(4) 36
(5) 40

This problem contains a great deal of information. Because you need to picture the garden and juggle numbers at the same time, the problem may seem confusing. One strategy that will make your work easier is to draw a rough sketch of the information in the problem. Follow these steps:

Step 1: Draw the shape indicated in the problem. Remember that you might need to combine shapes in a particular way.

Step 2: Write the information given in the problem in the correct place. This will let you see what information you have, what is missing and what you can calculate from the information.

Step 3: Perform whatever operations are indicated to convert the information to a usable form.

Step 4: Identify what you are being asked to find.

Step 5: Solve the problem.

Apply these steps to the example above:

Step 1: Draw a rectangle to represent the garden.

Step 2: The problem states that the width is 4 feet. Write 4 feet next to both ends of the rectangle.

Step 3: Now find the length. The problem states that the length is 12 inches (which converts to 1 foot) less than 3 times the width.

$3 \times \text{width} - 12 \text{ in.} =$
$3 \times 4 \text{ ft.} \quad - 1 \text{ ft.} \quad = 11 \text{ ft.}$

The labeled rectangle should look like this.

4 ft. | 11 ft. / 11 ft. | 4 ft.

Step 4: You are asked to find the perimeter.

Step 5: Solve

$$P = 2l + 2w$$
$$= 2(11) + 2(4)$$
$$= 22 + 8$$
$$= 30 \text{ ft.}$$

The correct choice is **(3) 30.**

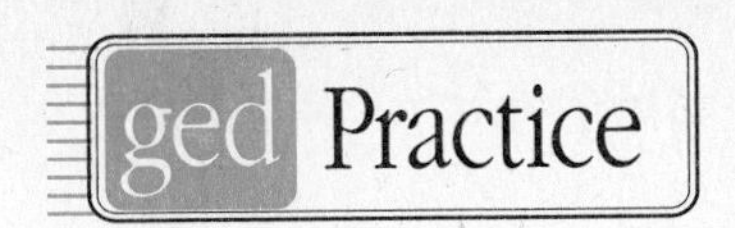

Solving Word Problems

<u>Directions</u>: Choose the <u>best answer</u> to each item.

1. A rectangular garden is 30 feet long and half as wide. A 3-foot diameter circular statue is placed in the center of the garden. What is the area to the nearest square foot of the garden space available for planting?

   (1) 413
   (2) 421
   (3) 443
   (4) 463
   (5) 503

2. A circular pool is surrounded by a 3-foot wide circular walkway. If the radius of the pool is 4 feet, what is the area of the walkway to the nearest square foot?

   (1) 80
   (2) 86
   (3) 88
   (4) 98
   (5) 104

3. A parallelogram is 6 feet on its short side and 15 feet on its long side. The height between the two long sides is $4\frac{1}{2}$ feet. What is its area in square feet?

   (1) 65
   (2) $67\frac{1}{2}$
   (3) $72\frac{1}{2}$
   (4) 75
   (5) $78\frac{1}{2}$

4. A large rectangular mural is 28 feet long. The width is 2 feet less than half the length. What is the distance in feet around the mural?

   (1) 74
   (2) 76
   (3) 80
   (4) 84
   (5) 336

5. A school playground is rectangular in shape. The entire playground measures 200 feet by 100 feet. A square blacktop area with a perimeter of 200 feet will be in the center. The rest of the playground will be grass. What is the area in square feet of the grassy portion of the playground?

   (1) 12,000
   (2) 15,500
   (3) 17,500
   (4) 19,000
   (5) 20,500

6. One side of the square base of a milk carton measures 4 inches. If the carton can be filled to a height of 8 inches, how many cubic inches can the carton hold?

   (1) 112
   (2) 128
   (3) 132
   (4) 138
   (5) 146

Answers are on page 396.

Triangles and Quadrilaterals

Triangles

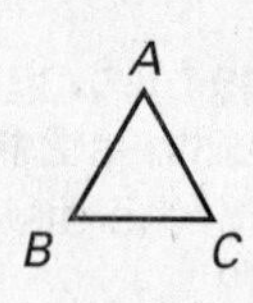

A geometric figure with three sides and three angles is called a **triangle.** A triangle is named by writing the vertices in any order. Triangles are identified in two ways: by the lengths of their sides and by the measure of their angles.

By the length of their sides: are any two sides equal, or congruent?

Note: The symbol ll is used to denote which sides are congruent. Sometimes the marks l or lll are also used.

Equilateral Triangle

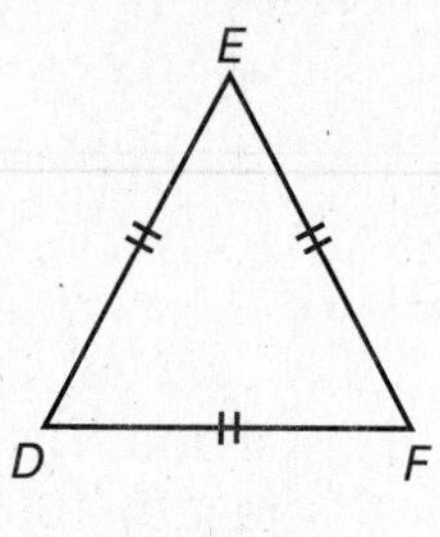

All sides are congruent.

Isosceles Triangle

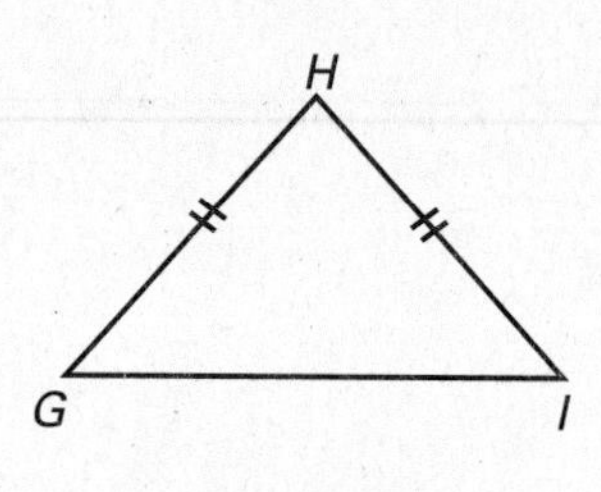

Two sides are congruent.

Scalene Triangle

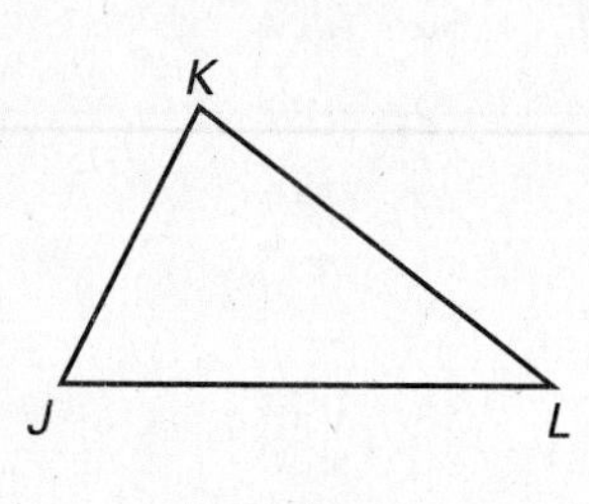

No sides are congruent.

A side is named by the two letters that form the ends of the side. In $\triangle JKL$, *JK, KL,* and *JL* are the three sides.

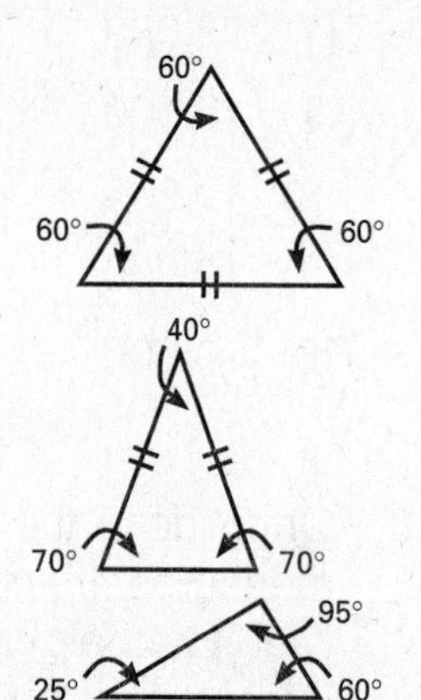

The angles in an equilateral triangle have equal measure. They all measure 60° $\left(\frac{180°}{3}\right)$.

An isosceles triangle has two angles of equal measure. They are opposite the two congruent sides.

A scalene triangle has no angles of equal measure. Remember it also has no congruent sides.

By the measure of their angles:

Right Triangle

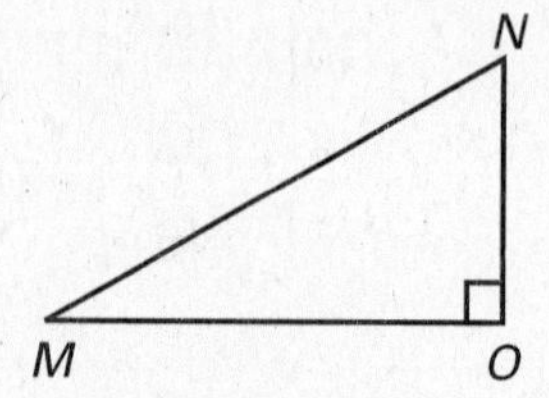

One angle is a right angle (90°).

Acute Triangle

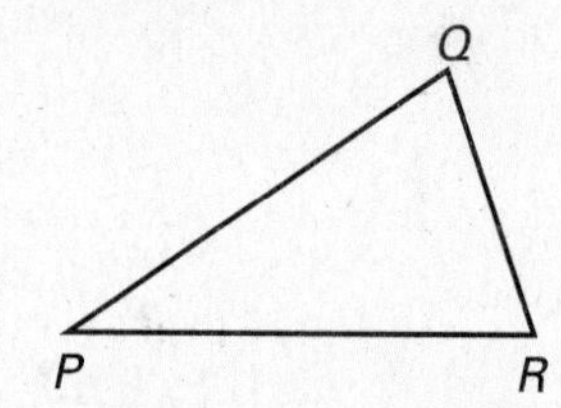

All three angles are acute angles (less than 90°).

Obtuse Triangle

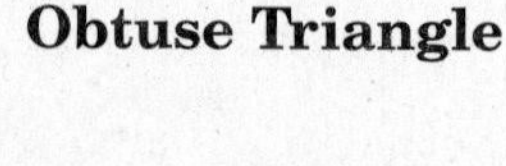
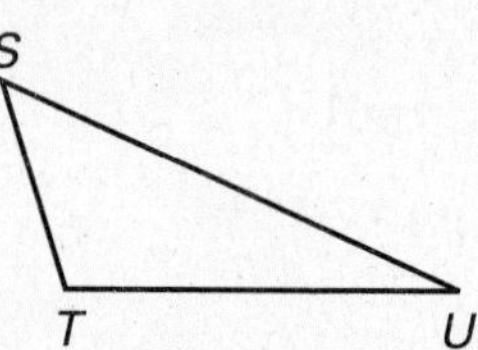

One angle is an obtuse angle (greater than 90°).

Recall that an angle is named by one letter, number, or the three letters that form it. In $\triangle MNO$, for example, one of the angles can be called $\angle O$ or $\angle MON$ or $\angle NOM$. Remember that the middle letter is the vertex point.

In any triangle:
$m\angle 1 + m\angle 2 + m\angle 3 = 180°$
In a right triangle:
$m\angle 1 = 90°$
$m\angle 2 + m\angle 3 = 90°$

In any triangle, the sum of the measures of the angles is 180°. To show this, cut out a triangle, tear off the corners, and arrange the corners to make a straight line. We can use this fact to find certain missing angle measures of a triangle.

Example: If the larger acute angle of a right triangle is twice the measure of the smaller acute angle, what is the measure of the smaller acute angle?

Draw a picture. Remember: a right triangle has a right angle (90°).

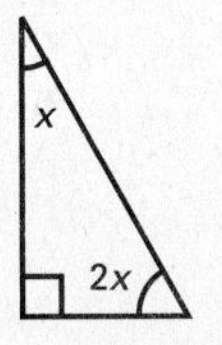

Let x = measure of the smaller acute angle
Then $2x$ = measure of the larger acute angle

$$x + 2x + 90° = 180°$$
$$3x = 90°$$
$$x = 30°$$

The smaller acute angle is **30°.**

Quadrilaterals

Any geometric figure with four sides is called a **quadrilateral.** A line segment drawn between the vertices of two non-adjacent sides is called a **diagonal** of the quadrilateral. The sum of the four angle measures of a quadrilateral is 360°. Quadrilaterals have special names.

Parallelogram

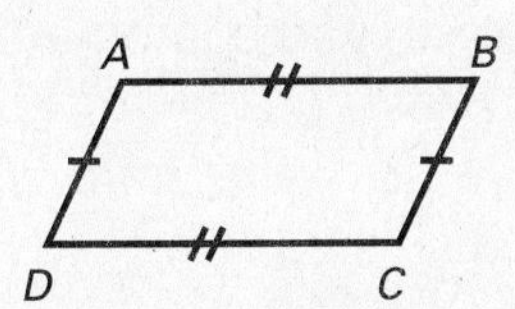

A quadrilateral with opposite sides parallel and congruent and opposite angles equal.

Rectangle

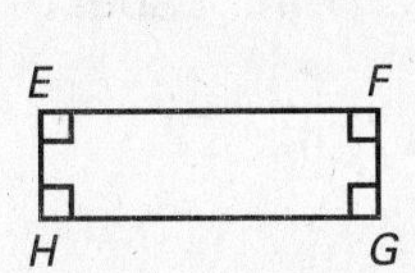

A special parallelogram with four right angles.

Rhombus

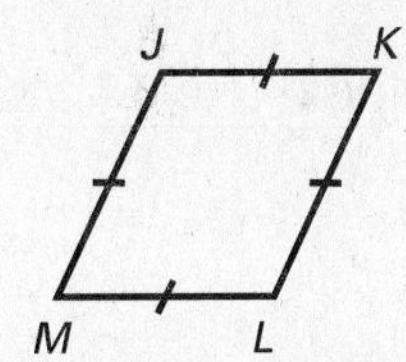

A parallelogram with four sides of equal length.

Square

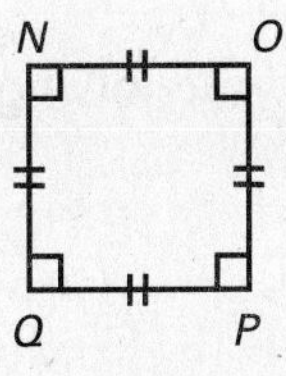

A rectangle with sides of equal length or a rhombus with right angles.

Trapezoid

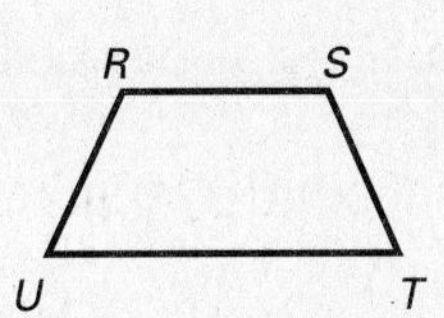

A quadrilateral with only one pair of parallel sides; the parallel sides are called **bases.**

Lesson 25

Directions: Provide the correct answer to each item.

Items 1 to 5 refer to the figure at the right. (Hint: There are 5 triangles.)

1. Name an equilateral triangle.
2. Name an isosceles triangle.
3. Name an obtuse triangle.
4. Name an acute triangle.
5. Name a scalene triangle.

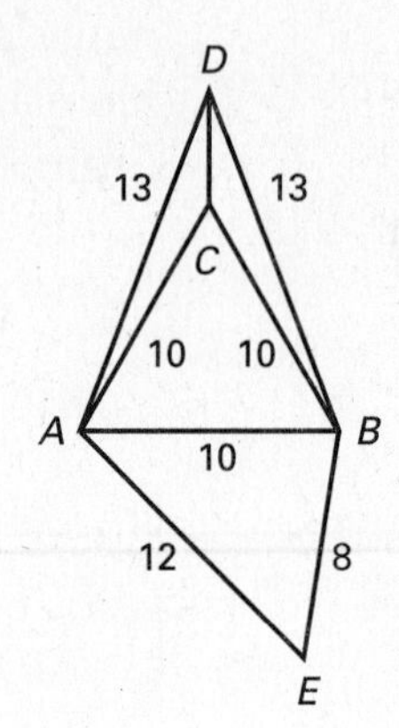

Items 6 to 14 refer to the figure at the right.

6. Name the triangle that has a right angle at *B*.
7. Name two triangles that have right angles at *D*.
8. Name two triangles that have right angles at *C*.

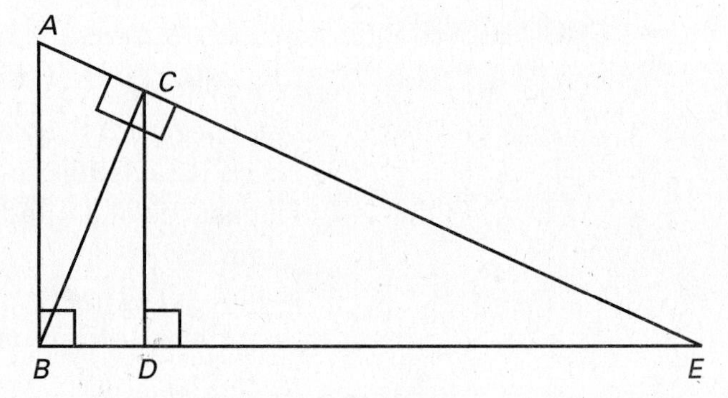

For items 9 to 14, $m\angle A = 55°$. (Hint: Label the angles as you compute their measures.)

9. If $m\angle A = 55°$, then what is $m\angle E$?
10. What is $m\angle ABC$?
11. What is $m\angle DCE$?
12. What is $m\angle CBD$?
13. What is $m\angle BCD$?
14. How are the four triangles in the figure similar?

Directions: Give all possibilities for the names of the following four-sided figures.

15. All sides have lengths of 10 inches.
16. All corners are right angles.
17. Opposite sides are parallel.
18. Only two sides are parallel.
19. There are no right angles.
20. Only one pair of opposite sides is equal.

Directions: Choose the best answer to each item.

21. In a right triangle the measure of one acute angle is 4 times the measure of the other acute angle. What is the measure of the smaller angle?

(1) less than 10°
(2) between 10° and 15°
(3) between 15° and 20°
(4) between 20° and 25°
(5) greater than 25°

22. A four-sided figure has sides in order of 8, 12, 8, 12. There are no right angles. What is the figure?

(1) triangle
(2) square
(3) trapezoid
(4) rhombus
(5) parallelogram

23. Which statement is true of the following figure?

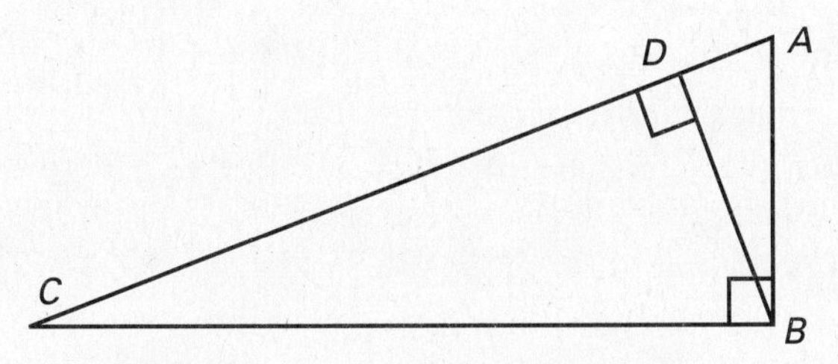

(1) $\triangle ABC$ is an equilateral triangle.
(2) $\triangle ABC$ is an obtuse triangle.
(3) $\triangle ABC$ is an acute triangle.
(4) There are exactly two right triangles.
(5) There are exactly three right triangles.

24. A triangle has sides 10, 10 and 8. What type of triangle is it?

(1) isosceles
(2) scalene
(3) equilateral
(4) obtuse
(5) right

25. In a triangle the measure of one angle is 15° more than the measure of the smallest, and the measure of the third is twice the measure of the smallest. If x is the measure of the smallest angle, then what is the equation to find the three angle measures?

(1) $(x) + (x + 15°) = 2x$
(2) $(x) + (x + 15°) + (2x) = 180°$
(3) $(x) + (15°) + (2 \cdot 15°) = 180°$
(4) $(x) + (x - 15°) + (2x) = 180°$
(5) $(15°) + (x + 15°) + (2x) = 180°$

26. How many triangles are in the figure?

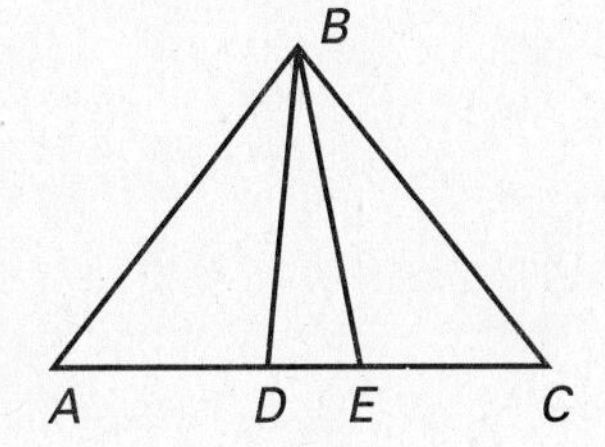

(1) 1
(2) 3
(3) 4
(4) 6
(5) 7

27. How many isosceles triangles are there in the figure?

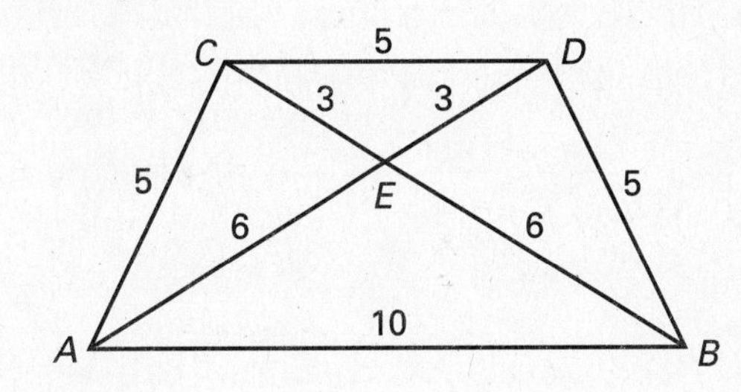

(1) 1
(2) 2
(3) 3
(4) 4
(5) 5

Answers are on page 397.

Strategies for Solving Word Problems

Use Logical Reasoning

Some items on the GED test must be solved using logic. Based on a few basic statements called **assumptions,** you will be asked to decide which conclusion is true.

Example: The sum of the measures of angles 1 and 2 is 180 degrees. The sum of the measures of angles 2 and 3 is 180 degrees. Based on this information, which of the following statements is true?

(1) $m\angle 1 = 120°$
(2) $m\angle 2 = 60°$
(3) $m\angle 3 = 100°$
(4) $m\angle 1 > m\angle 2$
(5) $m\angle 1 = m\angle 3$

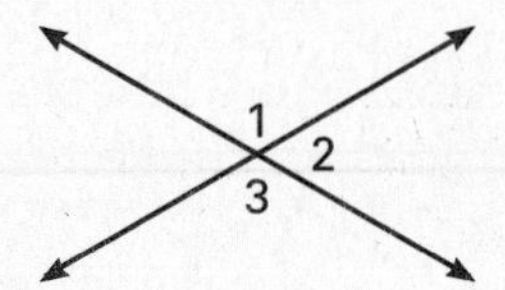

Options (1), (2), (3) and (4) are incorrect because the assumptions given do not provide any information about the measures of each angle. Although angle 1 looks greater than angle 2 in the figure, we cannot conclude that it is actually greater. Use only the information in the problem and any labels on the figure. Do not "jump to any conclusions."

Option 5 is correct. Angles 1 and 3 are congruent (equal).

Consider the facts: $m\angle 1 + m\angle 2 = 180°$
$m\angle 2 + m\angle 3 = 180°$

Then: $m\angle 1 + m\angle 2 = m\angle 2 + m\angle 3$

Subtract $m\angle 2$ from both sides: $m\angle 1 = m\angle 3$

The same principles you used to solve algebra are useful in solving geometry problems. In the example above, we subtracted the same quantity from both sides of an equation. Algebra can also be used to solve problems involving complementary and supplementary angles.

Example: The measure of an angle is twice the measure of its complement. What is the measure of the smaller angle?

Let x equal the smaller angle and $2x$ equal the larger angle. We know the sum of the measures of the two angles is 90° because the angles are complementary.

$x + 2x = 90°$
$3x = 90°$
$x = 30°$ The smaller angle measures **30°.**

Solving Word Problems

Directions: Choose the best answer to each item.

1. If $\angle ABC$ is a straight angle, which statement is true?

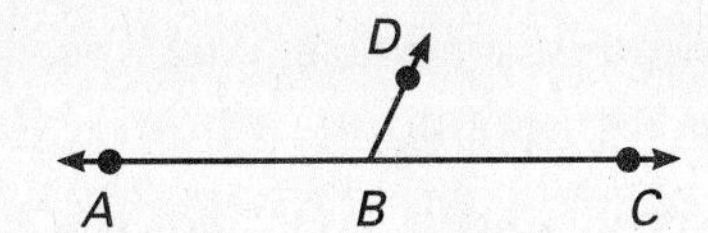

(1) $m\angle DBC = 90°$
(2) $\angle ABD$ and $\angle DBC$ are supplementary angles.
(3) $m\angle ABD = 90°$
(4) $\angle ABD$ and $\angle DBC$ are complementary angles.
(5) $m\angle ABC = 90°$

2. If angle 1 is congruent to angle 5, which statement must be true?

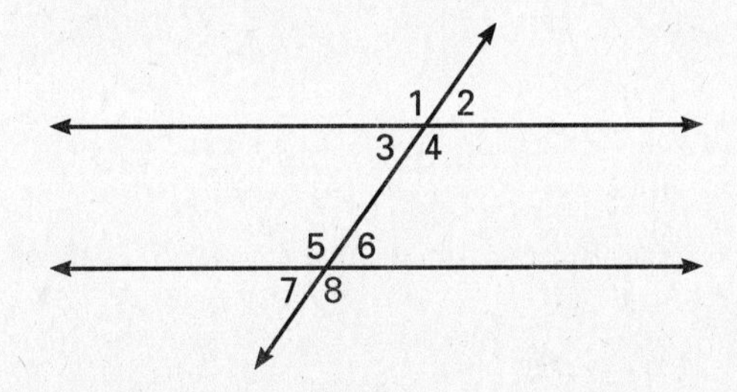

(1) Angle 1 is congruent to angle 2.
(2) Angle 3 is congruent to angle 4.
(3) Angle 3 is congruent to angle 7.
(4) Angle 3 is congruent to angle 8.
(5) Angle 7 is congruent to angle 8.

3. The measure of one angle is 12° less than the measure of its complement. If the measure of the larger angle is x, then

(1) $x + (x - 12°) = 180°$
(2) $x + (x - 12°) = 90°$
(3) $x + (12° - x) = 90°$
(4) $x + (x + 12°) = 180°$
(5) $x - 90° = x + 12°$

4. If $\angle ACB = 80°$, the sum of the measures of $\angle A$ and $\angle B$ together must be 100°. Why?

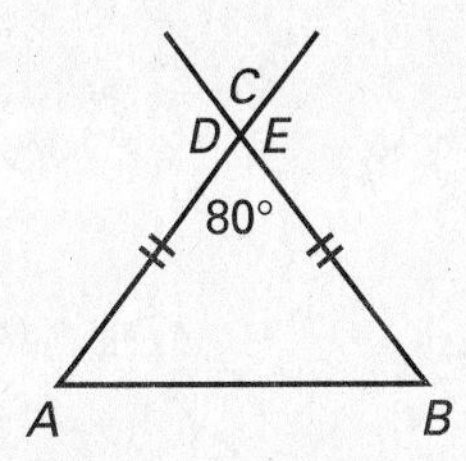

(1) They are congruent.
(2) They are adjacent.
(3) The sum of the measure of $\angle A$, $\angle B$, and $\angle ACB$ is 180°.
(4) Vertical angles at C are congruent.
(5) $\angle A$ and $\angle B$ are adjacent interior angles.

5. Which conclusion is false?

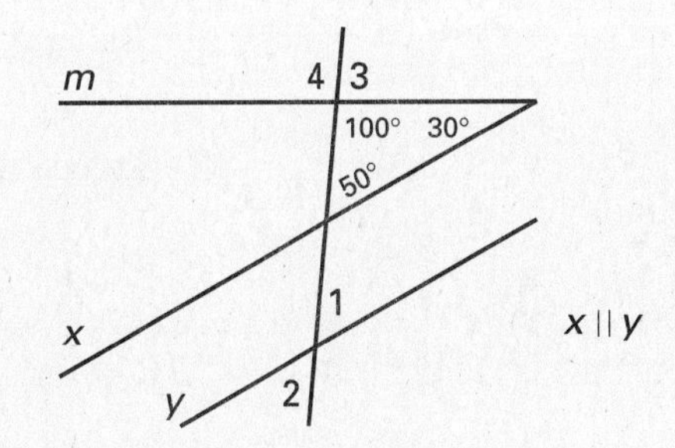

(1) $m\angle 3 = 80°$ because of the "supplementary angles" rule.
(2) $m\angle 1 = 50°$ because of the "corresponding angles" rule.
(3) $m\angle 2 = 50°$ because of the "vertical angles" rule.
(4) $m\angle 4 = 100°$ because of the "vertical angles" rule.
(5) $m\angle 4 = m\angle 3$ because of the "corresponding angles" rule.

Answers are on page 398.

Congruence and Similarity

Congruence

Congruent figures are the same shape and size. One figure will fit perfectly on top of the other. Triangles that are exactly the same shape and size are called **congruent triangles** (symbol: $\cong$). Congruent triangles have matching or corresponding vertices, congruent corresponding sides, and congruent corresponding angles. Marks show which parts correspond. Below are three ways to show that two triangles are congruent.

(1) Three sides are congruent (SSS).

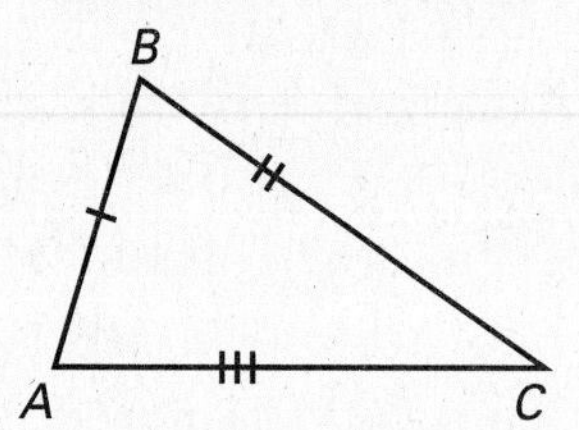

$\overline{AB} \cong \overline{DE}$
$\overline{BC} \cong \overline{EF}$
$\overline{CA} \cong \overline{FD}$
$\triangle ABC \cong \triangle DEF$

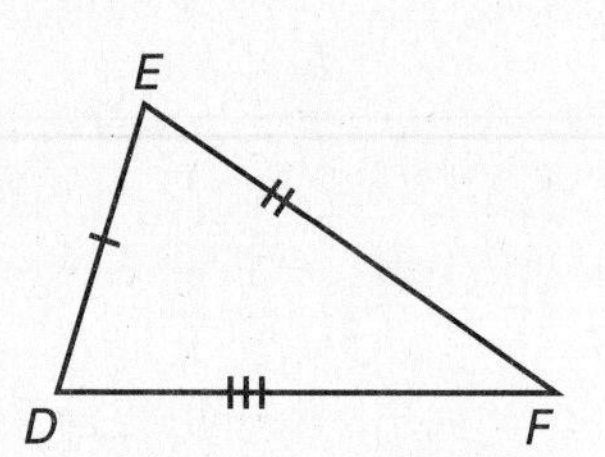

(2) Two sides and the angle between them are congruent (SAS).

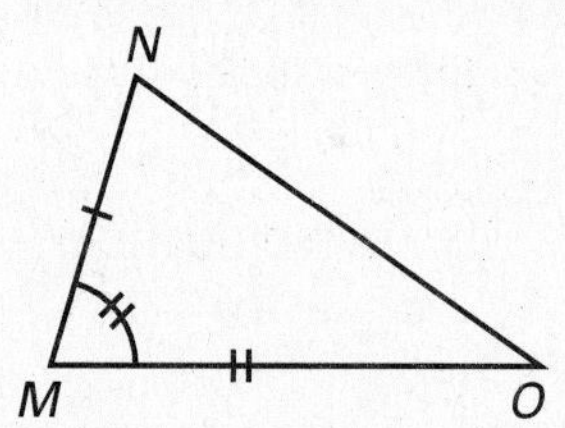

$\overline{MN} \cong \overline{HI}$
$\overline{MO} \cong \overline{HJ}$
$\angle M \cong \angle H$
$\triangle MNO \cong \triangle HIJ$

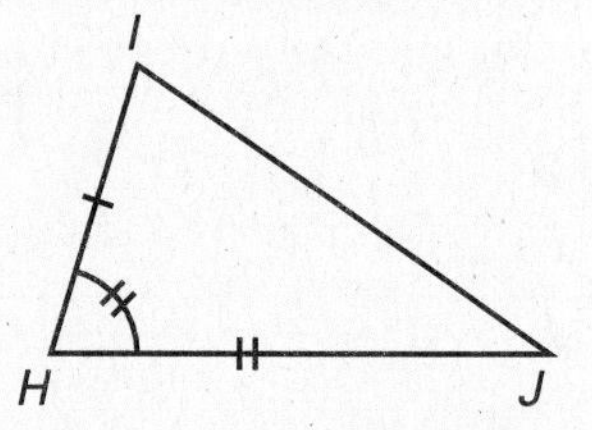

(3) Two angles and the side between them are congruent (ASA).

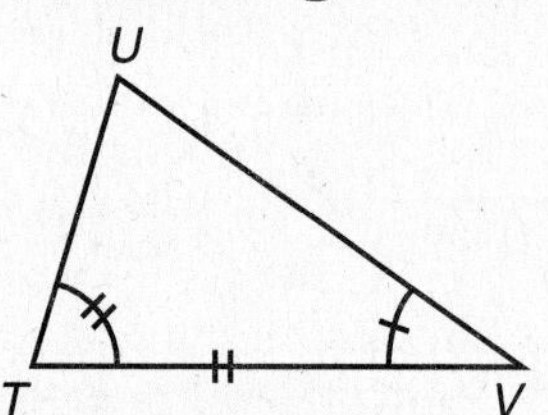

$\angle T \cong \angle Q$
$\angle V \cong \angle S$
$\overline{TV} \cong \overline{QS}$
$\triangle TUV \cong \triangle QRS$

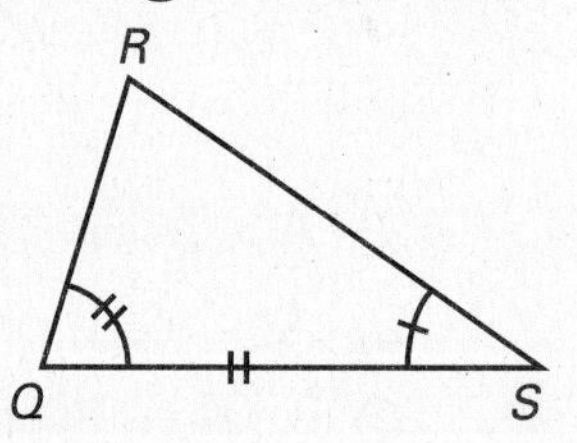

Using the explanations above, study why these pairs of triangles are congruent.

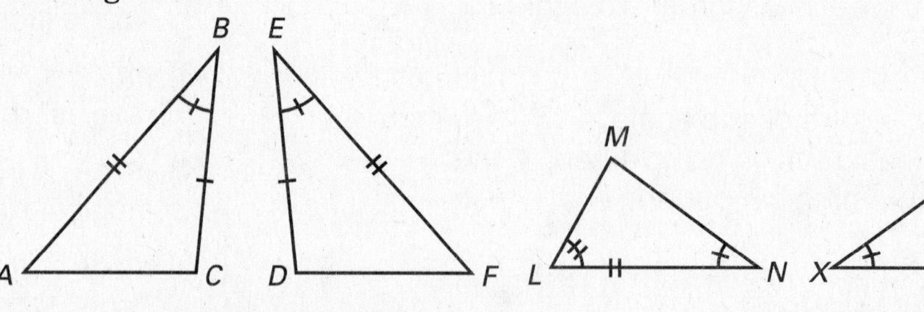

$\triangle ABC \cong \triangle FED$
Using Rule 2 or SAS,
$\overline{AB} \cong \overline{FE}$
$\angle B \cong \angle E$
$\overline{BC} \cong \overline{ED}$

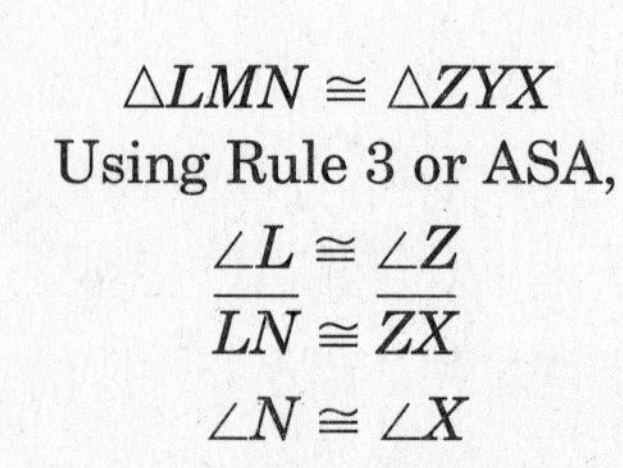

$\triangle LMN \cong \triangle ZYX$
Using Rule 3 or ASA,
$\angle L \cong \angle Z$
$\overline{LN} \cong \overline{ZX}$
$\angle N \cong \angle X$

Lesson 26

Directions: Name the corresponding parts of the two congruent triangles by completing the tables.

$\triangle ABE \cong \triangle CBD$

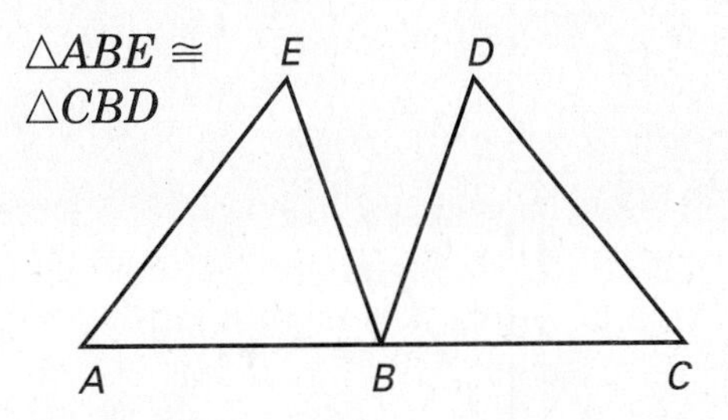

	In $\triangle ABE$	In $\triangle CBD$
1.	$\angle A$	?
2.	$\angle ABE$	?
3.	?	$\angle D$

	In $\triangle ABE$	In $\triangle CBD$
4.	?	$\overline{BC}$
5.	$\overline{AE}$	?
6.	$\overline{BE}$	?

Directions: For items 7 to 12, note on a sheet of paper whether each pair of triangles is congruent or not, and provide an explanation for each response.

7.

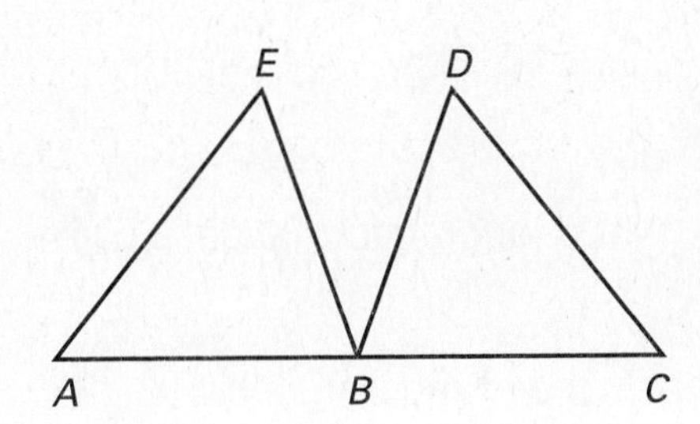

10.

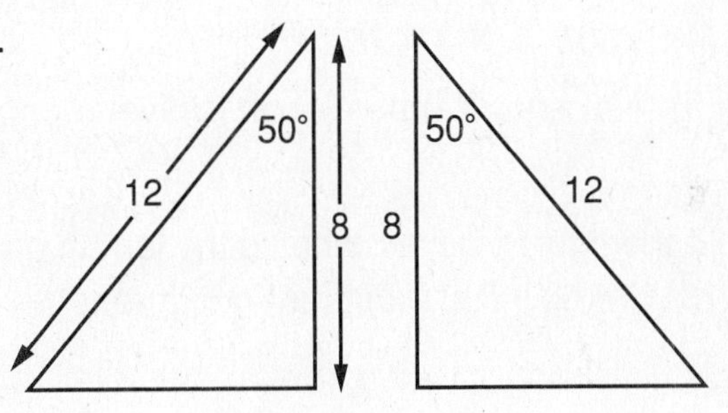

8.

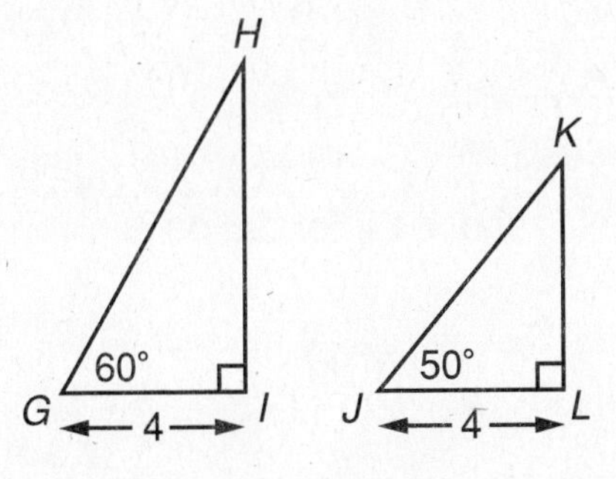

11.

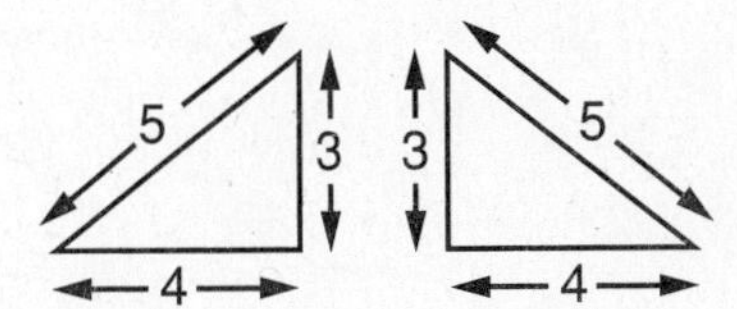

9.

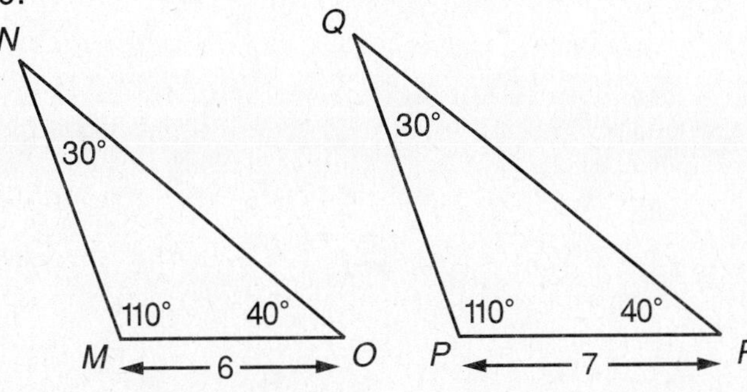

12.

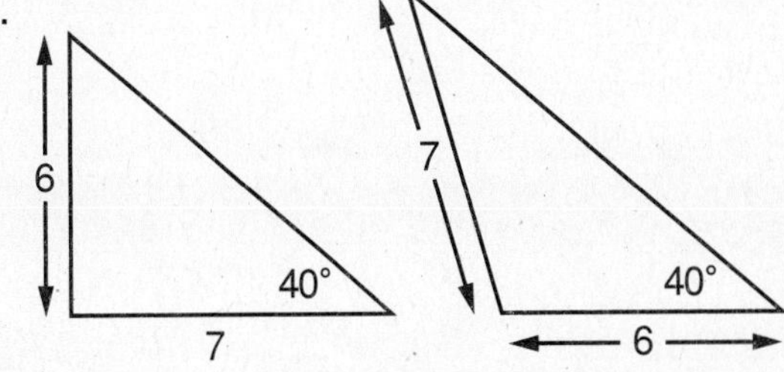

Directions: Choose the best answer to each item.

13. The two triangles in the figures below are congruent. What is the measurement of $\angle D$?

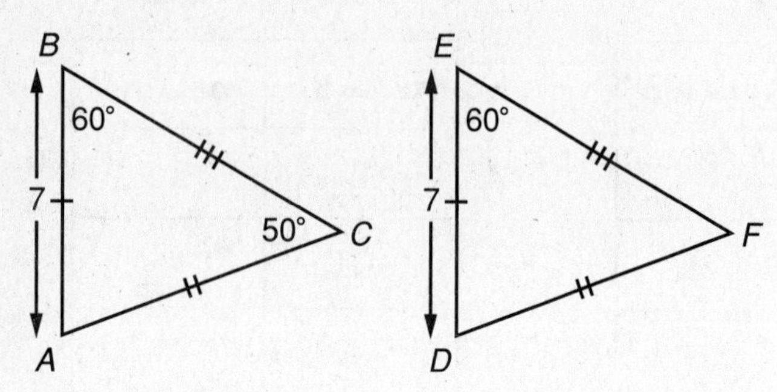

(1) 50°

(2) 70°

(3) 110°

(4) 180°

(5) Not enough information is given.

14. In the figures below, what is the length of $\overline{KL}$, if $\triangle GHI$ and $\triangle JKL$ are congruent?

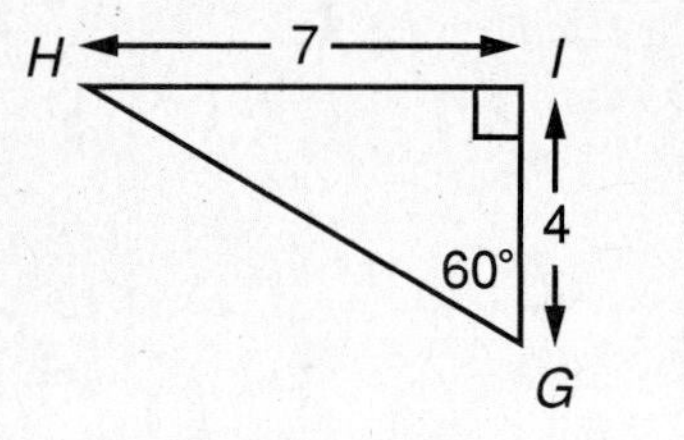

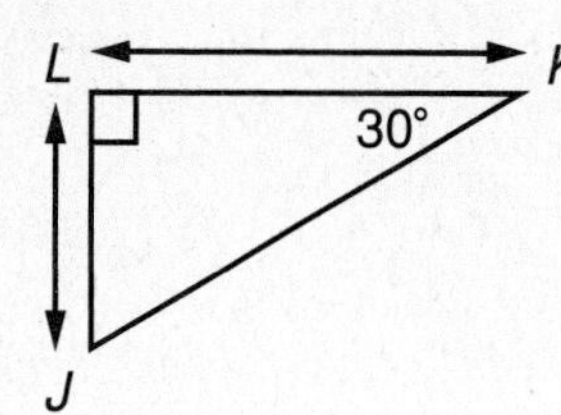

(1) 4

(2) 7

(3) 30

(4) 45

(5) 60

Items 15 and 16 refer to the following diagram.

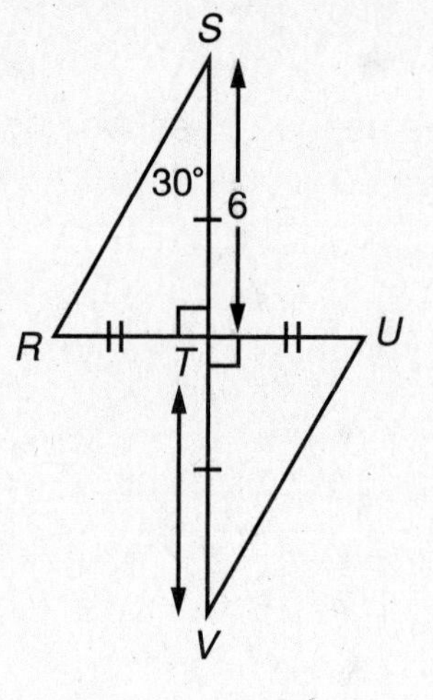

15. What is the measurement of $\angle V$?

(1) 6°

(2) 30°

(3) 60°

(4) 90°

(5) Not enough information is given.

16. What is the measurement of $\angle U$?

(1) 6°

(2) 30°

(3) 60°

(4) 90°

(5) Not enough information is given.

Directions: Choose the best answer to each item.

17. For the following triangles, what is the measurement of $\angle L$?

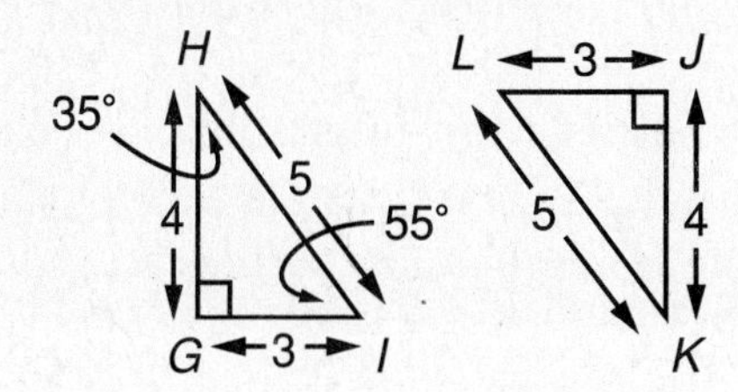

(1) 35°
(2) 45°
(3) 55°
(4) 90°
(5) Not enough information is given.

18. For the triangles below, what is the measurement of $\angle Q$?

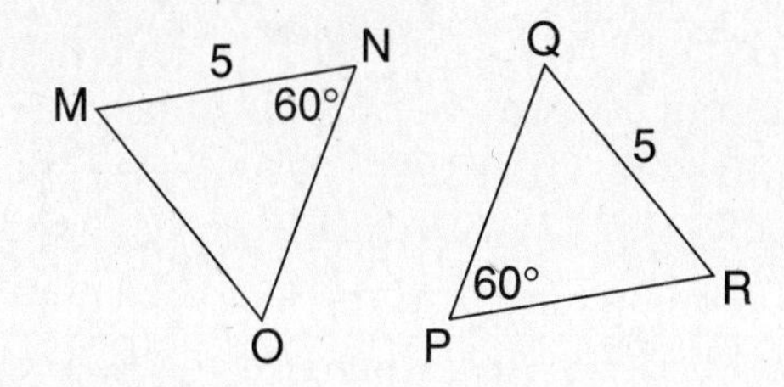

(1) 5°
(2) 30°
(3) 60°
(4) 120°
(5) Not enough information is given.

19. What is the length of $\overline{EF}$ in the triangles below?

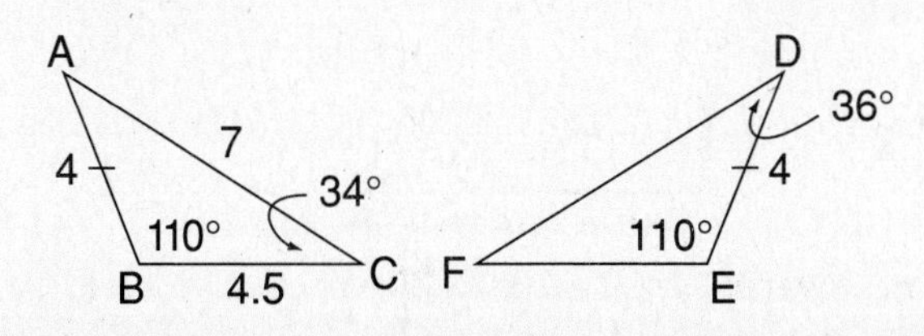

(1) 3
(2) 4
(3) 4.5
(4) 7
(5) 11

Items 20 and 21 refer to the following diagram.

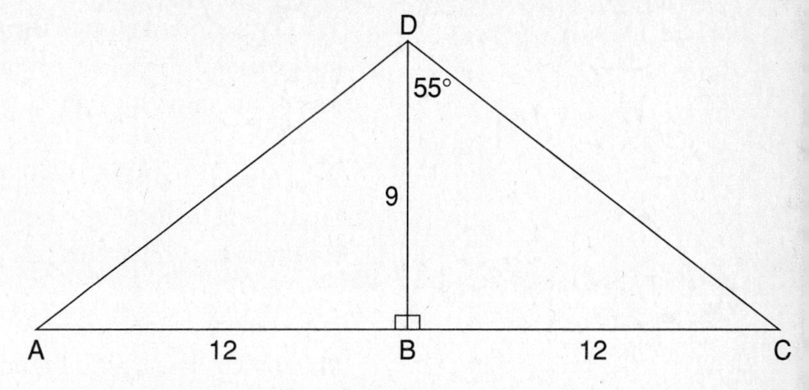

20. What is the measurement of $\angle A$?

(1) 9°
(2) 12°
(3) 35°
(4) 55°
(5) Not enough information is given.

21. What is the measurement of $\angle ADB$?

(1) 9°
(2) 35°
(3) 55°
(4) 90°
(5) Not enough information is given.

22. For the following triangles, what is the length of $\overline{DE}$?

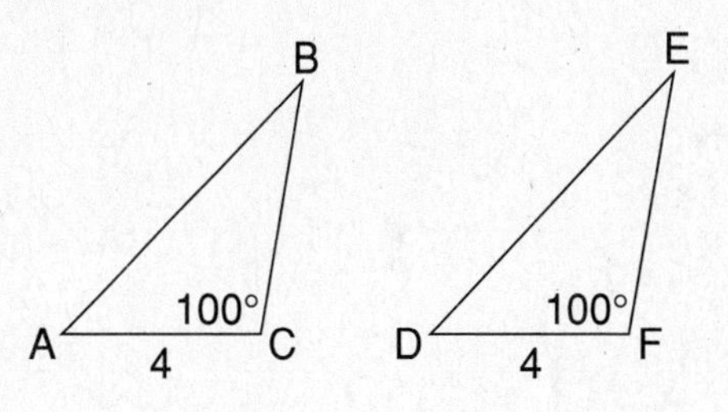

(1) 4
(2) 8
(3) 25
(4) 100
(5) Not enough information is given.

Answers are on page 399.

Two triangles are **similar** (symbol: ~) if the corresponding angles have equal measure and the corresponding sides are in proportion. They have the same shape, but they are *not necessarily* the same size.

When comparing triangles to find whether they are similar, you can look for relationships between the angles of one triangle and another triangle and between the sides of one triangle and another triangle.

If the angle measurement of each of two or more angles of one triangle is equal to the angle measurement of each of two or more angles of the other triangle, the two triangles are **similar.**

Example:

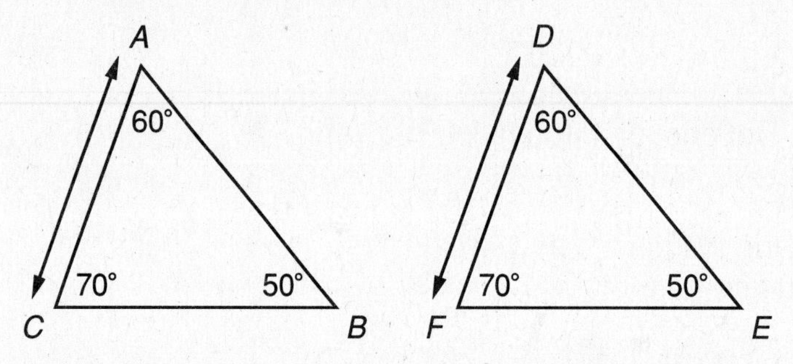

In $\triangle ABC$, $\angle A$ measures 60°. In $\triangle DEF$, $\angle D$ measures 60°. In $\triangle ABC$, $\angle B$ measures 50°. In $\triangle DEF$, $\angle E$ measures 50°. $\triangle ABC$ is **similar** to $\triangle DEF$ because at least two angles of $\triangle ABC$ are equal in measurement to each of two angles of $\triangle DEF$.

If the length of each of the sides of one triangle is proportional to the length of each of the sides of the other triangle, the two triangles are similar.

Example:

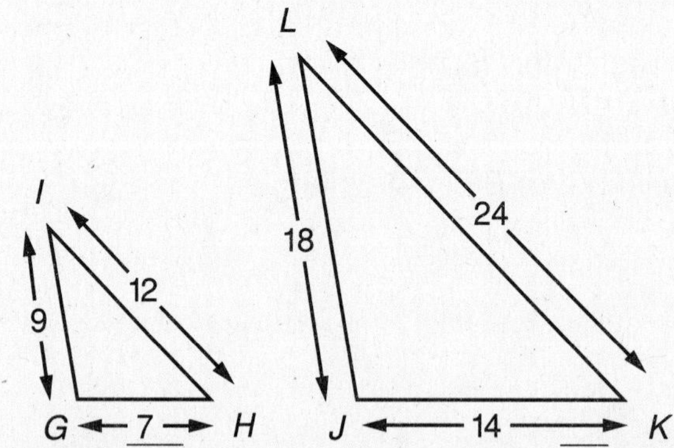

In $\triangle GHI$, $\overline{GH}$ measures 7. In $\triangle JKL$, $\overline{JK}$ measures 14.
In $\triangle GHI$, $\overline{GI}$ measures 9. In $\triangle JKL$, $\overline{JL}$ measures 18.
In $\triangle GHI$, $\overline{HI}$ measures 12. In $\triangle JKL$, $\overline{KL}$ measures 24.

$$\frac{7}{14} = \frac{9}{18} = \frac{12}{24}$$

$\triangle GHI$ is **similar** to $\triangle JKL$ because the length of each of the sides of $\triangle GHI$ is proportional to the length of each of the sides of $\triangle JKL$.

When two figures are similar, the smallest angle of one "corresponds" to the smallest angle of the other, and the longest side of one "corresponds" to the longest side of the other, and so on. To identify corresponding parts, match up the parts of the figures that are the same.

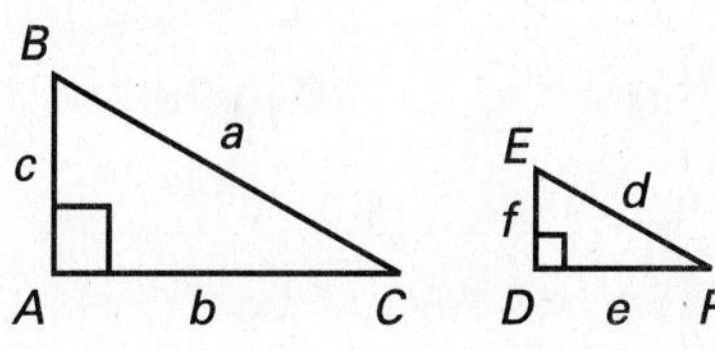

We have seen from previous examples the following relationships of similar (~) triangles:

(1) They have the same shape. $\triangle ABC \sim \triangle DEF$

(2) Corresponding angles are congruent (equal). $\angle A \cong \angle D \quad \angle B \cong \angle E \quad \angle C \cong \angle F$

(3) Corresponding sides are proportional. (The ratio of the lengths of the sides are in proportion.) $\frac{a}{d} = \frac{b}{e} = \frac{c}{f}$

A variety of practical problems can be solved using similar triangles. Similar triangles are often used when there is no way to actually perform the measurement of the missing length.

Example: What is the height of the flagpole?

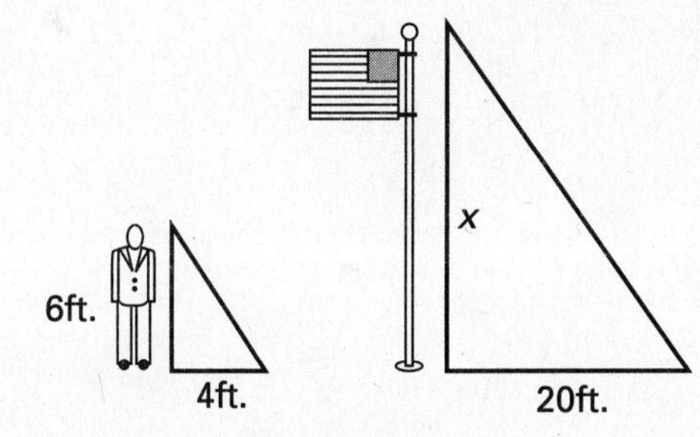

We can find the height of the flagpole in the diagram at the left. The diagram shows that the flagpole casts a shadow of 20 feet while a six-foot person casts a shadow of 4 feet. We use the fact that the triangles formed by the objects and their shadows are similar. We set up a proportion and solve.

$$\frac{\text{height of the person}}{\text{shadow of the person}} = \frac{\text{height of the flagpole}}{\text{shadow of the flagpole}}$$

$$\frac{6}{4} = \frac{x}{20}$$

$$4x = 6(20)$$

$$4x = 120$$

$$x = 30$$

The height of the flagpole is **30 feet.**

Lesson 26

Directions: Complete the following corresponding parts and ratios.

Items 1 to 9 refer to the diagram below.

$\triangle AEC \sim \triangle BDG$

1. $\angle A \cong$ ___?___
2. $\angle ACE \cong$ ___?___
3. $\angle CEA \cong$ ___?___
4. $\overline{AC}$ corresponds to ___?___.

5. $\overline{AE}$ corresponds to ___?___.
6. $\overline{CE}$ corresponds to ___?___.
7. $\frac{\overline{AC}}{\overline{BD}} = \frac{CE}{?}$
8. $\frac{\overline{AE}}{\overline{BG}} = \frac{AC}{?}$
9. $\frac{\overline{CE}}{\overline{DG}} = \frac{AE}{?}$

Items 10 and 11 refer to the following figure.

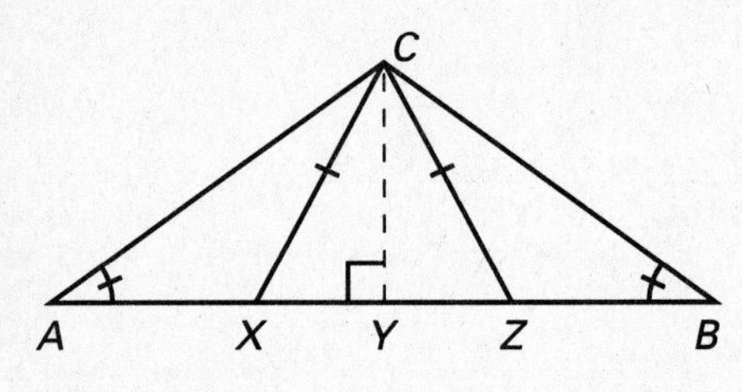

10. $\triangle ACB$ is isoceles.
Therefore, $\overline{AC} \cong$ ___?___
$\angle B \cong$ ___?___
$\overline{AY} \cong$ ___?___
$\angle ACY \cong$ ___?___

11. $\triangle XCZ$ is isoceles.
Therefore, $\overline{XC} \cong$ ___?___
$\angle CXY \cong$ ___?___
$\overline{XY} \cong$ ___?___
$\angle XCY \cong$ ___?___

Directions: At any given time, the ratios of the length of an object to the length of its shadow are equal. Complete the following table.

	Length of Object	Length of Shadow	Length of Flagpole	Length of Shadow
12.	6 ft.	6 ft.	?	14.5 ft.
13.	6 ft.	9 ft.	?	33 ft.
14.	5 ft.	3 ft.	?	21 ft.
15.	?	8 ft.	20 ft.	40 ft.
16.	5 ft.	12 ft.	20 ft.	?

Directions: Choose the best answer to each item.

17. In the figure below, $\triangle ABC$ and $\triangle DEC$ are similar. What is the length of $\overline{AB}$?

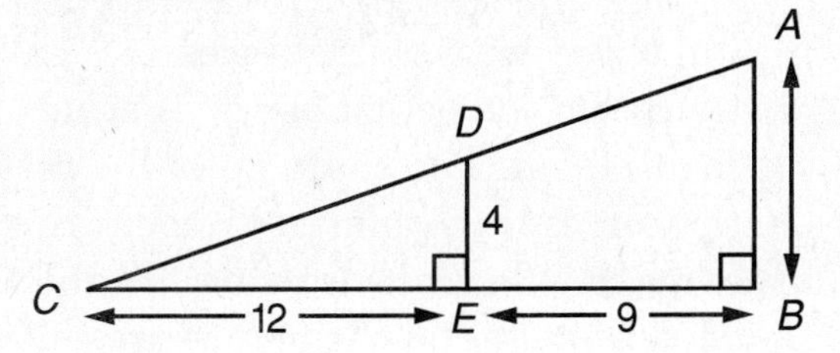

(1) 5
(2) 7
(3) 21
(4) 63
(5) Not enough information is given.

18. In the figure below, $\triangle GHI$ and $\triangle GJK$ are similar. What is the length of $\overline{GJ}$?

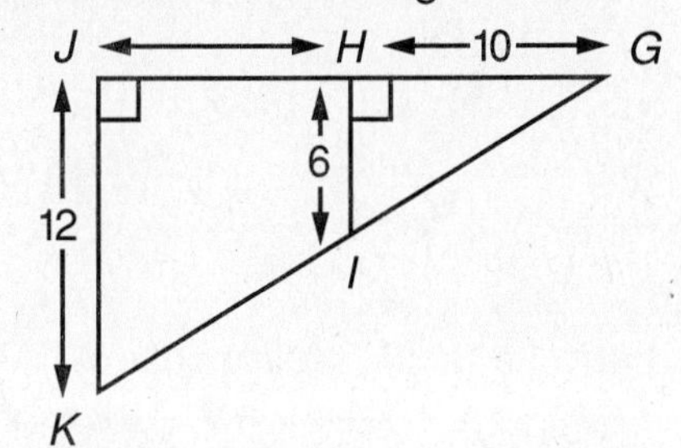

(1) 7.2
(2) 10
(3) 16
(4) 18
(5) 20

19. What is the height of the tree in the diagram below?

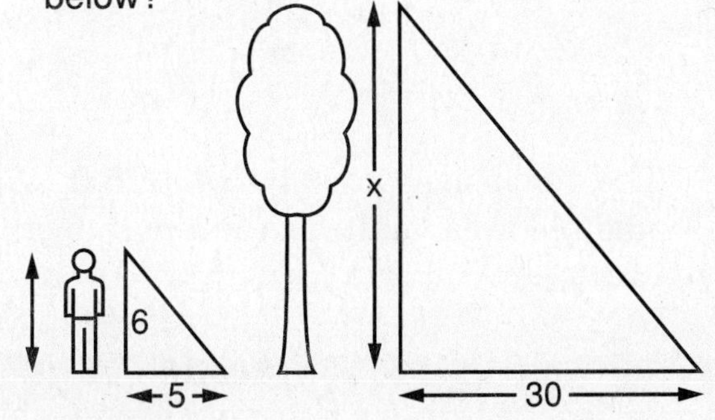

(1) 20
(2) 25
(3) 30
(4) 36
(5) 42

20. What is the length across the longest part of the lake shown in the diagram below?

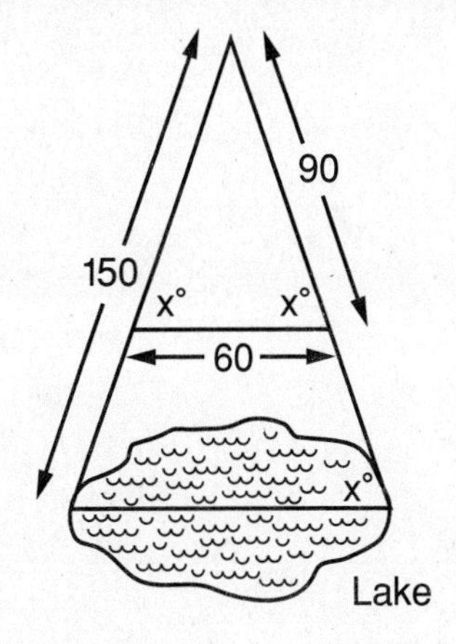

(1) 300
(2) 225
(3) 100
(4) 60
(5) 36

21. In the figure below, what is the length of $\overline{PQ}$?

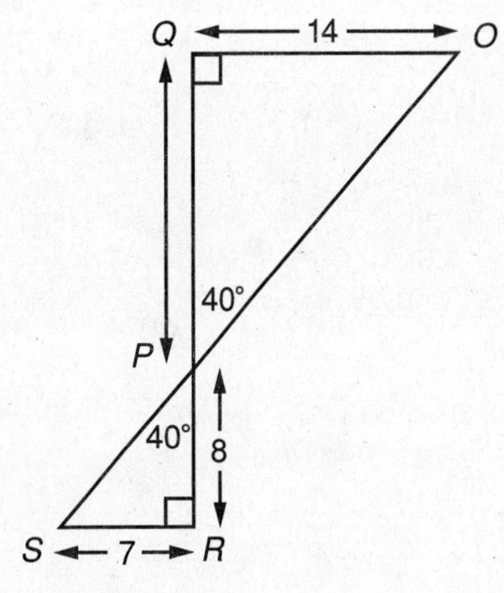

(1) 4
(2) 12.25
(3) 16
(4) 29
(5) Not enough information is given.

Answers are on page 400.

Strategies for Solving Word Problems

Use Proportion to Solve Indirect Measurement Problems

A **scale drawing** is a sketch of an object with all distances in proportion to corresponding distances on the actual object. A scale drawing is a form of indirect measurement. The **scale** gives the ratio of the sketch measurements to the corresponding measurements on the actual object. A map is one example of a scale drawing. On the map below, the scale shows that 1 centimeter on the map represents 5 kilometers on the land being measured.

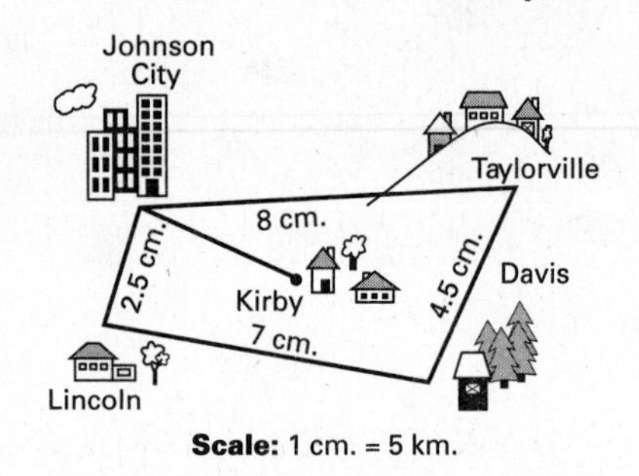

Example: The distance on the map between Taylorville and Davis is 4.5 cm. What is the actual distance between the two towns?

Use the scale and write a proportion using the ratio of map distance to actual distance.

$$\frac{\text{map distance}}{\text{actual distance}} \quad \frac{1 \text{ cm.}}{5 \text{ km.}} = \frac{4.5 \text{ cm.}}{x \text{ km.}}$$

$$\frac{1}{5} = \frac{4.5}{x}$$

$$x = (5)(4.5) = 22.5 \text{ km.}$$

The actual distance is **22.5 kilometers.**

A floor plan is another way to use indirect measurement.

Example: In the floor plan of the Martin's cabin, the living room measures 2 inches by $4\frac{1}{2}$ inches. Every 2 inches of the floor plan represents 9 actual feet. How many square feet of carpet are needed to cover the living room floor?

First, we use proportions to find the dimensions of the living room.

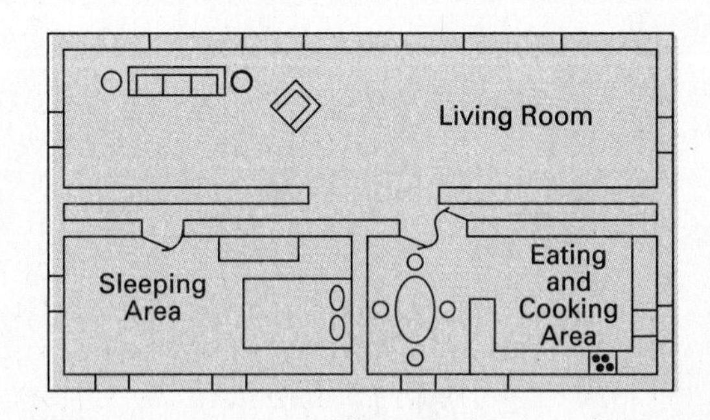

$$\frac{2''}{9'} = \frac{2''}{x'} \qquad \frac{2''}{9'} = \frac{4.5''}{x'}$$

$$2x = 18 \qquad 2x = 40.5$$

$$x = 9 \qquad x = 20.25$$

The dimensions are 9 ft. by 20.25 ft. To find the area, multiply the length by the width.

$$9 \times 20.25 = 182.25$$

The living room floor will need **$182\frac{1}{4}$ square feet** of carpet.

Solving Word Problems

Directions: Choose the best answer to each item.

Items 1 and 2 refer to the following drawing.

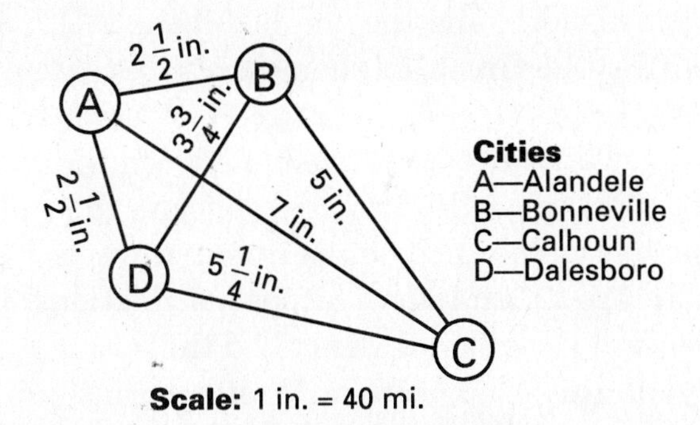

1. The distance on the map between Bonneville and Dalesboro is $3\frac{3}{4}$ inches. What is the actual distance in miles between the two cities?

   (1) 120
   (2) 150
   (3) 160
   (4) 180
   (5) 200

2. Ramona drove from Calhoun to Alandele, then from Alandele to Bonneville, and back to Calhoun. How many miles did she drive?

   (1) 460
   (2) 480
   (3) 540
   (4) 580
   (5) 620

3. On a map Sunnyview is 2.5 inches from Taylor. Actually these towns are 50 miles apart. What is the scale on this map?

   (1) Scale: 1 in. = $\frac{1}{5}$ mi.
   (2) Scale: 1 in. = 2 mi.
   (3) Scale: 1 in. = 2.5 mi.
   (4) Scale: 1 in. = 20 mi.
   (5) Scale: 1 in. = 25 mi.

4. The three sides of a triangle are 12 feet, 16 feet and 20 feet in length. The short side of a second triangle that is similar to the first is 15 feet long. What are the lengths of the other two sides of the second triangle?

   (1) 18 ft. and 23 ft.
   (2) 18 ft. and 25 ft.
   (3) 19 ft. and 23 ft.
   (4) 20 ft. and 24 ft.
   (5) 20 ft. and 25 ft.

5. The floor plan of a family room is drawn to a scale of 1 inch = 8 feet. It shows that one wall of this room is $1\frac{3}{4}$ inches long.The owner wants to place bookshelves that are 14 feet long on this wall. Will they fit?

   (1) No, the bookshelves are about 2 ft. too long.
   (2) No, the bookshelves are about 1 ft. too long.
   (3) Yes, the bookshelves fit exactly into this space.
   (4) Yes, the bookshelves will fit and leave about 1 ft. of extra space.
   (5) Yes, the bookshelves will fit and leave about 2 ft. of extra space.

6. A map shows the details of several counties. The scale is 1 inch = 1.8 miles. If the distance from one town to the city center is 3.5 inches on the map, how far is the actual distance?

   (1) 0.5 mi.
   (2) 1.9 mi.
   (3) 3.5 mi.
   (4) 6.3 mi.
   (5) 35 mi.

Answers are on page 400.

Pythagorean Relationships

Pythagorean Theorem

The ancient Egyptians used an indirect method to resurvey their fields after the yearly flooding of the Nile River. The ancient Greeks proved the method will always work, and they named it the **Pythagorean Theorem** after the Greek mathematician Pythagoras.

hypotenuse

legs

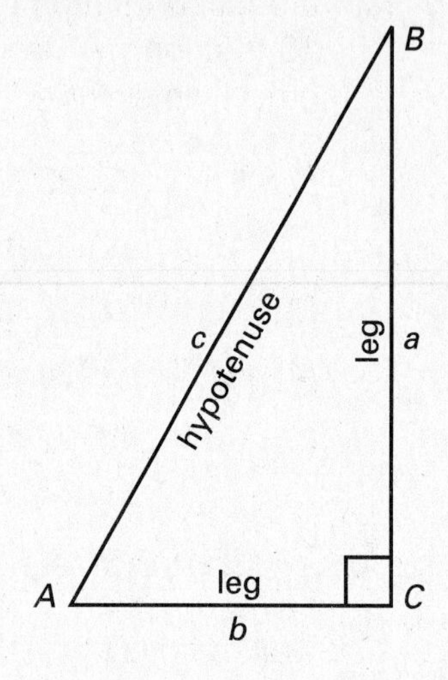

A special relationship exists between the sides of a right triangle. The side opposite the right angle is called the **hypotenuse.** The other sides are called **legs.** The special relationship, proved by the Greeks, is called the **Pythagorean Theorem.** It states that in a right triangle, the square of the length of the hypotenuse is equal to the sum of the squares of the lengths of the legs.

Pythagorean Theorem
$(\text{the hypotenuse})^2 = (\text{leg } 1)^2 + (\text{leg } 2)^2$
$c^2 = a^2 + b^2$

Remember, the square of a number is the number multiplied by itself.

Example: Joan places a 13-foot ladder so that the top of the ladder touches the roof of the house. Joan measures the distance from the house to the bottom of the ladder and finds it to be 5 feet. What is the distance from the ground to the roof?

Sometimes the problem and the picture say nothing about whether the triangle is a right triangle. You have to recognize the triangle is a right triangle.

Now think of Joan's ladder as the hypotenuse, the distance from the house as one leg, and the height to the roof as the other leg.

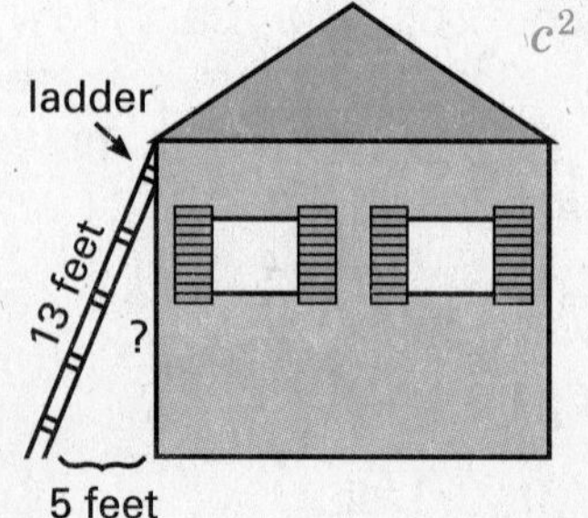

$c^2 = a^2 + b^2$, or $a^2 + b^2 = c^2$

$$(5)^2 + b^2 = (13)^2$$
$$25 + b^2 = 169$$
$$b^2 = 169 - 25$$
$$b^2 = 144$$
$$b = 12$$

The roof is **12 feet** from the ground.

The Pythagorean Theorem is often used in construction to "square up" a corner. A tape measure is used to measure two perpendicular segments that are 3 feet and 4 feet, and they are joined by a hypotenuse of 5 feet. The angle opposite the hypotenuse will be 90°, or a square corner. A diagonal is sometimes placed across a rectangular wall to give the wall added strength.

Example: If a wall is 8 ft. by 6 ft., what size diagonal board is needed?

Draw a picture.

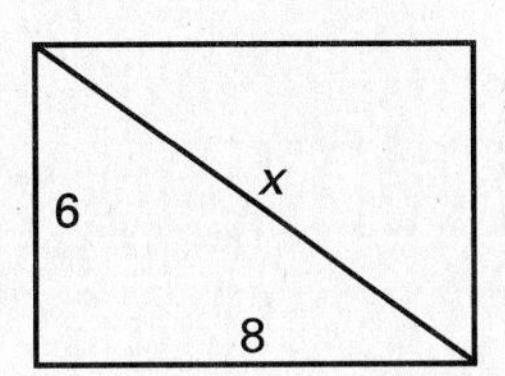

Use the Pythagorean Theorem.

$$c^2 = a^2 + b^2$$
$$x^2 = (6)^2 + (8)^2$$
$$x^2 = 36 + 64$$
$$x^2 = 100$$
$$x = 10$$

The board must be **10 feet long.**

Using algebra, the Pythagorean Theorem can be used to solve many problems. For example, if you are given the lengths of the sides of a triangle, you can determine whether the triangle is a right triangle or not.

Example: The lengths of the sides of a triangle are 3 in., 3.5 in., and 6 in. Is the triangle a right triangle?

If the triangle is a right triangle, the longest side must be the hypotenuse. Substitute the 3 in. for a and 3.5 in. for b. Solve for c. If the solution is 6, the triangle is a right triangle.

$$c^2 = a^2 + b^2$$
$$c^2 = 3^2 + 3.5^2$$
$$c^2 = 9 + 12.25$$
$$c^2 = 21.25$$
$$c = \sqrt{21.25}$$
$$c = 4.6097722 \text{ which rounds to } 4.61$$

Therefore, the triangle in the example is **not** a right triangle.

Lesson 27

Directions: The table gives lengths of sides of triangles. Complete the table to find which triangles are right triangles.

Triangle	Length of Sides							
	a	*b*	*c*	a^2	b^2	$a^2 + b^2$	c^2	Right Triangle?
1.	2	3	4					
2.	3	4	5					
3.	2	2	3					
4.	3	3	5					
5.	5	12	13					
6.	6	8	10					
7.	7	24	25					
8.	25	60	65					

Items 9 and 10 refer to the following diagram.

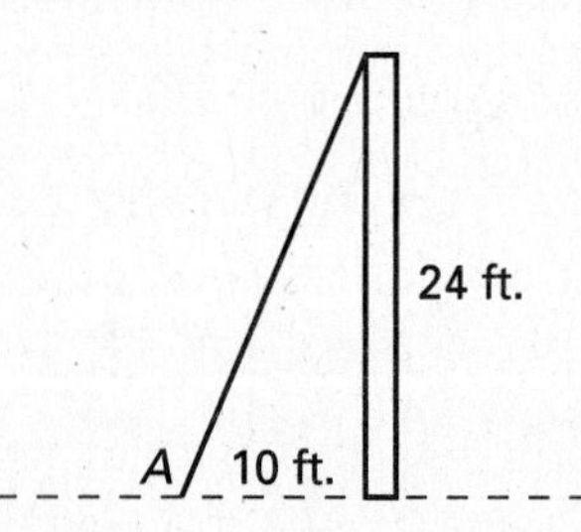

9. How many feet of wire are needed to reach from the top of the pole to point *A*?

10. Now point *A* is moved 8 feet further away from the pole. How much more wire will be needed?

Items 11 and 12 refer to the following diagram.

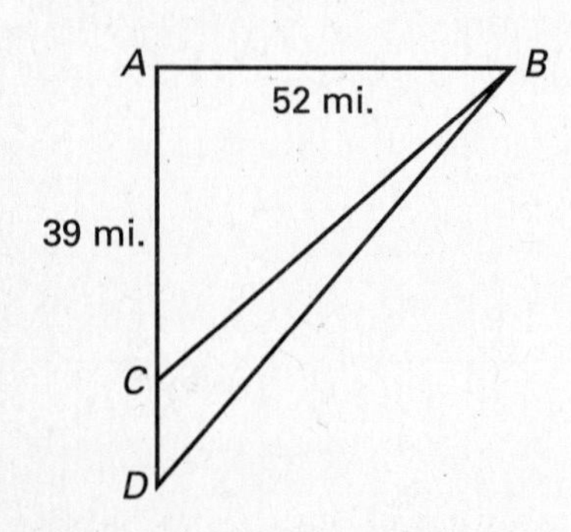

11. How far is it from Point *C* to Point *B*?

12. Point *D* is 11 miles from Point *C*. To the nearest whole mile, how far is it from Point *D* to Point *B*?

Directions: Choose the best answer to each item.

13. In a right triangle the short leg is 6 and the hypotenuse is 20. The other leg can be found by completing which calculation?

(1) $\sqrt{20 + 6}$
(2) $\sqrt{20 - 6}$
(3) $\sqrt{20 + 36}$
(4) $\sqrt{400 - 36}$
(5) $\sqrt{400 + 36}$

14. Which of the following is a right triangle?

(1) triangle with sides of 4, 5, and 6
(2) triangle with sides of 5, 7, and 9
(3) triangle with sides of 5, 12, and 13
(4) triangle with sides of 6, 8, and 11
(5) triangle with sides of 7, 9, and 12

15. Which is closest to the distance from A to B?

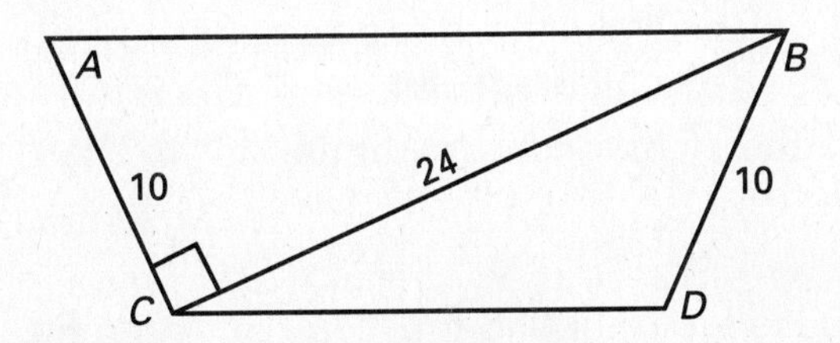

(1) 27
(2) 26
(3) 25
(4) 24
(5) 23

16. In a right triangle, the hypotenuse measures 10 meters and one leg measures 6 meters. What is the length in meters of the other leg?

(1) 5
(2) 6
(3) 7
(4) 8
(5) 9

17. Bill wants to find the width of a pond. He places stakes at A and B and then finds C so that C is a right triangle. This makes $\triangle ABC$ a right triangle. If $\overline{AC} = 60$ ft. and $\overline{BC} = 80$ ft., how far is it from A to B?

(1) less than 85 ft.
(2) between 85 and 95 ft.
(3) between 95 and 105 ft.
(4) between 105 and 115 ft.
(5) more than 115 ft.

18. A ladder is placed against the side of a building and reaches the second-story window. The ladder is 15 feet long. To find how far the window is from the ground, you must

(1) find 15^2 and 8^2 and add.
(2) find 15^2 and 8^2 and subtract.
(3) find 15^2 and 8^2 and take the square root of each.
(4) find 15^2 and 8^2 and subtract, then find the square root of the answer.
(5) find 15^2 and 8^2 and add, then find the square root of the answer.

19. One leg of a right triangle measures 24 feet. The hypotenuse measures 30 feet. What is the length of the other leg?

(1) 9
(2) 12
(3) 18
(4) 22
(5) 24

Answers are on page 401.

Strategies for Solving Word Problems

Use Mental Math

Using mental math means solving problems without pencil and paper or a calculator. It involves taking short cuts in computing to make solving problems easier.

Squares and Roots

You can use mental arithmetic to estimate the **squares** or the **square roots** of numbers from 1 to 10,000.

Example: What number when multiplied times itself gives a product of 4,096? In other words, find the square root of 4,096.

The first step involves a short cut for finding the square of numbers ending in 0. To find the square of a two-digit number ending in 0, square the tens digit and add two zeros.

$5^2 = 25$ so $50^2 = 2{,}500$
$6^2 = 36$ so $60^2 = 3{,}600$
$7^2 = 49$ so $70^2 = 4{,}900$

From our first step, we know the square root of 4,096 is between 60 and 70 because 4,096 is between 3,600 and 4,900.

Another shortcut allows us to find the square of a two-digit number ending in 5, from 15 to 95, such as 65^2. Multiply the ten's digit by the next higher consecutive number. Then add the digits "25" to the answer.

We have found that the square root of 4,096 must fall between 60 and 70.
Using the short cut, what is the square of 65?

65^2 Think, 6×7 next higher consecutive number

42 add the digits 25 = 4,225, so $65^2 = 4{,}225$.

The square root of 4,096 is greater than 60 and less than 65.

The square root must be either 61, 62, 63, or 64. Look at the last digit of each of the numbers. Which digit in the ones place when multiplied times itself will give you the last digit in the number 4,096? $4 \times 4 = 16$. We try 64^2 and find that $64 \times 64 = 4{,}096$. So **64** is the answer.

Solving Word Problems

Directions: Choose the best answer to each item.

1. What number times itself equals 3,025?

   (1) 35
   (2) 45
   (3) 55
   (4) 65
   (5) 75

2. Find the whole number closest to $\sqrt{2,650}$.

   (1) 49
   (2) 50
   (3) 51
   (4) 55
   (5) 60

3. Find the square root of 676.

   (1) 24
   (2) 25
   (3) 26
   (4) 27
   (5) 28

4. What number times itself equals 7,225?

   (1) 55
   (2) 65
   (3) 75
   (4) 85
   (5) 95

5. Find the square root of 3,136.

   (1) 36
   (2) 46
   (3) 56
   (4) 66
   (5) 76

6. What number times itself equals 4,624?

   (1) 62
   (2) 64
   (3) 66
   (4) 68
   (5) 70

7. Find the whole number closest to $\sqrt{1,205}$.

   (1) 30
   (2) 35
   (3) 40
   (4) 45
   (5) 50

8. Which of the following sets of numbers are in order from least to greatest?

   (1) $\sqrt{100}$, $\sqrt{144}$, 13
   (2) $\sqrt{100}$, 13, $\sqrt{144}$
   (3) 13, $\sqrt{100}$, $\sqrt{144}$
   (4) 13, $\sqrt{144}$, $\sqrt{100}$
   (5) $\sqrt{144}$, 13, $\sqrt{100}$

9. Which of the following sets of numbers are in order from greatest to least?

   (1) 2,300, 45^2, 49^2
   (2) 49^2, 2,300, 45^2
   (3) 45^2, 2,300, 49^2
   (4) 49^2, 45^2, 2,300
   (5) 2,300, 49^2, 45^2

Answers are on page 402.

Lessons 23–27

Directions: Choose the best answer to each item.

1. Find the volume of a cube (in cubic centimeters) when one side measures 6 cm.

   (1) 36
   (2) 108
   (3) 180
   (4) 216
   (5) 1,296

2. Find the approximate volume of the cylinder if it is filled to a height of 7 inches.

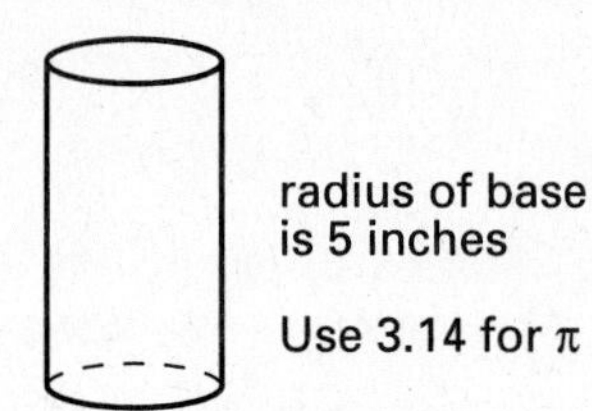

   (1) less than 350 cu. in.
   (2) from 350 to 425 cu. in.
   (3) from 425 to 500 cu. in.
   (4) from 500 to 575 cu. in.
   (5) over 575 cu. in.

3. What is the area in square feet of a rectangle that measures 2 yards by 2 feet?

   (1) 4
   (2) 12
   (3) 14
   (4) 15
   (5) 16

4. What is the area in square centimeters of the triangle in the figure?

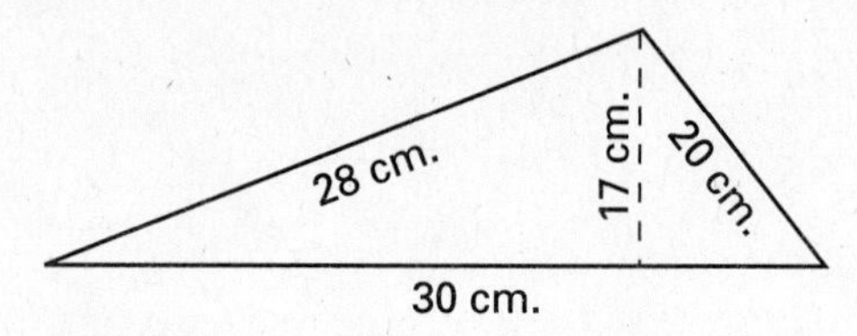

   (1) 65
   (2) 225
   (3) 255
   (4) 330
   (5) 720

5. A triangle has three sides that measure 6, 6, and 6. The triangle is

   (1) isosceles.
   (2) scalene.
   (3) equilateral.
   (4) right.
   (5) obtuse.

6. $\angle A$ and $\angle B$ are supplementary. If $m\angle A$ is twice $m\angle B$, what is the measure of $\angle B$?

   (1) 40°
   (2) 60°
   (3) 90°
   (4) 120°
   (5) 130°

Items 7 to 9 refer to the following figure.

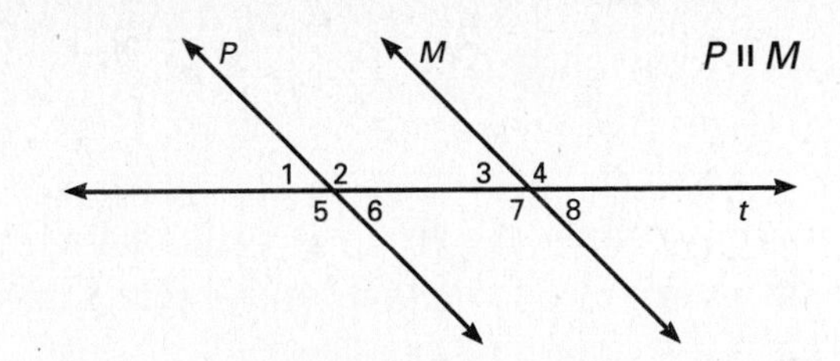

7. Which of the following statements is true about angles 4 and 5?

   (1) They are alternate exterior angles.
   (2) They are supplementary angles.
   (3) They are complementary angles.
   (4) They are corresponding angles.
   (5) They are alternate interior angles.

8. If $m\angle 8 = 45°$, which of the following statements is false?

   (1) $m\angle 8 = m\angle 1$
   (2) $m\angle 4 = 135°$
   (3) $m\angle 6 = 45°$
   (4) $m\angle 7 = 135°$
   (5) $m\angle 3 = 135°$

9. Which angle corresponds to $\angle 5$?

   (1) $\angle 3$
   (2) $\angle 4$
   (3) $\angle 6$
   (4) $\angle 7$
   (5) $\angle 8$

10. A 3-ft. post casts a $4\frac{1}{2}$-ft. shadow at the same time that a telephone pole casts a shadow of 33 ft. What is the length of the telephone pole?

   (1) 18
   (2) 22
   (3) 28
   (4) 33
   (5) 99

11. The distance between two towns on a map is $\frac{3}{4}$ of an inch. If the map scale is 1 inch = 40 miles, how far is the actual distance between the towns?

   (1) 15 mi.
   (2) 20 mi.
   (3) 25 mi.
   (4) 30 mi.
   (5) 35 mi.

Items 12 and 13 refer to the following diagram.

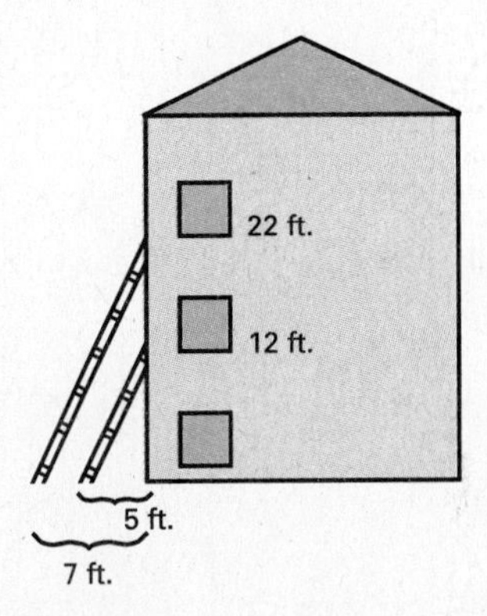

12. At an apartment building, the distance from the ground to the second-floor windows is 12 ft. If a ladder is positioned 5 ft. out from the building, about how long would the ladder need to be to reach a second-floor window?

   (1) 15 ft.
   (2) 14 ft.
   (3) 13 ft.
   (4) 12 ft.
   (5) 11 ft.

13. The third-floor windows are 22 ft. from the ground. If the ladder is positioned 7 ft. out from the building, about how long would the ladder need to be to reach a third-floor window?

   (1) 18 ft.
   (2) 20 ft.
   (3) 22 ft.
   (4) 23 ft.
   (5) 25 ft.

Answers are on page 402.

Irregular Figures

Area

A figure may be made up of several shapes. To find the area or volume of a combined figure, find the area or volume of each of the parts and add the results.

Example: Mr. Whittier plans to carpet the L-shaped office shown below. What is the area of the office in square feet?

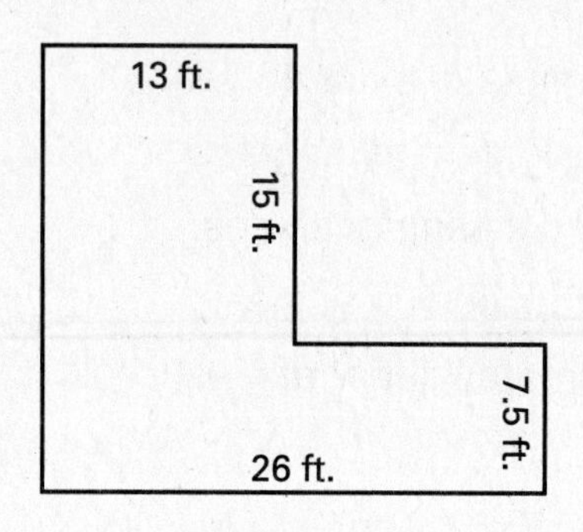

Break the figure into two rectangles.

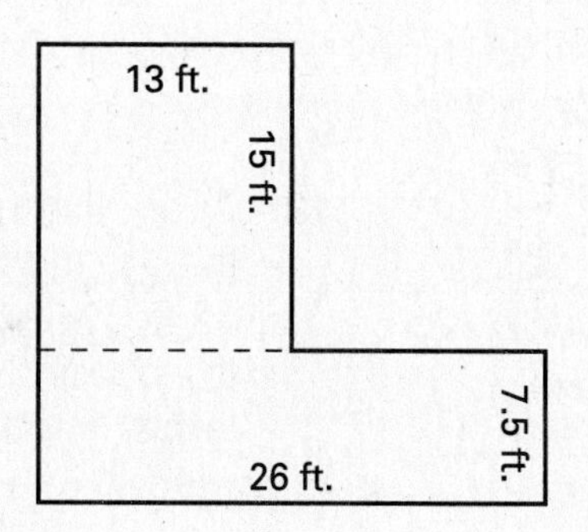

Find the area of each.

$A = lw$
$= 15(13)$
$= 195$ square feet

$A = lw$
$= 26(7.5)$
$= 195$ square feet.

Add the areas of the two rectangles:
$195 + 195 =$ **390 square feet.**

Example: What is the approximate area of the tabletop below?

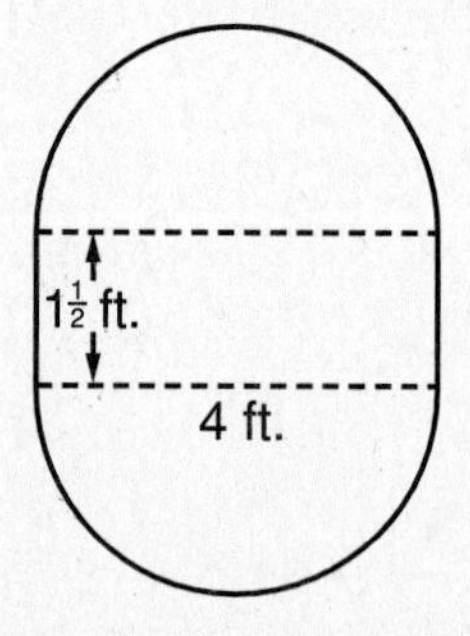

Think of each end as a half-circle. Find the circle's area.

$A = \pi r^2$
$= 3.14(2^2) \quad = 12.56$
which rounds to 13 square feet

Find the area of the rectangle.

$A = lw$
$= 4(1\frac{1}{2})$
$= 6$ square feet

Add the areas:
$13 + 6 =$ **19 square feet**

Volume

Example: Find the volume of the object shown below.

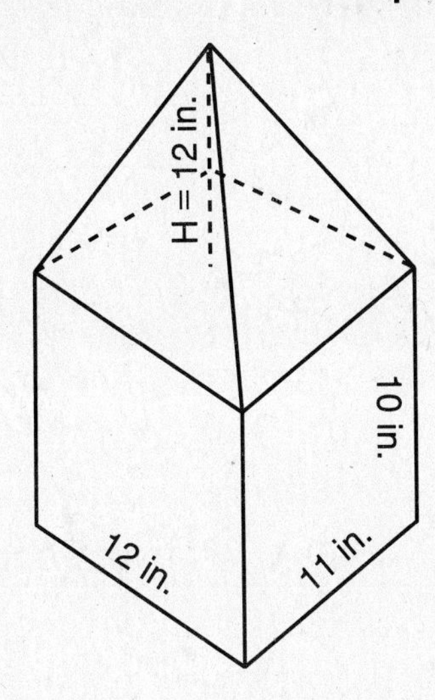

Find the volume of the pyramid.
First, find the area of the base:

$A = lw$
$= 12(11)$
$= 132$ square inches.

Next find the volume of the pyramid:

$V = \frac{1}{3}Ah$
$= \frac{1}{3}(132)(12)$
$= 44(12)$
$= 528$ cubic inches.

Then find the volume of the rectangular solid:

$V = lwh$
$= 12(11)(10)$
$= 132(10)$
$= 1{,}320$ cubic inches.

Add the results: $528 + 1{,}320 =$ **1,848 cubic inches.**

Example: Find the volume of the object shown below.

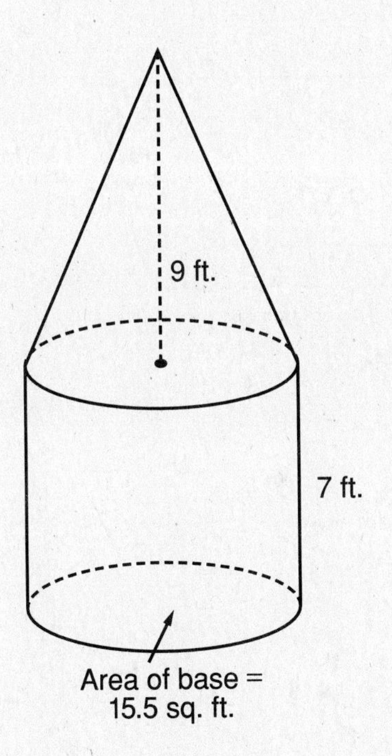

Find the volume of the cone.
The area of the base of the cylinder (15.5 square feet) is also the area of the base of the cone.

$V = \frac{1}{3}Ah$
$= \frac{1}{\not{3}}(15.5)(\not{9}^{3})$
$= 3(15.5)$
$= 46.5$ cubic feet

Find the volume of the cylinder.

$V = Ah$
$= (15.5)(7)$
$= 108.5$ cubic feet

Add the volumes:

$46.5 + 108.5 =$ **155 cubic feet**

Lesson 28

Directions: For items 1 to 5, find the area of each figure.

1.

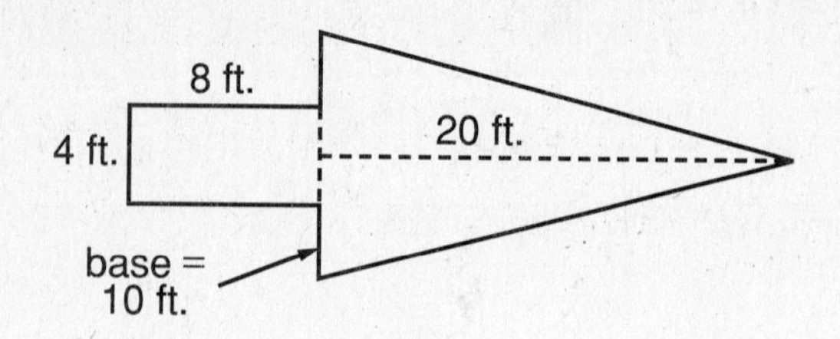

2.

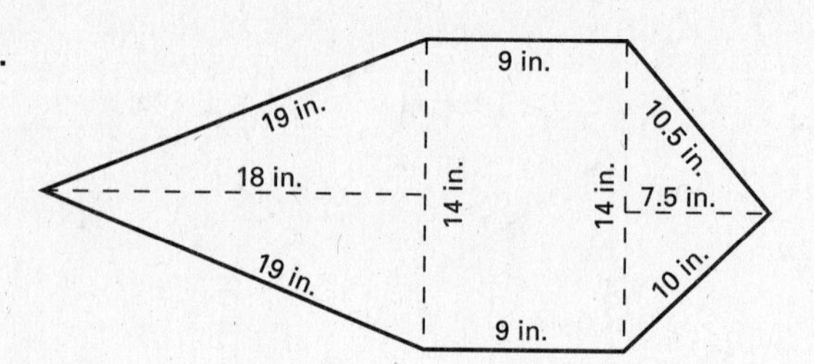

3.

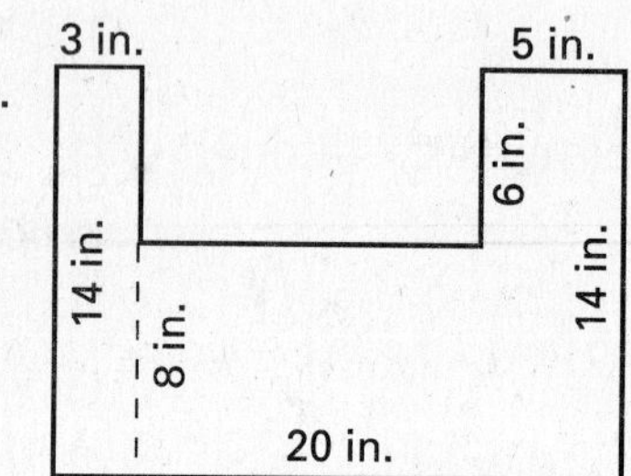

4.

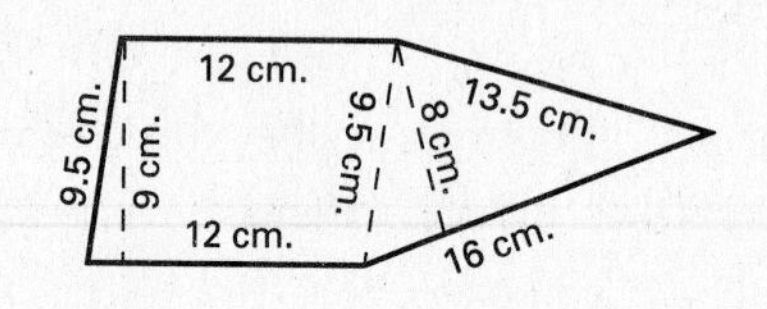

5.

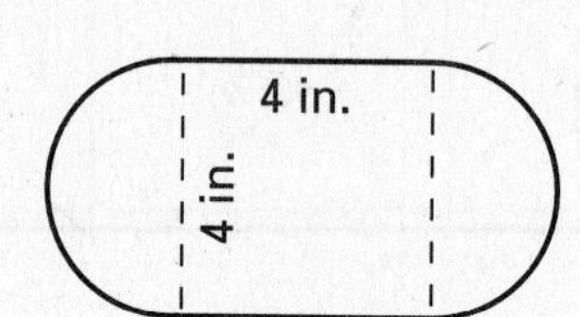

(Hint: Think of each end as a half-circle.)

Directions: For items 6 to 10, find the volume of the figure.

6.

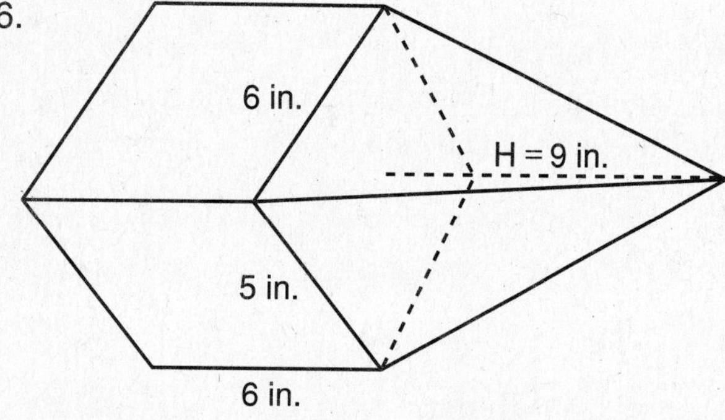

7.

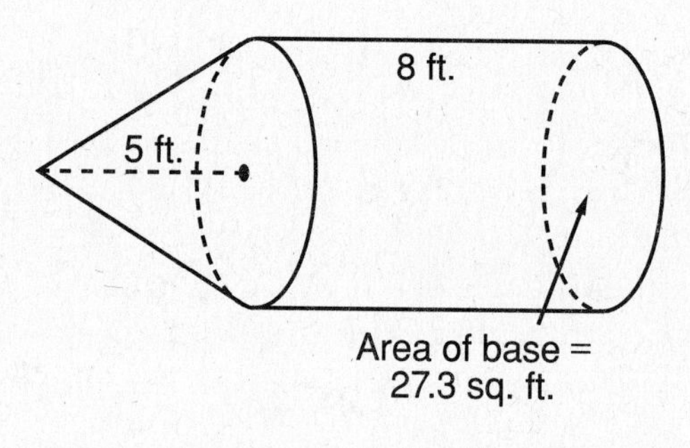

Directions: Each object below is made of two or more solid shapes combined into one. Name the shapes and find the volume of the object.

8.

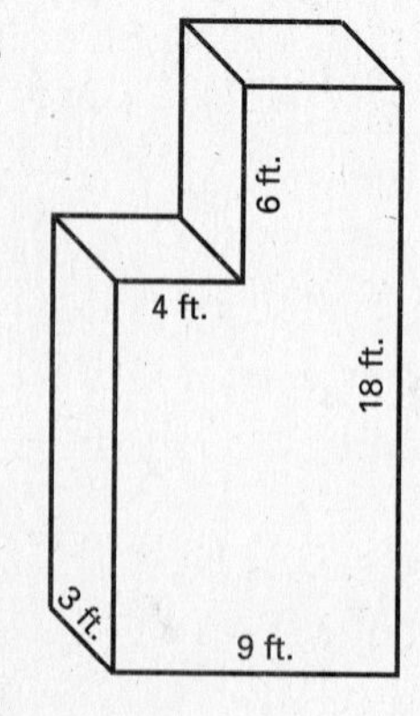

9.

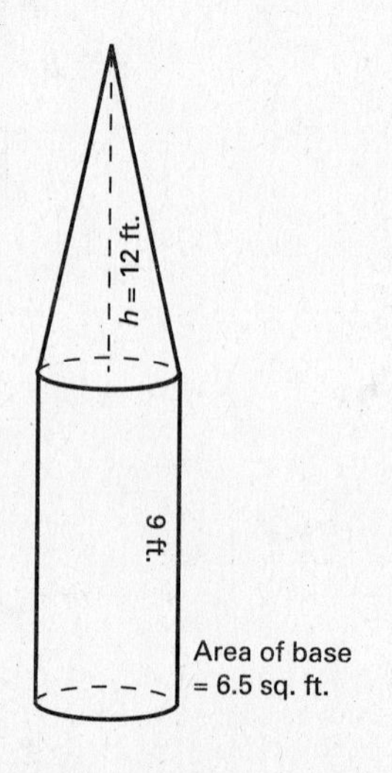

10.

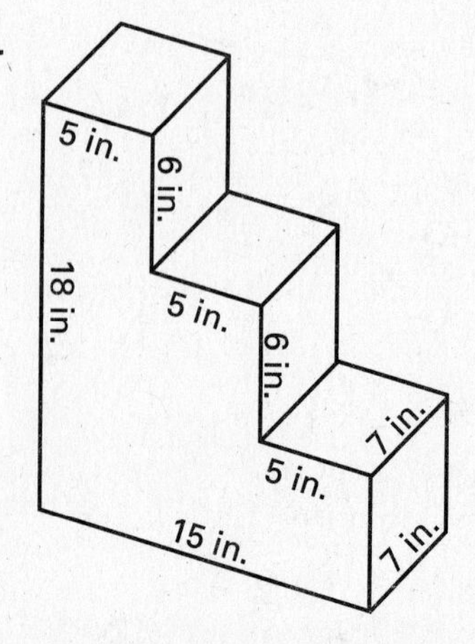

Directions: Choose the best answer to each item.

11. What is the area in square inches of the figure shown?

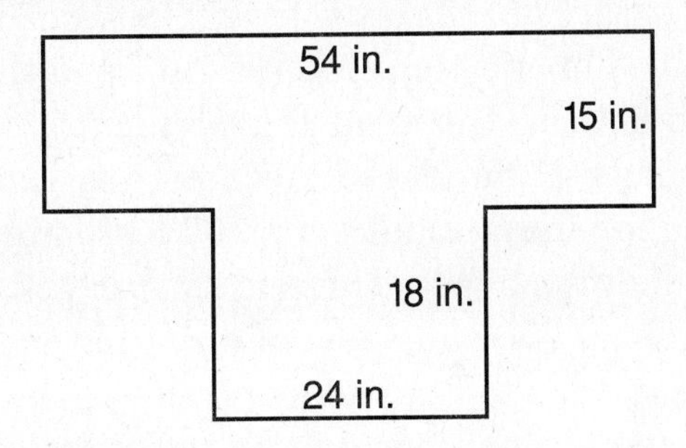

(1) 1,242
(2) 1,328
(3) 1,404
(4) 1,486
(5) 1,590

12. What is the volume in cubic feet of the figure shown?

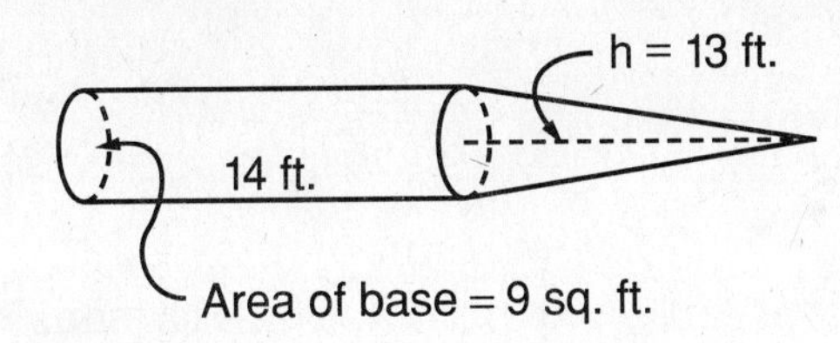

(1) 39
(2) 81
(3) 117
(4) 126
(5) 165

13. What is the area in square centimeters of the figure shown?

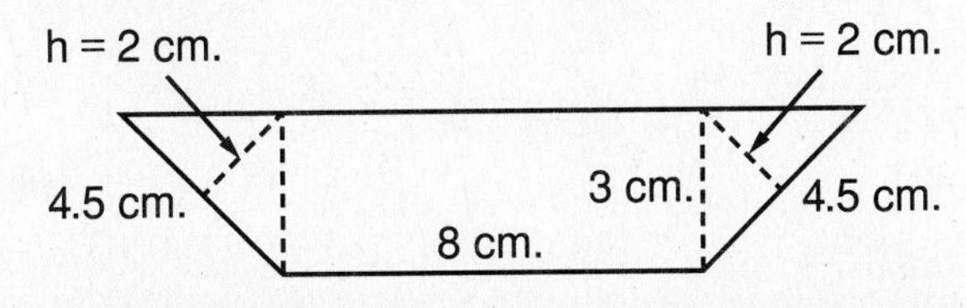

(1) 27
(2) 28.5
(3) 33
(4) 37.5
(5) 42

14. What is the area in square feet of the figure shown?

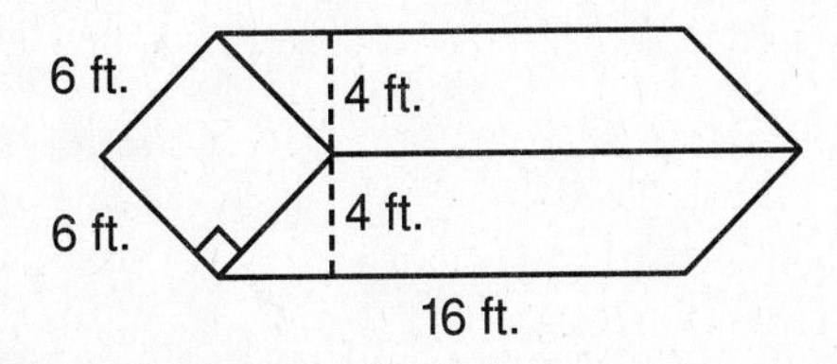

(1) 41
(2) 82
(3) 100
(4) 128
(5) 164

15. What is the volume in cubic centimeters of the figure shown?

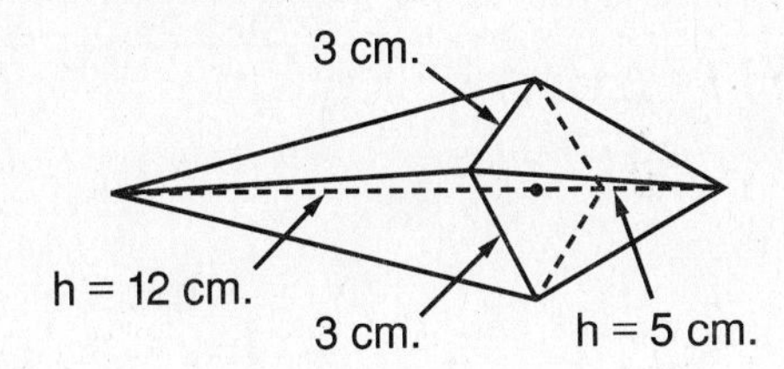

(1) 36
(2) 45
(3) 51
(4) 108
(5) 153

16. What is the area in square inches of the figure shown?

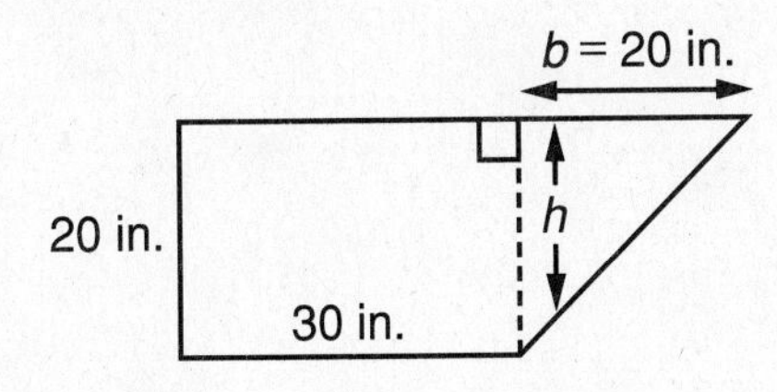

(1) 200
(2) 400
(3) 600
(4) 800
(5) 1,000

Answers are on page 402.

Strategies for Solving Word Problems

Solving a Multi-Step Problem

When a problem cannot be solved in one step, you need to read the problem several times and think about the steps for solving it.

Example: The city has contracted to build a new community center. The hole being dug for the foundation is 100 feet long by 35 feet wide. The depth is 6 feet. The trucks that will haul away the dirt each hold 250 cubic feet of dirt. How many truck loads will be needed to haul away all the dirt?

Think: First you need to find how many cubic feet of dirt will be removed. Then you can find the number of truck loads.

Step 1: The hole for the foundation is shaped like a rectangular solid. The amount of dirt to be taken from the hole is the same as the volume of the rectangular solid.

$V = lwh$
$= (100)(35)(6)$
$= 21{,}000$ cubic feet

Step 2: Find how many truck loads there are in 21,000 cubic feet. Each truck load equals 250 cubic feet. Divide the total number of cubic feet by 250.

$21{,}000 \div 250 = 84$

It will take **84** truck loads to haul away all the dirt.

Example: The diagram below shows the measurements of the Jacksons' living room and dining room. The Jacksons want to carpet both rooms with carpeting that costs $9.95 a square yard. What is the cost of carpeting the two rooms?

Think: First you need to find how many square yards of carpeting are needed. Then you can find the total cost.

Step 1: Find the areas of the two rooms in square feet.

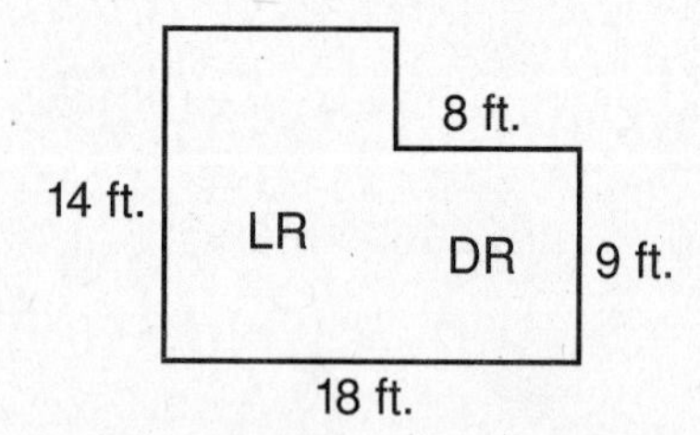

The living room is 14 feet by 18 − 8 feet.

$A = lw$
$= 14(10)$
$= 140$ square feet

The dining room is 9 feet by 8 feet.

$A = lw$
$= 9(8)$
$= 72$ square feet

Add the areas:

$140 + 72 = 212$ square feet

Step 2: Convert the square feet to square yards. Divide the total number of square feet by the number of square feet per square yard.

$212 \div 9 = 23.5$, which rounds to 24

The total amount of carpet needed is 24 square yards.

Step 3: Find the cost of the carpeting. Multiply the total number of square yards of carpet by the cost per square yard.

$24 \times \$9.95 = \$238.80$

The total cost of carpeting the two rooms is **$238.80.**

Example: The cylindrical water tank on top of Mark's apartment building is 12 feet high. The diameter of its base is 8 feet. There are 7.5 gallons of water in 1 cubic foot. Approximately how many gallons of water does the tank hold?

Think: First you need to find the volume of the tank in cubic feet. Then you can find the total number of gallons of water.

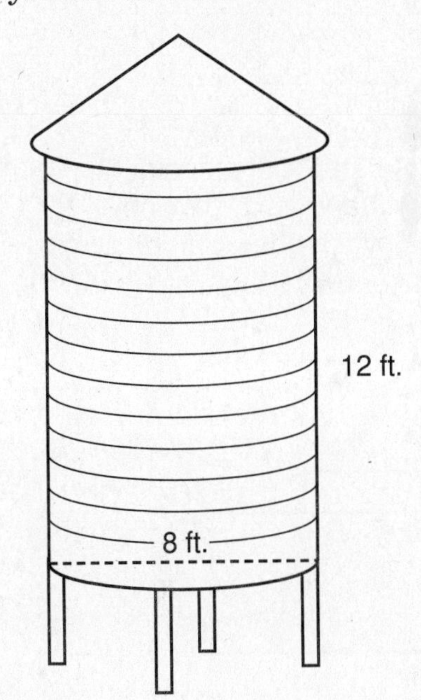

Step 1: Find the area of the base of the cylinder. The radius is $\frac{1}{2}$ the diameter, or 4 feet.

$$\begin{aligned} A &= \pi r^2 \\ &= 3.14(4^2) \\ &= 3.14(16) \\ &= 50.24 \text{ square feet} \end{aligned}$$

Step 2: Find the volume of the cylinder.

$$\begin{aligned} V &= Ah \\ &= (50.24)(12) \\ &= 602.88 \text{ cubic feet} \end{aligned}$$

Step 3: Multiply the total number of cubic feet by the number of gallons of water in 1 cubic foot.

$7.5 \times 602.88 = 4{,}521.6$ gallons

The tank holds approximately **4,522** gallons of water.

Solving Word Problems

Directions: Solve each problem.

1. In the following figure, the shaded part shows the walkway around a garden planted by the Neighborhood Garden Association. The garden measures 36 feet by 20 feet. How many square feet are on the surface of the walkway?

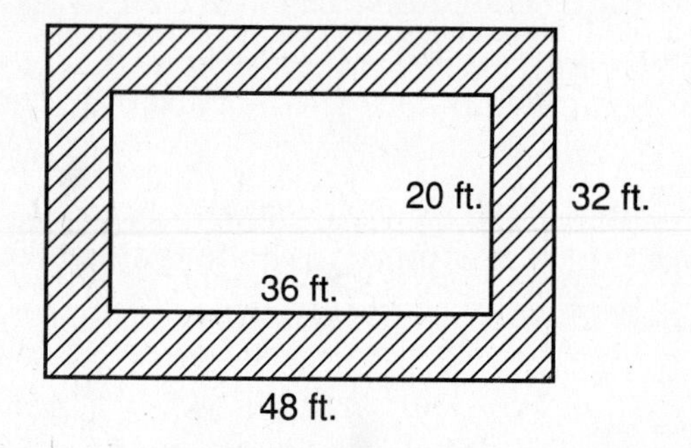

2. Martin is packing boxes that contain paint sets into cartons. How many boxes, which are 1 foot by 8 inches by 2 inches, can be fit into the carton shown below?

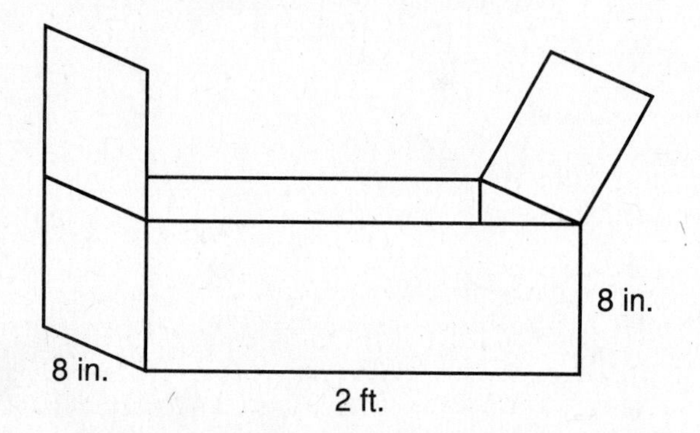

3. The following diagram shows the plan of a new neighborhood pool. The construction workers will line the edge of the pool with 6-inch ceramic tile strips. How many strips will they need?

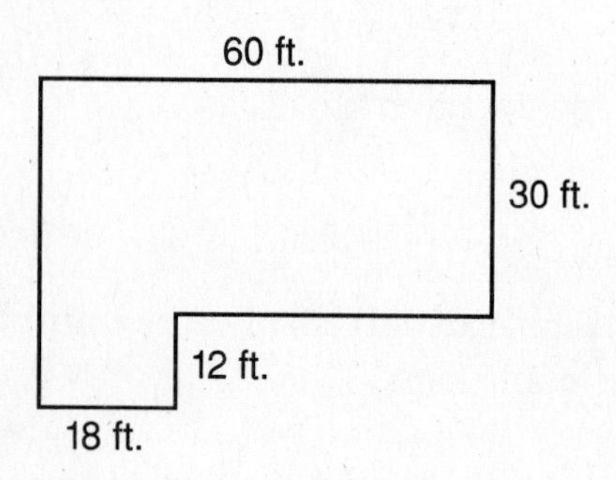

4. The following figure shows the dimensions of a refrigerator. It is advertised as an 18-cubic-foot refrigerator. How much space is taken up by the motor, insulation, and other such things?

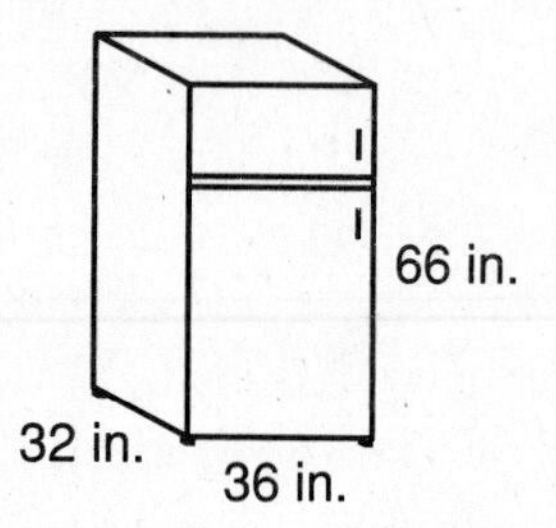

5. The following figure shows the plan for the lawn area of the Garcia's new home. How many cubic feet of soil will be needed to prepare the lawn area if it is spread to a depth of 6 inches?

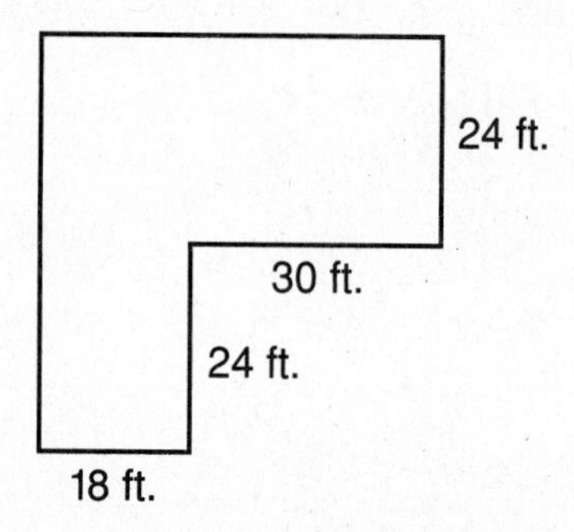

6. Tanya is ordering shingles for the roof of the Children's Center, shown in the following figure. A bundle of shingles covers 3 square yards of roof. How many bundles of shingles does Tanya need to order?

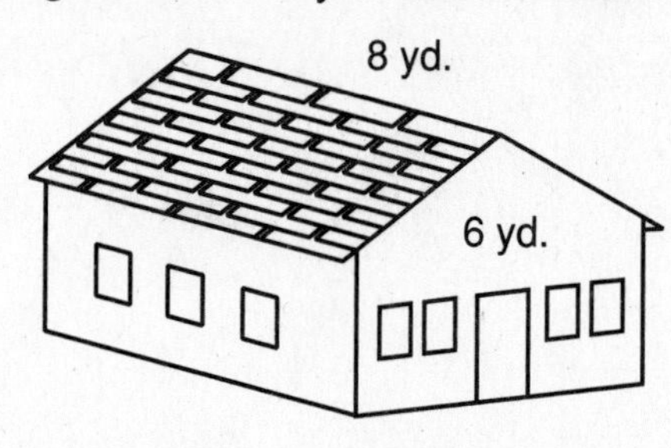

Directions: Choose the best answer for each item.

7. A toy maker is using cylindrical containers to hold a new toy. Each cylinder has a base area of 80 square inches and a height of 9 inches. If the toy is in the shape of a cube 8 inches on a side, how much of the cylinder in cubic inches needs to be filled with packing?

(1) 208
(2) 512
(3) 656
(4) 720
(5) 5,760

8. The members of the After-School Study Group want to paint the floor of their computer room. The floor is 45 feet by 60 feet. A gallon of paint covers 300 square feet. If the paint costs $21.95 a gallon, how much is the cost of paint for the floor?

(1) $ 131.70
(2) $ 197.55
(3) $ 658.50
(4) $1,317.00
(5) $1,975.50

9. A contractor building a group of houses has a large amount of trash to remove. Trash containers are 12 feet long, 8 feet wide and 6 feet high. If there are 27 cubic feet in 1 cubic yard, how many containers will be needed to remove 128 cubic yards of trash?

(1) less than 5
(2) 5
(3) 6
(4) 7
(5) more than 7

10. Mrs. Turner plans to tile her kitchen floor, which measures 6 feet by 9 feet. She will buy 6-inch square tiles. If the tiles cost $1.99 each, what is the total cost of the tiles she needs?

(1) $ 47.76
(2) $ 71.64
(3) $429.84
(4) $477.60
(5) $716.40

11. Shawn is paving a rectangular driveway with dimensions of 24 feet by 64 feet. He will put 3 inches of concrete on the driveway. Concrete is ordered by the cubic yard, and he cannot order part of a cubic yard. How many cubic yards of concrete should Shawn order?

(1) 11
(2) 12
(3) 13
(4) 14
(5) 15

12. Mrs. Miller needs to order new fencing for the preschool playground. The playground measures 36 feet by 90 feet. Fencing will be needed for all except the 6-foot gate. How many yards of fencing should Mrs. Miller order?

(1) 82
(2) 84
(3) 120
(4) 246
(5) 252

Answers are on page 405.

Area Relationships

Some problems are based on the relationship between the areas of two figures. You can use area formulas and what you know about solving equations to solve such problems.

Example: The two figures below have the same area. What is the side of the square?

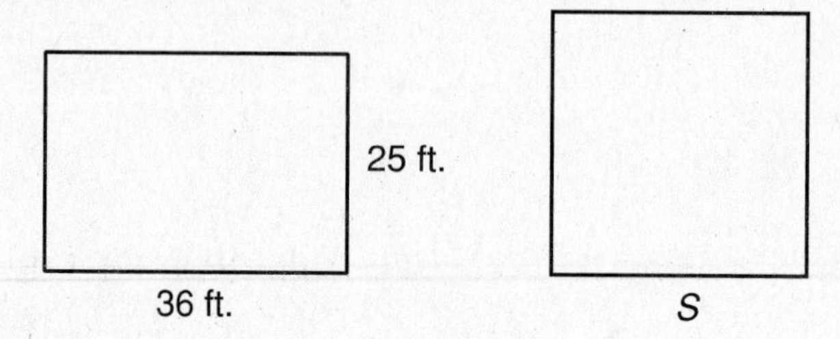

First: Find the area of the rectangle.

$$A = lw$$
$$= 36(25)$$
$$= 900$$

The area of the rectangle is 900 square feet.

To find the side of the square, use the formula for area of a square.

$$A = s^2$$
$$900 = s^2$$
$$\sqrt{900} = s$$
$$30 = s$$

The side of the square is **30 feet.**

Example: The two figures below have the same area. What is the length of side $\overline{BC}$?

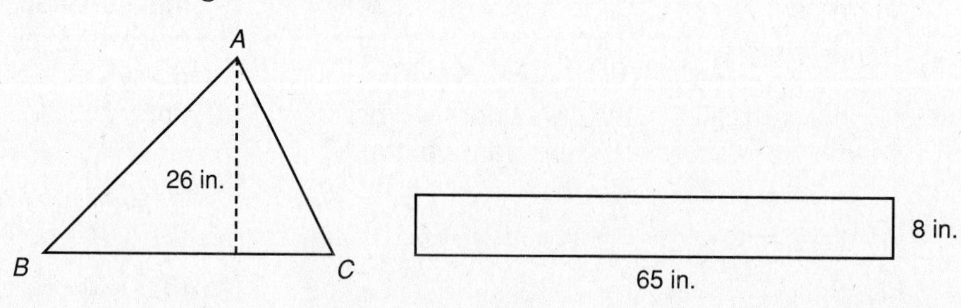

Find the area of the rectangle.

$$\begin{aligned} A &= lw \\ &= 65(8) \\ &= 520 \end{aligned}$$

The area of the rectangle is 520 square inches.

Side $\overline{BC}$ is the base of the triangle. To find the length of side $\overline{BC}$, use the formula for area of a triangle.

$$\begin{aligned} A &= \frac{1}{2}bh \\ 520 &= \frac{1}{2}b(26) \\ 520 &= \frac{1}{\cancel{2}_1}b(\cancel{26}^{13}) \\ \frac{520}{13} &= b \\ 40 &= b \end{aligned}$$

The length of side BC is **40 inches.**

Example: The two figures below have the same area. What is the base of the parallelogram?

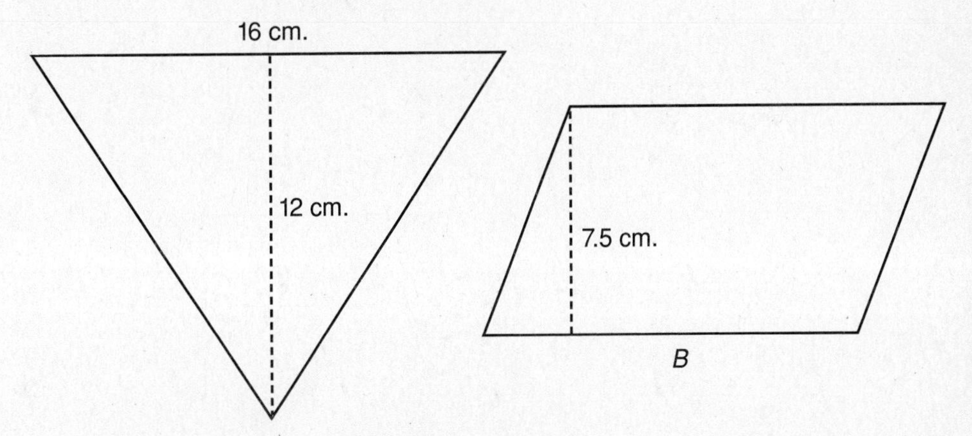

Find the area of the triangle.

$$\begin{aligned} A &= \frac{1}{2}bh \\ &= \frac{1}{2}(16)(12) \\ &= \frac{1}{\cancel{2}_1}(\cancel{16}^{8})(12) \\ &= 96 \end{aligned}$$

The area of the triangle is 96 square centimeters.
Use the formula for area of a parallelogram.

$$\begin{aligned} A &= bh \\ 96 &= b(7.5) \\ \frac{96}{7.5} &= b \\ 12.8 &= b \end{aligned}$$

The base of the parallelogram is **12.8 centimeters.**

Lesson 29

Directions: In items 1 to 4, the two figures in each pair have the same area. Find the unknown measure.

1. What is the side of the square?

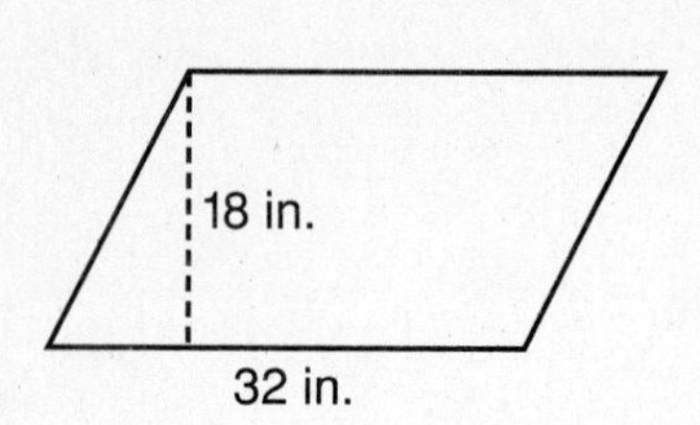

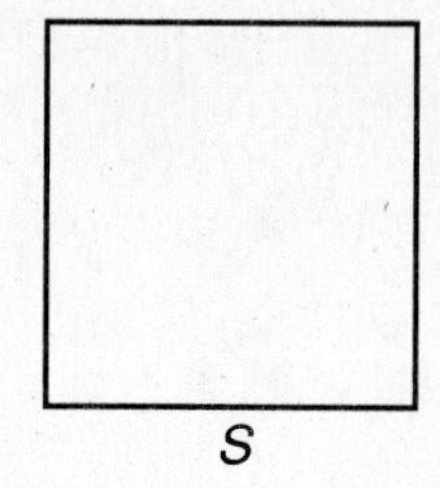

2. What is the height of the triangle?

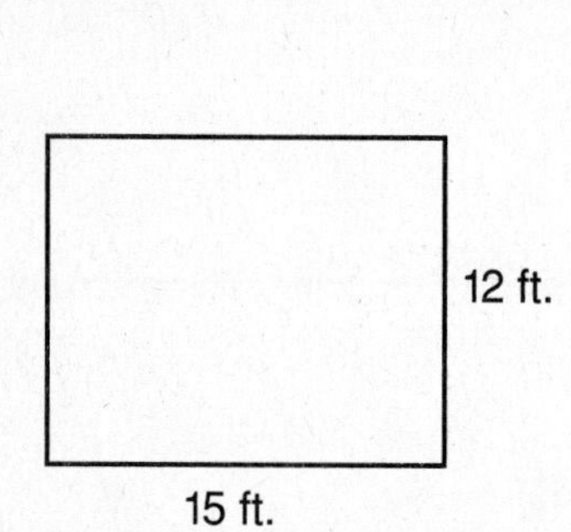

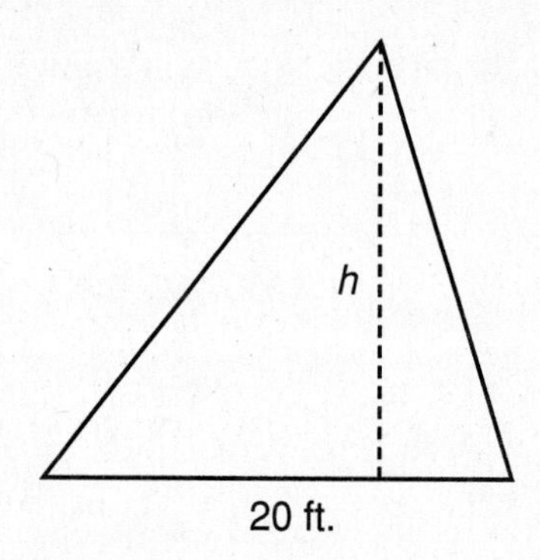

3. What is the base of the parallelogram?

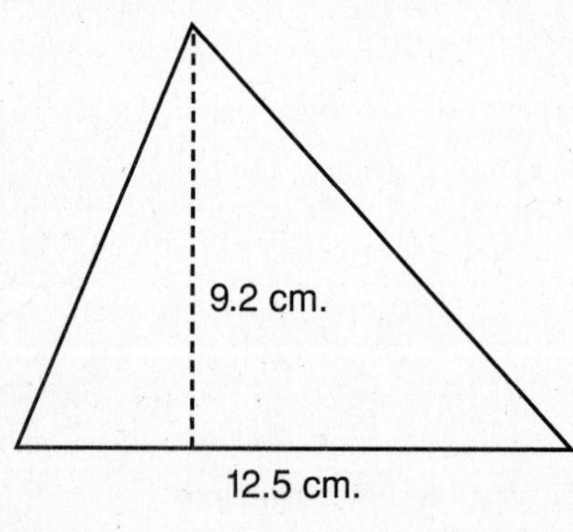

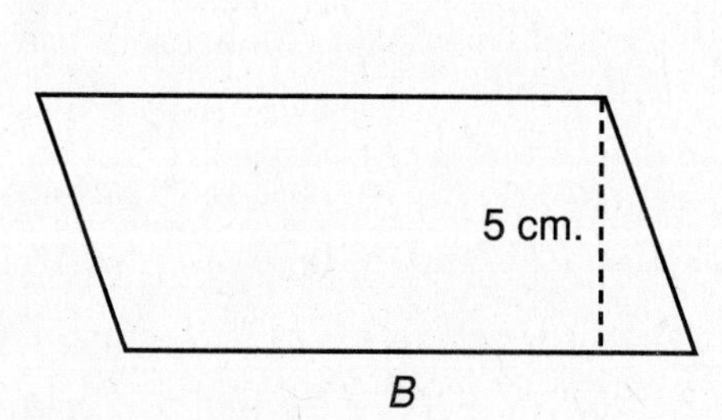

4. What is the side of the square?

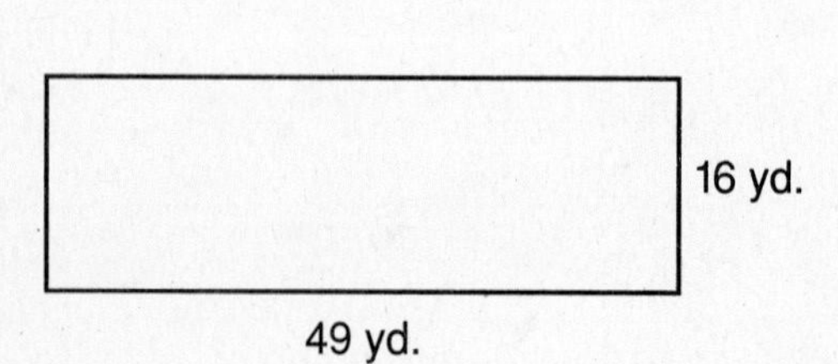

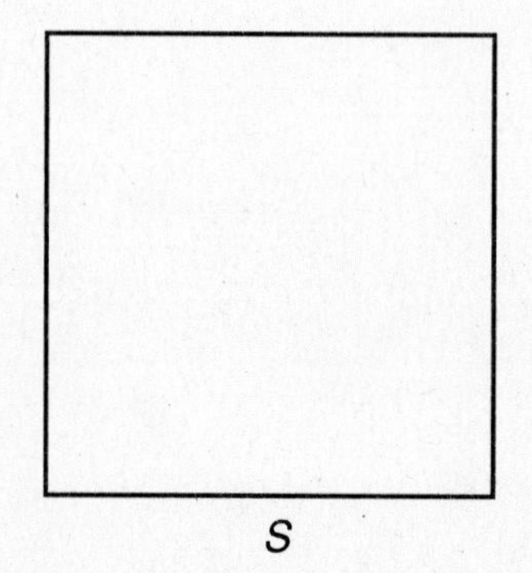

Directions: Choose the best answer to each item.

5. The length of a rectangle is 25 inches and the width is 16 inches. What is the side of a square that has the same area as the rectangle?

(1) 16 inches
(2) 18 inches
(3) 20 inches
(4) 200 inches
(5) Not enough information is given.

6. The following figures have the same area. What is the height of the parallelogram?

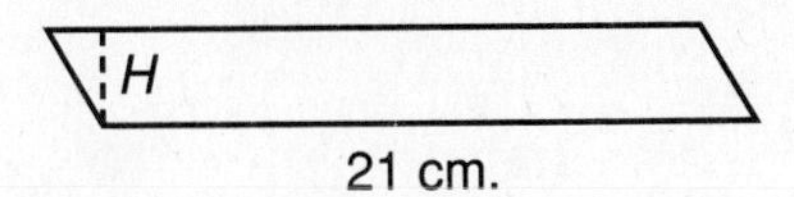

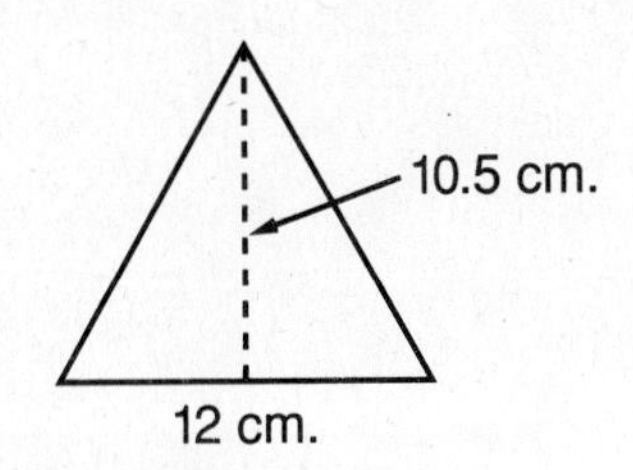

(1) 3.0 centimeters
(2) 3.5 centimeters
(3) 4.0 centimeters
(4) 9.0 centimeters
(5) 9.5 centimeters

7. A rectangle with the length 90 feet has the width $\frac{1}{2}$ foot. A triangle with the same area as the rectangle has a base of 18 feet. What is the height of the triangle?

(1) 1 foot
(2) 5 feet
(3) 9 feet
(4) 10 feet
(5) $10\frac{1}{2}$ feet

8. The following figures have the same area. What is the side of the square?

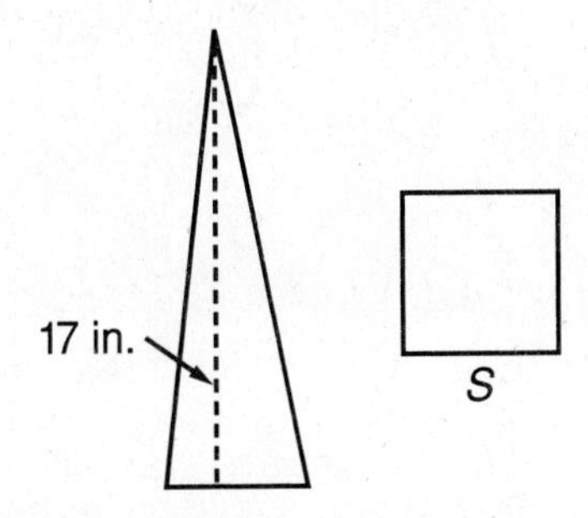

(1) $4\frac{1}{4}$ inches
(2) $8\frac{1}{2}$ inches
(3) $10\frac{1}{2}$ inches
(4) 17 inches
(5) Not enough information is given.

9. A square has the same area as a rectangle with length 64 centimeters and width 16 centimeters. What is the side of the square?

(1) 18 centimeters
(2) 20 centimeters
(3) 24 centimeters
(4) 32 centimeters
(5) Not enough information is given.

10. The following figures have the same area. What is the width of the rectangle?

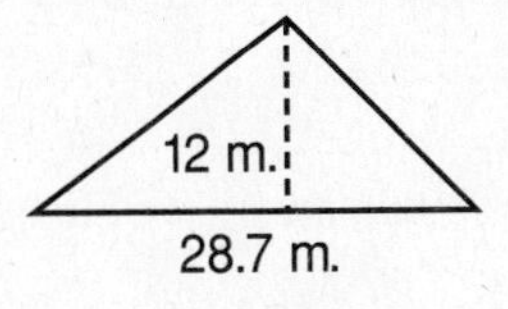

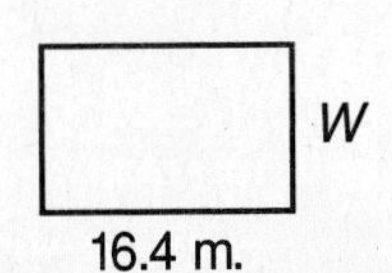

(1) 21.0 meters
(2) 20.5 meters
(3) 19.5 meters
(4) 11.5 meters
(5) 10.5 meters

Answers are on page 406.

Strategies for Solving Word Problems

Writing and Solving an Equation

You can use algebra to solve some complex problems in geometry. Use what you know about writing and solving equations.

Example: The perimeter of a rectangle is 44 inches. The length of the rectangle is 8 inches more than the width. What is the length of the rectangle?

Step 1: Since the length is described in terms of the width, start with the width. Let w stand for the width.

width $= w$

Step 2: Label the length in terms of w.

length $= w + 8$

Step 3: Write the equation. Use the formula for the perimeter of a rectangle.

$$P = 2l + 2w$$
$$44 = 2(w + 8) + 2w$$

Step 4: Solve for w.

$$44 = 2w + 16 + 2w$$
$$44 - 16 = 4w + 16 - 16$$
$$28 = 4w$$
$$\frac{28}{4} = \frac{4w}{4}$$
$$7 = w$$

The width is 7. The length is $7 + 8 =$ **15 inches.**

Step 5: Check your solution.

$P = 2l + 2w = 2(15) + 2(7) = 30 + 14 = 44$. The perimeter is 44 inches. The solution is correct.

Example: The area of a rectangle is three times the area of a square. The rectangle measures 27 feet by 25 feet. What is the side of the square?

Step 1: Let s stand for the side of the square.

side $= s$

Step 2: Write an equation to express the relationship between the areas. Use the formulas for area.

Area of rectangle $= 3 \times$ (Area of square)

$$lw = 3s^2$$
$$27 \times 25 = 3s^2$$

Step 3: Solve the equation for s.

$$27 \times 25 = 3s^2$$
$$675 = 3s^2$$
$$\frac{675}{3} = \frac{3s^2}{3}$$
$$225 = s^2$$
$$\sqrt{225} = s$$
$$15 = s$$

The side of the square is **15 feet.**

Step 4: Check the solution.

Area of square $= s^2 = 15 \times 15 = 225$

Area of rectangle $= 3 \times$ Area of square $= 3 \times 225 = 675 = 27 \times 25$.

The solution is correct.

Solving Word Problems

Directions: Choose the best answer to each item.

1. The perimeter of a rectangle is 48 feet. The length of the rectangle is 3 times the width. What is the width in feet?

   (1) 6
   (2) 8
   (3) 10
   (4) 12
   (5) 16

2. The area of a rectangle is 242 square feet. The length of the rectangle is twice as long as the width. What is the length in feet?

   (1) 10
   (2) 11
   (3) 18
   (4) 20
   (5) 22

3. The perimeter of a rectangle is 350 inches. The width of the rectangle is 10 inches less than the length. What is the length in inches?

   (1) 82.5
   (2) 92.5
   (3) 165
   (4) 175
   (5) 185

4. The area of a rectangle is half the area of a square. The rectangle measures 9 feet by 8 feet. What is the side of the square in inches?

   (1) 8
   (2) 9
   (3) 10
   (4) 11
   (5) 12

5. The height of a triangle is 4 times the base. The area of the triangle is 392 square inches. Find the base of the triangle.

   (1) 14 in.
   (2) 24 in.
   (3) 49 in.
   (4) 56 in.
   (5) 98 in.

6. A parallelogram with a base of 12 yards and a height of 21 yards has the same area as a triangle with a base of 24 yards. What is the height of the triangle in yards?

   (1) 12
   (2) 21
   (3) 24
   (4) 36
   (5) 42

7. The perimeter of a rectangle is 45 inches. The width is half the length. Find the length of the rectangle in inches.

   (1) 5
   (2) 5.5
   (3) 7
   (4) 7.5
   (5) 15

8. The perimeter of a square is 20 feet longer than the perimeter of a rectangle. The rectangle measures 28 feet by 16 feet. What is the side of the square in feet?

   (1) 20
   (2) 24
   (3) 27
   (4) 36
   (5) 40

Answers are on page 407.

ged Mini-Test Lessons 28 and 29

Directions: Choose the best answer to each item.

1. The following figures have the same area. What is the side of the square in centimeters?

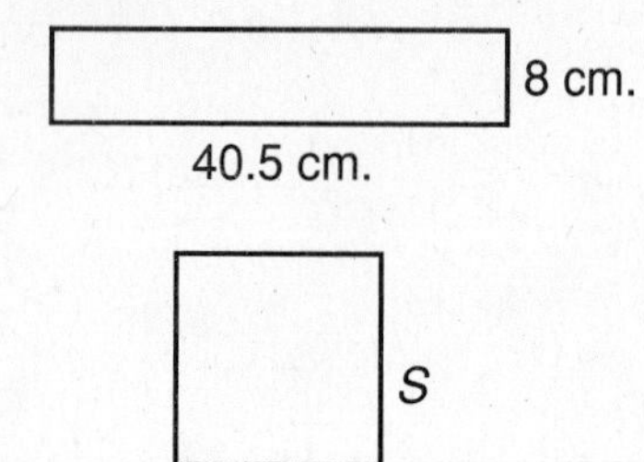

(1) 8
(2) 10
(3) 18
(4) 20.5
(5) 40.5

2. Mr. Robinson plans to put tiles on the floor of his son's bedroom. The bedroom is a rectangle with a length of 18 feet and a width of 9 feet. The tiles are 9-inch squares, and they cost $0.98 each. What is the total cost of the tiles he will need to buy?

(1) $ 28.22
(2) $ 31.36
(3) $162
(4) $233.28
(5) $282.24

3. Find the area in square inches of the following figure.

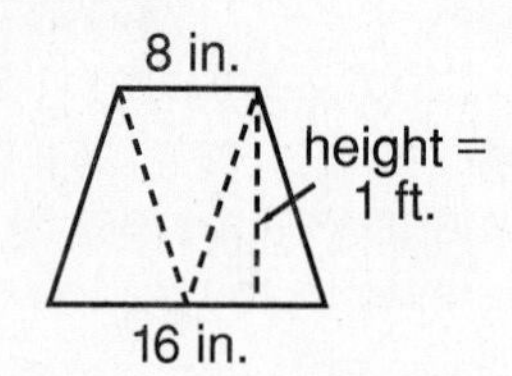

(1) 48
(2) 96
(3) 144
(4) 192
(5) 288

4. The perimeter of a rectangle is 210 feet. The length of the rectangle is 4 times the width. What is the length of the rectangle in feet?

(1) 21
(2) 42
(3) 63
(4) 84
(5) 105

5. Mr. Sikdar is planning to make a patio 20 feet by 16 feet in his backyard. How many cubic feet of concrete does he need if he plans to make the patio 6 inches deep?

(1) 160
(2) 320
(3) 456
(4) 480
(5) 1,920

6. The following figures have the same area. What is the height of the triangle in inches?

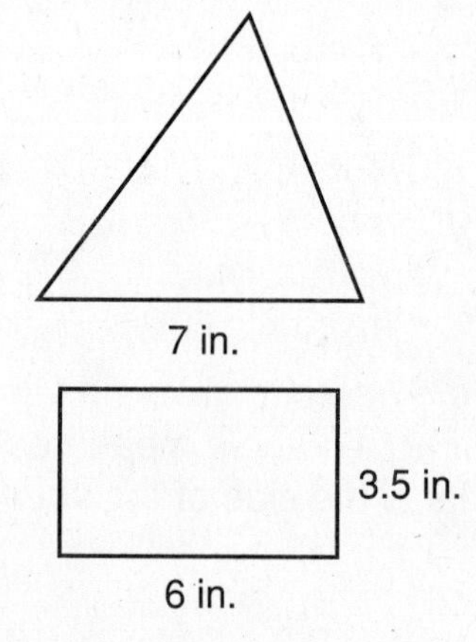

(1) 3.5
(2) 6
(3) 7
(4) 8
(5) 21

7. What is the approximate volume in cubic centimeters of the object shown below?

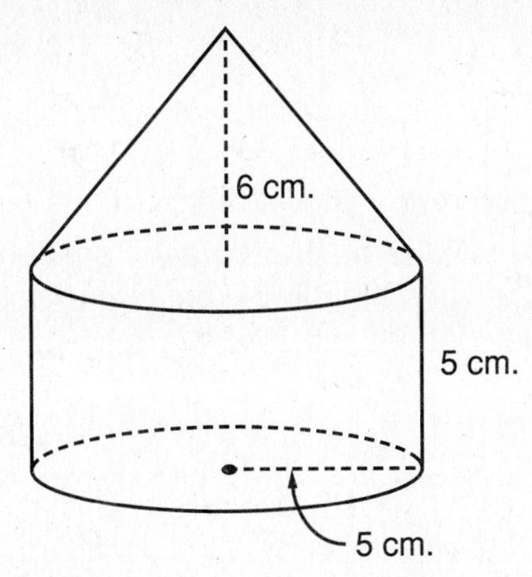

(1) 78.5
(2) 157
(3) 392.5
(4) 471
(5) 549.5

8. A circular pool 24 feet in diameter has a gravel walk around it that is 3 feet wide. Approximately how many square feet is the gravel walk?

(1) 144
(2) 225
(3) 254
(4) 452
(5) 706

9. What is the area in square feet of the following figure?

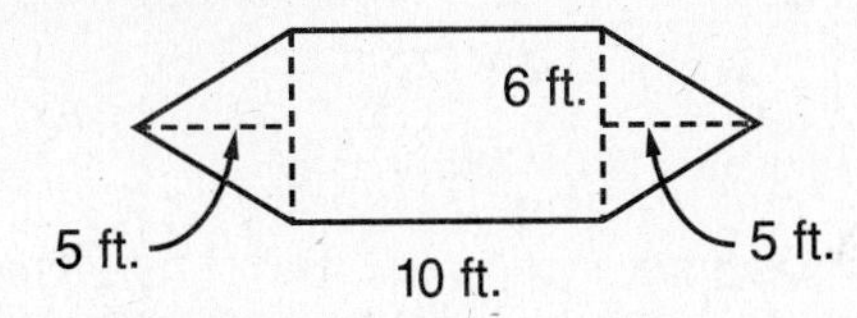

(1) 30
(2) 60
(3) 90
(4) 120
(5) 180

10. The perimeter of a rectangle is 26 yards. The width of the rectangle is 5 yards less than the length. What is the length in yards?

(1) 4
(2) 9
(3) 10
(4) 16
(5) 20

11. Find the volume in cubic inches of the following object.

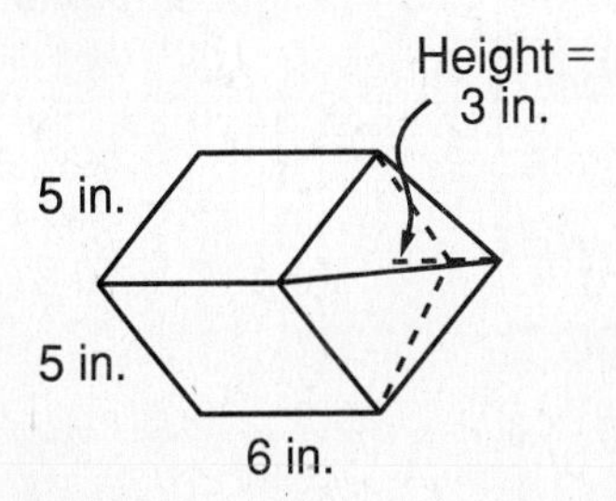

(1) 75
(2) 150
(3) 175
(4) 225
(5) 600

12. The area of a rectangle is 192 square centimeters. The length of the rectangle is 3 times the width. What is the width of the rectangle in centimeters?

(1) 6
(2) 8
(3) 24
(4) 48
(5) 64

13. A rectangular lawn measures 100 feet by 30 feet. What is the cost of seeding the lawn if 1 pound of seed costs $6.85 and covers 150 square feet?

(1) $1,370
(2) $ 685
(3) $ 137
(4) $ 68.50
(5) $ 13.70

Answers are on page 408.

ged Cumulative Review **Geometry**

Directions: Choose the best answer to each item.

Items 1 to 4 refer to the following figure.

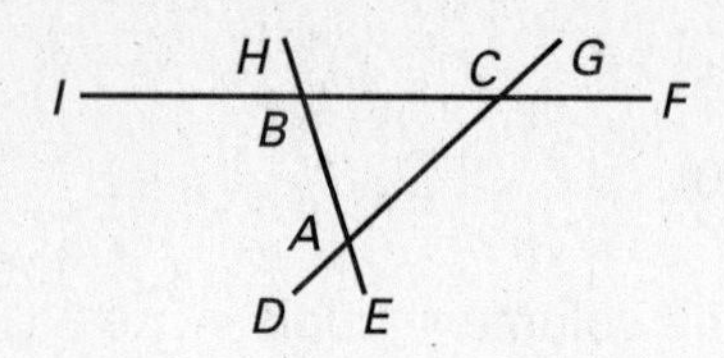

1. An example of a straight angle is
   (1) ∠*IBH*
   (2) ∠*GCF*
   (3) ∠*HBE*
   (4) ∠*DCF*
   (5) ∠*DCI*

2. Which angle forms a pair of vertical angles with ∠*DAH*?
   (1) ∠*DAE*
   (2) ∠*EAG*
   (3) ∠*GCF*
   (4) ∠*HAG*
   (5) ∠*IBH*

3. Which angle is supplementary to ∠*CBA*?
   (1) ∠*CBH*
   (2) ∠*HBI*
   (3) ∠*BCA*
   (4) ∠*CAB*
   (5) ∠*DAE*

4. Which angle forms a pair of vertical angles with ∠*BCA*?
   (1) ∠*DAB*
   (2) ∠*IBH*
   (3) ∠*IBA*
   (4) ∠*GCF*
   (5) ∠*DAE*

5. A photograph shows Mel to be $3\frac{1}{2}$ inches tall. In the same photograph Tom, who is 6 feet tall and standing next to Mel, appears to be 4 inches tall. How tall is Mel?
   (1) 5 ft. 10 in.
   (2) 5 ft. 8 in.
   (3) 5 ft. 6 in.
   (4) 5 ft. 4 in.
   (5) 5 ft. 3 in.

6. A sign 4 meters tall casts a shadow 4.8 meters long. At the same time a flagpole casts a shadow 24 meters long. How tall is the flagpole?
   (1) 10 m.
   (2) 15 m.
   (3) 20 m.
   (4) 24 m.
   (5) 25 m.

7. What is the perimeter of the figure shown?

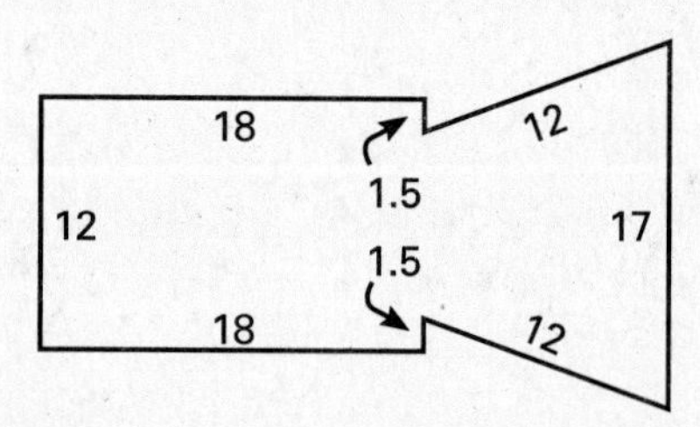

   (1) 72
   (2) 75
   (3) 90
   (4) 92
   (5) 101

8. The measure of $\angle CDE$ is 160°. $\overline{CF} \parallel \overline{BG}$. Which other angle has the same measure as $\angle CDE$?

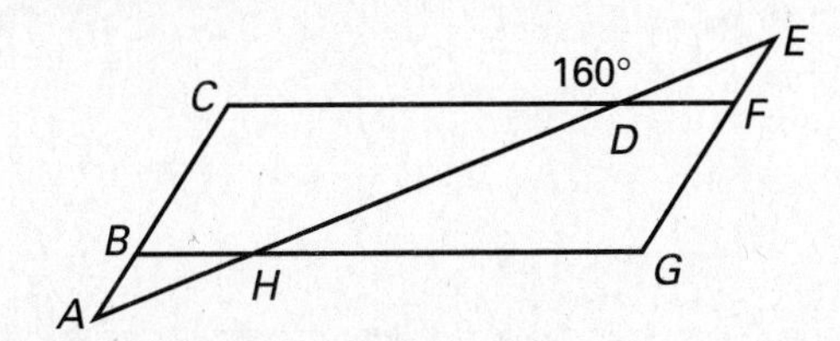

(1) $\angle BCD$
(2) $\angle EGH$
(3) $\angle AHG$
(4) $\angle ABH$
(5) $\angle DFE$

9. Which triangle is *not* isosceles?

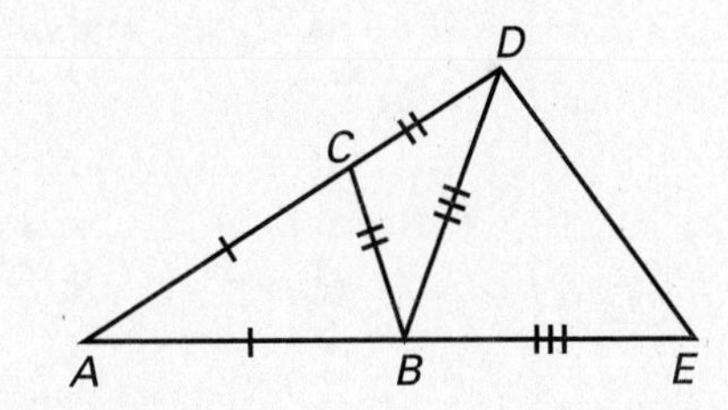

(1) $\triangle ABC$
(2) $\triangle BED$
(3) $\triangle BCD$
(4) $\triangle ABD$
(5) All triangles are isosceles.

10. The Pythagorean Theorem states that if one leg of a right triangle is *a* units, the other leg is *b* units and the hypotenuse is *c* units, then the length of the hypotenuse can be found by using this formula:

(1) $a^2 + b^2$
(2) $c^2 - b^2$
(3) $c^2 = a^2 + b^2$
(4) $\sqrt{c^2 - a^2}$
(5) $\sqrt{a^2 + c^2}$

11. $\angle ADB$ and $\angle C$ are right angles. How long is $\overline{AB}$?

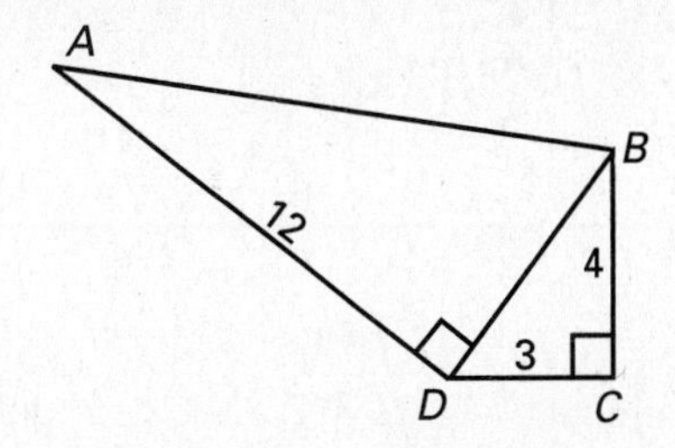

(1) 5
(2) 10
(3) 13
(4) 16
(5) 17

12. If *C* is the circumference, *d* is the diameter and *r* is the radius, which formula is false?

(1) $d = 2r$
(2) $C = \pi d$
(3) $C = 2\pi r$
(4) $\frac{C}{d} = \pi$
(5) $\pi = \frac{C}{r}$

13. What is the circumference of a circle with a radius of 3 cm.?

(1) 3π cm.
(2) 6π cm.
(3) 9π cm.
(4) 12π cm.
(5) 36π cm.

14. What is the area of a circle with a radius of 3 inches?

(1) 3π sq. in.
(2) 6π sq. in.
(3) 9π sq. in.
(4) 12π sq. in.
(5) 36π sq. in.

Items 15 and 16 refer to the following figure.

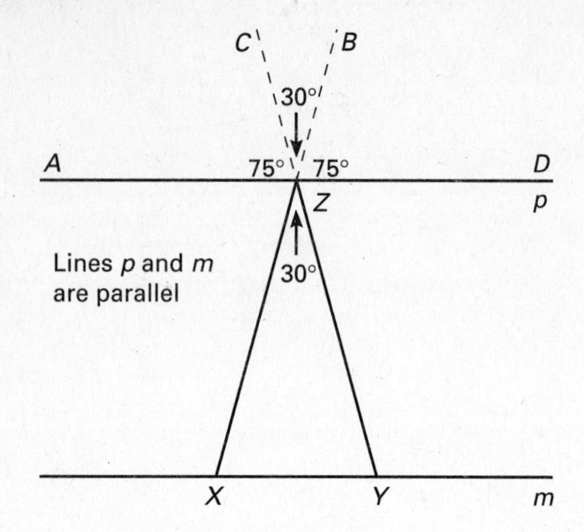

15. $m\angle ZXY = 75°$. Why?

    (1) $\angle ZXY$ and $\angle BZA$ are corresponding angles that are congruent.
    (2) $\angle ZXY$ and $\angle BZD$ are corresponding angles that are congruent.
    (3) $\angle ZXY$ and $\angle AZX$ are vertical angles.
    (4) $\angle AZB$ and $\angle CZD$ are vertical angles.
    (5) $\angle AZB$ and $\angle BZA$ are opposite angles.

16. $m\angle ZYX = 75°$. Why?

    (1) $\angle ZYX$ and $\angle CZA$ are corresponding angles.
    (2) $\angle ZYX$ and $\angle CZA$ are alternate interior angles.
    (3) $\angle Y$ is an interior angle with $\angle X$.
    (4) $\angle X$ and $\angle Y$ are corresponding angles.
    (5) $\angle X$ and $\angle A$ are corresponding angles.

17. The inside volume of a refrigerator is 19.5 cubic feet. It is 5.2 feet high and 1.5 feet deep, what is its width?

    (1) 2.5 ft.
    (2) 2.9 ft.
    (3) 3.75 ft.
    (4) 13 ft.
    (5) 15.2 ft.

18. Triangle *ECD* is a right triangle. Which angle is obtuse?

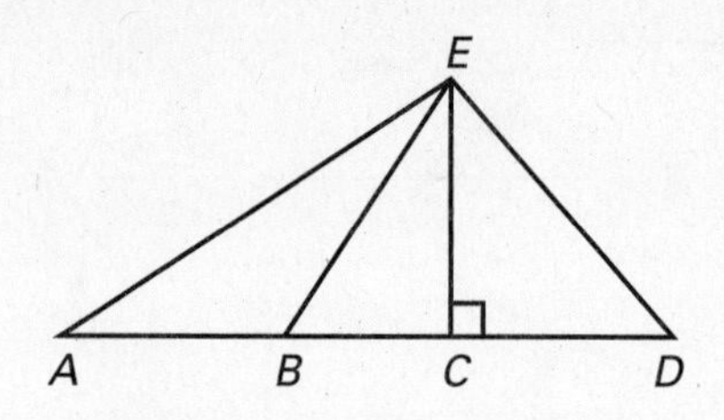

    (1) $\angle AED$
    (2) $\angle ACE$
    (3) $\angle BCE$
    (4) $\angle DAE$
    (5) Not enough information is given.

19. The measure of $\angle DAB$ is 70°. Which statement is true?

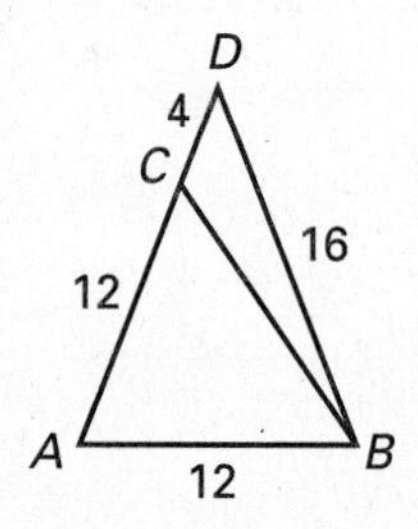

    (1) $m\angle ABC = 55°$
    (2) $m\angle D = 30°$
    (3) $m\angle ABC = 70°$
    (4) $m\angle ACB = 70°$
    (5) $m\angle CBD = 30°$

20. A post is 12 ft. high. A wire is secured to the top of the post and anchored to the ground 5 ft. from the base of the post. How long is the wire?

    (1) 9 ft.
    (2) 10 ft.
    (3) 12 ft.
    (4) 13 ft.
    (5) 17 ft.

21. What is the perimeter of this figure?

(1) $16 + 2\pi$
(2) $10 + 2\pi$
(3) $10 + 4\pi$
(4) $16 + 4\pi$
(5) $24 + 4\pi$

22. Rod, the manager of Cine-Max Theater, is changing the size of the popcorn box. He has made three choices shown below. Which statement is true about these three choices?

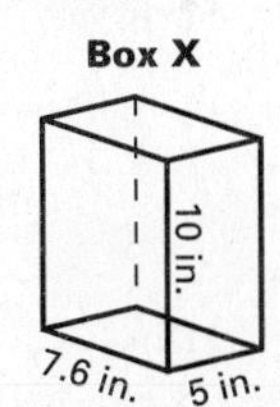

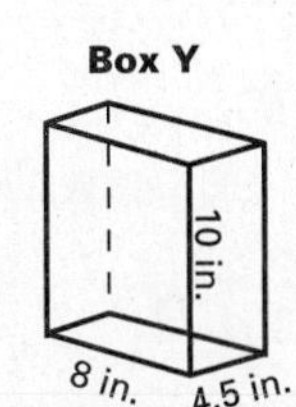

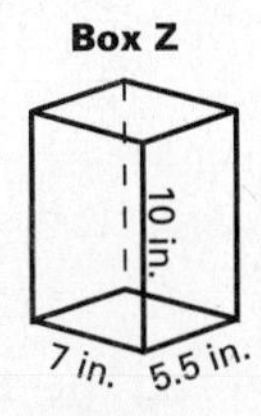

(1) Box X has the greatest volume.
(2) Box Y has the greatest volume.
(3) Box Z has the greatest volume.
(4) Box X and Box Z have the same volume.
(5) All three boxes have the same volume.

23. A triangle has sides measuring 8, 8, and 5. It is

(1) isosceles.
(2) scalene.
(3) equilateral.
(4) obtuse.
(5) right.

24. A triangle has sides measuring 9, 12, and 15. It is

(1) isosceles.
(2) acute.
(3) equilateral.
(4) obtuse.
(5) right.

Items 25 to 27 refer to the following figure.

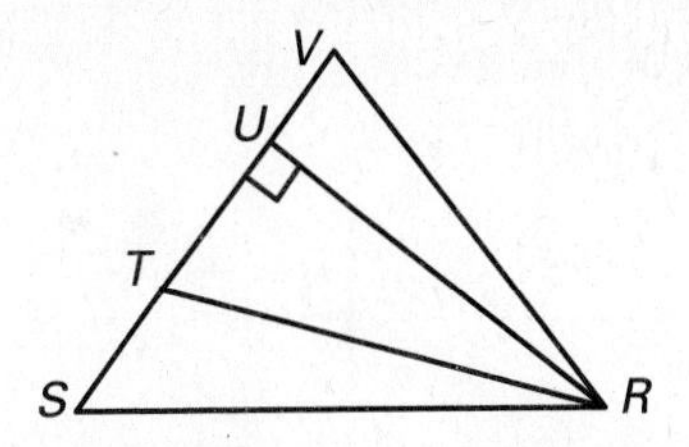

25. Which triangle is a right triangle?

(1) $\triangle TRS$
(2) $\triangle TUR$
(3) $\triangle SRV$
(4) $\triangle VRT$
(5) $\triangle STR$

26. $\triangle RTS$ is

(1) an acute triangle.
(2) a right triangle.
(3) a scalene triangle.
(4) an isosceles triangle.
(5) an equilateral triangle.

27. How many right triangles can be found?

(1) one
(2) two
(3) three
(4) four
(5) five

28. A perfume manufacturer has a gift box in the shape of a pyramid with a square base that is 3 inches on each side. The total volume of the box is 21 cubic inches. How high is the tallest bottle that can be packaged in this box?

(1) < 4 in. high
(2) $> 4 < 5$ in. high
(3) $> 5 < 6$ in. high
(4) $> 6 \leq 7$ in. high
(5) > 7 in. high

29. Similar marks on line segments and angles show congruent parts. $\triangle ABC \cong \triangle DCB$ Which of the following are not corresponding parts?

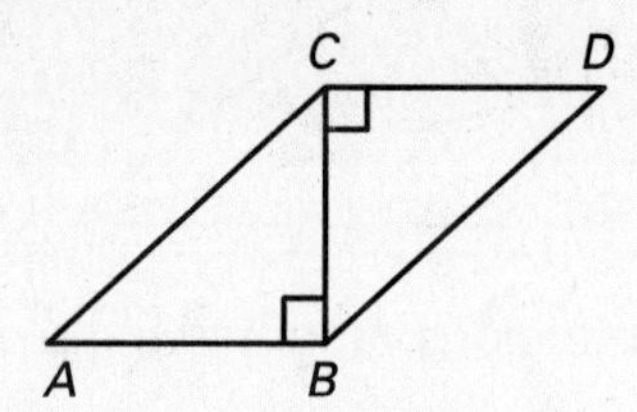

(1) $\overline{AB}$ and $\overline{DC}$
(2) $\triangle DCB$ and $\triangle ABC$
(3) $\angle A$ and $\angle D$
(4) $\overline{AC}$ and $\overline{DC}$
(5) $\overline{DB}$ and $\overline{CA}$

Items 30 and 31 refer to the following figure.

$\triangle ABC$ and $\triangle DEC$ are isosceles.

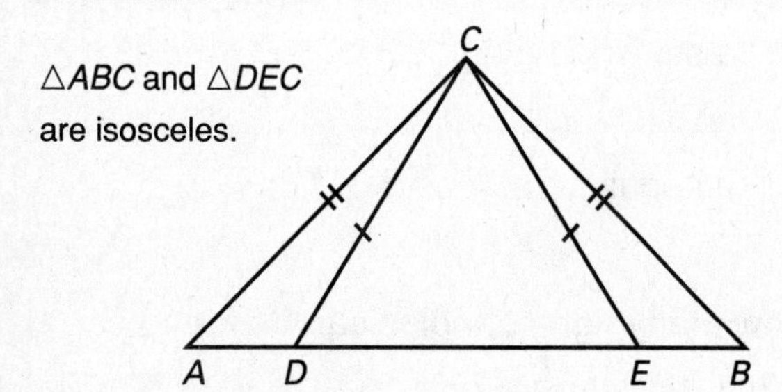

30. In order to prove $\triangle ADC \cong \triangle BEC$ by *SSS*, it is necessary to know that

(1) $\overline{AC} \cong \overline{DC}$
(2) $\overline{AE} \cong \overline{CE}$
(3) $\overline{AD} \cong \overline{BE}$
(4) $\angle ACD \cong \angle BCE$
(5) $\angle CAD \cong \angle CBE$

31. In order to prove $\triangle ACE \cong \triangle BCD$ by *SSS*, it is necessary to know that

(1) $\overline{AD} \cong \overline{BC}$
(2) $\overline{AE} \cong \overline{BD}$
(3) $\angle DAC \cong \angle EBC$
(4) $\angle CDE \cong \angle CED$
(5) $\overline{DE} \cong \overline{EB}$

32. Name the triangle that could possibly be congruent to $\triangle ADE$.

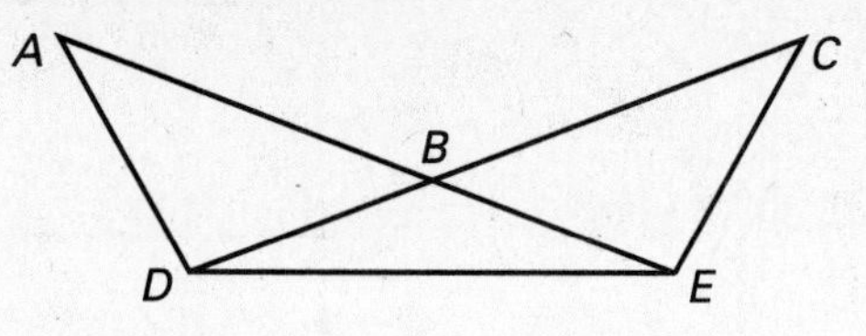

(1) $\triangle ADB$
(2) $\triangle DBE$
(3) $\triangle CEB$
(4) $\triangle CED$
(5) None could be congruent to $\triangle ADE$.

33. $\triangle AEC$ and $\triangle BDG$ are isosceles. What conclusion can be drawn?

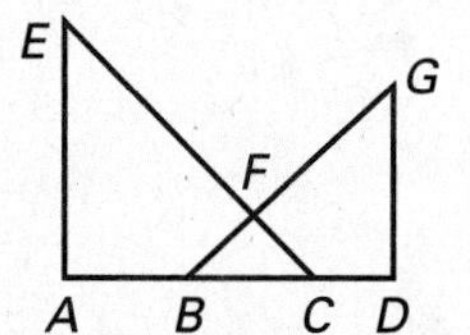

(1) $\overline{AB} \cong \overline{BC} \cong \overline{CD}$
(2) $\overline{AC} \cong \overline{AE}$
(3) $\overline{AE} \cong \overline{DG}$
(4) $\angle E \cong \angle G$
(5) $\overline{BF} \cong \overline{FC}$

34. An ancient Mexican pyramid has a volume of 82,600 cubic feet. It is 177 feet high and has a square base. What is the area of the base of the pyramid?

(1) 155.5 sq. ft.
(2) 466.6 sq. ft.
(3) 1,400 sq. ft.
(4) 4,200 sq. ft.
(5) More information is needed.

35. Which of the following is equal to $\sqrt{256}$?

(1) 14
(2) 15
(3) 16
(4) 17
(5) 18

Items 36 to 39 refer to the following figure.

Line a is parallel to line b.
$m\angle 3 = m\angle 4$

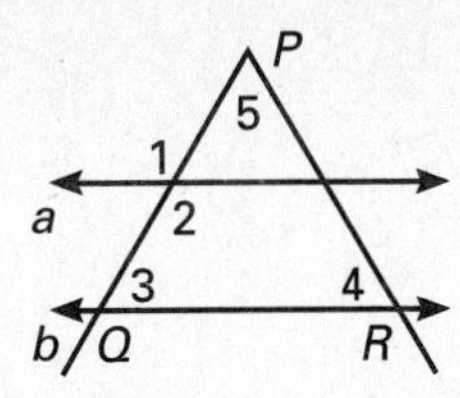

36. Which statement is true?

    (1) Angle 1 is an acute angle.
    (2) Angle 2 is a right angle.
    (3) Angle 4 is an obtuse angle.
    (4) Angle 3 is a right angle.
    (5) Angle 1 is an obtuse angle.

37. Which statement is true?

    (1) $m\angle 1 = m\angle 4$
    (2) $m\angle 2 + m\angle 3 = m\angle 4$
    (3) $m\angle 1 = m\angle 2$
    (4) $m\angle 1 + m\angle 3 = 90°$
    (5) $m\angle 1 = m\angle 3$

38. Which statement is true?

    (1) $\angle 3$ and $\angle 4$ are supplementary.
    (2) $\angle 1$ and $\angle 2$ are vertical angles.
    (3) $\angle 2$ and $\angle 3$ are complementary.
    (4) $\angle 1$ and $\angle 2$ are corresponding angles.
    (5) $\angle 2$ and $\angle 4$ are alternate interior angles.

39. Which statement is true if $m\angle 3 = 60°$?

    (1) $m\angle 3 = 45°$
    (2) $m\angle 4 = 120°$
    (3) $m\angle 2 = 60°$
    (4) Triangle PQR is an equilateral triangle.
    (5) $m\angle 1 = 135°$

40. If $\triangle ABF \sim \triangle ACG$, then which is a true proportion?

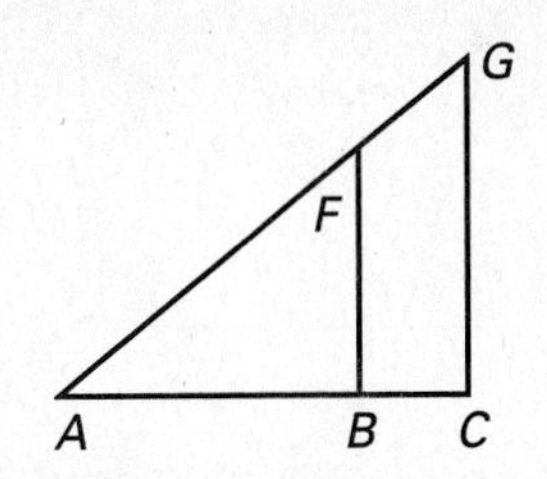

    (1) $\frac{\overline{AF}}{\overline{AB}} = \frac{\overline{AB}}{\overline{AC}}$
    (2) $\frac{\overline{AB}}{\overline{AC}} = \frac{\overline{FB}}{\overline{GC}}$
    (3) $\frac{\overline{AF}}{\overline{AC}} = \frac{\overline{AC}}{\overline{AB}}$
    (4) $\frac{\overline{AB}}{\overline{GC}} = \frac{\overline{AC}}{\overline{FB}}$
    (5) $\frac{\overline{AG}}{\overline{AC}} = \frac{\overline{AB}}{\overline{AF}}$

41. Find the length of $\overline{MN}$.

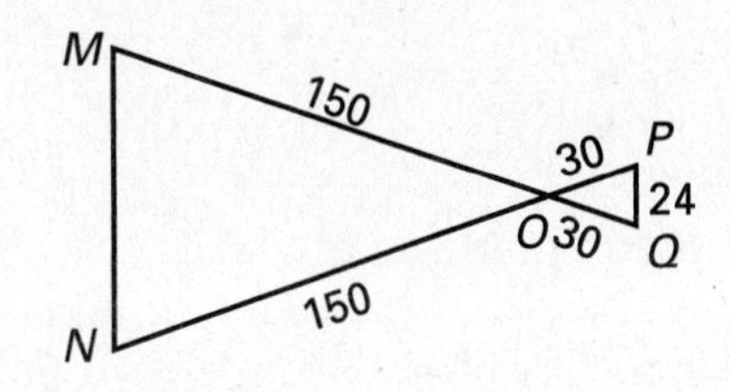

    (1) 48
    (2) 60
    (3) 90
    (4) 120
    (5) 150

42. Which of the following is equal to $\sqrt{289}$?

    (1) 16
    (2) 17
    (3) 18
    (4) 23
    (5) 27

Items 43 to 45 refer to the following figure.

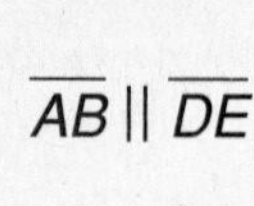

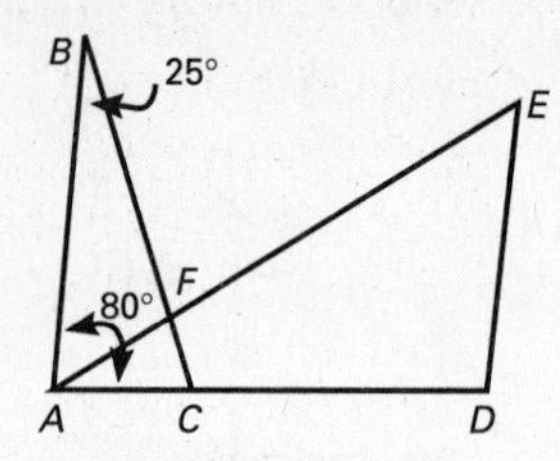

43. The measure of $\angle ACB$ is

(1) 20°
(2) 60°
(3) 75°
(4) 105°
(5) 110°

44. Angle D is

(1) the same size as $\angle BAC$
(2) the same size as $\angle BCD$
(3) the same size as $\angle B + \angle BAC$
(4) the supplement of $\angle BAC$
(5) the complement of $\angle B$

45. The measure of $\angle BCD$ is

(1) 105°
(2) 110°
(3) 120°
(4) 125°
(5) 130°

46. A cone-shaped candle mold has a base area of 4.5 square inches. The mold is 5 inches high. How many cubic inches of candle wax are needed to make 200 candles?

(1) 375 cu. in.
(2) 1,125 cu. in.
(3) 1,500 cu. in.
(4) 4,500 cu. in.
(5) 13,500 cu. in.

47. Which of the following is equal to 23?

(1) $\sqrt{529}$
(2) $\sqrt{523}$
(3) $\sqrt{519}$
(4) $\sqrt{513}$
(5) $\sqrt{509}$

Items 48 and 49 refer to the following figure.

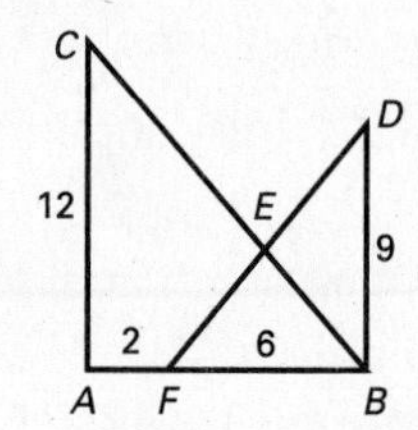

48. $\triangle ABC \sim \triangle BFD$. Which of the following must be true?

(1) $\angle C \cong \angle D$
(2) $\angle DFB \cong \angle C$
(3) $\angle CBA \cong \angle D$
(4) $\overline{BC} \cong \overline{BF}$
(5) $\overline{DB} \cong \overline{AB}$

49. $\triangle ABC \sim \triangle BFD$. If the length of $\overline{DF}$ is 12, then what is the length of $\overline{CB}$?

(1) 8
(2) 12
(3) 15
(4) 16
(5) 18

50. In a right triangle the short leg is 9 and the hypotenuse is 15. What is the measure of the other leg?

(1) 12
(2) 13
(3) 14
(4) 18
(5) 21

Items 51 and 52 refer to the following figure.

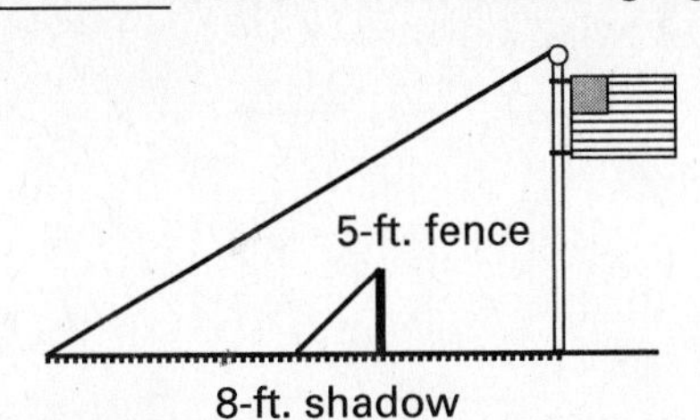

51. The flagpole casts a shadow 40 feet long. If the flagpole is x feet high, which proportion is used to find the height of the flagpole?

(1) $\frac{40}{5} = \frac{8}{x}$
(2) $40x = 8 \times 5$
(3) $40 \cdot x = 5 \cdot 8$
(4) $\frac{5}{x} = \frac{40}{8}$
(5) $\frac{5}{8} = \frac{x}{40}$

52. If the flagpole were 18 feet high, how long would its shadow be?

(1) about 20 ft. long
(2) about 25 ft. long
(3) about 30 ft. long
(4) about 35 ft. long
(5) about 40 ft. long

53. The scale on a house plan is 1 inch = 2 feet. The drawing of the kitchen is 5 in. by 7 in. What is the area of the actual kitchen?

(1) 12 sq. ft.
(2) 24 sq. ft.
(3) 35 sq. ft.
(4) 70 sq. ft.
(5) 140 sq. ft.

54. The perimeter of a rectangle is 20 yards. The width is $\frac{1}{4}$ of the length. Find the length.

(1) 2 yd.
(2) 8 yd.
(3) 10 yd.
(4) 12 yd.
(5) 14 yd.

55. A manufacturer of ice cream novelties is considering a new product. A cone is to be packed solid with a mixture of ice cream and nuts. How much more of the ice cream mixture does the larger cone (Sample B) hold than the smaller cone (Sample A)?

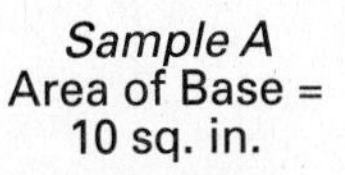

Sample B
Area of Base =
12 sq. in.

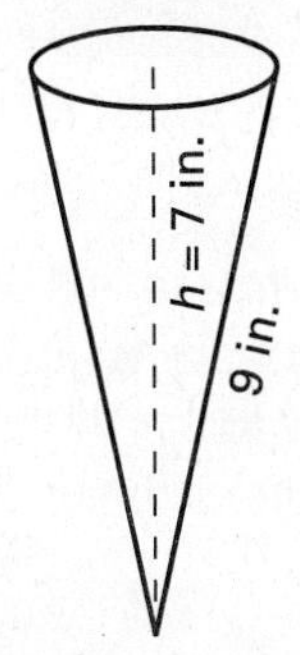

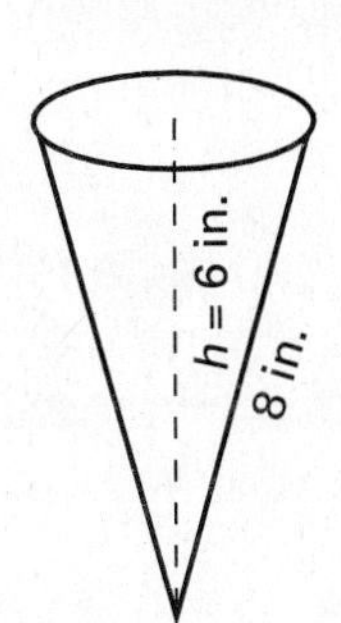

(1) $\frac{2}{3}$ cu. in.
(2) 3 cu. in.
(3) 5 cu. in.
(4) $6\frac{2}{3}$ cu. in.
(5) 9 cu. in.

56. If ∠8 measures 50°, name two angles that measure 130°.

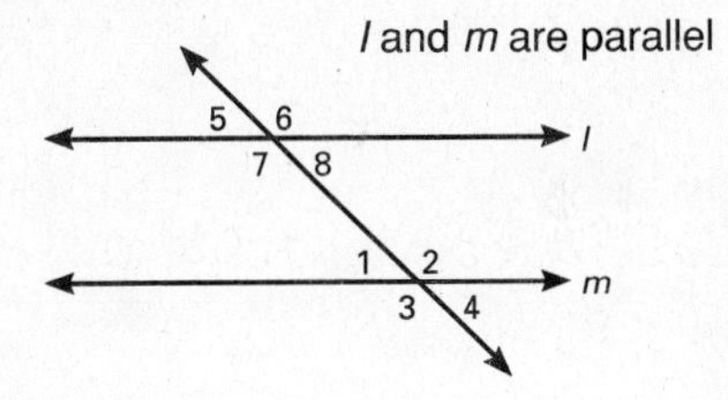

(1) ∠6 and ∠2
(2) ∠6 and ∠4
(3) ∠7 and ∠1
(4) ∠7 and ∠4
(5) ∠7 and ∠5

57. What is the approximate area of a circular pool with diameter 8 feet?

(1) 50 sq. ft.
(2) 48 sq. ft.
(3) 40 sq. ft.
(4) 20 sq. ft.
(5) 16 sq. ft.

58. The scale on a map is 1 in. = 1.5 mi. From Fenton to Prineville is actually 4 miles. How far apart are they on the map?

(1) less than 2 in.
(2) $2\frac{1}{3}$ in.
(3) $2\frac{2}{3}$ in.
(4) 3 in.
(5) 6 in.

59. A wooden rectangular frame is 15 by 25 inches. If diagonal braces are added to the back of the frame, how long (to the nearest whole inch) will each brace need to be?

(1) 18
(2) 23
(3) 27
(4) 29
(5) 32

60. In a right triangle, one leg is $7\frac{1}{2}$ feet and the other leg is 10 feet. What is the length in feet of the hypotenuse?

(1) $10\frac{1}{2}$
(2) $12\frac{1}{2}$
(3) 13
(4) $13\frac{1}{2}$
(5) 15

61. In a right triangle the measure of one acute angle is 5 times the measure of the other acute angle. The measure of the smaller angle is

(1) 10 degrees.
(2) 15 degrees.
(3) 20 degrees.
(4) 25 degrees.
(5) 30 degrees.

62. Find the approximate area in square feet of the following figure.

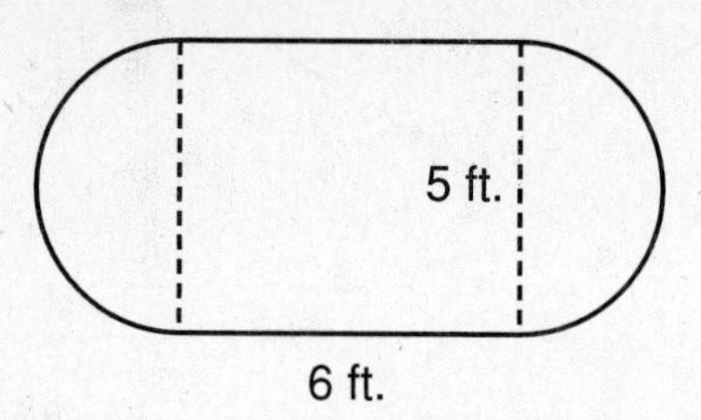

(1) 60
(2) 50
(3) 40
(4) 30
(5) 20

63. A parallelogram has a base of 32 yards and a height of 8 yards. What is the side of a square with the same area as the parallelogram?

(1) 8 yards
(2) 10 yards
(3) 12 yards
(4) 16 yards
(5) Not enough information is given.

64. The perimeter of a rectangle is 60 inches. The length is 6 inches longer than the width. Find the width in inches.

(1) 4
(2) 8
(3) 10
(4) 12
(5) 15

Answers are on page 409.

Performance Analysis
Unit 3 Cumulative Review: Geometry

Name: ______________________ **Class:** ______________ **Date:** ______________

Use the Answer Key on pages 409–412 to check your answers to the GED Unit 3 Cumulative Review: Geometry. Then use the chart to figure out the skill areas in which you need additional review. Circle on the chart the numbers of the test items you answered correctly. Then go back and review the lessons for the skill areas that are difficult for you. For additional review, see the *Steck-Vaughn GED Mathematics Exercise Book*, Unit 3: Geometry.

Use the chart below to find your strengths and weaknesses in application skills.

Concept	Item Numbers
Perimeter, Circumference, Area, and Volume pages 242–251, 284–299	**7**, 12, 13, 14, 17, **21**, **22**, 28, 34, 46, 54, **55**, 57, **62**, 63, 64
Angles pages 252–259	**1, 2, 3, 4, 8, 15, 16, 19, 36, 37, 38, 39, 43, 44, 45, 56**
Triangles and Quadilaterals pages 260–265	**9, 18,** 23, 24, **25, 26, 27,** 61
Congruence and Similarity pages 266–275	5, 6, **29, 30, 31, 32, 33, 40, 41, 48, 49, 51, 52,** 53, 58
Pythagorean Theorem, Squares, and Square Roots pages 276–283	10, **11,** 20, 35, 42, 47, 50, 59, 60

Boldfaced numbers indicate items based on charts, graphs, illustrations, and diagrams.

Posttest

MATHEMATICS

Directions

The Mathematics Test consists of multiple-choice questions intended to measure general mathematical skills and problem solving ability. The questions are based on short readings which often include a graph, chart, or figure.

You should spend no more than 115 minutes answering the questions. Work carefully, but do not spend too much time on any one question. Be sure you answer every question. You will not be penalized for incorrect answers.

Formulas you may need are given on page 427. Only some of the questions will require you to use a formula. Not all the formulas given will be needed.

Some questions contain more information than you will need to solve the problem. Other questions do not give enough information to solve the problem. If the question does not give enough information to solve the problem, the correct answer choice is "Not enough information is given."

The use of calculators is not allowed.

Record your answers on the separate answer sheet provided. Be sure all requested information is properly recorded on the answer sheet. You may make extra copies of page 428. To record your answers, mark the numbered space on the answer sheet beside the number that corresponds to the question in the test booklet. Mark only one answer space for each question; multiple answers will be scored as incorrect.

Example: If a grocery bill totaling $15.75 is paid with a $20.00 bill, how much change should be returned?

(1) $5.26
(2) $4.75
(3) $4.25
(4) $3.75
(5) $3.25

① ②  ④ ⑤

The correct answer is $4.25; therefore, answer space 3 would be marked on the answer sheet.

Do not rest the point of your pencil on the answer sheet while you are considering your answer. Make no stray or unnecessary marks. If you change an answer, erase your first mark completely. Mark only one answer for each question; multiple answers will be scored as incorrect. Do not fold or crease your answer sheet.

Directions: Choose the best answer to each item.

1. To make four seat covers, when each cover requires $4\frac{2}{3}$ yards of material, how many yards of material are needed?

   (1) $8\frac{2}{3}$
   (2) $13\frac{1}{3}$
   (3) $16\frac{1}{3}$
   (4) $16\frac{2}{3}$
   (5) $18\frac{2}{3}$

2. An early radar device (Model A) is operated at a maximum frequency of 3×10^3 megahertz. With improvements, a new device (Model B) was able to operate at 3×10^5 megahertz. Which statement is true?

   (1) The maximum frequency of Model B is less than that of Model A.
   (2) The maximum frequency of Model B is 2 times as great as that of Model A.
   (3) The maximum frequency of Model B is 3 times as great as that of Model A.
   (4) The maximum frequency of Model B is 10 times as great as that of Model A.
   (5) The maximum frequency of Model B is 100 times as great as that of Model A.

3. Janice works for a collection agency. She is paid a 5% commission on the amounts she collects. Last week she collected $1,900, $2,350, and $5,800. What is the total commission she earned for the week?

   (1) $ 177.50
   (2) $ 212.50
   (3) $ 290.50
   (4) $ 502.50
   (5) $1,550.50

4. Leonard gives swimming lessons at the community center. He is paid $5 per half-hour lesson and $9 per full-hour lesson. He has 20 students per week for a half-hour lesson and 12 students per week for a full-hour lesson. Which expression represents how many dollars Leonard will earn per week?

   (1) $20(5) + 12(9)$
   (2) $5(12) + 20\ (9)$
   (3) $12(20) + 5(9)$
   (4) $5(20) + 12(20)$
   (5) $12(9) + 20(9)$

5. In 1994 an airline hired an aircraft company to manufacture 80 new planes. The aircraft company can complete 16 planes per year. At this rate, in what year will the airline's order be completed?

   (1) 1995
   (2) 1997
   (3) 1999
   (4) 2001
   (5) 2004

6. Mr. Fayle owns 1,200 acres of forest land. He plans to donate 75% of the land to the county for a wildlife preserve. How many acres of land is he planning to donate?

   (1) 90
   (2) 450
   (3) 750
   (4) 900
   (5) 1,050

Item 7 refers to the following number line.

7. Which letter on the number line represents −5?

(1) *A*
(2) *B*
(3) *F*
(4) *K*
(5) *L*

Items 8 and 9 refer to the following information.

MARIA'S TAXES

Taxable Income	$19,000
TAX	**% OF INCOME**
Federal Tax Withheld	24
State Tax Withheld	3
Federal Tax Still Owed	4
State Tax Still Owed	1

8. After she pays all the tax she owes, what percent of her income will Maria have paid for federal and state income taxes?

(1) 34%
(2) 32%
(3) 27%
(4) 16%
(5) 8%

9. Maria must also pay a state disability tax of $475. What percent of her taxable income is the disability tax?

(1) 4%
(2) 3%
(3) 2.5%
(4) 1.5%
(5) 1%

10. Traveling at 55 miles per hour, approximately how long will it take Mona to drive 305 miles?

(1) between 2 and 3 hours
(2) between 3 and 4 hours
(3) between 4 and 5 hours
(4) between 5 and 6 hours
(5) between 6 and 7 hours

11. The square root of 14 is between which pair of numbers?

(1) 1.4 and 1.5
(2) 2 and 3
(3) 3 and 4
(4) 4 and 5
(5) 13 and 15

Item 12 is based on the following information.

THINGS TO DO LIST	
A. Complete work schedule	$\frac{1}{2}$ hour
B. Clean storeroom	2 hours
C. Order new supplies	$\frac{3}{4}$ hour
D. Staff meeting	$1\frac{1}{2}$ hours
E. Restock shelves	$1\frac{1}{4}$ hours
F. Type memo	$\frac{1}{2}$ hour

12. Cliff works 6 hours per day at a restaurant. The list shows the tasks he needs to get done and the amount of time each task will take. He has already spent two hours totaling last night's receipts and $2\frac{1}{2}$ hours completing the payroll. Which combination of tasks does he still have time to complete today?

(1) A and E
(2) B and E
(3) C and D
(4) C and F
(5) B only

13. A local market advertises tomatoes at 79 cents per pound. How much will it cost to buy 6 tomatoes?

(1) $3.16
(2) $3.88
(3) $4.74
(4) $5.82
(5) Not enough information is given.

Items 14 and 15 refer to the following information.

CARDSTON'S DEPARTMENT STORE EMPLOYEES	
Managers	15
Buyers	8
Accounting	12
Salespeople	45

14. What percent of Cardston's employees work in accounting?

(1) 12%
(2) 15%
(3) 26%
(4) 45%
(5) Not enough information is given.

15. Cardston's is holding a contest to improve employee morale. A computer will choose one employee at random to win a trip to Hawaii. What is the probability that a buyer will win the contest?

(1) $\frac{1}{2}$
(2) $\frac{1}{4}$
(3) $\frac{1}{8}$
(4) $\frac{1}{10}$
(5) $\frac{1}{80}$

16. On a map two cities are $3\frac{3}{4}$ inches apart. The map has a scale of 1 inch = 40 miles. What is the actual distance, in miles, between the two cities?

(1) 90
(2) 120
(3) 150
(4) 180
(5) 210

17. Preston bought an antique bookcase for $90 and refinished it. He sold the bookcase for 125% of the price he paid. How much did he get for the bookcase?

(1) $ 81.00
(2) $112.50
(3) $156.25
(4) $202.50
(5) Not enough information is given.

Item 18 refers to the following diagram.

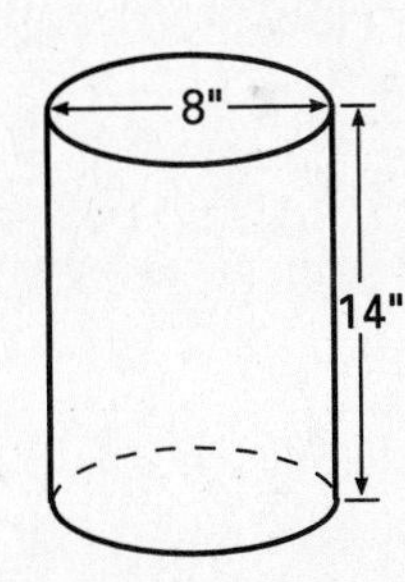

18. A can in the shape of a cylinder is 8 in. across and 14 in. tall. Which expression represents the approximate volume of the can in cubic inches?

(1) $(3.14)(4)(14)^2$
(2) $(3.14)(8)(14)^2$
(3) $(3.14)(8)^2(14)$
(4) $(3.14)(16)^2(14)$
(5) $(3.14)(4)^2(14)$

19. At Fabric Unlimited, adult patterns are $4 each and children's patterns are $3 each. A customer wants to buy 6 adult patterns and 2 children's patterns. Which expression represents how many dollars the customer will pay for the patterns?

(1) $(4 + 6)(2 + 3)$
(2) $2(6) + 3(4)$
(3) $2(4) + 3(6)$
(4) $3(2 + 4 + 6)$
(5) $6(4) + 2(3)$

20. Wayne can mow his back lawn in 45 minutes. What part of it can he mow in $\frac{1}{2}$ hour?

(1) $\frac{1}{4}$
(2) $\frac{1}{3}$
(3) $\frac{3}{8}$
(4) $\frac{2}{3}$
(5) $\frac{3}{4}$

21. A gallon of interior latex paint will cover 200 square feet. How many square feet will 8 gallons cover?

(1) 25
(2) 160
(3) 1,000
(4) 1,600
(5) 1,800

22. Which of the following expresses 2,062,493 in scientific notation?

(1) 2.062493×10^6
(2) 2.062493×10^7
(3) 20.62493×10^7
(4) 206.2493×10^6
(5) 2062.493×10^4

Items 23 and 24 refer to the following graph.

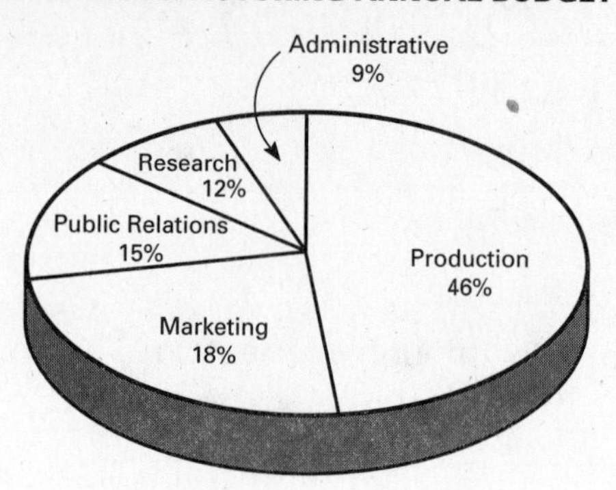

23. According to the graph, what percent of the company budget was spent on production and marketing?

(1) 33%
(2) 36%
(3) 57%
(4) 64%
(5) 67%

24. What percent of the budget is left after public relations and administrative expenses are paid?

(1) 91%
(2) 88%
(3) 85%
(4) 82%
(5) 76%

25. Max plays basketball. During the past four weeks, his team has scored 72, 66, 74, and 68 points. Which expression represents the team's mean score (average) for the month?

(1) $\frac{72+66+74+68}{4}$
(2) $4(72 + 66 + 74 + 68)$
(3) $\frac{(72)(66)(74)(68)}{4}$
(4) $\frac{4(72+66+74)}{68}$
(5) Not enough information is given.

26. A charity raises $500,000 to build shelters for the homeless. If 4% is spent on administrative costs, how much is left for building shelters?

(1) $200,000
(2) $260,000
(3) $420,000
(4) $450,000
(5) $480,000

27. An inspector at Watts-Up Light Bulb Company discovered that there were 5 defective light bulbs in every case of 150 light bulbs. What is the probability that a light bulb will be defective?

(1) $\frac{1}{3}$
(2) $\frac{1}{15}$
(3) $\frac{1}{30}$
(4) $\frac{1}{50}$
(5) $\frac{1}{100}$

28. A zoo built a circular platform with a 16-foot diameter for its sea lion exhibit. What is the approximate area (in square feet) of the platform?

(1) 50
(2) 201
(3) 252
(4) 804
(5) Not enough information is given.

29. A gardener plans to plant 12 tomato plants in each row of a garden. If there are 15 rows, how many plants will he need?

(1) 180
(2) 150
(3) 120
(4) 90
(5) 60

Item 30 refers to these diagrams.

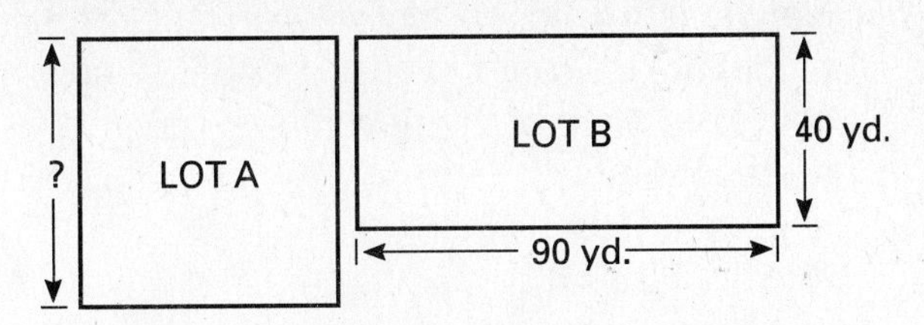

30. Lot A and Lot B have the same area. If Lot A is square, what is the length in yards of one of its sides?

(1) 40
(2) 50
(3) 60
(4) 90
(5) Not enough information is given.

31. A can of spaghetti sauce has 70 calories per serving. If 18 of the calories come from fat, to the nearest whole percent, what percent of the sauce's calories come from fat?

(1) 13%
(2) 26%
(3) 34%
(4) 52%
(5) Not enough information is given.

32. Arturo buys a computer desk for $398. If he pays $6\frac{1}{2}$ % sales tax on the purchase, what is the total cost of the desk?

(1) $258.70
(2) $404.50
(3) $423.87
(4) $433.24
(5) $656.70

33. A new computer monitor manufactured by Net-Co is priced 4% higher than last year's model. If last year's model cost \$500, how could the price of the new model be represented?

(1) 0.04(\$500)

(2) $\frac{\$500}{0.04}$

(3) 0.04(\$500) + 0.04

(4) 0.04 (\$500) + \$500

(5) $\frac{\$500}{0.04}$ + \$500

Item 34 refers to the following diagram.

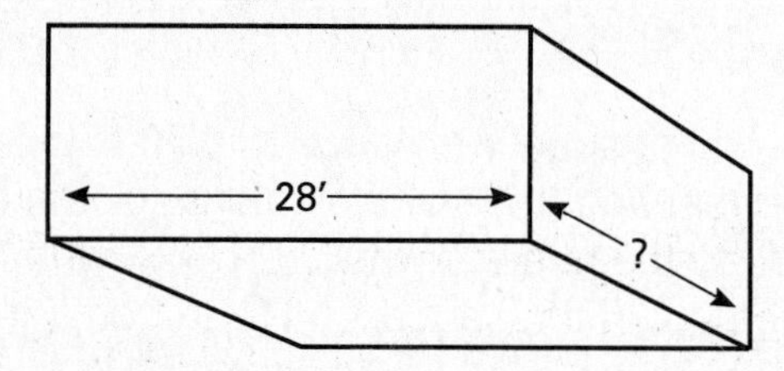

34. The rectangular storage tank has a volume of 3,920 cubic feet. If the length of the tank is 28 feet, what is the width of the tank?

(1) 7 feet

(2) 10 feet

(3) 14 feet

(4) 20 feet

(5) Not enough information is given.

35. Simplify: $6x + 2y - x + 3y$.

(1) $6x + 6y$

(2) $36xy$

(3) $5x + 5y$

(4) $7x^2 + 5y^2$

(5) $x^5 + y^5$

36. Oranges are priced at 6 for \$1.08. What is the cost of $3\frac{1}{2}$ dozen oranges?

(1) \$0.54

(2) \$3.78

(3) \$6.30

(4) \$6.48

(5) \$7.56

37. 620 clerks and cashiers register for a 2-day workshop on a new inventory system. Of the clerks and cashiers, 15% could not attend the banquet on the last day because they had to return to work. How many cashiers and clerks could not attend the banquet?

(1) 93

(2) 186

(3) 372

(4) 527

(5) 605

38. Given the formula $p = 3x(y + 2)$, find p if $y = 3$ and $x = 4$.

(1) 35

(2) 36

(3) 54

(4) 60

(5) 72

39. Nolan walks 5 miles in $1\frac{1}{4}$ hours. Which equation can be used to find how many miles Nolan can walk in 5 hours?

(1) $1\frac{1}{4}x = 25$

(2) $5x = 6\frac{1}{4}$

(3) $6\frac{1}{4}x = 5$

(4) $5x = 25$

(5) $25x = 5$

40. The perimeter of a rectangular garden is 74 feet. If the length of the garden is 22 feet, which equation could be used to find the width of the garden?

(1) $2w + 74 = 2(22)$
(2) $w^2 + 22^2 = 74$
(3) $\frac{74}{2(22)} = \frac{2}{w}$
(4) $\frac{w}{2} + 74 = 22$
(5) $74 - 2(22) = 2w$

41. Jaelle pays $12.80 per month to subscribe to the local newspaper. How much will it cost her to subscribe to the paper for 12 months?

(1) $ 76.80
(2) $128.00
(3) $153.60
(4) $230.40
(5) $256.00

42. Simplify: $2m + 2n + m - 3n$.

(1) $3m - n$
(2) $2m^2 - n^2$
(3) $m^3 - n^5$
(4) $12mn$
(5) $3m + 5n$

43. Aviva, an office manager, got a department store certificate from her staff. She wants to buy a shirt for $24.95 and a necklace for $18.65. If p represents the original amount of the gift certificate, how much will she have left to spend after buying the shirt and necklace?

(1) p + $24.95 − $18.65
(2) p − $24.95 − $18.65
(3) p − $24.95 + 18.65
(4) p + $24.95 + $18.65
(5) p($24.95 + $18.65)

44. For which value of z is the inequality $3z < 12$ true?

(1) 3
(2) 4
(3) 5
(4) 6
(5) 7

Item 45 is based on the following figure.

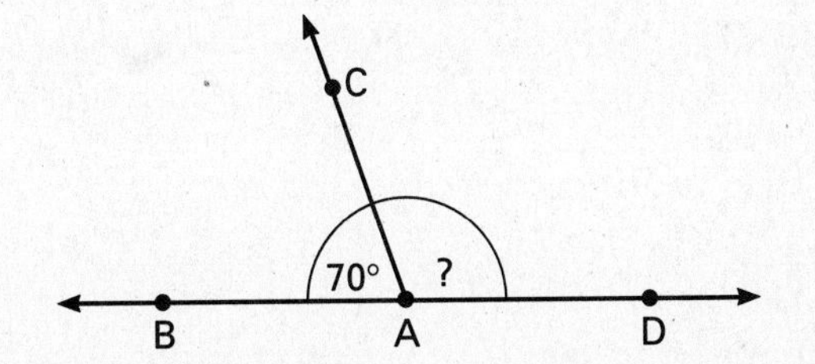

45. Angle BAC measures 70 degrees. How many degrees does angle CAD measure?

(1) 20°
(2) 50°
(3) 70°
(4) 90°
(5) 110°

Item 46 is based on the following figure.

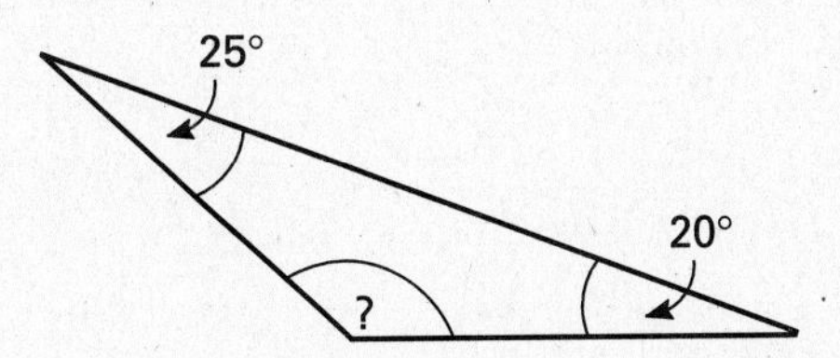

46. The triangle has three internal angles. One angle measures 25 degrees; another measures 20 degrees. How many degrees are in the third angle?

(1) 45°
(2) 105°
(3) 135°
(4) 315°
(5) Not enough information is given.

47. A pole 3 feet tall casts a shadow 5 feet long. At the same time, a building casts a shadow 120 feet long. How tall is the building?

(1) 72 ft.
(2) 150 ft.
(3) 175 ft.
(4) 200 ft.
(5) 1,800 ft.

48. The perimeter of a five-sided figure is 72 feet. The measurements of four of the sides are 16 feet, 15 feet, 20 feet, and 9 feet. What is the length in feet of the remaining side?

(1) 9
(2) 12
(3) 42
(4) 60
(5) Not enough information is given.

49. Bobbi walked 8 feet up a ramp and was 6 feet from the ground. How far from the ground was Bobbi after she walked 12 feet farther up the ramp?

(1) 12 ft.
(2) 18 ft.
(3) 14 ft.
(4) 15 ft.
(5) 4 ft.

Item 50 refers to the diagram below.

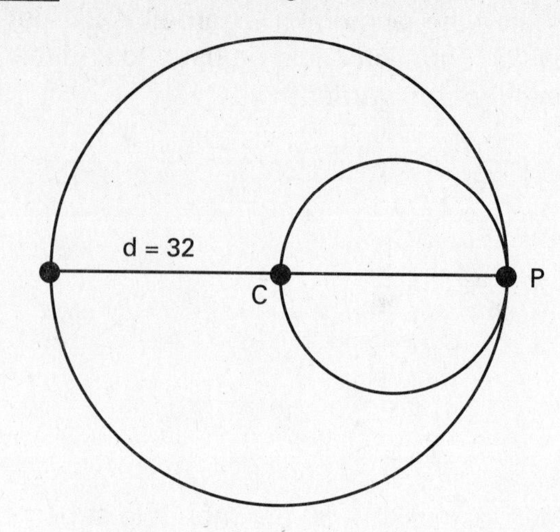

50. Two circles intersect at point *P*. If *C* is the center of the larger circle, and the diameter of the larger circle is 32, then the radius of the smaller circle must be

(1) 32.
(2) 24.
(3) 16.
(4) 8.
(5) Not enough information is given.

Item 51 refers to the following drawing.

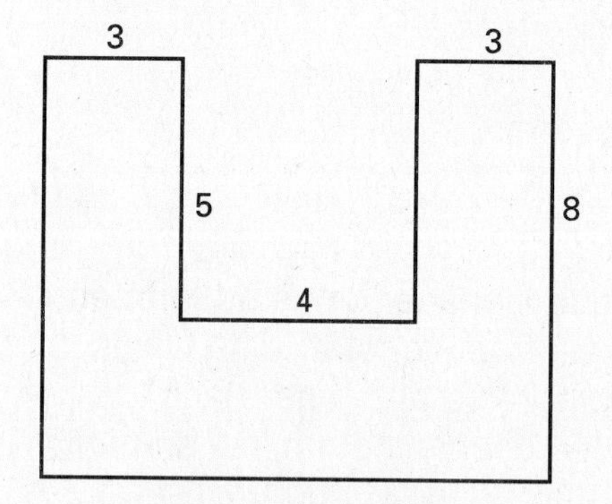

51. What is the perimeter of the figure?

(1) 18
(2) 23
(3) 36
(4) 41
(5) 46

Item 52 refers to the following drawing.

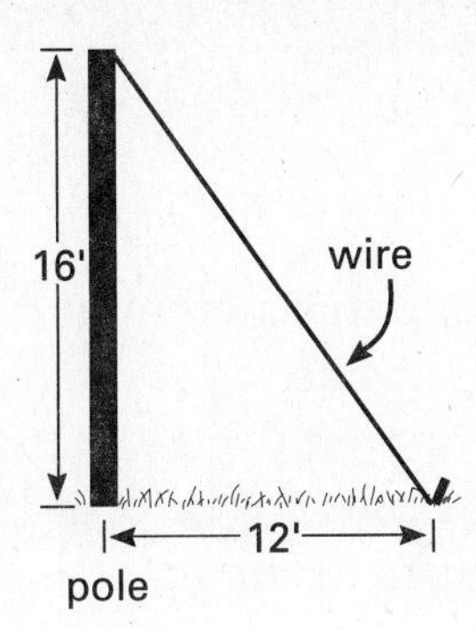

52. A wire is supporting a 16-foot pole. The wire is staked to the ground a distance of 12 feet from the bottom of the pole. How long is the wire?

(1) 4 ft.
(2) 8 ft.
(3) 10 ft.
(4) 16 ft.
(5) 20 ft.

Items 53 and 54 refer to the following figure.

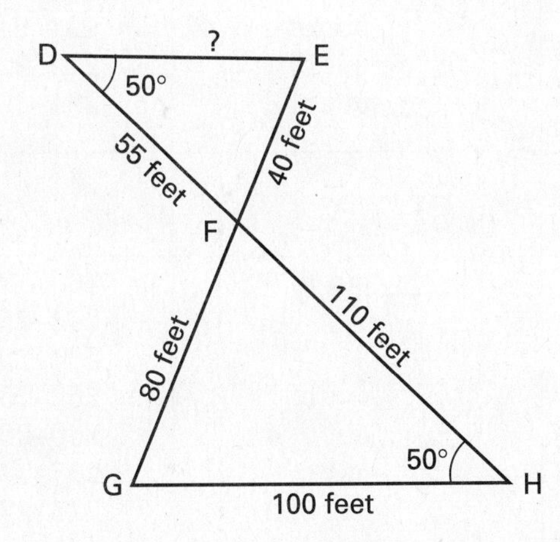

53. Triangles *DEF* and *FGH* are similar. Which side in triangle *FGH* corresponds to side *DF* in triangle *DEF?*

(1) side *EF*
(2) side *FG*
(3) side *FH*
(4) side *GH*
(5) Not enough information is given.

54. What is the length in feet of side *DE* in the figure?

(1) 20
(2) 33
(3) 40
(4) 50
(5) 55

Item 55 refers to the following figure.

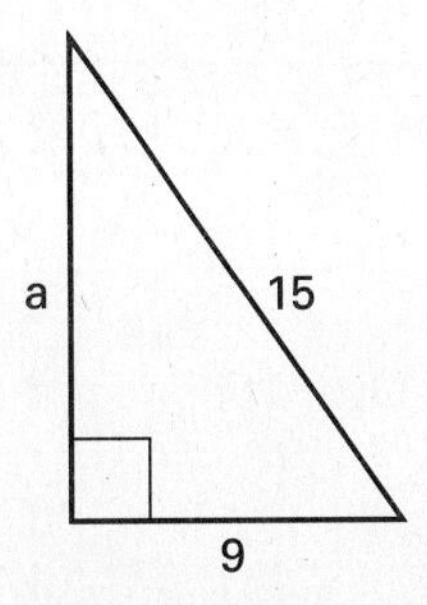

55. According to the Pythagorean Theorem, which statement is true for the triangle?

(1) $a^2 + 9^2 = 15^2$
(2) $a^2 - 9^2 = 15^2$
(3) $a^2 - 15^2 = 9^2$
(4) $a^2 + 15^2 = 9^2$
(5) $9^2 + 15^2 = a^2$

Item 56 refers to the following diagram.

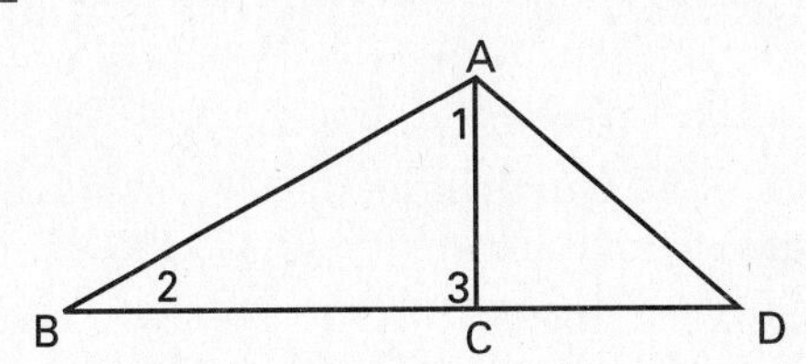

56. In triangle *ABC*, angle 2 is 35 degrees and angle 1 is 55 degrees. How many degrees is angle 3 of the same triangle?

(1) 145°
(2) 125°
(3) 100°
(4) 90°
(5) 55°

Answers are on page 412.

Posttest Correlation Chart: Mathematics

Name: ______________________ **Class:** ______________ **Date:** ______________

This chart can help you determine your strengths and weaknesses on the content and skill areas of the Mathematics GED Test. Use the Answer Key on pages 412–414 to check your answers to the test. Then circle on the chart the numbers of the test items you answered correctly. Put the total number correct for each content area and skill area in each row and column. Look at the total items correct in each column and row and decide which areas are difficult for you. Use the page references to study those areas. Use a copy of the Study Record Sheet on page 25 to guide your studying.

Content	Item Number	Total Correct	Page Ref.
Arithmetic *(pages 26–193)*			
Whole Number Operations	4, 10, 19, 29	_____out of 4	26–57
Fractions	1, 12	_____out of 2	58–87
Decimals	32, 36, 41	_____out of 3	88–113
Percent	3, 6, 8, 9, 14, 17, 26, 31, 33, 37	_____out of 10	114–145
Ratio, Proportion, Mean, Median, and Probability	5, 13, 15, 16, 20, 21, 25, 27	_____out of 8	146–165
Graphs and Charts	23, 24	_____out of 2	177–193
Algebra *(pages 194–241)*			
Equations	7, 35, 38, 39, 42, 43	_____out of 6	194–211
Graphs, Slope, Lines	44	_____out of 1	212–223
Powers and Roots	2, 11, 22	_____out of 3	224–229
Geometry *(pages 242–309)*			
Perimeter, Area, Volume	18, 28, 30, 34, 40, 48, 50, 51	_____out of 8	166–176 244–251 284–297
Angles	45	_____out of 1	252–257
Triangles and Quadrilaterals	46, 53, 54, 56	_____out of 4	260–265
Congruence and Similarity	47, 49	_____out of 2	266–273
Pythagorean Theorem	52, 55	_____out of 2	276–279

For additional help, see the Steck-Vaughn GED Mathematics Exercise Book.

SIMULATED TEST OF GENERAL EDUCATIONAL DEVELOPMENT

MATHEMATICS

Directions

The Mathematics Skills Simulated GED Test consists of multiple-choice questions intended to measure general mathematical skills and problem solving ability. The questions are based on short readings which often include a graph, chart, or figure.

You should spend no more than 90 minutes answering the questions. Work carefully, but do not spend too much time on any one question. Be sure you answer every question. You will not be penalized for incorrect answers.

Formulas you may need are given on page 427. Only some of the questions will require you to use a formula. Not all the formulas given will be needed.

Some questions contain more information than you will need to solve the problem. Other questions do not give enough information to solve the problem. If the question does not give enough information to solve the problem, the correct answer choice is "Not enough information is given."

The use of calculators is not allowed.

Record your answers on the separate answer sheet provided. Be sure all requested information is properly recorded on the answer sheet. You may make extra copies of page 428. To record your answers, mark the numbered space on the answer sheet beside the number that corresponds to the question in the test booklet. Mark only one answer space for each question; multiple answers will be scored as incorrect.

Example: If a grocery bill totaling $15.75 is paid with a $20.00 bill, how much change should be returned?

(1) $5.26

(2) $4.75

(3) $4.25

(4) $3.75

(5) $3.25

① ② ● ④ ⑤

The correct answer is $4.25; therefore, answer space 3 would be marked on the answer sheet.

Do not rest the point of your pencil on the answer sheet while you are considering your answer. Make no stray or unnecessary marks. If you change an answer, erase your first mark completely. Mark only one answer for each question; multiple answers will be scored as incorrect. Do not fold or crease your answer sheet.

Directions: Choose the best answer to each item.

1. A sign in the window at Keller's Stationers reads: "20% Off—All Items!" A customer wants to buy three items regularly priced at \$15.80, \$19.20, and \$11.50. What is the total sale price of the items?

   (1) \$ 9.30
   (2) \$18.60
   (3) \$37.20
   (4) \$46.50
   (5) \$55.80

2. Kenda earns \$10.75 per hour as a part-time bookkeeper. Which expression represents her total earnings if she works 4 hours on Monday, 6 hours on Tuesday, 1 hour on Wednesday, 2 hours on Thursday, and 7 hours on Friday?

   (1) $4 + 6 + 1 + 2 + 7$
   (2) $20 + 10.75$
   (3) $20(10.75)$
   (4) $40 + 10.75$
   (5) $40\ (10.75)$

Item 3 refers to the following number line.

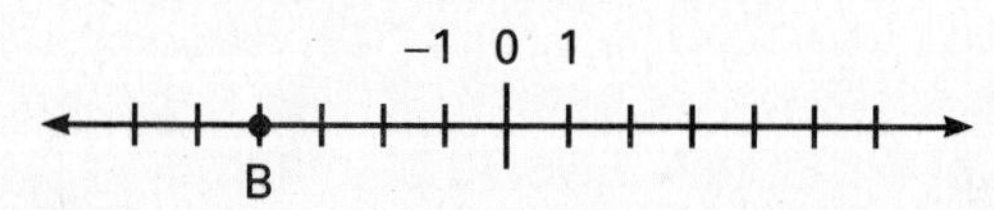

3. What number does Point *B* represent on the number line?

   (1) −4
   (2) −3
   (3) −1
   (4) 3
   (5) 4

4. Lee bought a package of 4 batteries with a 50-cents-off coupon. He also mailed in a proof-of-purchase seal for a company rebate of \$1.00. If *p* represents the original price, which expression represents the cost of the batteries to Lee?

   (1) $p + \$0.50 + \$1.00$
   (2) $p - \$0.50 - \$1.00$
   (3) $4p(\$1.00 - \$0.50)$
   (4) $p - 4(\$0.50 + \$1.00)$
   (5) $p - \$0.50 + \$1.00$

5. Rosa earns twice as much as her husband Mark. If their combined monthly income is \$3,900, how much does Rosa earn in one month?

   (1) \$ 975
   (2) \$1,300
   (3) \$1,950
   (4) \$2,600
   (5) Not enough information is given.

6. In a football game, two teams scored a total of 41 points. The winning team scored 13 more points than the losing team. How many points did the winning team score?

   (1) 14
   (2) 20
   (3) 27
   (4) 28
   (5) 38

Item 7 is based on the following figure.

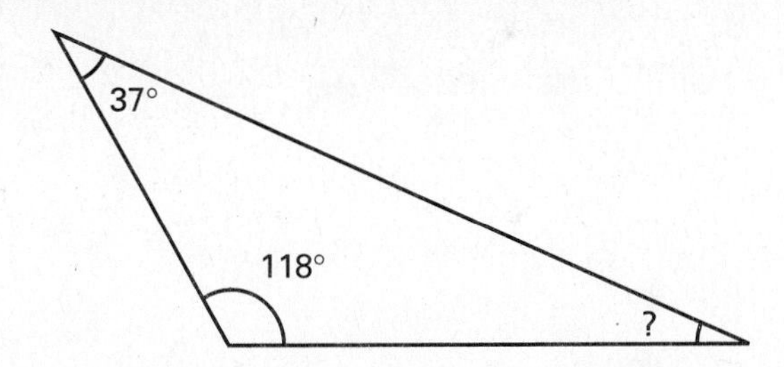

7. The triangle above has three internal angles. One angle measures 37 degrees. Another measures 118 degrees. How many degrees does the third angle measure?

   (1) 25°
   (2) 62°
   (3) 81°
   (4) 143°
   (5) Not enough information is given.

8. Wilson is selling subscriptions to DiskDrive Magazine. A one-year subscription costs \$25.80; a two-year subscription is \$41.76. Which expression shows how much Wilson will collect if he sells 15 one-year and 12 two-year subscriptions?

   (1) 12(\$25.80) + 15(\$41.76)
   (2) 15(\$25.80) + 12(\$41.76)
   (3) 12(\$25.76) + 15(\$41.80)
   (4) 27(\$67.56)
   (5) Not enough information is given.

9. Jeff wants to pour a rectangular concrete slab 10 feet long, 4 feet wide, and $\frac{1}{2}$ foot thick. How many cubic feet of concrete will he need?

   (1) 10
   (2) 20
   (3) 30
   (4) 40
   (5) Not enough information is given.

Items 10 and 11 refer to the following information.

Karen and Luis own the Sandwich Shoppe. They surveyed 240 people to find out what kind of sandwiches to serve as their lunch and dinner specials. The lunch special is \$3 and the dinner special is \$5.

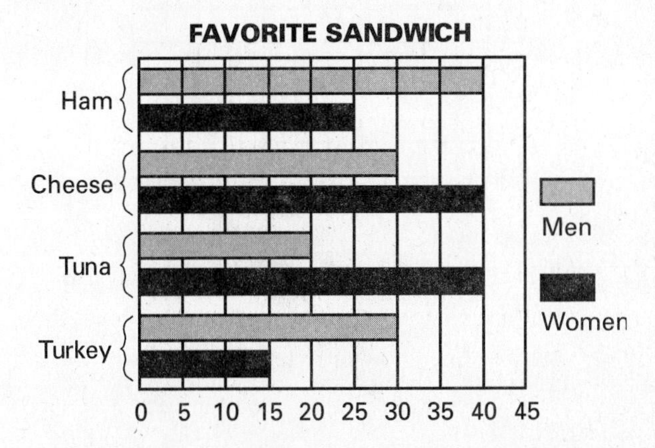

10. How many more men preferred ham than tuna?

    (1) 20
    (2) 15
    (3) 10
    (4) 5
    (5) Not enough information is given.

11. If Karen and Luis choose tuna for the lunch and dinner specials, how much money will they earn from specials in one day?

    (1) \$160
    (2) \$180
    (3) \$240
    (4) \$300
    (5) Not enough information is given.

Item 12 is based on the following graph.

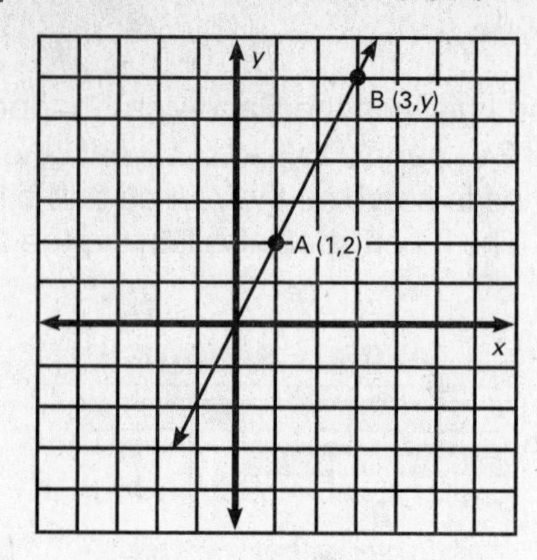

12. What is the y-coordinate of point B if the slope of line AB is $+2$?

(1) -3
(2) 4
(3) 6
(4) 8
(5) 12

13. Working as a store clerk, James earns $17,160 per year. If his boss gives him a 6% raise, what is his new annual salary?

(1) $17,760.00
(2) $18,189.60
(3) $20,660.20
(4) $22,243.40
(5) $27,456.00

14. A delivery truck is loaded with boxes of T-shirts. Of the 200 boxes on the truck, 150 contain large shirts and 50 contain extra-large. What is the probability that the first box taken off the truck will contain large shirts?

(1) $\frac{1}{200}$
(2) $\frac{1}{150}$
(3) $\frac{2}{3}$
(4) $\frac{1}{4}$
(5) $\frac{3}{4}$

15. To make 4 skirts, when each skirt requires $3\frac{3}{8}$ yards of material, how many yards of material are needed?

(1) $7\frac{3}{8}$
(2) $10\frac{1}{2}$
(3) $12\frac{3}{8}$
(4) $12\frac{7}{8}$
(5) $13\frac{1}{2}$

16. Given the formula $x = 3a(b^2 - 8)$, find x if $a = 5$ and $b = 4$.

(1) 0
(2) 24
(3) 40
(4) 72
(5) 120

17. Which of the following expresses 18,592 in scientific notation?

(1) 1.8592×10^3
(2) 1.8592×10^4
(3) 1.8592×10^5
(4) 18.592×10^2
(5) 18.592×10^4

18. Sandra's monthly salary is $1,660. If 24% is deducted for taxes, what is the amount of her take-home pay?

(1) $ 398.40
(2) $ 630.80
(3) $1,261.60
(4) $1,494.20
(5) $1,636.00

19. Evaluate $4x^3 - 2y$, if $x = 2$ and $y = 10$.

(1) 4
(2) 8
(3) 10
(4) 12
(5) 20

20. It takes $2\frac{1}{4}$ yards of material to make a child's jumpsuit. How many jumpsuits can be made from $22\frac{1}{2}$ yards of material?

(1) 6
(2) 8
(3) 10
(4) 11
(5) 13

Item 21 refers to the following diagram.

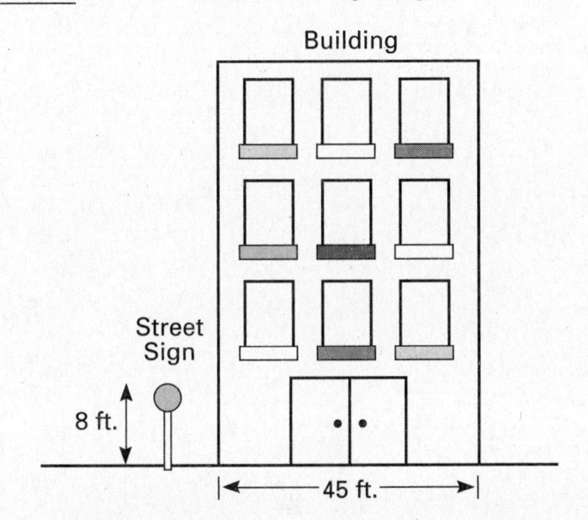

21. A landlord needs to find out the height of the apartment building. At 4 PM he takes several measurements. The shadow of the building is 30 feet long. He also determines the width of the building (45 feet) and the height of a street sign (8 feet). What additional measurement does he need to be able to determine the height of the building?

(1) the angle of the sunlight
(2) the distance from the building to the corner
(3) the volume of the building
(4) the width of the street
(5) the length of the shadow of the street sign

22. One angle of an isosceles triangle is 90 degrees. If the legs of the angle are equal in length, what are the measures (in degrees) of the two remaining angles?

(1) 30 and 30
(2) 30 and 60
(3) 45 and 45
(4) 60 and 60
(5) Not enough information is given.

23. A computer store charges its customers a 2% processing fee on all credit card purchases. If a customer makes a credit card purchase of $2,065, what is the amount of the processing fee?

(1) $ 4.13
(2) $ 8.26
(3) $ 41.30
(4) $ 82.60
(5) $413.00

24. For which value of y is the inequality $\frac{5}{y} > 1$ true?

(1) 4
(2) 5
(3) 6
(4) 7
(5) 8

25. Working 4 hours per day, Matt can inspect 92 computer monitors. At the same rate, which expression represents the number of monitors (N) Matt could inspect if he worked 7 hours per day?

(1) $N = \frac{4}{7}(92)$
(2) $N = \frac{7}{4}(92)$
(3) $N = 4(92)$
(4) $N = 7(92)$
(5) Not enough information is given.

26. The city of Whitesburg is 12 miles due east of Wellington. Fallsdale is 16 miles due south of Wellington. What is the straight line distance between Whitesburg and Fallsdale?

(1) 17 miles
(2) 20 miles
(3) 22 miles
(4) 25 miles
(5) 28 miles

Item 27 refers to the following information.

Toys and Games	Price
A. Funny Faces Game	$ 8.95
B. Doll	$16.75
C. Sing-Along Videotape	$ 9.99
D. Building Blocks	$18.60
E. Wading Pool	$ 9.45
F. Doll Furniture	$25.37

27. Linda has $50 to spend on birthday gifts for her daughter. She has already chosen two items: a dress-up gown ($14.50) and a toy helicopter ($16.80). Which combination of toys and games can she also buy without spending more than $50?

(1) A and C
(2) A and E
(3) B and E
(4) C and E
(5) F only

28. A cube has a volume of 216 cubic feet. Which expression shows the equation to use to find the length of a side *(s)*?

(1) $s^3 = 216$
(2) $s^2 = 216$
(3) $3s = 216$
(4) $s = \frac{216}{3}$
(5) Not enough information is given.

29. The Gordons have three children. What is the average monthly food expense of the family during a 4-month period in which they spent $593.60, $585.92, $602.35, and $608.25?

(1) $590.75
(2) $593.28
(3) $597.53
(4) $602.04
(5) $607.14

Item 30 refers to the following figure.

30. The pennant is an isosceles triangle. What is the perimeter (in inches) of the pennant if angles *A* and *C* are equal?

(1) 45
(2) 62
(3) 64
(4) 79
(5) 96

31. In a golf tournament, there are twice as many amateur golfers as professional golfers. Altogether, 96 golfers are in the tournament. How many of them are amateurs?

(1) 24
(2) 32
(3) 43
(4) 60
(5) 64

32. Which expression represents the minimum average amount Akayo must save each month to save $1,170 in five months?

(1) 5(1,170)
(2) 5(1,170) − 5
(3) 5(1,170) − 1,170
(4) $\frac{1,170}{5}$
(5) Not enough information is given.

Item 33 is based on the following drawing.

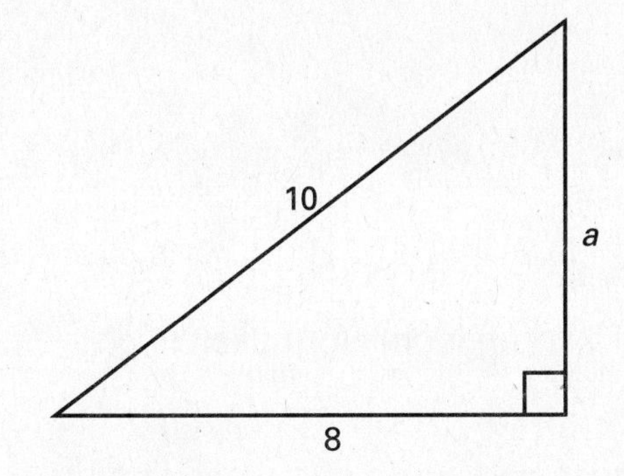

33. Which equation could be used to find the length of side *a* for the triangle?

(1) $8^2 + 10^2 = a^2$
(2) $a^2 - 10^2 = 5^2$
(3) $a^2 + 10^2 = 5^2$
(4) $a^2 - 8^2 = 10^2$
(5) $a^2 + 8^2 = 10^2$

Item 34 is based on the following diagram

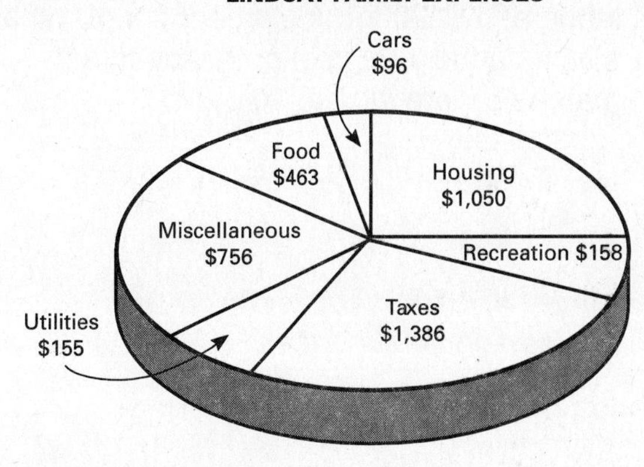

34. Approximately how many times more money does the Lindsay family spend on food than it spends on recreation?

(1) 2
(2) 3
(3) 5
(4) 8
(5) 12

35. Brenda manufactures cloth baby carriers in her home at a cost of $15 per carrier. If she sells the carriers for 35% more than her cost, how could the price of one carrier be represented?

(1) 0.35($15)
(2) $\frac{\$15}{0.35}$
(3) 0.35($15) + 0.35
(4) 0.35($15) + $15
(5) $\frac{\$15}{0.35} + \$15$

36. A plumber contracts to do a job. The costs for the job are $1,500 for materials, $600 for labor, and $220 for equipment rental. What would be the total cost of the job if the materials were increased by 25%?

(1) $ 680
(2) $1,055
(3) $1,875
(4) $2,320
(5) $2,695

37. Richenda wants to buy a new car priced at $9,500. If she makes a 20% down payment, which expression shows how much she will owe?

(1) (0.20)($9,500)
(2) $9,500 − (0.20)($9,500)
(3) (0.20)($9,500) − 0.20
(4) $9,500 − 0.20
(5) (0.20)($9,500) + $9,500

Item 38 refers to the following figure.

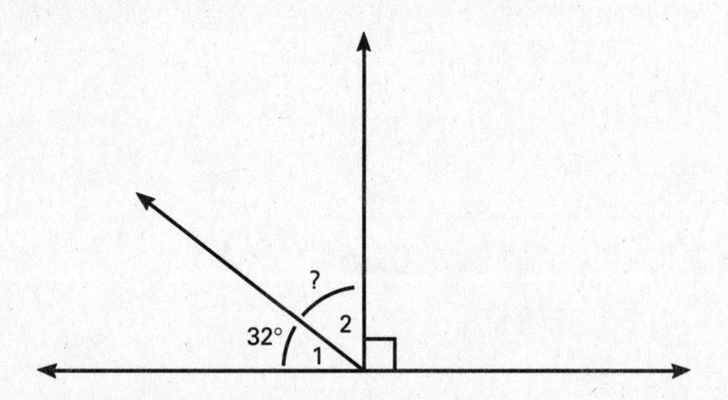

38. If angle 1 measures 32 degrees, what is the measure of angle 2?

(1) 32 degrees
(2) 45 degrees
(3) 58 degrees
(4) 90 degrees
(5) 148 degrees

39. The dosage instructions state that the patient should be given 5 milligrams of the drug for every 30 pounds of body weight. How many milligrams should be given to a patient who weighs 180 pounds?

(1) 15
(2) 30
(3) 90
(4) 150
(5) Not enough information is given.

40. The square root of 24 is between which pair of numbers?

(1) 2.4 and 2.5
(2) 3 and 4
(3) 4 and 5
(4) 5 and 6
(5) 12 and 13

41. Michela has worked for McKenzie Publishing 4 years longer than Stuart. Stuart has worked there 3 years less than Sofia. If Sofia has worked at the company for 8 years, how long has Michela worked there?

(1) 9
(2) 10
(3) 11
(4) 12
(5) Not enough information is given.

42. Teresa can wax a hardwood floor in $1\frac{1}{2}$ hours. What part of it can she wax in 45 minutes?

(1) $\frac{1}{4}$
(2) $\frac{1}{2}$
(3) $\frac{2}{3}$
(4) $\frac{3}{4}$
(5) $\frac{4}{5}$

43. A map has a scale of 1 inch = 50 miles. What is the actual distance, in miles, between two cities that are $2\frac{1}{2}$ inches apart on the map?

(1) 80
(2) 100
(3) 110
(4) 125
(5) 145

44. Doris is three times as old as her daughter Anne. If Doris is 33, which equation could be used to determine the age (a) of Anne?

(1) $\frac{a}{3} = 33$
(2) $a - 3 = 33$
(3) $3a = 33$
(4) $a + 3 = 33$
(5) $3(a + 3) = 33$

Item 45 refers to the following diagram.

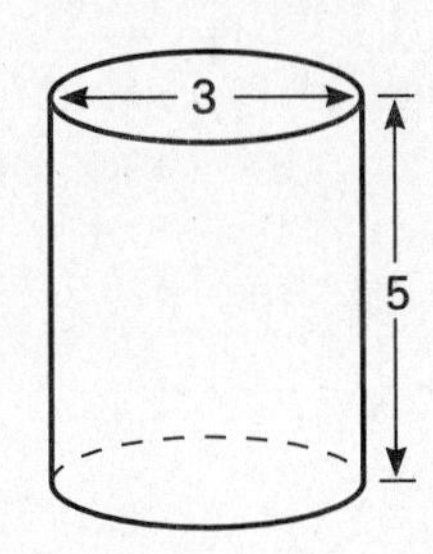

45. A cylindrical oil drum is 3 feet across and 5 feet tall. Which expression represents the approximate volume in cubic feet?

(1) $(3.14)(1.5)^2(5)$
(2) $(3.14)(1.5)(5)^2$
(3) $(3.14)(3)(5)$
(4) $(3.14)(3)^2(5)$
(5) $(3.14)(6)^2(5)$

Item 46 is based on the following graph.

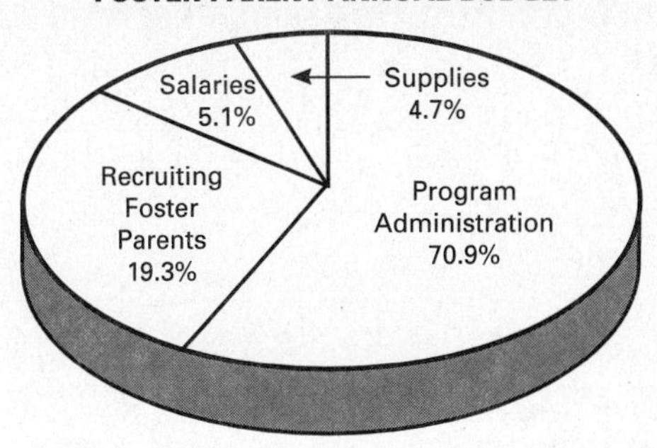

46. The graph above shows how a foster parent program spent its funds in a given year. According to the graph, what percent of the money was left after recruiting and salary expenses?

(1) 24.4%
(2) 29.1%
(3) 75.6%
(4) 80.7%
(5) 105.1%

47. Corene has two jobs. She earns $7 per hour as a clerk, and $9 per hour as a receptionist. Which expression represents how much money she will earn in a week if she works 15 hours as a receptionist and 22 hours as a clerk?

(1) 15(7) + 22 (9)
(2) 9(15) + 7 (15)
(3) 15(9) + 22 (7)
(4) 22(7) + 15 (7)
(5) 15(16) + 22 (16)

48. The volume of a rectangular storage tank is 900 cubic feet. If its length is 20 feet and its height is 9 feet, how many feet wide is it?

(1) 5
(2) 45
(3) 720
(4) 871
(5) Not enough information is given.

Item 49 refers to the following figure.

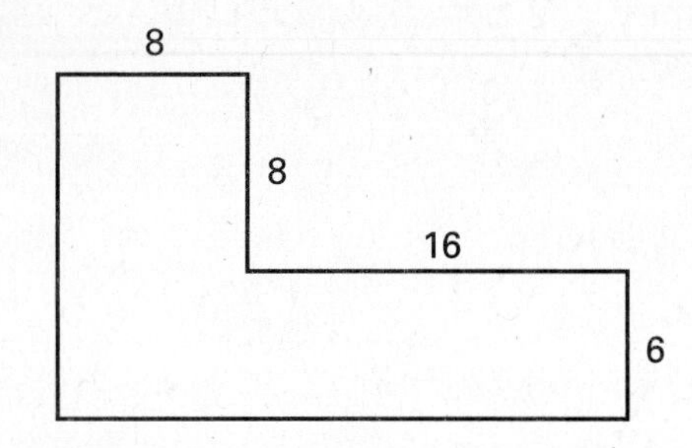

49. If all the internal angles of the figure are right angles, what is the area of the figure?

(1) 64
(2) 144
(3) 160
(4) 208
(5) Not enough information is given.

50. Traveling at 55 miles per hour, approximately how long will it take Arvin to drive 420 miles?

(1) between 4 and 5 hours
(2) between 5 and 6 hours
(3) between 6 and 7 hours
(4) between 7 and 8 hours
(5) between 8 and 9 hours

51. Karleen can drive 200 miles on a tank of gas. At this rate, how many tanks of gas will she use to drive from Los Angeles to Boston, a distance of 3,000 miles?

(1) 12
(2) 13
(3) 14
(4) 15
(5) 16

52. Land sonar equipment is used in construction for measuring distances and determining types of soil. Type A land sonar equipment has a range of 3.2×10^3 miles. Type B has a range of 4.8×10^4 miles. Which statement is true?

(1) The range of Type A sonar is 15 times greater than the range of Type B sonar.
(2) The range of Type B sonar is 15 times greater than the range of Type A sonar.
(3) The range of Type A sonar is 1.6 times greater than the range of Type B sonar.
(4) The range of Type B sonar is 1.6 times greater than the range of Type A sonar.
(5) The range of Type A sonar is 6 times greater than the range of Type B sonar.

Item 53 is based on the following diagram.

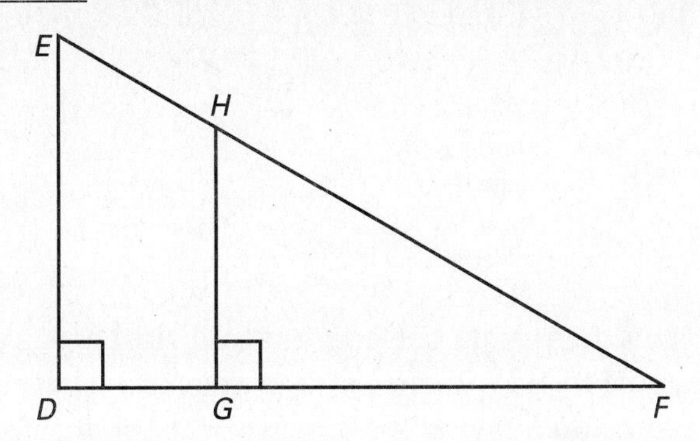

53. Triangles FGH and FDE are similar. If $\overline{FG}$ = 12 inches, $\overline{DG}$ = 6 inches, and $\overline{GH}$ = 8 inches, what is the measure (in inches) of $\overline{DE}$?

(1) 10

(2) 12

(3) 18

(4) 27

(5) Not enough information is given.

54. On a five-day trip, a trucker drove distances of 318, 315, 320, 298, and 264 miles. What was the mean (average) number of miles driven per day?

(1) 276

(2) 291

(3) 298

(4) 303

(5) 308

55. A triangle with a base of 28.7 meters and a height of 12 meters has the same area as a rectangle with a length of 16.4 meters. What is the width of the rectangle?

(1) 21.0 meters

(2) 20.5 meters

(3) 19.5 meters

(4) 11.5 meters

(5) 10.5 meters

56. Toni and Dan need to buy a weekly expense register, 50 file folders, 3 notebooks, and 6 adding machine paper rolls. The costs of the supplies are as follows:

notebooks	$4 each or 3 for $10
adding machine paper rolls	$1 each or 3 for $2
file folders	10 for $5
weekly expense register	$15

What is the least amount they can spend to buy the supplies they need?

(1) $34

(2) $42

(3) $48

(4) $54

(5) $58

Answers are on page 415.

Simulated GED Test Correlation Chart: Mathematics

Name: ______________________ **Class:** ____________ **Date:** ____________

This chart can help you determine your strengths and weaknesses on the content and skill areas of the Mathematics GED Test. Use the Answer Key on pages 415–417 to check your answers to the test. Then circle on the chart the numbers of the test items you answered correctly. Put the total number correct for each content area and skill area in each row and column. Look at the total items correct in each column and row and decide which areas are difficult for you. Use the page references to study those areas. Use a copy of the Study Record Sheet on page 25 to guide your studying.

Content	Item Number	Total Correct	Page Ref.
Arithmetic *(pages 26–193)*			
Whole Number Operations	5, 47, 56	_____out of 3	26–57
Fractions	15, 20	_____out of 2	58–87
Decimals	1, 2, 4, 8, 27, 50	_____out of 6	88–113
Percent	13, 18, 23, 35, 36, 37	_____out of 6	114–145
Ratio, Proportion, Mean, Median, and Probability	14, 25, 29, 32, 39, 42, 43, 51, 54	_____out of 9	146–165
Graphs and Charts	10, 11, 34, 46	_____out of 4	177–193
Algebra *(pages 194–241)*			
Equations	3, 6, 16, 19, 31, 41, 44	_____out of 7	194–211
Graphs, Slope, Line	12, 24	_____out of 2	212–223
Powers and Roots	17, 40, 52	_____out of 3	224–233
Geometry *(pages 242–309)*			
Perimeter, Area, Volume	9, 28, 45, 48, 49, 55	_____out of 6	166–176 244–251 284–297
Angles	7, 22, 38	_____out of 3	252–257
Triangles and Quadrilaterals	30	_____out of 1	260–265
Congruence and Similarity	21, 53	_____out of 2	266–273
Pythagorean Theorem	26, 33	_____out of 2	276–279

For additional help, see the Steck-Vaughn GED Mathematics Exercise Book.

Answers and Explanations

PRETEST (Pages 14–23)

All items relate to the skill of application.

1. **(2) 35(8)(10)** Multiply 35 by 8 to find the number of valves made in one day. Multiply by 10 to find the number made in 10 days.

2. **(3) between 720 and 760** Divide the monthly cost ($40) by the rate (0.054). Estimate: $40 ÷ 0.054 is equal to approximately 740.

3. **(1) 2.73815×10^5** Move the decimal point to the right 5 places to multiply by 10^5.

4. **(1) $\frac{230 + 212 + 257}{3}$** Use the formula for finding the mean. The problem is solved by adding the scores and dividing by the number of scores, 3.

5. **(2) B and E** Add the games he has already chosen ($15.25 + $34.95 = $50.20) and subtract from $90 to find out how much money he has left. $90 − $50.20 = $39.80 Estimate the cost of the combinations in the list. Option (2), B and E, is the only option which totals less than $39.80.

6. **(3) 3,600** Use the formula for the volume of a rectangular container:
$V = lwh = (40)(9)(10) = 3{,}600$.

7. **(2) between 4 and 5 percent** Divide the state tax ($124.33) by the total earnings ($2,759). Convert to a percent. Or round the numbers to estimate an answer: 130 ÷ 2,760.

8. **(2) The volume of *B* is one cubic foot smaller than the volume of *A*.** The volume of *A* is 64 cubic feet ($4 \times 4 \times 4 = 64$). The volume of *B* is 63 cubic feet ($3 \times 3.5 \times 6 = 63$).

9. **(4) 56** $\triangle ABC$ and $\triangle CDE$ are similar triangles and have proportional sides. Set up a proportion:

$$\frac{\text{Side } \overline{AC}}{\text{Side } \overline{CE}} = \frac{\text{Side } \overline{AB}}{\text{Side } \overline{DE}}$$

$$\frac{\cancel{28}^{14}}{\cancel{2}_1} = \frac{x}{4}$$

$$\frac{14}{1} = \frac{x}{4}$$

$$x = (14)(4)$$

$$x = 56$$

10. **(2) 4** Divide $13\frac{1}{2}$ yards by $3\frac{3}{8}$ yards. Change to improper fractions $\left(13\frac{1}{2} = \frac{27}{2}; 3\frac{3}{8} = \frac{27}{8}\right)$. Invert $\frac{27}{8}$ and multiply. $\frac{27}{2} \div \frac{27}{8} = \frac{^1\cancel{27}}{_1\cancel{2}} \times \frac{\cancel{8}^4}{\cancel{27}_1} = 4$

11. **(1) $(3.14)(5)^2(12)$** Use the formula $V = \pi r^2 h$. Divide the diameter (10) in half to find the radius.

12. **(3) $\frac{1}{50}$** Write as a fraction and reduce: $\frac{8}{400} = \frac{1}{50}$.

13. **(4) 300–350** Approximately 625 people attended on Saturday and 300 on Monday. Subtract Monday's attendance:
625 − 300 = 325.

14. **(2) 11** Approximately 550 people attended on Friday. Divide by 50 people per guard:
550 ÷ 50 = 11 guards.

15. **(2) 8** Divide 12 by $1\frac{1}{2}$.
$12 \div \frac{3}{2} = \cancel{12}^4 \times \frac{2}{\cancel{3}_1} = 8$.

16. **(2) 20** Find the area of Room *A*: $16 \times 25 = 400$. Find the square root of 400 to find one side of Room *B*. (See the formula page for finding the area of a square and a rectangle.)

17. **(5) 4** Divide 5 grains by $1\frac{1}{4}$ grains per tablet:
$1\frac{1}{4} = \frac{5}{4}. \frac{5}{1} \div \frac{5}{4} = \frac{\cancel{5}}{1} \times \frac{4}{\cancel{5}} = 4$

18. **(2) 750** Multiply the number of hours the company is open (5) by the rate (150).

19. **(1) $6^2 + B^2 = 10^2$** Use the Pythagorean Theorem. The hypotenuse *(c)* measures 10. The side given *(a)* measures 6.

20. **(4) 60 degrees** The sum of the interior angles of any triangle is 180°. Add the known angles and subtract from 180.
180 − (60 + 60) = 180 − 120 = 60.

21. **(4) 11 and 12** Estimate the square root of 125: $\sqrt{121} = 11; \sqrt{144} = 12$.

22. **(2) 110** Use the proportion:

$$\frac{1 \text{ in.}}{20 \text{ mi.}} = \frac{5\frac{1}{2} \text{ in.}}{x \text{ mi.}}$$

$$x = 5\tfrac{1}{2} \times 20$$

$$x = \frac{11}{\cancel{2}_1} \times \frac{\cancel{20}^{10}}{1} = 110$$

23. **(2) 47%** Add the food and rent rates (28% + 25% = 53%) and subtract from 100% (100% − 53% = 47%). Or add the percents for clothing, auto, recreation, and other (15% + 12% + 10% + 10% = 47%).

24. **(2) side $\overline{AC}$** When intersecting lines connect parallel line segments, similar triangles are formed. In the figure, $\angle B$ corresponds to $\angle D$ and $\angle A$ corresponds to $\angle E$. Therefore, side $\overline{BC}$ corresponds to side $\overline{CD}$ and side $\overline{CE}$ corresponds to side $\overline{AC}$.

25. **(3) 64** Similar triangles have proportionate sides. Set up a proportion using any pair of corresponding sides:

$$\frac{\text{Side } \overline{AC}}{\text{Side } \overline{CE}} = \frac{\text{Side } \overline{AB}}{\text{Side } \overline{DE}} = \frac{15}{60} = \frac{16}{x} = \frac{1}{4} = \frac{16}{x}$$

$x = 64$

26. **(2) $\frac{(16)(36)}{24}$** Set up a proportion and solve for length.

27. **(1) $\frac{252}{900(2)}$** Use the formula $i = prt$ and solve for rate: $r = \frac{i}{pt} = \frac{252}{(900)(2)}$.

28. **(3) $3x$ = $1.36** Let x equal the price in 1970. Then, $3x$ equals the current price.

29. **(5) 36**

$c = 2a^2(b - 4)$
$c = 2(3^2)(6 - 4)$
$c = 2(9)(2)$
$c = 36$

30. **(5) $8,970** They still owe 100% − 25% = 75% (75% = 0.75). 0.75 × $11,960 = $8,970.

31. **(4) between 11 and 12 feet** Use the Pythagorean Theorem:

$c^2 = a^2 + b^2$
$12^2 = 4^2 + b^2$
$144 = 16 + b^2$
$144 - 16 = b^2$
$128 = b^2$

Estimate the square root of 128. It falls between 11 and 12 ($11^2 = 121$; $12^2 = 144$.)

32. **(4) 24** Add the sides. The measurement of the bottom side is 8 (the horizontal distance is 2 + 2 + 2 + 2). The left side is 4 (the vertical distance is 1 + 1 + 1 + 1).
$P = 8 + 8 + 4 + 4$ or $P = 2(8) + 2(4)$
$P = 16 + 8 = 24$

33. **(4) WRLO's frequency is 2 times higher than KPRQ's frequency.** 21.0×10^5 is 2,100,000. 4.2×10^6 is 4,200,000. 2,100,000 × 2 = 4,200,000.

34. **(3) 372** Divide the total miles by the number of days: 1,860 ÷ 5 = 372.

35. **(5) Not enough information is given.** You need to know the map scale to solve the problem. For example, if the map represented 5 miles as one inch, then the circle would represent an area of 314 square miles, or $3.14\,[2(5)]^2$.

36. **(1) 5(18) = 3d** Use a proportion: $\frac{3 \text{ boxes}}{5 \text{ days}} = \frac{18 \text{ boxes}}{d \text{ days}}$ Therefore, $5(18) = 3d$.

37. **(2) 72** Use a proportion:

$$\frac{18 \text{ games won}}{40 \text{ games played}} = \frac{x \text{ games won}}{160 \text{ games played}}$$

$\frac{18(160)}{40} = 72.$

38. **(2) $63.00** The charity keeps 35% (0.35): 100% − 65%. (0.35)($180) = $63.00

39. **(4) F** Count four spaces to the right from zero.

40. **(3) $41.10** Find the total sales ($725 + $385 + $260 = $1,370) and multiply by the rate (3% = .03). $1,370 × 0.03 = $41.10.

41. **(4) 1999** To meet the goal the university needs 6,000 students (20,000 − 14,000 = 6,000). Divide 6,000 by 1,500. It will take four years: 1995 + 4 = 1999.

42. **(3) 4%** The raise equals x% of the base pay: ($15,080 − $14,500) = ($x$)($14,500) Divide the raise over the base pay and change to a percent.

$$x = \frac{(\$15{,}080 - \$14{,}500)}{\$14{,}500} = \frac{\$580}{\$14{,}500} = .04 = 4\%$$

43. **(4) 38** An isosceles triangle has two sides the same length and two angles of equal measure. The equal angles are opposite the equal (congruent) sides. Side $\overline{BC}$ must be 10 feet in length so the perimeter must be 38 feet (10 + 10 + 18).

44. **(5) 27** Use a proportion

$$\frac{\text{pole height } (x)}{\text{pole shadow } (36)} = \frac{\text{post height } (3)}{\text{post shadow } (4)}$$

$4x = (36)(3)$
$4x = 108$
$x = 27$

45. **(5) 280** Divide the number who responded (126) by the rate (45% = 0.45). Or set up a ratio: $\frac{45\%}{126} = \frac{100\%}{x}$

$12{,}600 = 45x$
$280 = x$

46. **(2) $1,435** Find the amount of discount, then the cost of the used parts ($1,400 × 0.60 = $840; $1,400 − $840 = $560). Add the cost of the used parts ($560) to the cost of labor ($875).

47. **(2) $6x + 2y$** Combine the x's ($4x + 2x = 6x$); combine the y's ($y + y = 2y$).

48. **(3) 60** If 85% voted for the proposal, 15% (100% − 85%) voted against (15% = 0.15). $400 \times 0.15 = 60$

49. **(3) 18** Substitute values in the formula for the perimeter of a triangle: $p = a + b + c$. Solve the equation for a.

$$86 = a + 32 + 36$$
$$86 = a + 68$$
$$86 - 68 = a + 68 - 68$$
$$18 = a$$

50. **(4) p − $10.00 − $5.50** You need to subtract both amounts from the original price of the toaster oven.

51. **(2) 27** $3(n^2 - z^4) =$
$$3(5^2 - 2^4) =$$
$$3(25 - 16) =$$
$$3(9) = 27$$

52. **(5) $90.25**
$$\$9.50(5 + 3 + 1\tfrac{1}{2}) =$$
$$\$9.50\left(9\tfrac{1}{2}\right) = \$90.25$$

53. **(4) 32** Let x represent those who chose Plan B and $2x$ represent those who chose Plan A. Use the equation:
$$x + 2x = 48$$
$$3x = 48$$
$$\frac{3x}{3} = \frac{48}{3}$$
$$x = 16$$
$$2x = 32$$

54. **(2) 60 yards** Add the areas of the patio and the pool to find the area of the fenced region. ($A = lw$) Divide by the length of the fenced region to find the width:
4,500 + 1,500 = 6,000
6,000 ÷ 100 = 60.

55. **(2) 64°** The sum of the interior angles of any triangle is 180°. Add the given angles (80 + 36 = 116) and subtract from 180. 180° − 116° = 64°

56. **(4) −3** Use the formula: $m = \frac{y_2 - y_1}{x_2 - x_1}$
$$3 = \frac{3 - y}{3 - 1}$$
$$3 = \frac{3 - y}{2}$$
$$6 = 3 - y$$
$$3 = -y$$
$$-3 = y$$

UNIT 1: ARITHMETIC

Lesson 1

GED Practice: Lesson 1 (Pages 30–31)

1. **five hundred**
2. **three million**
3. **six thousand**
4. **eight billion**
5. **seven hundred thousand**
6. **sixty thousand**
7. **eighty million**
8. **6** is in the **hundred thousands place.** $6 \times 100{,}000 =$ **600,000**
   7 is in the **ten thousands place.** $7 \times 10{,}000 =$ **70,000**
   9 is in the **thousands place.** $9 \times 1{,}000 =$ **9,000**
   3 is in the **hundreds place.** $3 \times 100 =$ **300**
   0 is in the **tens place.** $0 \times 10 =$ **0**
   8 is in the **ones place.** $8 \times 1 =$ **8**
9. **b.** Eight thousand, four hundred sixteen
10. **d.** Eight million, four hundred twenty thousand, one hundred six
11. **a.** Eighty-four million, two hundred thousand, one hundred sixty
12. **c.** Eight hundred forty-two thousand, sixteen
13. **c.** Ten million, two hundred fifty thousand, nine hundred
14. **b.** Twelve thousand, five hundred nine
15. **d.** One thousand, two hundred fifty-nine
16. **a.** One hundred twenty-five thousand, ninety
17. **1,305 < 1,503**
18. **34,000 > 29,989**
19. **102,667 > 102,657**
20. **5,690,185 < 5,690,185,100**
21. **7,650,300 > 7,649,950**
22. **875,438 = 875,438**
23. **3,492,012,558 = 3,492,012,558**
24. **75,390,000 < 75,391,540**

25. **9,500,000 < 9,500,000,000**

26. **45,100 > 45,099**

27. **7,456,795 < 7,500,000**

28. **319,002,110 = 319,002,110**

29. **8,600**

30. **700,000**

31. **5,000,000**

32. **50,000,000**

33. **11,000**

34. **50,000**

35. **60,000,000**

36. **74,000**

37. **Monday—$19,000**
    Tuesday—$12,000
    Wednesday—$14,000
    Thursday—$14,000
    Friday—$22,000
    Saturday—$29,000

38. **Wednesday.** $13,940 is greater than $13,772.

39. **Tuesday.** $12,316 is less than all the other amounts.

40. **Saturday.** $28,795 is greater than all the other amounts.

41. **$12,316; $13,772; $13,940; $18,756; $21,592; $28,795**

GED Practice: Solving Word Problems (Page 33)

1. **(2) $40 − $28** You are finding "how much is left."
2. **(3) $269 × 12** You need to add the same amount, $269, 12 times.
3. **(1) $137 + $124** You are finding the total of two costs.
4. **(2) $827 − $189** You are finding out how much is left.
5. **(4) 348 ÷ 6** You need to break 348 pages into 6 equal parts.
6. **(1) 168 + 154** You are finding the total of two amounts.
7. **(4) $114 ÷ 6** You need to break an amount into equal parts.
8. **(3) 68 × $7** You need to add the same amount, $7, 68 times.

Lesson 2

GED Practice: Lesson 2 (Pages 36–37)

1. **768**

$$\begin{array}{r} 305 \\ +463 \\ \hline 768 \end{array}$$

2. **8,682**

$$\begin{array}{r} 4{,}172 \\ +4{,}510 \\ \hline 8{,}682 \end{array}$$

3. **6,927**

$$\begin{array}{r} 6{,}795 \\ +\ 132 \\ \hline 6{,}927 \end{array}$$

4. **1,139**

$$\begin{array}{r} 193 \\ 317 \\ +629 \\ \hline 1{,}139 \end{array}$$

5. **61,235**

$$\begin{array}{r} 56{,}439 \\ +\ 4{,}796 \\ \hline 61{,}235 \end{array}$$

6. **244,967**

$$\begin{array}{r} 36{,}075 \\ 1{,}936 \\ 189{,}006 \\ +\ 17{,}950 \\ \hline 244{,}967 \end{array}$$

7. **54,263**

$$\begin{array}{r} 19{,}067 \\ +35{,}196 \\ \hline 54{,}263 \end{array}$$

8. **213,658**

$$\begin{array}{r} 65{,}196 \\ 6{,}725 \\ 114{,}021 \\ +\ 27{,}716 \\ \hline 213{,}658 \end{array}$$

9. **109,636**

$$\begin{array}{r} 81{,}427 \\ 3{,}584 \\ +24{,}625 \\ \hline 109{,}636 \end{array}$$

10. **134**

$$\begin{array}{r} 76 \\ +58 \\ \hline 134 \end{array}$$

11. **188**

$$\begin{array}{r} 36 \\ 9 \\ 74 \\ 48 \\ 6 \\ +15 \\ \hline 188 \end{array}$$

12. **758,450**

$$\begin{array}{r} 588{,}394 \\ 61{,}042 \\ +109{,}014 \\ \hline 758{,}450 \end{array}$$

13. **54,271**

$$\begin{array}{r} 54{,}095 \\ +\quad 176 \\ \hline 54{,}271 \end{array}$$

14. **91,923**

$$\begin{array}{r} 35{,}100 \\ 49{,}257 \\ +\ 7{,}566 \\ \hline 91{,}923 \end{array}$$

15. **1,414**

$$\begin{array}{r} 950 \\ 308 \\ 77 \\ 29 \\ +\ 50 \\ \hline 1{,}414 \end{array}$$

16. 208,115

6,019
85,200
+116,896
208,115

17. 35

86
−51
35

18. 31,313

51,964
−20,651
31,313

19. 327

494
−167
327

20. 963

1,258
− 295
963

21. 412

680
−268
412

22. 929

3,205
−2,276
929

23. 581

800
−219
581

24. 1,272

5,067
−3,795
1,272

25. 6,829

10,508
− 3,679
6,829

26. 1,305

5,000
−3,695
1,305

27. 406,985

419,003
− 12,018
406,985

28. 1,025

10,000
− 8,975
1,025

29. 3,433

20,000
−16,567
3,433

30. 188,575

375,000
−186,425
188,575

31. 505,232

510,000
− 4,768
505,232

32. 24,490

100,000
− 75,510
24,490

33. 864

432
× 2
864

34. 3,730

746
× 5
3,730

35. 125,304

15,663
× 8
125,304

36. 273,681

30,409
× 9
273,681

37. 828

36
×23
108
72
828

38. 386,384

5,084
× 76
30 504
355 88
386,384

39. 53,685

1,193
× 45
5 965
47 72
53,685

40. 78,624

3,276
× 24
13 104
65 52
78,624

41. 6,976,800

2,584
×2,700
0 000
00 00
1 808 8
5 168
6,976,800

42. 1,505,820

25,097
× 60
00 000
1 505 82
1,505,820

43. 4,097,450

8,050
× 509
72 450
00 00
4 025 0
4,097,450

44. 3,758,458

1,247
×3,014
4 988
12 47
000 0
3 741
3,758,458

45. 57,525

885
× 65
4 425
53 10
57,525

46. 415,340

2,186
× 190
0 000
196 74
218 6
415,340

47. 600,625

775
×775
3 875
54 25
542 5
600,625

48. 7,640,000

3,056
×2,500
0 000
00 00
1 528 0
6 112
7,640,000

49. 458

458
7) 3,206
2 8
40
35
56
56

50. 5,996

5,996
4) 23,984
20
3 9
3 6
38
36
24
24

UNIT 1

51. 42,363

```
     42,363
 6)254,178
   24
    14
    12
     2 1
     1 8
       37
       36
        18
        18
```

52. 8,132

```
      8,132
 16)130,112
    128
      2 1
      1 6
        51
        48
         32
         32
```

53. 583r4

```
    583r4
 6)3,502
   3 0
     50
     48
      22
      18
       4
```

54. 6,366r10

```
     6,366r10
 12)76,402
    72
     4 4
     3 6
       80
       72
        82
        72
        10
```

55. 7,692r4

```
      7,692r4
 13)100,000
     91
      9 0
      7 8
      1 20
      1 17
         30
         26
          4
```

56. 9,138r3

```
      9,138r3
 24)219,315
    216
      3 3
      2 4
        91
        72
        195
        192
          3
```

57. 1,098

```
     1,098
 35)38,430
    35
     3 43
     3 15
       280
       280
```

58. 659r7

```
      659r7
 46)30,321
    27 6
     2 72
     2 30
       421
       414
         7
```

59. 5,302r1

```
      5,302r1
 68)360,537
    340
     20 5
     20 4
        13
        00
        137
        136
          1
```

60. 680

```
        680
 205)139,400
     123 0
      16 40
      16 40
         00
         00
```

61. 1,026r5

```
       1,026r5
 155)159,035
     155
       4 0
       0 0
       4 03
       3 10
         935
         930
           5
```

62. 40,430

```
      40,430
 15)606,450
    60
     6 4
     6 0
       45
       45
```

63. 28,480

```
      28,480
 25)712,000
    50
    212
    200
     12 0
     10 0
      2 00
      2 00
```

64. 2,572r121

```
        2,572r121
 163)419,357
     326
      93 3
      81 5
      11 85
      11 41
         447
         326
         121
```

65. 4,236

```
        4,236
 120)508,320
     480
      28 3
      24 0
       4 32
       3 60
         720
         720
```

GED Practice: Solving Word Problems (Page 39)

1. (2) $12

```
  $40
 − 28
  $12
```

2. (5) $3,228

```
   $269
 ×   12
    538
   269
 $3,228
```

3. (4) $261

```
  $137
 + 124
  $261
```

4. **(5) $638**

$$\begin{array}{r} \$827 \\ -\ 189 \\ \hline \$638 \end{array}$$

5. **(3) 58**

$$\begin{array}{r} 58 \\ 6\overline{)348} \\ 30 \\ \hline 48 \\ 48 \\ \hline \end{array}$$

6. **(1) 322**

$$\begin{array}{r} 168 \\ +\ 154 \\ \hline 322 \end{array}$$

7. **(3) $19**

$$\begin{array}{r} \$\ 19 \\ 6\overline{)\$114} \\ 6 \\ \hline 54 \\ 54 \\ \hline \end{array}$$

8. **(5) $476**

$$\begin{array}{r} 68 \\ \times\ \$7 \\ \hline \$476 \end{array}$$

Lesson 3

All items relate to the skill of application.

GED Practice: Lesson 3 (Pages 42–45)

1. **(3) $1,090** When a deposit is made to an account, the numbers are added ($715 + $375 = $1,090).

2. **(5) Not enough information is given.** You need to know the number of people that can sit at each of the 12 tables to solve the problem.

3. **(3) between 1,650 and 1,750** Round the number of calories he has eaten to the nearest 100 (685 rounds to 700). Subtract 700 from 2,400 to get an estimate (2,400 – 700 = 1,700). The estimate 1,700 falls within the range in option 3.

4. **(3) $500** Divide the amount for the year by the number of months in a year ($6,000 ÷ 12 = $500).

5. **(2) $294** Subtract the deductions from the gross pay to find his take-home pay ($440 – $146 = $294). Check by addition ($146 + $294 = $440).

6. **(5) Not enough information is given.** To figure the cost of the job, you need to know how much the new parts will cost.

7. **(4) 18** Divide the total sales by the price of one set ($5,760 ÷ $320 = 18). The size of the television (20 inches) is not needed to solve the problem. Check by multiplication (18 × $320 = $5,760).

8. **(4) $188** Subtract the amount more from this week's amount ($236 – $48 = $188).

9. **(5) Not enough information is given.** You need to know how many miles were driven on the second day to solve the problem.

10. **(1) 90** Multiply the number of bags by the number of buns in a bag (15 × 6 = 90). You do not need the number of days Quan Le works to solve the problem.

11. **(2) $186** Subtract the amount she has from the amount she needs ($223 – $37 = $186). Check by addition ($37 + $186 = $223).

12. **(5) Not enough information is given.** You need to know how many pounds one bag weighs to solve the problem.

13. **(3) $206,000** Multiply the number of employees by the amount of each bonus (412 × $500 = $206,000).

14. **(2) 224** Multiply the number of offices by the number of people who could work in each office (112 × 2 = 224). You do not need the number of floors to solve the problem.

15. **(1) $6** Subtract the hourly amount spent on child care from her hourly wage ($8 – $2 = $6). You do not need to know how many hours she works to solve the problem.

16. **(5) Not enough information is given.** You need to know how many people are on the custodial staff. The number of rooms that can be cleaned in an hour is not necessary information.

17. **(4) $568,450** Add the individual department budgets to find the total budget amount ($180,500 + $92,750 + $225,125 + $18,725 + $51,350 = $568,450).

18. **(4) between $200,000 and $220,000** This problem asks for an estimate. Round the maintenance budget—the largest amount—to the nearest thousand ($225,125 rounds to $225,000). Round the recreation budget—the smallest amount—to the nearest ten thousand ($18,725 rounds to $19,000). Subtract ($225,000 – $19,000 = $206,000). The number $206,000 falls in the range in option 4. The exact solution, $206,400, falls in the same range.

19. **(3) $361,000** "To double" means multiply by 2. Therefore, multiply by 2 the amount on the chart for the police department ($180,500 × 2 = $361,000).

20. **(2) $3,745** Divide the recreation budget by the number of parks ($18,725 ÷ 5 = $3,745). Check by multiplication (5 × $3,745 = $18,725).

21. **(2) $44,850** Subtract the difference $6,500 from this year's budget for the office department ($51,350 − $6,500 = $44,850). Check by addition ($6,500 + $44,850 = $51,350).

22. **(1) $45,159** Divide the final budget amount by the number of months in a year ($541,908 ÷ 12). You do not need any information from the chart to solve this problem. Check by multiplication (12 × $45,159 = $541,908).

23. **(2) $720** Multiply the number of tiles by the cost of each tile (240 × $3 = $720). The amounts spent on cabinets and plumbing are not needed to solve the problem.

24. **(2) 515** Add the goals for each day of the week (100 + 95 + 120 + 110 + 90 = 515).

25. **(3) Wednesday** Review the data on the list. On every day except Wednesday (120 > 117) the actual number is greater than the goal number.

26. **(2) 58** Only the actual data for Thursday and Friday are needed for this problem. Subtract Thursday's actual number from Friday's (180 − 122 = 58). Check by addition (122 + 58 = 180).

27. **(5) Not enough information is given.** You need to know how many months Maya will be paying for the stereo to solve the problem.

28. **(5) Not enough information is given.** You need the score of the third game to solve the problem.

29. **(4) 820** Subtract the number of employees that will be laid off from the total number of employees (1,192 − 372 = 820). You do not need the number of branches to solve the problem. Check by addition (372 + 820 = 1,192).

30. **(3) $159,000** Multiply the number of seats by the cost of a ticket (26,500 × $6 = $159,000). You do not need to know the number of years the game has been at the coliseum to solve the problem.

Lesson 4

GED Practice: Lesson 4 (Pages 50–53)

1. **(4) multiplication and addition** You need to multiply the number of payments by the amount of the regular payment and add the final payment.

2. **(5) $1,082** Multiply $115 by 8 to get $920 and add $162. ($920 + $162 = $1,082).

3. **(3) $1,490** Multiply to find how much Hugh pays in one year ($312 × 2 = $624). Add Brenda's amount ($624 + $866 = $1,490).

4. **(4) multiplication and subtraction** Find the cost of the TV if you buy on credit by multiplying the payment amount by 12. Subtract the cash amount to find the difference.

5. **(3) $80** Multiply to find the cost of the TV if you buy on credit (12 × $45 = $540). Subtract the cost if you pay in cash ($540 − $460 = $80).

6. **(3) $1,412** Subtract the amount the insurance company refunded from the total cost of medical bills. ($3,794 − $2,382 = $1,412).

7. **(2) addition and division** Add to find the total number of apartments. Then divide the cost of the service by the number of apartments.

8. **(5) $72** There are 43 apartments (18 + 25 = 43). Divide the yearly cost by the number of apartments ($3,096 ÷ 43 = $72).

9. **(1) $280** Add to find the hours F. Blau worked (8 + 7 + 8 + 8 + 6 + 3 = 40). Then multiply by the hourly wage (40 × $7 = $280).

10. **(2) 12** Add to find the total hours M. Bodine worked (8 + 8 + 9 + 8 + 8 + 8 = 49). Find the hours T. Ortiz worked (6 + 6 + 8 + 9 + 8 = 37). Find the difference (49 – 37 = 12).

11. **(1) 2** Subtract to find the difference between the amount he earned and the amount he needs to earn ($252 − $240 = $12). Divide the result by his hourly wage ($12 ÷ $6/hr = 2 hr). R. Perez needs to work 2 more hours.

12. **(3) $370** T. Ortiz worked 37 hours which you originally found by adding in problem 10. Add the raise to the hourly wage ($8 + $2 = $10). Multiply the hours by the new hourly wage (37 × $10 = $370).

13. (2) $56 Multiply 6 days by 8 hours to find the total number of hours F. Blau worked the following week ($6 \times 8 = 48$). Add to find the total number of hours worked the given week ($8 + 7 + 8 + 8 + 6 + 3 = 40$). Subtract to find the difference between the number of hours worked ($48 - 40 = 8$). Multiply by the hourly wage to find the difference in earnings ($8 \times \$7 = \$56$).

14. (2) $360 Multiply the number of pairs of pants sold by the sale price for pants ($24 \times \$15 = \$360$).

15. (5) $52 Find out the amount they would spend for the items using the regular prices and the amount they spent using the sale prices.

12 T-shirts
Regular Price: $12 \times \$8 = \$96$
Sale Price: $12 \times \$7 = \$84$

5 jerseys
Regular Price: $5 \times \$10 = \$50$
Sale Price: $5 \times \$8 = \$40$

10 pairs of pants
Regular Price: $10 \times \$18 = \$180$
Sale Price: $10 \times \$15 = \$150$

Total at Regular Price:
$\$96 + \$50 + \$180 = \$326$

Total at Sale Price:
$\$84 + \$40 + \$150 = \$274$

Find the difference in the totals ($\$326 - \$274 = \$52$).

16. (2) $48 Find the new sale price ($\$15 - \$3 = \$12$). Find the cost of 4 pairs by multiplying ($\$12 \times 4 = \$48$). You do not need to know how many pairs they have in stock.

17. (1) $280 Multiply the number produced by the number of days ($56 \times 5 = 280$) and multiply the result by the amount paid per piece ($280 \times \$1 = \$280$).

18. (5) Not enough information is given. You need the amount Mr. Butler is paid per item and the number of items he produced to find out how much he earned.

19. (1) 10 Take the number she produced (50) and divide it by the number of hours she worked ($50 \div 5 = 10$).

20. (2) 414 Multiply to find the number of small diapers ($15 \times 66 = 990$). Multiply to find the number of large diapers ($24 \times 24 = 576$). Find the difference between the two numbers ($990 - 576 = 414$)

21. (3) between 35 and 40 days Find the total ordered. You already know the number of small and large diapers from problem 20. Find the number of medium diapers by multiplying ($18 \times 42 = 756$). The total diapers are found by adding ($990 + 756 + 576 = 2{,}322$). Since the problem calls for an estimate, round 2,322 to the nearest hundred, 2,300, and divide by 60. The answer is approximately 38 days.

22. (4) $627 Add to find the total number of boxes ordered ($15 + 18 + 24 = 57$). Multiply to find the total cost ($57 \times \$11 = \$627$).

23. (3) 171,217 Add the weekday numbers ($34{,}196 + 44{,}640 + 42{,}079 + 50{,}302 = 171{,}217$). Notice that there is no attendance for Monday.

24. (2) between 5,000 and 6,500 Estimate totals for the day games and the night games and find the difference. Round numbers to the nearest hundred.
Day game estimate:
$47{,}600 + 34{,}200 + 49{,}700 = 131{,}500$
Night game estimate:
$44{,}600 + 42{,}100 + 50{,}300 = 137{,}000$
The difference is 5,500
($137{,}000 - 131{,}500 = 5{,}500$)

25. (4) 850 Add the enrollment for the three known quarters ($650 + 624 + 705 = 1{,}979$). Subtract this amount from the year's enrollment ($2{,}829 - 1{,}979 = 850$).

GED Mini-Test: Lessons 1–4 (Pages 54–57)

All items relate to the skill of application.

1. (3) $15 × 10 Multiply to find the amount she will earn from haircuts ($\$15 \times 10$).

2. (4) $710 Find the amount spent on food by multiplying $9 by the number of employees ($65 \times \$9 = \$585$). Add the amount spent on the room ($\$585 + \$125 = \$710$).

3. (2) between 3,500 and 4,500 Round both figures to the nearest thousand and subtract ($10{,}000 - 6{,}000 = 4{,}000$).

4. (3) 26,000 Find the number of calculators per year by multiplying by 52, the number of weeks in a year ($500 \times 52 = 26{,}000$).

5. **(3) 50** Find how many medals the United States and China won by adding (108 + 54 = 162). Subtract the number won by the Unified Team (162 − 112 = 50).

6. **(1) 22** Add the number of medals for the three countries (31 + 29 + 30 = 90). Then subtract the total from the number won by the Unified Team (112 − 90 = 22). Check by addition (90 + 22 = 112).

7. **(3) 12 × $2** Multiply the number of parts by the rate per piece to find the amount Art earns in an hour (12 × $2).

8. **(3) 65,044** Add the numbers for the three months (13,042 + 25,907 + 26,095 = 65,044).

9. **(4) 26,095 − 6,521** The month with the most rentals is May; the month with the fewest is February. Subtract: 26,095 − 6,521.

10. **(3) about 4 times higher** Round the two numbers to the nearest hundred and divide: 26,100 ÷ 6,500 ≅ 4. Rounding 6,521 to 7,000 does not give you a clear answer. Remember that rounding to a smaller place value column gives you greater accuracy.

11. **(5) 60** Multiply the number of hours per day by the number of days per week (3 × 5 = 15). Multiply that product by 4 weeks (15 × 4 = 60).

12. **(3) $23** Find how much she spent:
2 books × $35 = $70; 3 books × $18 = $54;
1 notebook = $3. Add the amounts ($70 + $54 + $3 = $127). Subtract the amount she spent from $150 ($150 – $127 = $23).

13. **(5) Not enough information is given.** You need to know how many ribbons are in a box to solve the problem.

14. **(2) between 5 and 6** If you divide 900 by 150 you get 6, which is mentioned in two choices. Try multiplying 160 by 6; the result, 960, is too high. A better estimate is less than 6.

15. **(4) $140** Multiply the number of spaces in the garage by each price to find how much the garage owner will make at each price. Subtract to find the difference:
70 × $6 = $420; 70 × $8 = $560;
$560 − $420 = $140.

16. **(3) $17** Subtract the sale price from the original price of each item to find the savings. Add the savings for each of the items to find the total savings.
1 wool sweater: $45 − $36 = $9
1 dress shirt: $21 − $17 = $4
1 necktie: $19 − $15 = $4
$9 + $4 + $4 = $17

17. **(3) $64** Multiply the sale price of each item by the number of items, 2. Add the results together to find the total cost.
Two dress shirts: $17 × 2 = $34
Two neckties: $15 × 2 = $30
$34 + $30 = $64

18. **(2) $18** Subtract the sale price from the original price of the sweater to find the amount of savings. Multiply the result by the number of items to find the total savings.
$45 − $36 = $9
$9 × 2 = $18

19. **(3) 365** Multiply the average speed 55 mph by 3 hours. Multiply the average speed 50 mph by 4 hours. Add the products.
55 × 3 = 165 miles
50 × 4 = 200 miles
165 + 200 = 365 miles

20. **(1) 18** Divide the total length of rope, 900 feet, by equal lengths of 50 feet to find the number of packages needed.

$$\begin{array}{r} 18 \\ 50\overline{)900} \\ \underline{50} \\ 400 \\ \underline{400} \end{array}$$

21. **(2) 578** Subtract the number of writing pads used on Tuesday from the original number of pads. From the number left, subtract the number of pads used on Wednesday.

$$\begin{array}{r} 1{,}000 \\ \underline{-\ 188} \\ 812 \end{array} \qquad \begin{array}{r} 812 \\ \underline{-234} \\ 578 \end{array}$$

22. **(1) 25** Divide the total weight by the number of boxes. 1,800 ÷ 72 = 25

23. **(1) between 9,000 and 11,000** This problem calls for an estimate. Round the number of boxes to the nearest 10 (21 rounds to 20). Round the number of bolts per box to the nearest ten (506 rounds to 510).
510 × 20 = 10,200. The number of bolts on hand is not needed to solve the problem.

24. **(4) $1,236** Subtract the amount she actually owed from the amount she paid ($5,752 − $4,516 = $1,236). Check by addition ($4,516 + $1,236 = $5,752).

25. **(2) 1,988** Subtract the number of votes the winner won by from the winner's total number of votes to find the other candidate's number of votes.

$$\begin{array}{r} 2{,}035 \\ -\quad 47 \\ \hline 1{,}988 \end{array}$$

26. **(5) $175** You do not need to know the price of the cameras. Add to find the number of sales (2 + 3 = 5). Multiply the number of sales by $35 ($35 × 5 = $175).

27. **(2) 16,106** Subtract the attendance at Tuesday's game from the attendance at Friday night's game (50,302 − 34,196 = 16,106).

28. **(1) 12** Divide the total bill, $1,800, by equal payments of $150 to find the number of payments.

$$\begin{array}{r} 12 \\ \$150\overline{)\ \$1{,}800} \\ 1\,50 \\ \hline 300 \\ 300 \\ \hline \end{array}$$

29. **(1) 1,536** Subtract this year's enrollment from last year's enrollment to find how many children the enrollment grew by.

$$\begin{array}{r} 20{,}132 \\ -18{,}596 \\ \hline 1{,}536 \end{array}$$

Lesson 5

All items relate to the skill of application.

GED Practice: Lesson 5 (Pages 62–63)

1. $\frac{5}{8}$ The figure is divided into 8 equal parts, and 5 of the parts are shaded.

2. $\frac{2}{3}$ There are 3 circles in the group, and 2 are shaded.

3. $\frac{9}{10}$ The figure is divided into 10 equal parts, and 9 of the parts are shaded.

4. $\frac{1}{6}$ There are 6 figures in the group, and 1 is shaded.

5. $\frac{5}{9}$ The figure is divided into 9 equal parts, and 5 of the parts are shaded.

6. $\frac{9}{16}$ **7.** $\frac{7}{10}$ **8.** $\frac{19}{45}$ **9.** $\frac{3}{50}$ **10.** $\frac{9}{70}$ **11.** $\frac{43}{500}$

12. $\frac{8}{5}$ Each figure is divided into 5 equal parts. There are 8 parts shaded among the figures.

13. $\frac{7}{2}$ Each figure is divided into 2 equal parts. There are 7 parts shaded among the figures.

14. $\frac{8}{3}$ Each figure is divided into 3 equal parts. There are 8 parts shaded among the figures.

15. $2\frac{5}{6}$ Each figure is divided into 6 equal parts. Two figures are completely shaded, and 5 of the 6 parts of the last figure are shaded.

16. $1\frac{3}{8}$ Each figure is divided into 8 equal parts. One figure is completely shaded, and 3 of the 8 parts of the last figure are shaded.

17. $3\frac{1}{4}$ Each figure is divided into 4 equal parts. Three figures are completely shaded, and 1 of the 4 parts of the last figure is shaded.

18. $3\frac{1}{2}$ 7 ÷ 2 = 3 r1

19. $2\frac{2}{3}$ 8 ÷ 3 = 2 r2

20. $5\frac{7}{10}$ 57 ÷ 10 = 5 r7

21. 3 15 ÷ 5 = 3

22. $3\frac{1}{3}$ 10 ÷ 3 = 3 r1

23. $8\frac{5}{9}$ 77 ÷ 9 = 8 r5

24. $4\frac{3}{4}$ 19 ÷ 4 = 4 r3

25. $3\frac{7}{8}$ 31 ÷ 8 = 3 r7

26. $2\frac{5}{6}$ 17 ÷ 6 = 2 r5

27. $4\frac{1}{2}$ 9 ÷ 2 = 4 r1

28. 7 42 ÷ 6 = 7

29. $3\frac{5}{12}$ 41 ÷ 12 = 3 r5

30. $\frac{13}{2}$ 2 × 6 = 12, 12 + 1 = 13

31. $\frac{17}{2}$ 2 × 8 = 16, 16 + 1 = 17

32. $\frac{43}{8}$ 8 × 5 = 40, 40 + 3 = 43

33. $\frac{23}{6}$ 6 × 3 = 18, 18 + 5 = 23

34. $\frac{11}{5}$ 5 × 2 = 10, 10 + 1 = 11

35. $\frac{46}{7}$ 7 × 6 = 42, 42 + 4 = 46

36. $\frac{13}{8}$ 8 × 1 = 8, 8 + 5 = 13

37. $\frac{47}{4}$ $4 \times 11 = 44$, $44 + 3 = 47$

38. $\frac{21}{2}$ $2 \times 10 = 20$, $20 + 1 = 21$

39. $\frac{38}{9}$ $9 \times 4 = 36$, $36 + 2 = 38$

40. $\frac{61}{12}$ $12 \times 5 = 60$, $60 + 1 = 61$

41. $\frac{47}{5}$ $5 \times 9 = 45$, $45 + 2 = 47$

GED Practice: Solving Word Problems (Page 65)

1. **(2)** $5\frac{3}{4} - 2\frac{3}{8}$ You are comparing two quantities and finding the difference ("how much more").
2. **(1)** $8\frac{1}{8} + 4\frac{4}{5}$ You are combining.
3. **(3)** $15\frac{3}{4} \times \frac{1}{2}$ You are finding a fractional part "of" a whole amount.
4. **(4)** $7\frac{1}{2} \div \frac{3}{4}$ You are finding how many equal parts ($\frac{3}{4}$) are in the whole ($7\frac{1}{2}$).
5. **(1)** $18\frac{1}{10} + 12\frac{2}{5}$ You are finding a total.
6. **(3)** $\frac{5}{8} \times \frac{2}{5}$ You are finding the fractional part "of" the whole.
7. **(4)** $6\frac{1}{4} \div 2\frac{1}{2}$ You are finding how many equal parts ($2\frac{1}{2}$) are in the whole ($6\frac{1}{4}$).
8. **(2)** $3\frac{1}{4} - 2\frac{4}{5}$ You are finding the difference between two quantities.

Lesson 6

All items relate to the skill of application.

GED Practice: Lesson 6 (Pages 68–69)

1. $\frac{1}{2}$ Divide numerator and denominator by 2.
2. $\frac{2}{3}$ Divide numerator and denominator by 3.
3. $\frac{2}{5}$ Divide numerator and denominator by 5.
4. $\frac{3}{4}$ Divide numerator and denominator by 2.
5. $\frac{2}{5}$ Divide numerator and denominator by 3.
6. $\frac{2}{3}$ Divide numerator and denominator by 9.
7. $\frac{1}{4}$ Divide numerator and denominator by 5.
8. $\frac{1}{4}$ Divide numerator and denominator by 12.
9. $\frac{4}{5}$ Divide numerator and denominator by 2.
10. $\frac{2}{5}$ Divide numerator and denominator by 6.
11. $\frac{1}{6}$ Divide numerator and denominator by 7.
12. $\frac{2}{3}$ Divide numerator and denominator by 12.
13. $\frac{4}{8}$ $2 \times 4 = 8$ and $1 \times 4 = 4$.
14. $\frac{8}{12}$ $3 \times 4 = 12$ and $2 \times 4 = 8$.
15. $\frac{6}{21}$ $7 \times 3 = 21$ and $2 \times 3 = 6$.
16. $\frac{20}{25}$ $5 \times 5 = 25$ and $4 \times 5 = 20$.
17. $\frac{20}{32}$ $8 \times 4 = 32$ and $5 \times 4 = 20$.
18. $\frac{49}{63}$ $9 \times 7 = 63$ and $7 \times 7 = 49$.
19. $\frac{36}{120}$ $10 \times 12 = 120$ and $3 \times 12 = 36$.
20. $\frac{36}{96}$ $8 \times 12 = 96$ and $3 \times 12 = 36$.
21. $\frac{27}{36}$ $4 \times 9 = 36$ and $3 \times 9 = 27$.
22. $\frac{36}{81}$ $9 \times 9 = 81$ and $4 \times 9 = 36$.
23. $\frac{27}{150}$ $50 \times 3 = 150$ and $9 \times 3 = 27$.
24. $\frac{48}{100}$ $25 \times 4 = 100$ and $12 \times 4 = 48$.
25. $\frac{1}{3} > \frac{1}{4}$ because $\frac{4}{12} > \frac{3}{12}$.
26. $\frac{3}{4} < \frac{7}{8}$ because $\frac{6}{8} < \frac{7}{8}$.
27. $\frac{3}{9} = \frac{1}{3}$ because $\frac{3}{9}$ reduces to $\frac{1}{3}$.
28. $\frac{2}{3} > \frac{1}{2}$ because $\frac{4}{6} > \frac{3}{6}$.
29. $\frac{1}{3} < \frac{2}{5}$ because $\frac{5}{15} < \frac{6}{15}$.
30. $\frac{5}{6} = \frac{15}{18}$ because $\frac{15}{18}$ reduces to $\frac{5}{6}$.
31. $\frac{9}{12} = \frac{3}{4}$ because $\frac{9}{12}$ reduces to $\frac{3}{4}$.
32. $\frac{7}{10} > \frac{2}{3}$ because $\frac{21}{30} > \frac{20}{30}$.
33. $\frac{7}{15} > \frac{2}{5}$ because $\frac{7}{15} > \frac{6}{15}$.
34. $\frac{3}{4} < \frac{17}{20}$ because $\frac{15}{20} < \frac{17}{20}$.
35. $\frac{8}{9} = \frac{16}{18}$ because $\frac{16}{18}$ reduces to $\frac{8}{9}$.
36. $\frac{5}{6} > \frac{3}{4}$ because $\frac{10}{12} > \frac{9}{12}$.

UNIT 1

37. **(1) A greater fraction of shirts passed inspection at 10 A.M.** Compare $\frac{9}{10}$ and $\frac{7}{8}$. Change both to equal fractions with a denominator of 40: $\frac{9}{10} = \frac{36}{40}$; $\frac{7}{8} = \frac{35}{40}$. Compare numerators: 36 is greater than 35, so $\frac{9}{10}$ is greater than $\frac{7}{8}$.

38. **(3) The same fraction of customers saw the ad each day.** The fractions $\frac{4}{12}$ and $\frac{6}{18}$ are equal fractions. Both reduce to $\frac{1}{3}$.

39. **(3) Team C** Change each fraction to an equal fraction with a denominator of 20, the least common denominator.

Team A = $\frac{8}{20}$

Team B = $\frac{10}{20}$

Team C = $\frac{5}{20}$

Team D = $\frac{12}{20}$

Team E = $\frac{15}{20}$

Team C is lowest because 5 is lower than any of the other numerators.

40. **(5) Team E** Team E is highest because 15 is greater than any of the other numerators (see explanation for item 39).

41. **(5) Teams D and E** Change $\frac{1}{2}$ to an equal fraction with a denominator of 20. $\frac{1}{2} = \frac{10}{20}$. Only $\frac{12}{20}$ and $\frac{15}{20}$ are greater than $\frac{10}{20}$.

42. **(2) Teams B, D, and E** Team B's work = $\frac{1}{2}$; Teams D and E each completed more than $\frac{1}{2}$.

43. **(1) Team F completed less than Team E.** Compare $\frac{2}{3}$ and $\frac{3}{4}$ (Team E's work). Change to equal fractions with a denominator of 12, the lowest common denominator: $\frac{2}{3} = \frac{8}{12}$ and $\frac{3}{4} = \frac{9}{12}$. $\frac{2}{3}$ is less than $\frac{3}{4}$.

GED Practice: Estimation and Fractions (Pages 71–73)

1. $\frac{1}{3} < \frac{1}{2}$ $\quad \frac{1}{3} = \frac{2}{6}$; $\frac{1}{2} = \frac{3}{6}$; $\frac{2}{6} < \frac{3}{6}$

2. $\frac{5}{6} > \frac{1}{2}$ $\quad \frac{1}{2} = \frac{3}{6}$; $\frac{5}{6} > \frac{3}{6}$

3. $\frac{2}{3} > \frac{1}{2}$ $\quad \frac{2}{3} = \frac{4}{6}$; $\frac{1}{2} = \frac{3}{6}$; $\frac{4}{6} > \frac{3}{6}$

4. $\frac{1}{4} < \frac{1}{2}$ $\quad \frac{1}{2} = \frac{2}{4}$; $\frac{1}{4} < \frac{2}{4}$

5. $\frac{5}{8} > \frac{1}{2}$ $\quad \frac{1}{2} = \frac{4}{8}$; $\frac{5}{8} > \frac{4}{8}$

6. $\frac{4}{5} > \frac{1}{2}$ $\quad \frac{4}{5} = \frac{8}{10}$; $\frac{1}{2} = \frac{5}{10}$; $\frac{8}{10} > \frac{5}{10}$

7. $\frac{1}{8} < \frac{1}{2}$ $\quad \frac{1}{2} = \frac{4}{8}$; $\frac{1}{8} < \frac{4}{8}$

8. $\frac{3}{6} = \frac{1}{2}$ $\quad \frac{1}{2} = \frac{3}{6}$; $\frac{3}{6} = \frac{3}{6}$

9. $\frac{3}{4} > \frac{1}{2}$ $\quad \frac{1}{2} = \frac{2}{4}$; $\frac{3}{4} > \frac{2}{4}$

10. $\frac{2}{7} < \frac{1}{2}$ $\quad \frac{2}{7} = \frac{4}{14}$; $\frac{1}{2} = \frac{7}{14}$; $\frac{4}{14} < \frac{7}{14}$

11. $\frac{2}{5} < \frac{1}{2}$ $\quad \frac{2}{5} = \frac{4}{10}$; $\frac{1}{2} = \frac{5}{10}$; $\frac{4}{10} < \frac{5}{10}$

12. $\frac{7}{10} > \frac{1}{2}$ $\quad \frac{1}{2} = \frac{5}{10}$; $\frac{7}{10} > \frac{5}{10}$

13. $\frac{4}{9} < \frac{1}{2}$ $\quad \frac{4}{9} = \frac{8}{18}$; $\frac{1}{2} = \frac{9}{18}$; $\frac{8}{18} < \frac{9}{18}$

14. $\frac{3}{7} < \frac{1}{2}$ $\quad \frac{3}{7} = \frac{6}{14}$; $\frac{1}{2} = \frac{7}{14}$; $\frac{6}{14} < \frac{7}{14}$

15. $\frac{3}{5} > \frac{1}{2}$ $\quad \frac{3}{5} = \frac{6}{10}$; $\frac{1}{2} = \frac{5}{10}$; $\frac{6}{10} > \frac{5}{10}$

16. **4** $\frac{3}{4}$ is greater than $\frac{1}{2}$ because $\frac{3}{4} > \frac{2}{4}$. The mixed number $3\frac{3}{4}$ rounds to the whole number 4.

17. **5** $\frac{1}{3}$ is less than $\frac{1}{2}$ because $\frac{2}{6} < \frac{3}{6}$. The mixed number $5\frac{1}{3}$ rounds to the whole number 5.

18. **1** $\frac{2}{3}$ is greater than $\frac{1}{2}$ because $\frac{4}{6} > \frac{3}{6}$. The fraction $\frac{2}{3}$ rounds to the whole number 1.

19. **0** $\frac{1}{4}$ is less than $\frac{1}{2}$ because $\frac{1}{4} < \frac{2}{4}$. The fraction $\frac{1}{4}$ rounds to the whole number 0.

20. **6** $\frac{3}{5}$ is greater than $\frac{1}{2}$ because $\frac{6}{10} > \frac{5}{10}$. The mixed number $5\frac{3}{5}$ rounds to the whole number 6.

21. **2** $\frac{1}{6}$ is less than $\frac{1}{2}$ because $\frac{1}{6} < \frac{3}{6}$. The mixed number $2\frac{1}{6}$ rounds to the whole number 2.

22. **8** $\frac{7}{8}$ is greater than $\frac{1}{2}$ because $\frac{7}{8} > \frac{4}{8}$. The mixed number $7\frac{7}{8}$ rounds to the whole number 8.

23. **2** $\frac{1}{2}$ is equal to $\frac{1}{2}$. The mixed number $1\frac{1}{2}$ rounds to the whole number 2.

24. **4** $\frac{2}{5}$ is less than $\frac{1}{2}$ because $\frac{4}{10} < \frac{5}{10}$. The mixed number $4\frac{2}{5}$ rounds to the whole number 4.

25. **7** $\frac{7}{10}$ is greater than $\frac{1}{2}$ because $\frac{7}{10} > \frac{5}{10}$. The mixed number $6\frac{7}{10}$ rounds to the whole number 7.

26. **1** $\frac{11}{20}$ is greater than $\frac{1}{2}$ because $\frac{11}{20} > \frac{10}{20}$. The fraction $\frac{11}{20}$ rounds to the whole number 1.

27. **3** $\frac{3}{7}$ is less than $\frac{1}{2}$ because $\frac{6}{14} < \frac{7}{14}$. The mixed number $3\frac{3}{7}$ rounds to the whole number 3.

28. 11 $\frac{5}{6}$ is greater than $\frac{1}{2}$ because $\frac{5}{6} > \frac{3}{6}$. The mixed number $10\frac{5}{6}$ rounds to the whole number 11.

29. 8 $\frac{3}{8}$ is less than $\frac{1}{2}$ because $\frac{3}{8} < \frac{4}{8}$. The mixed number $8\frac{3}{8}$ rounds to the whole number 8.

30. 0 $\frac{7}{15}$ is less than $\frac{1}{2}$ because $\frac{14}{30} < \frac{15}{30}$. The fraction $\frac{7}{15}$ rounds to the whole number 0.

31. (3) 30 gallons
Round the amounts: $14\frac{1}{3}$ rounds to 14
$6\frac{3}{4}$ rounds to 7
$9\frac{1}{4}$ rounds to 9
Add the rounded amounts: $14 + 7 + 9 = 30$

32. (4) 78
Round the amounts: $9\frac{5}{8}$ rounds to 10
$27\frac{1}{4}$ rounds to 27
$4\frac{2}{3}$ rounds to 5
$36\frac{3}{8}$ rounds to 36
Add the rounded amounts:
$10 + 27 + 5 + 36 = 78$

33. (2) 22
Round the amounts: $10\frac{2}{5}$ rounds to 10
$12\frac{1}{6}$ rounds to 12
Add the rounded amounts: $10 + 12 = 22$

34. (2) 2
Round the amounts:
Peanuts in Mix B $4\frac{1}{5}$ rounds to 4
Peanuts in Mix A $2\frac{3}{8}$ rounds to 2
Subtract the rounded amounts: $4 - 2 = 2$

35. (4) 5
Round the amounts:
cashews $2\frac{2}{3}$ rounds to 3
Brazil nuts $2\frac{1}{8}$ rounds to 2
Add the rounded amounts: $3 + 2 = 5$

36. (4) 4
Estimate the number of pounds in Mix A.
Round the amounts: $2\frac{2}{3}$ rounds to 3
$2\frac{3}{8}$ rounds to 2
$3\frac{1}{2}$ rounds to 4
$2\frac{1}{8}$ rounds to 2
Add the rounded amounts: $3 + 2 + 4 + 2 = 11$
Estimate the number of pounds in Mix B.
Round the amounts: $6\frac{1}{2}$ rounds to 7
$3\frac{7}{8}$ rounds to 4
$4\frac{1}{5}$ rounds to 4
Add the rounded amounts: $7 + 4 + 4 = 15$
Find the difference: $15 - 11 = 4$

37. (1) 3 To find if $5\frac{1}{3}$ is nearer to 5 or 6, compare $\frac{1}{3}$ and $\frac{1}{2}$ using common denominators.
$\frac{1}{3} = \frac{2}{6}$ $\frac{1}{2} = \frac{3}{6}$ $\frac{2}{6} < \frac{3}{6}$
Since $\frac{2}{6} < \frac{1}{2}$ the whole number, 5, stays the same. $5\frac{1}{3}$ rounds to 5.
To find if $1\frac{7}{8}$ is nearer to 1 or 2, compare $\frac{7}{8}$ and $\frac{1}{2}$ using common denominators.
$\frac{1}{2} = \frac{4}{8}$ $\frac{7}{8} > \frac{4}{8}$
Since $\frac{7}{8} > \frac{1}{2}$, add 1 to the whole number. $1\frac{7}{8}$ rounds to 2.
Subtract to find approximately how many feet of board will be left. $5 - 2 = 3$

38. (2) 2 To find if $2\frac{2}{3}$ is nearer to 2 or 3, compare $\frac{2}{3}$ and $\frac{1}{2}$ using common denominators.
$\frac{2}{3} = \frac{4}{6}$ $\frac{1}{2} = \frac{3}{6}$ $\frac{4}{6} > \frac{3}{6}$
Since $\frac{4}{6} > \frac{1}{2}$ add 1 to the whole number. $2\frac{2}{3}$ rounds to 3.
To find if $\frac{3}{4}$ is nearer to 0 or 1, compare $\frac{3}{4}$ and $\frac{1}{2}$ using common denominators.
$\frac{1}{2} = \frac{2}{4}$ $\frac{3}{4} > \frac{2}{4}$
Since $\frac{3}{4} > \frac{1}{2}$, add 1 to the whole number. $\frac{3}{4}$ rounds to 1.
Subtract to find about how many cups of dressing are left. $3 - 1 = 2$

39. (4) 4 To find if $\frac{1}{4}$ is nearer to 1 or 2, compare $\frac{1}{4}$ and $\frac{1}{2}$ using common denominators.
$\frac{1}{2} = \frac{2}{4}$ $\frac{1}{4} < \frac{2}{4}$
Since $\frac{1}{4} < \frac{1}{2}$ the whole number, 0, stays the same. $\frac{1}{4}$ rounds to 0.
To find if $\frac{3}{4}$ is nearer to 0 or 1, compare $\frac{3}{4}$ and $\frac{1}{2}$ using common denominators.
$\frac{1}{2} = \frac{2}{4}$ $\frac{3}{4} > \frac{2}{4}$
Since $\frac{3}{4} > \frac{1}{2}$, add 1 to the whole number. $\frac{3}{4}$ rounds to 1.
Add to find about how many pies are already made. $3 + 0 + 1 = 4$

40. **(2) 2** To find if $1\frac{1}{3}$ is nearer to 1 or 2, compare $\frac{1}{3}$ and $\frac{1}{2}$ using common denominators.

$\frac{1}{3} = \frac{2}{6}$ $\quad \frac{1}{2} = \frac{3}{6}$ $\quad \frac{2}{6} < \frac{3}{6}$

Since $\frac{1}{3} < \frac{1}{2}$, the whole number, 1, stays the same. $1\frac{1}{3}$ rounds to 1.

To find if $1\frac{1}{8}$ is nearer to 1 or 2, compare $\frac{1}{8}$ and $\frac{1}{2}$ using common denominators.

$\frac{1}{2} = \frac{4}{8}$ $\quad \frac{1}{8} < \frac{4}{8}$

Since $\frac{1}{8} < \frac{1}{2}$, the whole number, 1, stays the same. $1\frac{1}{8}$ rounds to 1.

Add to find about how many cups of milk are needed. $1 + 1 = 2$

41. **(3) 6** To find if $2\frac{1}{4}$ is nearer to 2 or 3, compare $\frac{1}{4}$ and $\frac{1}{2}$ using common denominators.

$\frac{1}{2} = \frac{2}{4}$ $\quad \frac{1}{4} < \frac{2}{4}$

Since $\frac{1}{4} < \frac{1}{2}$, the whole number, 2, stays the same. $2\frac{1}{4}$ rounds to 2.

To find if $1\frac{7}{8}$ is nearer to 1 or 2, compare $\frac{7}{8}$ and $\frac{1}{2}$ using common denominators.

$\frac{1}{2} = \frac{4}{8}$ $\quad \frac{7}{8} > \frac{4}{8}$

Since $\frac{7}{8} > \frac{1}{2}$, add 1 to the whole number. $1\frac{7}{8}$ rounds to 2.

Add to find the approximate total number of miles Tara walked. $2 + 2 + 2 = 6$

42. **(3) 4** To find if $1\frac{1}{3}$ is nearer to 1 or 2, compare $\frac{1}{3}$ and $\frac{1}{2}$ using common denominators.

$\frac{1}{3} = \frac{2}{6}$ $\quad \frac{1}{2} = \frac{3}{6}$ $\quad \frac{2}{6} < \frac{3}{6}$

Since $\frac{2}{6} < \frac{1}{2}$, the whole number, 1, stays the same. $1\frac{1}{3}$ rounds to 1.

To find if $4\frac{7}{8}$ is nearer to 4 or 5, compare $\frac{7}{8}$ and $\frac{1}{2}$.

$\frac{1}{2} = \frac{4}{8}$ $\quad \frac{7}{8} > \frac{4}{8}$

Since $\frac{7}{8} > \frac{4}{8}$, add 1 to the whole number. $4\frac{7}{8}$ rounds to 5.

Subtract to find about how many more yards of fabric she needs. $5 - 1 = 4$

43. **(4) 5** To find if $3\frac{3}{4}$ is nearer to 3 or 4, compare $\frac{3}{4}$ and $\frac{1}{2}$ using common denominators.

$\frac{1}{2} = \frac{2}{4}$ $\quad \frac{3}{4} > \frac{2}{4}$

Since $\frac{3}{4} > \frac{1}{2}$, add 1 to the whole number. $3\frac{3}{4}$ rounds to 4.

To find if $1\frac{1}{3}$ is nearer to 1 or 2, compare $\frac{1}{3}$ and $\frac{1}{2}$ using common denominators.

$\frac{1}{3} = \frac{2}{6}$ $\quad \frac{1}{2} = \frac{3}{6}$ $\quad \frac{2}{6} < \frac{3}{6}$

Since $\frac{2}{6} < \frac{1}{2}$, the whole number, 1, stays the same. $1\frac{1}{3}$ rounds to 1.

Add to find about how many miles it is to the doctor's office. $4 + 1 = 5$

44. **(4) 4** To find if $1\frac{3}{4}$ is nearer to 1 or 2, compare $\frac{3}{4}$ and $\frac{1}{2}$ using common denominators.

$\frac{1}{2} = \frac{2}{4}$ $\quad \frac{3}{4} > \frac{2}{4}$

Since $\frac{3}{4} > \frac{1}{2}$, add 1 to the whole number. $1\frac{3}{4}$ rounds to 2.

To find out if $2\frac{1}{8}$ is nearer to 2 or 3, compare $\frac{1}{8}$ and $\frac{1}{2}$ using common denominators.

$\frac{1}{2} = \frac{4}{8}$ $\quad \frac{1}{8} < \frac{4}{8}$

Since $\frac{1}{8} < \frac{1}{2}$, the whole number, 2, stays the same. $2\frac{1}{8}$ rounds to 2.

Add to find about how many yards of silk are needed. $2 + 2 = 4$

Lesson 7

All items relate to the skill of application.

GED Practice: Lesson 7 (Pages 76–77)

1. **1** $\quad \frac{2}{3} + \frac{1}{3} = \frac{3}{3} = 1$

2. $\mathbf{\frac{2}{3}}$ $\quad \frac{1}{6} + \frac{3}{6} = \frac{4}{6}$; $\frac{4}{6} \div \frac{2}{2} = \frac{2}{3}$

3. $\mathbf{\frac{7}{8}}$ $\quad \frac{3}{4} = \frac{6}{8}$; $+\frac{1}{8} = \frac{1}{8}$; total $\frac{7}{8}$

4. $\mathbf{\frac{2}{3}}$ $\quad \frac{1}{6} = \frac{1}{6}$; $+\frac{1}{2} = \frac{3}{6}$; total $\frac{4}{6}$; $\frac{4}{6} \div \frac{2}{2} = \frac{2}{3}$

5. $\mathbf{\frac{23}{40}}$ $\quad \frac{3}{8} = \frac{15}{40}$; $+\frac{1}{5} = \frac{8}{40}$; total $\frac{23}{40}$

6. $\mathbf{\frac{11}{15}}$ $\quad \frac{2}{5} = \frac{6}{15}$; $+\frac{1}{3} = \frac{5}{15}$; total $\frac{11}{15}$

7. $\mathbf{7\frac{7}{8}}$ $\quad 2\frac{1}{4} = 2\frac{2}{8}$; $+5\frac{5}{8} = 5\frac{5}{8}$; total $7\frac{7}{8}$

8. $\mathbf{14\frac{1}{2}}$ $\quad 3\frac{5}{6} = 3\frac{5}{6}$; $+10\frac{2}{3} = 10\frac{4}{6}$; total $13\frac{9}{6}$

Reduce:
$13\frac{9}{6} = 13 + 1\frac{3}{6} = 14\frac{3}{6} = 14\frac{1}{2}$

UNIT 1

UNIT 1

9. $11\frac{1}{10}$

$$\begin{array}{r} 2\frac{4}{5} = 2\frac{8}{10} \\ +8\frac{3}{10} = 8\frac{3}{10} \\ \hline 10\frac{11}{10} \end{array}$$

Reduce:
$10\frac{11}{10} = 10 + 1\frac{1}{10} = 11\frac{1}{10}$

10. $42\frac{5}{6}$

$$\begin{array}{r} 18\frac{1}{6} = 18\frac{1}{6} \\ +24\frac{2}{3} = 24\frac{4}{6} \\ \hline 42\frac{5}{6} \end{array}$$

11. $8\frac{1}{12}$

$$\begin{array}{r} 3\frac{3}{4} = 3\frac{9}{12} \\ +4\frac{1}{3} = 4\frac{4}{12} \\ \hline 7\frac{13}{12} \end{array}$$

Reduce:
$7\frac{13}{12} = 7 + 1\frac{1}{12} = 8\frac{1}{12}$

12. $7\frac{1}{8}$

$$\begin{array}{r} 1\frac{1}{2} = 1\frac{4}{8} \\ +5\frac{5}{8} = 5\frac{5}{8} \\ \hline 6\frac{9}{8} \end{array}$$

Reduce:
$6\frac{9}{8} = 6 + 1\frac{1}{8} = 7\frac{1}{8}$

13. $28\frac{14}{15}$

$$\begin{array}{r} 12\frac{3}{5} = 12\frac{9}{15} \\ +16\frac{1}{3} = 16\frac{5}{15} \\ \hline 28\frac{14}{15} \end{array}$$

14. $55\frac{13}{24}$

$$\begin{array}{r} 16\frac{7}{8} = 16\frac{21}{24} \\ +38\frac{2}{3} = 38\frac{16}{24} \\ \hline 54\frac{37}{24} \end{array}$$

Reduce:
$54\frac{37}{24} = 54 + 1\frac{13}{24} = 55\frac{13}{24}$

15. $39\frac{9}{10}$

$$\begin{array}{r} 20\frac{1}{5} = 20\frac{2}{10} \\ +19\frac{7}{10} = 19\frac{7}{10} \\ \hline 39\frac{9}{10} \end{array}$$

16. $31\frac{7}{9}$

$$\begin{array}{r} 10\frac{1}{9} = 10\frac{1}{9} \\ +21\frac{2}{3} = 21\frac{6}{9} \\ \hline 31\frac{7}{9} \end{array}$$

17. $17\frac{14}{15}$

$$\begin{array}{r} 5\frac{3}{5} = 5\frac{9}{15} \\ +12\frac{1}{3} = 12\frac{5}{15} \\ \hline 17\frac{14}{15} \end{array}$$

18. $39\frac{1}{2}$

$$\begin{array}{r} 22\frac{1}{10} = 22\frac{1}{10} \\ +17\frac{2}{5} = 17\frac{4}{10} \\ \hline 39\frac{5}{10} \end{array}$$

Reduce:
$39\frac{5}{10} = 39\frac{1}{2}$

19. $56\frac{3}{8}$

$$\begin{array}{r} 31\frac{5}{8} = 31\frac{5}{8} \\ +24\frac{3}{4} = 24\frac{6}{8} \\ \hline 55\frac{11}{8} \end{array}$$

Reduce:
$55\frac{11}{8} = 55 + 1\frac{3}{8} = 56\frac{3}{8}$

20. $36\frac{3}{10}$

$$\begin{array}{r} 17\frac{1}{2} = 17\frac{5}{10} \\ +18\frac{4}{5} = 18\frac{8}{10} \\ \hline 35\frac{13}{10} \end{array}$$

Reduce:
$35\frac{13}{10} = 35 + 1\frac{3}{10} = 36\frac{3}{10}$

21. $1\frac{3}{10}$

$$\begin{array}{r} \frac{3}{4} = \frac{15}{20} \\ \frac{3}{20} = \frac{3}{20} \\ +\frac{2}{5} = \frac{8}{20} \\ \hline \frac{26}{20} \end{array}$$

Reduce:
$\frac{26}{20} = 1\frac{6}{20} = 1\frac{3}{10}$

22. $1\frac{1}{24}$

$$\begin{array}{r} \frac{5}{8} = \frac{15}{24} \\ \frac{1}{6} = \frac{4}{24} \\ +\frac{1}{4} = \frac{6}{24} \\ \hline \frac{25}{24} = 1\frac{1}{24} \end{array}$$

23. $10\frac{1}{60}$

$$\begin{array}{r} 1\frac{1}{4} = 1\frac{15}{60} \\ 3\frac{1}{10} = 3\frac{6}{60} \\ +5\frac{2}{3} = 5\frac{40}{60} \\ \hline 9\frac{61}{60} \end{array}$$

Reduce:
$9\frac{61}{60} = 9 + 1\frac{1}{60} = 10\frac{1}{60}$

24. $7\frac{1}{2}$

$$\begin{array}{r} 3\frac{1}{2} = 3\frac{4}{8} \\ 1\frac{2}{8} = 1\frac{2}{8} \\ +2\frac{3}{4} = 2\frac{6}{8} \\ \hline 6\frac{12}{8} \end{array}$$

Reduce:
$6\frac{12}{8} = 6 + 1\frac{4}{8} = 7\frac{4}{8} = 7\frac{1}{2}$

25. $6\frac{11}{24}$

$$\begin{array}{r} 1\frac{3}{4} = 1\frac{18}{24} \\ \frac{7}{8} = \frac{21}{24} \\ +3\frac{5}{6} = 3\frac{20}{24} \\ \hline 4\frac{59}{24} \end{array}$$

Reduce:
$4\frac{59}{24} = 4 + 2\frac{11}{24} = 6\frac{11}{24}$

26. $9\frac{7}{18}$

$$\begin{array}{r} \frac{5}{9} = \frac{10}{18} \\ 3\frac{2}{3} = 3\frac{12}{18} \\ +5\frac{1}{6} = 5\frac{3}{18} \\ \hline 8\frac{25}{18} \end{array}$$

Reduce:
$8\frac{25}{18} = 8 + 1\frac{7}{18} = 9\frac{7}{18}$

27. 8

$$\begin{array}{r} 1\frac{7}{20} = 1\frac{7}{20} \\ 2\frac{1}{4} = 2\frac{5}{20} \\ +4\frac{2}{5} = 4\frac{8}{20} \\ \hline 7\frac{20}{20} \end{array}$$

Reduce:
$7\frac{20}{20} = 7 + 1 = 8$

28. $15\frac{13}{20}$

$$\begin{array}{r} 3\frac{9}{10} = 3\frac{18}{20} \\ 5\frac{1}{2} = 5\frac{10}{20} \\ +6\frac{1}{4} = 6\frac{5}{20} \\ \hline 14\frac{33}{20} \end{array}$$

Reduce:
$14\frac{33}{20} = 14 + 1\frac{13}{20} = 15\frac{13}{20}$

29. $11\frac{13}{30}$

$$\begin{array}{r} 2\frac{1}{3} = 2\frac{10}{30} \\ 7\frac{2}{5} = 7\frac{12}{30} \\ +1\frac{7}{10} = 1\frac{21}{30} \\ \hline 10\frac{43}{30} \end{array}$$

Reduce:
$10\frac{43}{30} = 10 + 1\frac{13}{30} = 11\frac{13}{30}$

30. $9\frac{1}{4}$

$$\begin{array}{r} 4\frac{1}{6} = 4\frac{2}{12} \\ 3\frac{2}{3} = 3\frac{8}{12} \\ +1\frac{5}{12} = 1\frac{5}{12} \\ \hline 8\frac{15}{12} \end{array}$$

Reduce:
$8\frac{15}{12} = 8 + 1\frac{3}{12} = 9\frac{3}{12} = 9\frac{1}{4}$

31. $\frac{1}{4}$

$$\begin{array}{r} \frac{3}{4} = \frac{3}{4} \\ -\frac{1}{2} = \frac{2}{4} \\ \hline \frac{1}{4} \end{array}$$

32. $\frac{1}{2}$

$$\begin{array}{r} \frac{2}{3} = \frac{4}{6} \\ -\frac{1}{6} = \frac{1}{6} \\ \hline \frac{3}{6} \div \frac{3}{3} = \frac{1}{2} \end{array}$$

33. $\frac{1}{3}$

$$\begin{array}{r} \frac{5}{6} = \frac{5}{6} \\ -\frac{1}{2} = \frac{3}{6} \\ \hline \frac{2}{6} \div \frac{2}{2} = \frac{1}{3} \end{array}$$

34. $\frac{5}{12}$

$$\begin{array}{r} \frac{3}{4} = \frac{9}{12} \\ -\frac{1}{3} = \frac{4}{12} \\ \hline \frac{5}{12} \end{array}$$

35. $\frac{2}{5}$

$$\begin{array}{r} \frac{9}{10} = \frac{9}{10} \\ -\frac{1}{2} = \frac{5}{10} \\ \hline \frac{4}{10} \div \frac{2}{2} = \frac{2}{5} \end{array}$$

36. $\frac{17}{36}$

$$\begin{array}{r} \frac{5}{9} = \frac{20}{36} \\ -\frac{1}{12} = \frac{3}{36} \\ \hline \frac{17}{36} \end{array}$$

37. $10\frac{1}{2}$

$$\begin{array}{r} 14 = 13\frac{2}{2} \\ -\ 3\frac{1}{2} = 3\frac{1}{2} \\ \hline 10\frac{1}{2} \end{array}$$

38. $11\frac{7}{8}$

$$\begin{array}{r} 27\frac{1}{4} = 27\frac{2}{8} = 26\frac{10}{8} \\ -15\frac{3}{8} = 15\frac{3}{8} = 15\frac{3}{8} \\ \hline 11\frac{7}{8} \end{array}$$

39. $1\frac{2}{5}$

$$\begin{array}{r} 3\frac{1}{5} = 2\frac{6}{5} \\ -1\frac{4}{5} = 1\frac{4}{5} \\ \hline 1\frac{2}{5} \end{array}$$

40. $7\frac{7}{8}$

$$\begin{array}{r} 26\frac{1}{4} = 26\frac{2}{8} = 25\frac{10}{8} \\ -18\frac{3}{8} = 18\frac{3}{8} = 18\frac{3}{8} \\ \hline 7\frac{7}{8} \end{array}$$

41. $6\frac{3}{4}$

$$\begin{array}{r} 10\frac{1}{2} = 10\frac{2}{4} = 9\frac{6}{4} \\ -\ 3\frac{3}{4} = 3\frac{3}{4} = 3\frac{3}{4} \\ \hline 6\frac{3}{4} \end{array}$$

42. $8\frac{5}{12}$

$$\begin{array}{r} 15\frac{2}{3} = 15\frac{8}{12} \\ -\ 7\frac{1}{4} = 7\frac{3}{12} \\ \hline 8\frac{5}{12} \end{array}$$

43. $7\frac{1}{6}$

$$\begin{array}{r} 20 = 19\frac{6}{6} \\ -12\frac{5}{6} = 12\frac{5}{6} \\ \hline 7\frac{1}{6} \end{array}$$

44. $25\frac{1}{8}$

$$\begin{array}{r} 40\frac{7}{8} = 40\frac{7}{8} \\ -15\frac{3}{4} = 15\frac{6}{8} \\ \hline 25\frac{1}{8} \end{array}$$

45. $7\frac{8}{15}$

$$\begin{array}{r} 35\frac{1}{3} = 35\frac{5}{15} = 34\frac{20}{15} \\ -27\frac{4}{5} = 27\frac{12}{15} = 27\frac{12}{15} \\ \hline 7\frac{8}{15} \end{array}$$

46. $4\frac{1}{12}$

$$\begin{array}{r} 20\frac{3}{4} = 20\frac{9}{12} \\ -16\frac{2}{3} = 16\frac{8}{12} \\ \hline 4\frac{1}{12} \end{array}$$

47. $7\frac{1}{2}$

$$\begin{array}{r} 18\frac{3}{10} = 18\frac{3}{10} = 17\frac{13}{10} \\ -10\frac{4}{5} = 10\frac{8}{10} = 10\frac{8}{10} \\ \hline 7\frac{5}{10} = 7\frac{1}{2} \end{array}$$

48. $12\frac{7}{8}$

$$\begin{array}{r} 16\frac{1}{8} = 16\frac{1}{8} = 15\frac{9}{8} \\ -\ 3\frac{1}{4} = 3\frac{2}{8} = 3\frac{2}{8} \\ \hline 12\frac{7}{8} \end{array}$$

49. $1\frac{1}{8}$

$$\begin{array}{r} 12 = 11\frac{8}{8} \\ -10\frac{7}{8} = 10\frac{7}{8} \\ \hline 1\frac{1}{8} \end{array}$$

50. $11\frac{13}{20}$

$$\begin{array}{r} 25\frac{1}{4} = 25\frac{5}{20} = 24\frac{25}{20} \\ -13\frac{3}{5} = 13\frac{12}{20} = 13\frac{12}{20} \\ \hline 11\frac{13}{20} \end{array}$$

51. $\frac{9}{20}$

$$\begin{array}{r} \frac{7}{10} = \frac{14}{20} \\ -\frac{1}{4} = \frac{5}{20} \\ \hline \frac{9}{20} \end{array}$$

52. $2\frac{5}{12}$

$$\begin{array}{r} 3\frac{2}{3} = 3\frac{8}{12} \\ -1\frac{1}{4} = 1\frac{3}{12} \\ \hline 2\frac{5}{12} \end{array}$$

53. $1\frac{1}{5}$

$$\begin{array}{r} 4 = 3\frac{5}{5} \\ -2\frac{4}{5} = 2\frac{4}{5} \\ \hline 1\frac{1}{5} \end{array}$$

54. $6\frac{7}{8}$

$$\begin{array}{r} 9\frac{1}{2} = 9\frac{4}{8} = 8\frac{12}{8} \\ -2\frac{5}{8} = 2\frac{5}{8} = 2\frac{5}{8} \\ \hline 6\frac{7}{8} \end{array}$$

55. $3\frac{2}{3}$

$$\begin{array}{r} 15\frac{1}{9} = 14\frac{10}{9} \\ -11\frac{4}{9} = 11\frac{4}{9} \\ \hline 3\frac{6}{9} = 3\frac{2}{3} \end{array}$$

56. $2\frac{13}{15}$

$$\begin{array}{r} 6\frac{1}{5} = 6\frac{3}{15} = 5\frac{18}{15} \\ -3\frac{1}{3} = 3\frac{5}{15} = 3\frac{5}{15} \\ \hline 2\frac{13}{15} \end{array}$$

57. $6\frac{7}{12}$

$$\begin{array}{r} 8\frac{5}{6} = 8\frac{10}{12} \\ -2\frac{1}{4} = 2\frac{3}{12} \\ \hline 6\frac{7}{12} \end{array}$$

58. $10\frac{3}{10}$

$$\begin{array}{r} 10\frac{9}{10} = 10\frac{9}{10} \\ -\ \frac{3}{5} = \frac{6}{10} \\ \hline 10\frac{3}{10} \end{array}$$

59. $11\frac{1}{8}$

$$\begin{array}{r} 12 = 11\frac{8}{8} \\ -\ \frac{7}{8} = \frac{7}{8} \\ \hline 11\frac{1}{8} \end{array}$$

60. $12\frac{13}{16}$

$$\begin{array}{r} 13\frac{3}{4} = 13\frac{12}{16} = 12\frac{28}{16} \\ -\ \frac{15}{16} = \frac{15}{16} = \frac{15}{16} \\ \hline 12\frac{13}{16} \end{array}$$

61. $19\frac{1}{3}$

$$\begin{array}{r} 20 = 19\frac{3}{3} \\ -\ \frac{2}{3} = \frac{2}{3} \\ \hline 19\frac{1}{3} \end{array}$$

62. $8\frac{7}{8}$

$$\begin{array}{r} 18\frac{3}{4} = 18\frac{6}{8} = 17\frac{14}{8} \\ -\ 9\frac{7}{8} = 9\frac{7}{8} = 9\frac{7}{8} \\ \hline 8\frac{7}{8} \end{array}$$

63. $7\frac{3}{5}$

$$\begin{array}{r} 11 = 10\frac{5}{5} \\ -\ 3\frac{2}{5} = 3\frac{2}{5} \\ \hline 7\frac{3}{5} \end{array}$$

64. $4\frac{1}{12}$

$$\begin{array}{r} 16\frac{1}{4} = 16\frac{3}{12} \\ -12\frac{1}{6} = 12\frac{2}{12} \\ \hline 4\frac{1}{12} \end{array}$$

UNIT 1

65. $4\frac{5}{7}$

$$\begin{array}{r} 10 = 9\frac{7}{7} \\ -\ 5\frac{2}{7} = 5\frac{2}{7} \\ \hline 4\frac{5}{7} \end{array}$$

GED Practice: Solving Word Problems (Page 79)

1. **(5) $33\frac{1}{2}$** Add to find the total number of gallons.

$$\begin{array}{r} 8\frac{5}{10} \\ 9\frac{3}{10} \\ 8 \\ +\ 7\frac{7}{10} \\ \hline 32\frac{15}{10} = 33\frac{5}{10} = 33\frac{1}{2} \end{array}$$

2. **(2) $5\frac{1}{2}$** Subtract to find how many days he "has left."

$$\begin{array}{r} 10\ = 9\frac{2}{2} \\ -4\frac{1}{2} = 4\frac{1}{2} \\ \hline 5\frac{1}{2} \end{array}$$

3. **(1) $1\frac{3}{4}$** Subtract to find "how many more."

$$\begin{array}{r} 3\frac{1}{2} = 3\frac{2}{4} = 2\frac{6}{4} \\ -1\frac{3}{4} = 1\frac{3}{4} = 1\frac{3}{4} \\ \hline 1\frac{3}{4} \end{array}$$

4. **(3) $28\frac{1}{8}$** Add to find the total yardage.

$$\begin{array}{r} 12\frac{1}{2} = 12\frac{4}{8} \\ 8\frac{7}{8} = \ 8\frac{7}{8} \\ +\ 6\frac{3}{4} = \ 6\frac{6}{8} \\ \hline 26\frac{17}{8} = \\ 26 + 2\frac{1}{8} = 28\frac{1}{8} \end{array}$$

5. **(4) $20\frac{7}{10}$** Add to find the total hours.

$$\begin{array}{r} 5\frac{2}{5} = 5\frac{4}{10} \\ 6\frac{1}{2} = 6\frac{5}{10} \\ +8\frac{4}{5} = 8\frac{8}{10} \\ \hline 19\frac{17}{10} = \\ 19 + 1\frac{7}{10} = 20\frac{7}{10} \end{array}$$

6. **(1) $2\frac{7}{12}$ feet** Subtract to find the difference between the two lengths.

$$\begin{array}{r} 8\frac{1}{3} = 8\frac{4}{12} = 7\frac{16}{12} \\ -5\frac{3}{4} = 5\frac{9}{12} = 5\frac{9}{12} \\ \hline 2\frac{7}{12} \end{array}$$

Lesson 8

GED Practice: Lesson 8 (Pages 82–83)

1. $\mathbf{\frac{1}{3}}$ $\frac{1}{\cancel{2}_{1}}\times\frac{\cancel{2}^{1}}{3}=\frac{1}{3}$

2. $\mathbf{\frac{7}{10}}$ $\frac{7}{\cancel{8}_{2}}\times\frac{\cancel{4}^{1}}{5}=\frac{7}{10}$

3. $\mathbf{\frac{9}{28}}$ $\frac{3}{4}\times\frac{3}{7}=\frac{9}{28}$

4. $\mathbf{\frac{1}{20}}$ $\frac{\cancel{3}^{1}}{\cancel{8}_{4}}\times\frac{\cancel{2}^{1}}{\cancel{15}_{5}}=\frac{1}{20}$

5. $\mathbf{\frac{20}{27}}$ $\frac{5}{\cancel{6}_{3}}\times\frac{\cancel{8}^{4}}{9}=\frac{20}{27}$

6. $\mathbf{\frac{3}{8}}$ $\frac{\cancel{9}^{3}}{\cancel{10}_{2}}\times\frac{\cancel{5}^{1}}{\cancel{12}_{4}}=\frac{3}{8}$

7. **2** $6\times\frac{1}{3}=\frac{\cancel{6}^{2}}{1}\times\frac{1}{\cancel{3}_{1}}=\frac{2}{1}=2$

8. **9** $15\times\frac{3}{5}=\frac{\cancel{15}^{3}}{1}\times\frac{3}{\cancel{5}_{1}}=\frac{9}{1}=9$

9. **12**
$18\times\frac{2}{3}=\frac{\cancel{18}^{6}}{1}\times\frac{2}{\cancel{3}_{1}}=\frac{12}{1}=12$

10. **21**
$\frac{7}{8}\times 24=\frac{7}{\cancel{8}_{1}}\times\frac{\cancel{24}^{3}}{1}=\frac{21}{1}=21$

11. $\mathbf{5\frac{1}{2}}$
$11\times\frac{1}{2}=\frac{11}{1}\times\frac{1}{2}=\frac{11}{2}=5\frac{1}{2}$

12. $\mathbf{5\frac{1}{3}}$
$24\times\frac{2}{9}=\frac{\cancel{24}^{8}}{1}\times\frac{2}{\cancel{9}_{3}}=\frac{16}{3}=5\frac{1}{3}$

13. $\mathbf{1\frac{1}{3}}$
$\frac{4}{5}\times 1\frac{2}{3}=\frac{4}{\cancel{5}_{1}}\times\frac{\cancel{5}^{1}}{3}=\frac{4}{3}=1\frac{1}{3}$

14. $\mathbf{2\frac{1}{4}}$
$1\frac{1}{2}\times 1\frac{1}{2}=\frac{3}{2}\times\frac{3}{2}=\frac{9}{4}=2\frac{1}{4}$

15. **48**
$9\times 5\frac{1}{3}=\frac{\cancel{9}^{3}}{1}\times\frac{16}{\cancel{3}_{1}}=\frac{48}{1}=48$

16. $\mathbf{\frac{7}{16}}$ $\frac{3}{8}\times 1\frac{1}{6}=\frac{\cancel{3}^{1}}{8}\times\frac{7}{\cancel{6}_{2}}=\frac{7}{16}$

17. **6**
$2\frac{2}{5}\times 2\frac{1}{2}=\frac{\cancel{12}^{6}}{\cancel{5}_{1}}\times\frac{\cancel{5}^{1}}{\cancel{2}_{1}}=\frac{6}{1}=6$

18. $\mathbf{13\frac{3}{4}}$ $3\frac{1}{3}\times 4\frac{1}{8}=\frac{\cancel{10}^{5}}{\cancel{3}_{1}}\times\frac{\cancel{33}^{11}}{\cancel{8}_{4}}=$
$\frac{55}{4}=13\frac{3}{4}$

19. $\mathbf{3\frac{3}{4}}$ $\frac{5}{\cancel{6}_{2}}\times\frac{\cancel{9}^{3}}{2}=\frac{15}{4}=3\frac{3}{4}$

20. $\mathbf{16\frac{1}{2}}$ $2\frac{3}{4}\times 6=\frac{11}{\cancel{4}_{2}}\times\frac{\cancel{6}^{3}}{1}=$
$\frac{33}{2}=16\frac{1}{2}$

21. $\mathbf{15\frac{3}{4}}$ $3\times 5\frac{1}{4}=\frac{3}{1}\times\frac{21}{4}=$
$\frac{63}{4}=15\frac{3}{4}$

22. $\mathbf{11\frac{1}{3}}$ $2\frac{5}{6}\times 4=\frac{17}{\cancel{6}_{3}}\times\frac{\cancel{4}^{2}}{1}=$
$\frac{34}{3}=11\frac{1}{3}$

23. $\mathbf{5\frac{4}{9}}$ $2\frac{1}{3}\times 2\frac{1}{3}=\frac{7}{3}\times\frac{7}{3}=$
$\frac{49}{9}=5\frac{4}{9}$

24. $\mathbf{1\frac{3}{25}}$
$\frac{4}{5}\times 1\frac{2}{5}=\frac{4}{5}\times\frac{7}{5}=\frac{28}{25}=1\frac{3}{25}$

25. $\mathbf{2\frac{3}{16}}$ $1\frac{3}{4}\times 1\frac{1}{4}=\frac{7}{4}\times\frac{5}{4}=$
$\frac{35}{16}=2\frac{3}{16}$

26. $\mathbf{9\frac{1}{2}}$
$2\frac{3}{8}\times 4=\frac{19}{\cancel{8}_{2}}\times\frac{\cancel{4}^{1}}{1}=\frac{19}{2}=9\frac{1}{2}$

27. $\mathbf{10\frac{2}{3}}$ $3\frac{1}{3}\times 3\frac{1}{5}=\frac{\cancel{10}^{2}}{3}\times\frac{16}{\cancel{5}_{1}}=$
$\frac{32}{3}=10\frac{2}{3}$

28. $\mathbf{7\frac{1}{2}}$
$4\times 1\frac{7}{8}=\frac{\cancel{4}^{1}}{1}\times\frac{15}{\cancel{8}_{2}}=\frac{15}{2}=7\frac{1}{2}$

29. **4** $2\frac{2}{3}\times 1\frac{1}{2}=\frac{\cancel{8}^{4}}{\cancel{3}_{1}}\times\frac{\cancel{3}^{1}}{\cancel{2}_{1}}=4$

30. **7**
$3\frac{1}{9}\times 2\frac{1}{4}=\frac{\cancel{28}^{7}}{\cancel{9}_{1}}\times\frac{\cancel{9}^{1}}{\cancel{4}_{1}}=\frac{7}{1}=7$

31. **$\frac{2}{5}$** $\frac{1}{3} \div \frac{5}{6} = \frac{1}{\cancel{3}_1} \times \frac{\cancel{6}^2}{5} = \frac{2}{5}$

32. **$1\frac{2}{3}$** $\frac{2}{3} \div \frac{2}{5} = \frac{\cancel{2}^1}{3} \times \frac{5}{\cancel{2}_1} = \frac{5}{3} = 1\frac{2}{3}$

33. **$\frac{7}{20}$** $\frac{7}{10} \div 2 = \frac{7}{10} \div \frac{2}{1}$
$= \frac{7}{10} \times \frac{1}{2} = \frac{7}{20}$

34. **3** $\frac{5}{8} \div \frac{5}{24} = \frac{\cancel{5}^1}{\cancel{8}_1} \times \frac{\cancel{24}^3}{\cancel{5}_1} = \frac{3}{1} = 3$

35. **$\frac{2}{7}$**
$\frac{6}{7} \div 3 = \frac{6}{7} \div \frac{3}{1} = \frac{\cancel{6}^2}{7} \times \frac{1}{\cancel{3}_1} = \frac{2}{7}$

36. **$\frac{2}{3}$** $\frac{4}{9} \div \frac{2}{3} = \frac{\cancel{4}^2}{\cancel{9}_3} \times \frac{\cancel{3}^1}{\cancel{2}_1} = \frac{2}{3}$

37. **$3\frac{1}{2}$** $\frac{7}{8} \div \frac{1}{4} = \frac{7}{\cancel{8}_2} \times \frac{\cancel{4}^1}{1} = \frac{7}{2} = 3\frac{1}{2}$

38. **36** $4\frac{1}{2} \div \frac{1}{8} = \frac{9}{2} \div \frac{1}{8} =$
$\frac{9}{\cancel{2}_1} \times \frac{\cancel{8}^4}{1} = \frac{36}{1} = 36$

39. **45** $15 \div \frac{1}{3} = \frac{15}{1} \div \frac{1}{3} =$
$\frac{15}{1} \times \frac{3}{1} = \frac{45}{1} = 45$

40. **$1\frac{3}{5}$** $\frac{4}{5} \div \frac{1}{2} = \frac{4}{5} \times \frac{2}{1} = \frac{8}{5} = 1\frac{3}{5}$

41. **8** $12 \div 1\frac{1}{2} = \frac{12}{1} \div \frac{3}{2} =$
$\frac{\cancel{12}^4}{1} \times \frac{2}{\cancel{3}_1} = \frac{8}{1} = 8$

42. **$2\frac{1}{4}$** $3\frac{3}{4} \div 1\frac{2}{3} = \frac{15}{4} \div \frac{5}{3} =$
$\frac{\cancel{15}^3}{4} \times \frac{3}{\cancel{5}_1} = \frac{9}{4} = 2\frac{1}{4}$

43. **26** $6\frac{1}{2} \div \frac{1}{4} = \frac{13}{2} \div \frac{1}{4} =$
$\frac{13}{\cancel{2}_1} \times \frac{\cancel{4}^2}{1} = \frac{26}{1} = 26$

44. **$1\frac{1}{2}$** $2\frac{1}{4} \div 1\frac{1}{2} = \frac{9}{4} \div \frac{3}{2} =$
$\frac{\cancel{9}^3}{\cancel{4}_2} \times \frac{\cancel{2}^1}{\cancel{3}_1} = \frac{3}{2} = 1\frac{1}{2}$

45. **27** $18 \div \frac{2}{3} = \frac{18}{1} \div \frac{2}{3} =$
$\frac{\cancel{18}^9}{1} \times \frac{3}{\cancel{2}_1} = \frac{27}{1} = 27$

46. **10** $2\frac{2}{5} \div \frac{6}{25} = \frac{12}{5} \div \frac{6}{25} =$
$\frac{\cancel{12}^2}{\cancel{5}_1} \times \frac{\cancel{25}^5}{\cancel{6}_1} = \frac{10}{1} = 10$

47. **$4\frac{1}{5}$** $4\frac{9}{10} \div 1\frac{1}{6} = \frac{49}{10} \div \frac{7}{6} =$
$\frac{\cancel{49}^7}{\cancel{10}_5} \times \frac{\cancel{6}^3}{\cancel{7}_1} = \frac{21}{5} = 4\frac{1}{5}$

48. **$3\frac{1}{3}$** $6\frac{1}{9} \div 1\frac{5}{6} = \frac{55}{9} \div \frac{11}{6} =$
$\frac{\cancel{55}^5}{\cancel{9}_3} \times \frac{\cancel{6}^2}{\cancel{11}_1} = \frac{10}{3} = 3\frac{1}{3}$

49. **$2\frac{2}{3}$** $6 \div 2\frac{1}{4} = \frac{6}{1} \div \frac{9}{4} =$
$\frac{\cancel{6}^2}{1} \times \frac{4}{\cancel{9}_3} = \frac{8}{3} = 2\frac{2}{3}$

50. **$\frac{29}{32}$** $3\frac{5}{8} \div 4 = \frac{29}{8} \div \frac{4}{1} =$
$\frac{29}{8} \times \frac{1}{4} = \frac{29}{32}$

51. **$\frac{1}{2}$** $\frac{3}{4} \div 1\frac{1}{2} = \frac{3}{4} \div \frac{3}{2} =$
$\frac{\cancel{3}^1}{\cancel{4}_2} \times \frac{\cancel{2}^1}{\cancel{3}_1} = \frac{1}{2}$

52. **8** $2\frac{2}{3} \div \frac{1}{3} = \frac{8}{3} \div \frac{1}{3} =$
$\frac{8}{\cancel{3}_1} \times \frac{\cancel{3}^1}{1} = 8$

53. **$3\frac{1}{5}$** $4 \div 1\frac{1}{4} = \frac{4}{1} \div \frac{5}{4} =$
$\frac{4}{1} \times \frac{4}{5} = \frac{16}{5} = 3\frac{1}{5}$

54. **$3\frac{2}{3}$** $5\frac{1}{2} \div 1\frac{1}{2} = \frac{11}{2} \div \frac{3}{2} =$
$\frac{11}{\cancel{2}_1} \times \frac{\cancel{2}^1}{3} = \frac{11}{3} = 3\frac{2}{3}$

55. **$8\frac{1}{3}$** $10 \div 1\frac{1}{5} = \frac{10}{1} \div \frac{6}{5} =$
$\frac{\cancel{10}^5}{1} \times \frac{5}{\cancel{6}_3} = \frac{25}{3} = 8\frac{1}{3}$

56. **35** $8\frac{3}{4} \div \frac{1}{4} = \frac{35}{4} \div \frac{1}{4} =$
$\frac{35}{\cancel{4}_1} \times \frac{\cancel{4}^1}{1} = \frac{35}{1} = 35$

57. **27** $12 \div \frac{4}{9} = \frac{12}{1} \div \frac{4}{9} =$
$\frac{\cancel{12}^3}{1} \times \frac{9}{\cancel{4}_1} = \frac{27}{1} = 27$

58. **20** $16 \div \frac{4}{5} = \frac{16}{1} \div \frac{4}{5} =$
$\frac{\cancel{16}^4}{1} \times \frac{5}{\cancel{4}_1} = \frac{20}{1} = 20$

59. **$5\frac{19}{40}$** $9\frac{1}{8} \div 1\frac{2}{3} = \frac{73}{8} \div \frac{5}{3} =$
$\frac{73}{8} \times \frac{3}{5} = \frac{219}{40} = 5\frac{19}{40}$

60. **$1\frac{9}{11}$** $4 \div 2\frac{1}{5} = \frac{4}{1} \div \frac{11}{5} =$
$\frac{4}{1} \times \frac{5}{11} = \frac{20}{11} = 1\frac{9}{11}$

GED Practice: Solving Word Problems (Page 85)

1. **(3) $7\frac{7}{8}$** Multiply $15\frac{3}{4}$ by $\frac{1}{2}$ because you need to find $\frac{1}{2}$ "of" the order.
$15\frac{3}{4} \times \frac{1}{2} = \frac{63}{4} \times \frac{1}{2} = \frac{63}{8} = 7\frac{7}{8}$

2. **(4) 45** You need to find how many $\frac{1}{3}$ parts are in 15. Divide 15 by $\frac{1}{3}$.
$15 \div \frac{1}{3} = \frac{15}{1} \div \frac{1}{3} = \frac{15}{1} \times \frac{3}{1} = \frac{45}{1} = 45$

3. **(4) 48** The word "of" means multiply. Multiply the number of animals, 60, by $\frac{4}{5}$ (the fraction that is dogs).
$60 \times \frac{4}{5} = \frac{\cancel{60}^{12}}{1} \times \frac{4}{\cancel{5}_1} = \frac{48}{1} = 48$

4. **(3) $2\frac{1}{2}$** Add the three amounts.
$\frac{1}{2} + \frac{3}{4} + 1\frac{1}{4} = \frac{2}{4} + \frac{3}{4} + \frac{5}{4} = \frac{10}{4} = 2\frac{2}{4} = 2\frac{1}{2}$

5. **(5) Not enough information is given.** You need to know Elio's present salary to solve the problem.

6. **(3) 32** Divide the height of the stack (24 inches) by the thickness of one board $\left(\frac{3}{4}\text{ inch}\right)$. You do not need the length of the boards to solve the problem.
$24 \div \frac{3}{4} = \frac{24}{1} \div \frac{3}{4} = \frac{\cancel{24}^8}{1} \times \frac{4}{\cancel{3}_1} = \frac{32}{1} = 32$

7. **(2) $2\frac{1}{2}$** Divide $6\frac{1}{4}$ (the total length of the trail) by $2\frac{1}{2}$ (the distance Steve can cover in one hour). You need to find how many $2\frac{1}{2}$ parts are in $6\frac{1}{4}$.

$$6\frac{1}{4} \div 2\frac{1}{2} = \frac{25}{4} \div \frac{5}{2} = \frac{\overset{5}{\cancel{25}}}{\underset{2}{\cancel{4}}} \times \frac{\overset{1}{\cancel{2}}}{\underset{1}{\cancel{5}}} = \frac{5}{2} = 2\frac{1}{2}$$

8. **(2) $1\frac{1}{8}$** Subtract $\frac{7}{8}$ from 3.

$$\begin{array}{r} 3 = 2\frac{8}{8} \\ -1\frac{1}{8} = 1\frac{7}{8} \\ \hline 1\frac{1}{8} \end{array}$$

GED Mini-Test: Lessons 5–8 (Pages 86–87)

All items relate to the skill of application.

1. **(3) $250 \times \frac{3}{5}$** The problem states that $\frac{3}{5}$ of the cars are speeding. Since 250 cars are driving, then $\frac{3}{5}$ of 250 is the expression used to solve the problem. It can be written as $\frac{3}{5} \times 250$ or $250 \times \frac{3}{5}$.

2. **(3) $3\frac{5}{12}$** First find the total square yards of carpeting he needs using addition.

$$\begin{array}{r} 12\frac{1}{2} = 12\frac{6}{12} \\ 20\frac{3}{4} = 20\frac{9}{12} \\ +13\frac{1}{3} = 13\frac{4}{12} \\ \hline 45\frac{19}{12} = 45 + 1\frac{7}{12} = 46\frac{7}{12} \end{array}$$

Then subtract the result from 50 square yards (the amount on hand).

$$\begin{array}{r} 50 = 49\frac{12}{12} \\ -46\frac{7}{12} = 46\frac{7}{12} \\ \hline 3\frac{5}{12} \end{array}$$

3. **(4) $14\frac{13}{16}$** Subtract to find the difference.

$$\begin{array}{r} 18\frac{1}{4} = 18\frac{4}{16} = 17\frac{20}{16} \\ -\ 3\frac{7}{16} = 3\frac{7}{16} = 3\frac{7}{16} \\ \hline 14\frac{13}{16} \end{array}$$

4. **(2) golden oak** Compare the five fractions on the list by changing $\frac{1}{8}$, $\frac{3}{8}$, and $\frac{1}{4}$ to equal fractions with the common denominator of 16.

$\frac{1}{8} = \frac{2}{16}$ $\frac{3}{8} = \frac{6}{16}$ $\frac{1}{4} = \frac{4}{16}$

Compare with the other two on the list: $\frac{3}{16}$ and $\frac{1}{16}$. The fraction $\frac{3}{8}$ represents the most customers.

5. **(3) $\frac{13}{16}$** Add $\frac{3}{8}$, $\frac{1}{4}$, and $\frac{3}{16}$.

$$\begin{array}{r} \frac{3}{8} = \frac{6}{16} \\ \frac{1}{4} = \frac{4}{16} \\ +\frac{3}{16} = \frac{3}{16} \\ \hline \frac{13}{16} \end{array}$$

6. **(1) 25** Find $\frac{1}{8}$ of 200 using multiplication.

$$200 \times \frac{1}{8} = \frac{\overset{25}{\cancel{200}}}{1} \times \frac{1}{\underset{1}{\cancel{8}}} = \frac{25}{1} = 25$$

7. **(1) $\frac{1}{5} + \frac{1}{8}$** Add $\frac{1}{5}$ and $\frac{1}{8}$ to find the total fraction spent on food and clothes.

8. **(2) \$520** Find $\frac{1}{4}$ of their monthly income of \$2,080 using multiplication.

$$\$2{,}080 \times \frac{1}{4} = \frac{\overset{520}{\cancel{\$2{,}080}}}{1} \times \frac{1}{\underset{1}{\cancel{4}}} = \frac{520}{1} = \$520$$

9. **(5) Not enough information is given.** The chart tells that they spend $\frac{1}{20}$ of their income on two items: savings and miscellaneous. You need to know the part of their income that is spent on savings alone.

10. **(1) $\frac{1}{25}$** Write as a fraction and reduce to lowest terms.

$$\frac{4}{100} \div \frac{4}{4} = \frac{1}{25}$$

11. **(2) 8** Divide 10 (the number of hours the store is open) by $1\frac{1}{4}$ (the number of hours each tape lasts).

$$10 \div 1\frac{1}{4} = \frac{10}{1} \div \frac{5}{4} = \frac{\overset{2}{\cancel{10}}}{1} \times \frac{4}{\underset{1}{\cancel{5}}} = \frac{8}{1} = 8$$

12. **(4) $7\frac{1}{12}$** Add the three distances: $1\frac{2}{3}$ (to work), $1\frac{2}{3}$ (from work), and $3\frac{3}{4}$ (her evening walk).

$$\begin{array}{r} 1\frac{2}{3} = 1\frac{8}{12} \\ 1\frac{2}{3} = 1\frac{8}{12} \\ +3\frac{3}{4} = 3\frac{9}{12} \\ \hline 5\frac{25}{12} = 5 + 2\frac{1}{12} = 7\frac{1}{12} \end{array}$$

13. **(3) $2\frac{1}{2} \times 5\frac{1}{2}$** Multiply $5\frac{1}{2}$ (the rate of growth) by $2\frac{1}{2}$ (the number of weeks).

Lesson 9

All items relate to the skill of application.

GED Practice: Lesson 9 (Pages 90–91)

1. **five thousandths** The 5 is in the thousandths place.

2. **eight tenths** The 8 is in the tenths place.

3. **seven hundredths** The 7 is in the hundredths place.

4. **nine ten-thousandths** The 9 is in the ten-thousandths place.

5. **one tenth** The 1 is in the tenths place.

6. **nine hundredths** The 9 is in the hundredths place.

7. 1 is in the tenths place **$1 \times 0.1 = 0.1$**
8 is in the hundredths place. **$8 \times 0.01 = 0.08$**
3 is in the thousandths place. **$3 \times 0.001 = 0.003$**
6 is in the ten-thousandths place.
$6 \times 0.0001 = 0.0006$
2 is in the hundred-thousandths place.
$2 \times 0.00001 = 0.00002$

8. **c. one and thirty-five hundredths**

9. **b. one and thirty-five thousandths**

10. **a. one and three hundred five thousandths**

11. **d. one and three hundred five ten-thousandths**

12. **d. one and two tenths**

13. **a. twelve hundredths**

14. **c. twelve ten-thousandths**

15. **b. twelve thousandths**

16. **five and twenty-five hundredths**

17. **six and eight thousandths**

18. **thirty-seven hundredths**

19. **one and one hundredth**

20. **two and five thousandths**

21. **four and five hundredths**

22. **three and nine tenths**

23. **eight hundredths**

24. **three and four thousandths**

25. **twelve and six tenths**

26. **$0.32 > 0.3109$** Add zeros: $0.3200 > 0.3109$ The number 3,200 is greater than 3,109.

27. **$0.98 < 1.9$** The first number, 0.98, does not have a whole number part; the second number, 1.9, has a whole number part of 1, so it is larger.

28. **$0.5 = 0.50$** The 0 after the 5 in 0.50 does not change the value of the number. Both have the same value: five tenths.

29. **$0.006 < 0.06$** Add zero to the second number: 0.060. The first number, 0.006, is less because 6 is less than 60.

30. **$1.075 < 1.57$** Both have the same whole number part. Add a zero to the second number: 1.570. The first number, 1.075, is smaller because 75 is less than 570.

31. **$0.18 > 0.108$** Add a zero to the first number: 0.180. The first number, 0.18, is greater because 180 is greater than 108.

32. **$2.38 < 2.83$** Both have the same whole number part. The first number is less because 38 is less than 83.

33. **$1.09 > 1.009$** Both have the same whole number part. Add a zero to the first number: 1.090. The first number is greater because 90 is greater than 9.

34. **$3.60 = 3.600$** Both have the same whole number part. The zeros after the 6 in both numbers do not change the value. Both have the same value.

35. **3.09 3.4 3.9 3.901** Add zeros so that each number has three decimal places, and compare: 3.400, 3.090, 3.900, 3.901.

36. **0.08 0.8 0.89** Add zeros so that each number has two decimal places, and compare: 0.08, 0.80, 0.89.

37. **0.9054 0.95 0.954** Add zeros so that each number has four decimal places, and compare: 0.9500, 0.9540, 0.9054.

38. **12.001 12.04 12.608 12.8** Add zeros so that each number has three decimal places, and compare: 12.608, 12.001, 12.800, 12.040.

39. **3.6** The number to the right of the tenths place is 5 or more: 3.5719. Add 1 to the tenths place and drop the remaining digits to the right.

40. **125.070** The number to the right of the thousandths place is 5 or more: 125.0699. Since there is a 9 in the thousandths place, add 1 to the hundredths place and drop the remaining digits.

41. **5.13** The number to the right of the hundredths place is less than 5: 5.132. Drop the remaining digit to the right.

42. **17.1** The number to the right of the tenths place is 5 or more: 17.0813. Add 1 to the tenths place and drop the remaining digits to the right.

43. **0.64** The number to the right of the hundredths place is less than 5: 0.6415. Drop the remaining digits to the right.

44. **4** The number to the right of the ones place is 5 or more: 3.8301. Add 1 and drop the remaining digits to the right.

45. **1** The number to the right of the ones place is 5 or more: 0.543. Add 1 and drop the remaining digits to the right.

46. **4.62** The number to the right of the hundredths place is 5 or more: 4.617. Add 1 to the hundredths place and drop the remaining digits.

47. **1.2** The number to the right of the tenths place is less than 5: 1.2491. Drop the remaining digits.

48. **0.510** The number to the right of the thousandths place is less than 5: 0.5104. Drop that digit.

49. **3** The number to the right of the ones place is less than 5: 3.389. Drop the remaining digits.

50. A. 0.3619 rounds to **0.4 g**
B. 0.7082 rounds to **0.7 g**
C. 0.0561 rounds to **0.1 g**
D. 0.9357 rounds to **0.9 g**

GED Practice: Estimation and Money (Pages 93–95)

1. **(4) between $30 and $33**
Estimate the cost of the three games:

	Sale Price	Estimate
Fast Pitch	$8.99	$ 9
Crown of Power	10.09	10
Dugout Derby	12.78	13
Total estimate:		$32

2. **(5) She cannot afford any other game.**
Find the total of the estimate:

	Sale Price	Estimate
Par 4	$6.29	$ 6
Batwing	7.98	8
Total estimate:		$14

Subtract from $20: $20 − $14 = $6
None of the other games among the choices costs $6 or less.

3. **(2) $8**
Estimate the cost of the games at the regular price:

	Regular Price	Estimate
Dugout Derby	$17.25	$17
Crown of Power	13.72	14
Total estimate:		$31

Estimate the cost of the games at the sale price:

	Sale Price	Estimate
Dugout Derby	$12.78	$13
Crown of Power	10.09	10
Total estimate:		$23

Find the difference between the two estimates: $31 − $23 = $8

4. **(3) between $19 and $21**
Estimate the cost of the items:

	Estimate
4 Bosco specials: $2.95 rounds to $3. $3 × 4 = 12	$12
4 small fries: $1.05 rounds to $1. $1 × 4 = 4	4
4 medium drinks: $1.19 rounds to $1. $1 × 4 = 4	+ 4
Total estimate	$20

5. **(5) $24**
Estimate the cost of the items:

	Estimate
No-frills burger: $1.19 rounds to $1.	$1
small fries: $1.05 rounds to $1.	1
large drink: $1.99 rounds to $2.	+ 2
Total the estimate:	$4

Multiply the total by 6: $4 × 6 = $24

6. **(3) $7** Round the cost of each item to the nearest whole dollar and add to find the total of the estimates.
$1.29 rounds to $1.
$2.99 rounds to $3.
$1.59 rounds to $2.
$1.19 rounds to $1.
$1 + $3 + $2 + $1 = $7

7. **(4) $27** Round the cost of each item to the nearest whole dollar and add to find the total of the estimates. Be sure to add the amounts for multiple items.
$2.97 rounds to $3.
$6.47 rounds to $6.
$14.99 rounds to $15.
$3 + $3 + $6 + $15 = $27

8. **(3) $4** Round the cost of each item to the nearest whole dollar, add to find the total cost of the estimates, and subtract the total cost from $10.
$1.99 rounds to $2.
$2.19 rounds to $2.
$1.14 rounds to $1.
$0.85 rounds to $1.
$2 + $2 + $1 + $1 = $6
$10 − $6 = $4

9. **(5) $8** Round the cost of 1 tape to the nearest whole dollar, multiply to find the approximate cost of 8 tapes, and subtract from $40 to find the approximate change.
$3.98 rounds to $4
8 × $4 = $32
$40 − 32 = $8

10. **(4) between $14 and $16** Round the cost of each item to the nearest whole dollar, multiply rounded costs for any multiple items, and add to find the total of the estimates.
a $\frac{1}{2}$-chicken: $5.99 rounds to $6.
2 large orders of stuffing: $2.79 rounds to $3.
$3 × 2 = $6
3 medium drinks: $1.09 rounds to $1.
$1 × 3 = $3
$6 + $6 + $3 = $15

11. **(2) between $7 and $9** Round the cost of each item to the nearest whole dollar, and add to find the total of the estimates.
a $\frac{1}{4}$-chicken: $3.99 rounds to $4.
1 small order of vegetables: $1.59 rounds to $2.
1 small order of potatoes: $1.29 rounds to $1.
1 small soda: $0.79 rounds to $1.
$4 + $2 + $1 + $1 = $8

12. **(4) between $27 and $37** Round the cost of each item to the nearest whole dollar, multiply rounded costs for any multiple items, and add to find the total of the estimates.
Three $\frac{1}{2}$ chickens: $5.99 rounds to $6.
$6 × 3 = $18
3 large orders of stuffing:
$2.79 rounds to $3.
$3 × 3 = $9
3 large orders of vegetables:
$3.19 rounds to $3.
$3 × 3 = $9
$18 + $9 + $9 = $36

13. **(3) 5** Round the cost of a 3-inch potted plant to the nearest whole dollar. Divide $10 by the estimated cost.
$1.79 rounds to $2.
$10 ÷ $2 = 5

14. **(3) 3** Round amounts to the nearest whole dollar. Add to find the cost of Ernesto's purchase, so far. Subtract the cost so far from $20. Divide the difference by the rounded cost of the 4-inch plant.
Two 5-inch potted plants: $3.69 rounds to $4.
$4 × 2 = $8
One watering can: $1.89 rounds to $2.
$8 + $2 = $10
$20 − $10 = $10
4-inch potted plants: $2.89 rounds to $3.
$10 ÷ $3 = 3r1

15. **(4) $4** Round amount of one bag of potting soil and multiply by 3. Round amount of one watering can and add to total. Subtract total from $15 to get the approximate change.
$3.19 rounds to $3.
3 × $3 = $9
$1.89 rounds to $2.
$9 + $2 = $11
$15 − $11 = $4

16. **(4) $51** Round the cost of each item to the nearest whole dollar and add to find the total of the estimates.
$38.82 rounds to $39.
$12.39 rounds to $12.
$39 + $12 = $51

17. **(5) $36** Round the cost of a package of nails to the nearest whole dollar. Multiply to find the total cost of six of the items.
$5.75 rounds to $6.
$6 × 6 = $36

18. **(4) $232** Round the cost of each item to the nearest whole dollar, and add to find the total of the estimates.
$185.60 rounds to $186.
$46.36 rounds to $46.
$186 + $46 = $232

19. **(5) $600** Round amounts to the nearest hundred dollars. Subtract to find the amount still needed.
$685 rounds to $700.
$110 rounds to $100.
$700 − $100 = $600

Lesson 10

All items relate to the skill of application.

GED Practice: Lesson 10 (Pages 98–100)

1. 2.63
0.03
+2.60
2.63

2. 5.4
1.35
+4.05
5.40

3. 15.103
7.100
+8.003
15.103

4. 5.547
9
8 1 1
6.900
−1.353
5.547

5. 2.925
4 1
5.075
−2.150
2.925

6. 4.175
9
2 1 1
10.300
−6.125
4.175

7. 4.81
3.61
+1.20
4.81

8. 11.78
9
5 1 1
16.05
− 4.27
11.78

9. 82.887
1.850
0.030
19.007
+62.000
82.887

10. 44.155
12.400
11.080
16.100
+ 4.575
44.155

11. 9,032
1 9
5 1 1
16,004.1
− 6,972.1
9,032.0

12. 2.794
9
7 1 1
3.800
−1.006
2.794

13. 2.947
1 1 6 1
12.870
− 9.923
2.947

14. 17.105
1
1 2 1 6 1
23.070
− 5.965
17.105

15. 22.668
14.010
8.600
+ 0.058
22.668

16. 31.85
1
5 7 1
56.80
−24.95
31.85

17. 6.701
0.950
1.843
3.008
+0.900
6.701

18. 0.2593
5 9 9 1
0.6000
−0.3407
0.2593

19. 10.316
3.150
2.816
4.050
+0.300
10.316

20. 23.35
8 1
39.05
−15.70
23.35

21. 5.295
0.125
1.400
3.760
+0.010
5.295

22. 12.75
1
4 5 1
25.60
−12.85
12.75

23. 3.4 $8.5 \times 0.4 = 3.40$ There are 2 decimal places in the problem.

24. 0.024 $0.04 \times 0.6 = 0.024$ There are 3 decimal places in the problem. You need to write a zero as a placeholder.

25. 0.0112 $5.6 \times 0.002 = 0.0112$ There are 4 decimal places in the problem.

26. 1.02 $3.4 \times 0.3 = 1.02$ There are 2 decimal places in the problem.

27. 36.72 $12 \times 3.06 = 36.72$ There are 2 decimal places in the problem.

28. 310.17 $21.1 \times 14.7 = 310.17$ There are 2 decimal places in the problem.

29. 0.096 $0.008 \times 12 = 0.096$ There are 3 decimal places in the problem.

30. 0.6 $0.04 \times 15 = 0.60$ There are 2 decimal places in the problem.

31. 0.084 $1.2 \times 0.07 = 0.084$ There are 3 decimal places in the problem.

32. 60.0 $25 \times 2.4 = 60.0$ There is 1 decimal place in the problem.

33. 12.84
1.07
× 12
2 14
10 7
12.84

34. 0.549
0.09
× 6.1
00 9
54
0.54 9

35. 4.086
2.27
× 1.8
1 816
2 27
4.086

36. 75.6

5.04
× 15
25 20
50 4
75.60

37. 0.02

0.008
× 2.5
0040
0 016
0.0200

38. 0.1155

1.05
× 0.11
105
10 5
0 00
0.11 55

39. 0.144

0.012
× 12
0 024
00 12
00 0
00.144

40. 0.2145

7.15
× 0.03
21 45
00 0
0 00
0.21 45

41. 18.375

12.25
× 1.5
612 5
1225
18.375

42. 2.56

2.56
8) 20.48
16
4 4
4 0
48
48

43. 1.0972

1.0972
3) 3.2916
3
29
27
21
21
6
6

44. 0.136

0.136
7) 0.952
7
25
21
42
42

45. 0.418

0.418
5) 2.090
2 0
9
5
40
40

46. 0.0035

0.0035
6) 0.0210
18
30
30

47. 62

62.
0.07) 4.34.
4 2
14
14

48. 0.025

0.025
1.6.) 0.0.400
32
80
80

49. 6.094

6.094
1.05.) 6.39.870
6 30
9 87
9 45
420
420

50. 2.14

2.14
3.6) 7.7.04
7 2
5 0
3 6
1 44
1 44

51. 0.04

0.04
0.9) 00.36
36

52. 0.3

0.3
1.5) 0.4.5
4 5

53. 0.12

0.12
2.25) 0.27.00
22 5
4 50
4 50

54. 0.072

0.072
1.2) 0.0.864
84
24
24

55. 146

146.
0.03) 4.38.
3
1 3
1 2
18
18

56. 0.248

0.248
0.15) 0.03.720
3 0
72
60
120
120

57. 0.29 0.285 rounds to 0.29

7) 2.000
1 4
60
56
40
35
5

58. 0.27 0.272 rounds to 0.27

11) 3.000
2 2
80
77
30
22
8

59. 0.46 0.461 rounds to 0.46

13) 6.000
5 2
80
78
20
13
7

UNIT 1

60. 0.42 0.416 rounds to 0.42

$$\begin{array}{r} 0.416 \\ 12\overline{)5.000} \\ \underline{4\,8} \\ 20 \\ \underline{12} \\ 80 \\ \underline{72} \\ 8 \end{array}$$

61. 0.83 0.833 rounds to 0.83

$$\begin{array}{r} 0.833 \\ 6\overline{)5.000} \\ \underline{4\,8} \\ 20 \\ \underline{18} \\ 20 \\ \underline{18} \\ 2 \end{array}$$

62. 2.22 2.222 rounds to 2.22

$$\begin{array}{r} 2.222 \\ 9\overline{)20.000} \\ \underline{18} \\ 2\,0 \\ \underline{1\,8} \\ 20 \\ \underline{18} \\ 20 \end{array}$$

GED Practice: Solving Word Problems (Page 103)

1. **(2) $35 – ($12.98 + $10.67 + $5.98)** You need to add the prices for the three items together and then subtract the total from $35. The parentheses around the three prices in option (2) tell you to do that step first, then subtract.

2. **(4) 5.37** Carry out the operations:
$35 – ($12.98 + $10.67 + $5.98) =
$35 – $29.63 = $5.37

3. **(1) 25($1.05 – $0.89)** The best way to work the problem is to find the difference between the two prices for one diskette. Then multiply the difference by 25 to find the total savings. You can get the same answer by multiplying each price by 25 and finding the difference: 25($1.05) – 25($0.89).

4. **(4) $4.00** Carry out the operations:
25($1.05 – $0.89) =
25($0.16) = $4

5. **(1) 2($45.79) + $18.25** Multiply the cost of one tire by 2 and add the amount for the oil change.

6. **(3) $109.83** Carry out the operations:
2($45.79) + $18.25 =
$91.58 + $18.25 = $109.83

7. **(1) $202** Subtract to find the annual increase. Divide the annual increase by 12, the number of months in a year.
$21,000 – $18,575 = $2,425
$2,425 ÷ 12 = $202.083, which rounds to $202.

Lesson 11

All items relate to the skill of application.

GED Practice: Lesson 11 (Pages 106–107)

1. $\frac{1}{4}$ Write 25 over 100 and reduce to lowest terms.
$\frac{25}{100} \div \frac{25}{25} = \frac{1}{4}$

2. $\frac{2}{5}$ Write 4 over 10 and reduce to lowest terms.
$\frac{4}{10} \div \frac{2}{2} = \frac{2}{5}$

3. $\frac{7}{20}$ Write 35 over 100 and reduce to lowest terms.
$\frac{35}{100} \div \frac{5}{5} = \frac{7}{20}$

4. $\frac{7}{8}$ Write 875 over 1,000 and reduce to lowest terms.
$\frac{875}{1{,}000} \div \frac{125}{125} = \frac{7}{8}$

5. $\frac{3}{8}$ Write 375 over 1,000 and reduce to lowest terms.
$\frac{375}{1{,}000} \div \frac{125}{125} = \frac{3}{8}$

6. $\frac{19}{25}$ Write 76 over 100 and reduce to lowest terms.
$\frac{76}{100} \div \frac{4}{4} = \frac{19}{25}$

7. $\frac{16}{125}$ Write 128 over 1,000 and reduce to lowest terms.
$\frac{128}{1{,}000} \div \frac{8}{8} = \frac{16}{125}$

8. $\frac{1}{20}$ Write 5 over 100 and reduce to lowest terms.
$\frac{5}{100} \div \frac{5}{5} = \frac{1}{20}$

9. $\frac{5}{16}$ $\frac{31\frac{1}{4}}{100} = 31\frac{1}{4} \div 100 =$
$\frac{\overset{5}{\cancel{125}}}{4} \times \frac{1}{\underset{4}{\cancel{100}}} = \frac{5}{16}$

10. $\frac{1}{12}$ $\frac{8\frac{1}{3}}{100} = 8\frac{1}{3} \div 100 =$
$\frac{\overset{1}{\cancel{25}}}{3} \times \frac{1}{\underset{4}{\cancel{100}}} = \frac{1}{12}$

11. $\frac{7}{15}$ $\frac{46\frac{2}{3}}{100} = 46\frac{2}{3} \div 100 =$
$\frac{\overset{7}{\cancel{140}}}{3} \times \frac{1}{\underset{5}{\cancel{100}}} = \frac{7}{15}$

12. $\frac{15}{16}$ $\frac{93\frac{3}{4}}{100} = 93\frac{3}{4} \div 100 =$
$\frac{\overset{15}{\cancel{375}}}{4} \times \frac{1}{\underset{4}{\cancel{100}}} = \frac{15}{16}$

13. $\frac{11}{12}$ $\frac{91\frac{2}{3}}{100} = 91\frac{2}{3} \div 100 =$
$\frac{\overset{11}{\cancel{275}}}{3} \times \frac{1}{\underset{4}{\cancel{100}}} = \frac{11}{12}$

14. $\frac{1}{6}$ $\frac{16\frac{2}{3}}{100} = 16\frac{2}{3} \div 100 =$
$\frac{\overset{1}{\cancel{50}}}{3} \times \frac{1}{\underset{2}{\cancel{100}}} = \frac{1}{6}$

15. $\frac{1}{15}$ $\quad \frac{6\frac{2}{3}}{100} = 6\frac{2}{3} \div 100 =$

$\frac{\cancel{20}^{1}}{3} \times \frac{1}{\cancel{100}_{5}} = \frac{1}{15}$

16. $\frac{19}{80}$ $\quad 0.23\frac{3}{4} = 23\frac{3}{4} \div 100 =$

$\frac{\cancel{95}^{19}}{4} \times \frac{1}{\cancel{100}_{20}} = \frac{19}{80}$

17. $\frac{27}{40}$ $\quad \frac{675}{1{,}000} \div \frac{25}{25} = \frac{27}{40}$

18. $\frac{49}{600}$ $\quad \frac{8\frac{1}{6}}{100} = 8\frac{1}{6} \div 100 =$

$\frac{49}{6} \times \frac{1}{100} = \frac{49}{600}$

19. $\frac{1}{8}$ $\quad \frac{125}{1{,}000} \div \frac{125}{125} = \frac{1}{8}$

20. $\frac{13}{120}$ $\quad \frac{10\frac{5}{6}}{100} = 10\frac{5}{6} \div 100 =$

$\frac{\cancel{65}^{13}}{6} \times \frac{1}{\cancel{100}_{20}} = \frac{13}{120}$

21. $\frac{11}{20}$ $\quad \frac{55}{100} \div \frac{5}{5} = \frac{11}{20}$

22. $\frac{4}{75}$ $\quad \frac{5\frac{1}{3}}{100} = 5\frac{1}{3} \div 100 =$

$\frac{\cancel{16}^{4}}{3} \times \frac{1}{\cancel{100}_{25}} = \frac{4}{75}$

23. $\frac{7}{25}$ $\quad \frac{28}{100} \div \frac{4}{4} = \frac{7}{25}$

24. $\frac{51}{400}$ $\quad \frac{12\frac{3}{4}}{100} = 12\frac{3}{4} \div 100 =$

$\frac{51}{4} \times \frac{1}{100} = \frac{51}{400}$

25. 0.8 Divide 4 by 5.

```
   0.8
5) 4.0
   4 0
```

26. 0.375 Divide 3 by 8.

```
   0.375
8) 3.000
   2 4
     60
     56
      40
      40
```

27. 0.667 Divide 2 by 3. Divide to four decimal places and round to the thousandths place.

```
   0.6666 rounds to 0.667
3) 2.0000
   1 8
     20
     18
      20
      18
       20
       18
        2
```

28. 0.417 Divide 5 by 12 to four decimal places and round.

```
    0.4166 rounds to 0.417
12) 5.0000
    4 8
      20
      12
       80
       72
        80
        72
         8
```

29. $0.83\frac{1}{3}$ $\quad 0.83\frac{2}{6} = 0.83\frac{1}{3}$

```
   0.83 2/6
6) 5.00
   4 8
     20
     18
      2
```

30. $0.88\frac{8}{9}$ $\quad 0.88\frac{8}{9}$

```
   0.88 8/9
9) 8.00
   7 2
     80
     72
      8
```

31. $0.46\frac{2}{3}$ $\quad 0.46\frac{10}{15} = 0.46\frac{2}{3}$

```
    0.46 10/15
15) 7.00
    6 0
    1 00
      90
      10
```

32. $0.06\frac{1}{4}$ $\quad 0.06\frac{4}{16} = 0.06\frac{1}{4}$

```
    0.06 4/16
16) 1.00
      96
       4
```

33. (3) $23\frac{1}{2}$ **cents** Divide \$2.82 by 12 to two decimal places and write the remainder as a fraction.

$0.23\frac{6}{12} = 0.23\frac{1}{2}$

```
    0.23 6/12
12) $2.82
     2 4
       42
       36
        6
```

34. (5) \$2.20 Multiply $13\frac{3}{4}$ cents (or 0.1375) by 16.

$13\frac{3}{4} \times 16 = \frac{55}{\cancel{4}_{1}} \times \frac{\cancel{16}^{4}}{1} = \frac{220}{1}$

cents or \$2.20

or

```
   $0.1375
 ×      16
    8250
   1 375
   $2.2000
```

35. (2) \$1.09 Multiply $15\frac{1}{2}$ cents (or 0.155) by 7 and round to the nearest whole cent.

$15\frac{1}{2} \times 7 = \frac{31}{2} \times \frac{7}{1} = \frac{217}{2}$

$= 108.5$ cents

```
   108.5
2) 217.0
   2
    17
    16
     1 0
     1 0
```

or

```
   $0.155
 ×      7
   $1.085 rounds to $1.09
```

36. (3) $14\frac{3}{4}$ **cents** Divide \$0.59 by 4 to two decimal places and write the remainder as a fraction.

```
   $0.14 3/4
4) $0.59
    0 4
      19
      16
       3
```

UNIT 1

GED Practice: Solving Word Problems (Page 109)

1. **(3) 0.003** Since the answer options are written as decimals, change $\frac{1}{100}$ to a decimal: $\frac{1}{100} = 0.01$.
   Subtract: $0.010 - 0.007 = 0.003$

2. **(2) $\frac{1}{8}$** Since the answer options are written as fractions, change 0.75 to a fraction: $0.75 = \frac{3}{4}$.
   Subtract: $\frac{7}{8} = \frac{7}{8}$, $-\frac{3}{4} = -\frac{6}{8}$; result $\frac{1}{8}$

3. **(4) $17.00** Since the answer options are written as decimals, change $12\frac{1}{2}$ to a decimal: $12\frac{1}{2} = 12.5$.
   Multiply: 12.5 × 1.36
   7 50
   37 5
   125
   17.000 = $17.00

4. **(1) 10** To find the difference of $\frac{3}{8}$ and 0.4 solve the problem using decimals as the faster method. Convert $\frac{3}{8}$ to a decimal: $\frac{3}{8} = 0.375$.
   Subtract: $0.400 - 0.375 = 0.025$
   Now multiply the number of people surveyed by 0.025. 400 × 0.025
   2 000
   8 00
   10.000

5. **(2) 11** You need to divide $35\frac{3}{4}$ by 3.25. You can use either method. Both methods are shown:
   Fractions: $(3.25 = 3\frac{1}{4})$
   $$35\frac{3}{4} \div 3\frac{1}{4} = \frac{143}{4} \div \frac{13}{4} = \frac{\overset{11}{\cancel{143}}}{\underset{1}{\cancel{4}}} \times \frac{\overset{1}{\cancel{4}}}{\underset{1}{\cancel{13}}} = \frac{11}{1} = 11$$
   Decimals: $(35\frac{3}{4} = 35.75)$
   3.25)35.75 = 11
   32 5
   3 25
   3 25

6. **(2) $16\frac{1}{4}$** Since the answer options are fractions, change 2.75 and 3.375 to fractions. Then add and reduce.
   $$2.75 = 2 + \frac{75 \div 25}{100 \div 25} = 2\frac{3}{4}$$
   $$3.375 = 3 + \frac{375 \div 125}{1{,}000 \div 125} = 3\frac{3}{8}$$
   Add:
   $2\frac{3}{4} = 2\frac{6}{8}$
   $3\frac{1}{2} = 3\frac{4}{8}$
   $3\frac{7}{8} = 3\frac{7}{8}$
   $2\frac{3}{4} = 2\frac{6}{8}$
   $+3\frac{3}{8} = +3\frac{3}{8}$
   $13\frac{26}{8}$
   Reduce: $13\frac{26}{8} = 13 + 3\frac{2}{8} = 16\frac{2}{8} = 16\frac{1}{4}$

Lesson 11

All items relate to the skill of application.

GED Practice: Lesson 11 (Page 111)

1. **(3) $\frac{3}{4}$** Subtract $6\frac{3}{4}$ from $7\frac{1}{2}$.
   $7\frac{1}{2} = 7\frac{2}{4} = 6\frac{6}{4}$
   $-6\frac{3}{4} = 6\frac{3}{4} = 6\frac{3}{4}$
   $\frac{3}{4}$

2. **(2) $0.72** To find the actual answer, subtract to find the difference between the two prices for one gallon. Multiply the difference by 18 to find the total difference.
   18($1.39 − $1.35)
   18($0.04) = $0.72
   To use estimation to narrow down the choices, round 18 to the nearest ten: 20. Work the problem using the rounded number.
   20($1.39 − $1.35)
   20($0.04) = $0.80 estimated solution
   Options (1), (3), (4), and (5) are far from the estimated solution. Option (2) is the most likely answer.

3. **(4) $12.75** To find the actual answer, multiply $1.25 by 5 half-hours because you are putting together the same amount, $1.25, 5 times. Add the result to $6.50 because you are finding the total of two costs.
$5(\$1.25) + (6.50)$
$\$6.25 + \$6.50 = \$12.75$
To use estimation to narrow down the choices, round $1.25 to the nearest dollar: $1.00. Work the problem using the rounded number.
$5(\$1.00) + 1(6.50)$
$\$5.00 + 6.50 = \$11.50$ estimated solution
Options (1), (2), and (5) are far from the estimated solution. Since you rounded down $1.25, your estimated answer will be less than the actual answer. Therefore, option (3) is eliminated. Option (4) is the most likely answer.

4. **(5) 36** To find the actual answer, divide 12 by $\frac{1}{3}$ because you are breaking a whole into equal parts of $\frac{1}{3}$ each. To divide by a fraction, invert the fraction and multiply.
$12 \div \frac{1}{3} = 12 \times \frac{3}{1} = 36$
In this item, options (3) and (4) are not far enough apart in value to use estimation as a strategy.

5. **(3) (35 × $6) + (4 × $9)** Multiply the number of hours worked at each dollar amount because for each dollar amount, you are putting together the same amount several times. Then add the two results.
In this item, appropriate operations rather than the actual answer is requested. Using estimation to narrow down options would not apply.

6. **(1) 100** To find the actual answer, multiply the number of square feet per carton by the number of cartons because you need to put together the same amount,
$12\frac{1}{2}$ square feet, 8 times.
$12\frac{1}{2} \times 8$
$\frac{25}{2} \times 8 = 100$
To use estimation to narrow down the choices, round each number to the greatest place. $12\frac{1}{2}$ rounds to 10. 8 rounds to 10. Work the problem using the rounded numbers.
$10 \times 10 = 100$
Options (2), (3), (4), and (5) are far from the estimated solution. Option (1) is the most likely answer.

7. **(3) 7,988** To find the actual answer, add the number of people who attended on Saturday to the number of people who attended on Sunday because you need to put together the numbers to find the total.
$$\begin{array}{r} 5{,}136 \\ +2{,}852 \\ \hline 7{,}988 \end{array}$$
In this item, options (3) and (4) are not far enough apart in value to use estimation as a strategy.

8. **(2) $1,080** To find the actual answer, subtract the sale price from the wholesale price to find the difference in the two prices. Multiply the difference by 120 sweaters to find how much money Tran made.
$120(\$18 - \$9)$
$120(\$9) = \$1{,}080$
To use estimation to narrow down the choices, round each number: 120 rounded to the nearest hundred is 100; $18 rounded to the nearest ten is $20; $9 rounded to the nearest ten is $10. Work the problem using the rounded numbers.
$100(\$20 - \$10)$
$100(\$10) = \$1{,}000$ estimated solution
Options (1), (3), (4), and (5) are far from the estimated solution. Option (2) is the most likely answer.

GED Mini-Test: Lessons 9–11 (Pages 112–113)

All items relate to the skill of application.

1. **(4) 58.7 miles** Add the miles for each trip to find a total: $12.3 + 15.8 + 21.7 + 8.9 = 58.7$.

2. **(2) 0.52 − (0.06 + 0.1)** First you need to find how much of ingredient A and B are in the conditioner: add 0.06 and 0.1 grams. Then subtract that result from the total needed: 0.52 grams. The parentheses in option 2 tell you to do the addition step first.

3. **(3) 3.75** Divide $6.98 by $1.86. Divide to the thousandths place and round to the hundredths place.
3.752 rounds to 3.75.
$$\begin{array}{r} 3.752 \\ \$1.86\overline{)\,\$6.98.000} \\ 5\,58 \\ \hline 1\,40\,0 \\ 1\,30\,2 \\ \hline 9\,80 \\ 9\,30 \\ \hline 500 \\ 372 \\ \hline 128 \end{array}$$

UNIT 1

4. **(1) 263.8 miles** Add the miles for each day to find a total:

87.6	Line up the decimal points.
17.3	Write zeros as needed, then
14.0	add.
102.2	
+ 42.7	
263.8	

5. **(3) $12.26** Multiply Monday's mileage by $0.14

87.6	Since there are three decimal
×$0.14	places in the problem, there
3 504	must be three decimal places in
8 76	the answer. Round to the
$12.264	nearest whole cent: $12.264 rounds to $12.26.

6. **(2) $6.16** Multiply $3.85 by 1.6.

$3.85	Since there are 3 decimal places
× 1.6	in the problem, there are 3 decimal
2 31 0	places in the answer.
3 85	
$6.16 0	Drop the final zero. $6.160 = $6.16

7. **(5) 0.009** Subtract the thickness of the gold from the desired thickness:

$$\begin{array}{r} 0.0\overset{2}{\not{3}}\overset{1}{0} \\ -0.021 \\ \hline 0.009 \end{array}$$

8. **(4) between $75 and $80** Round each amount to the nearest whole number and find the difference:

$127.76 rounds to $128	$128
$51.04 rounds to $51	− 51
	$77

The difference is approximately $77.

9. **(1) ($82.37 + $136.29) − ($51.04 + $127.76)** Find the total earned on Saturday, $82.37 + $136.29, and the total earned on Sunday, $51.04 + $127.76. Subtract Sunday's total from Saturday's. The second set of parentheses is subtracted from the first.

10. **(3) 0.015 inch** Add the two thicknesses: 0.005 and 0.01.

$$\begin{array}{r} 0.005 \\ +0.010 \\ \hline 0.015 \end{array}$$

11. **(2) $438.81** Multiply $2,507.47 by 0.175. Round to the nearest whole cent.

$ 2,507.47	Since there are 5 decimal places
× 0.175	in the problem, there are 5
12 53735	decimal places in the answer.
175 5229	Round to the hundredths place.
250 747	
$438.80725	$438.80725 rounds to $438.81.

12. **(5) 0.1($226.40) + $226.40** First find the amount of the raise by multiplying 0.1 by his present salary. This step can be written 0.1($226.40). Then add the result to the present salary to find the new salary: 0.1($226.40) is added to $226.40.

Lesson 12

All items relate to the skill of application.

GED Practice: Lesson 12 (Pages 115–117)

1. **0.6** 60% = .60. = 0.6
2. **0.048** 4.8% = .04.8 = 0.048
3. **0.04** 4% = .04. = 0.04
4. **$0.05\frac{1}{2}$ or 0.055** $5\frac{1}{2}\%$ = $.05.\frac{1}{2}$ = $0.05\frac{1}{2}$
5. **2** 200% = 2.00. = 2
6. **1.3** 130% = 1.30. = 1.3
7. **$0.09\frac{1}{4}$ or 0.0925** $9\frac{1}{4}\%$ = $.09.\frac{1}{4}$ = $0.09\frac{1}{4}$
8. **0.056** 5.6% = .05.6 = 0.056
9. **2.25** 225% = 2.25. = 2.25
10. **85%** 0.85 = 0.85. = 85%
11. **36%** 0.36 = 0.36. = 36%
12. **14.4%** 0.144 = 0.14.4 = 14.4%
13. **40%** 0.4 = 0.40. = 40%
14. **450%** 4.5 = 4.50. = 450%
15. **$16\frac{2}{3}\%$** $0.16\frac{2}{3}$ = $0.16.\frac{2}{3}$ = $16\frac{2}{3}\%$
16. **875%** 8.75 = 8.75. = 875%
17. **37.5%** 0.37.5 = 37.5%
18. **$7\frac{1}{3}\%$** $0.07.\frac{1}{3}$ = $7\frac{1}{3}\%$
19. **150%** 1.50.
20. **17.5%** 0.17.5
21. **9%** 0.09.
22. **40%** $\frac{2}{\not{5}_1} \times \frac{\not{100}^{20}}{1} = \frac{40}{1} = 40\%$
23. **75%** $\frac{3}{\not{4}_1} \times \frac{\not{100}^{25}}{1} = \frac{75}{1} = 75\%$
24. **$62\frac{1}{2}\%$ or 62.5%** $\frac{5}{\not{8}_2} \times \frac{\not{100}^{25}}{1} = \frac{125}{2} = 62\frac{1}{2}\%$

25. **$33\frac{1}{3}\%$** $\frac{1}{3} \times \frac{100}{1} = \frac{100}{3} = 33\frac{1}{3}\%$

26. **235%** $2 \times 100 = 200\%$
$\frac{7}{\cancel{20}_1} \times \frac{\cancel{100}^5}{1} = \frac{35}{1} = 35\%$
Add the two percents.
$200\% + 35\% = 235\%$

27. **$347\frac{1}{2}\%$ or 347.5%** $3 \times 100 = 300\%$
$\frac{19}{\cancel{40}_2} \times \frac{\cancel{100}^5}{1} = \frac{95}{2} = 47\frac{1}{2} = 47\frac{1}{2}\%$
Add the percents.
$300\% + 47\frac{1}{2}\% = 347\frac{1}{2}\%$

28. **$162\frac{1}{2}\%$ or 162.5%** $1 \times 100 = 100\%$
$\frac{5}{\cancel{8}_2} \times \frac{\cancel{100}^{25}}{1} = \frac{125}{2} = 62\frac{1}{2} = 62\frac{1}{2}\%$
Add the percents.
$100\% + 62\frac{1}{2}\% = 162\frac{1}{2}\%$

29. **$466\frac{2}{3}\%$** $4 \times 100 = 400\%$
$\frac{2}{3} \times \frac{100}{1} = \frac{200}{3} = 66\frac{2}{3} = 66\frac{2}{3}\%$
Add the percents.
$400\% + 66\frac{2}{3}\% = 466\frac{2}{3}\%$

30. **$236\frac{2}{3}\%$** $2 \times 100 = 200\%$
$\frac{11}{\cancel{30}_3} \times \frac{\cancel{100}^{10}}{1} = \frac{110}{3} = 36\frac{2}{3} = 36\frac{2}{3}\%$
Add the percents.
$200\% + 36\frac{2}{3}\% = 236\frac{2}{3}\%$

31. **187.5%** $1 \times 100 = 100\%$
$\frac{7}{\cancel{8}_2} \times \frac{\cancel{100}^{25}}{1} = 87.5\%$
Add the percents.
$100\% + 87.5\% = 187.5\%$

32. **310%** $3 \times 100 = 300\%$
$\frac{1}{\cancel{10}_1} \times \frac{\cancel{100}^{10}}{1} = 10\%$
Add the percents.
$300\% + 10\% = 310\%$

33. **420%** $4 \times 100 = 400\%$
$\frac{1}{\cancel{5}_1} \times \frac{\cancel{100}^{20}}{1} = 20\%$
Add the percents.
$400\% + 20\% = 420\%$

34. $\mathbf{\frac{1}{5}}$ $\frac{20}{100} \div \frac{20}{20} = \frac{1}{5}$

35. $\mathbf{\frac{13}{20}}$ $\frac{65}{100} \div \frac{5}{5} = \frac{13}{20}$

36. $\mathbf{\frac{21}{25}}$ $\frac{84}{100} \div \frac{4}{4} = \frac{21}{25}$

37. $\mathbf{\frac{3}{25}}$ $\frac{12}{100} \div \frac{4}{4} = \frac{3}{25}$

38. $\mathbf{1\frac{2}{5}}$ $\frac{140}{100} \div \frac{20}{20} = \frac{7}{5} = 1\frac{2}{5}$

39. $\mathbf{2\frac{3}{4}}$ $\frac{275}{100} \div \frac{25}{25} = \frac{11}{4} = 2\frac{3}{4}$

40. $\mathbf{\frac{39}{100}}$

41. $\mathbf{1\frac{1}{5}}$ $\frac{120}{100} \div \frac{20}{20} = \frac{6}{5} = 1\frac{1}{5}$

42. $\mathbf{3\frac{1}{4}}$ $\frac{325}{100} \div \frac{25}{25} = \frac{13}{4} = 3\frac{1}{4}$

43.

Decimal	Fraction	Percent
0.1	$\frac{10}{100} = \mathbf{\frac{1}{10}}$	**10%**
0.2	$\frac{20}{100} = \mathbf{\frac{1}{5}}$	20%
0.25	$\frac{25}{100} = \frac{1}{4}$	**25%**
0.3	$\frac{30}{100} = \mathbf{\frac{3}{10}}$	**30%**
$0.33\frac{1}{3}$	$\frac{33\frac{1}{3}}{100} = \frac{1}{3}$	**$33\frac{1}{3}\%$**
0.4	$\frac{40}{100} = \mathbf{\frac{2}{5}}$	40%
0.5	$\frac{50}{100} = \mathbf{\frac{1}{2}}$	**50%**
0.6	$\frac{60}{100} = \mathbf{\frac{3}{5}}$	60%
$0.66\frac{2}{3}$	$\frac{66\frac{2}{3}}{100} = \mathbf{\frac{2}{3}}$	**$66\frac{2}{3}\%$**
0.7	$\frac{70}{100} = \frac{7}{10}$	**70%**
0.75	$\frac{75}{100} = \frac{3}{4}$	**75%**
0.8	$\frac{80}{100} = \mathbf{\frac{4}{5}}$	80%
0.9	$\frac{90}{100} = \mathbf{\frac{9}{10}}$	**90%**

GED Practice: Solving Word Problems (Page 119)

1. **(4) $600** $33\frac{1}{3}\%$ is $\frac{1}{3}$. Solve by multiplying by $\frac{1}{3}$ or dividing by 3. $\$1,800 \div 3 = \$600$

2. **(1) $1.10** 20% is $\frac{1}{5}$. Solve by multiplying by $\frac{1}{5}$ or dividing by 5. $\$5.50 \div 5 = \$1.10$

3. **(2) $6.00** $66\frac{2}{3}\%$ is $\frac{2}{3}$. Solve by multiplying by $\frac{2}{3}$. $\$9 \times \frac{2}{3} = \frac{\cancel{9}^3}{1} \times \frac{2}{\cancel{3}_1} = \frac{6}{1} = \$6$

4. **(1) 100** $83\frac{1}{3}\%$ is $\frac{5}{6}$. Solve by multiplying by $\frac{5}{6}$.

$120 \times \frac{5}{6} = \frac{\overset{20}{\cancel{120}}}{1} \times \frac{5}{\underset{1}{\cancel{6}}} = \frac{100}{1} = 100$

5. **(3) 175** $62\frac{1}{2}\%$ is $\frac{5}{8}$. Solve by multiplying by $\frac{5}{8}$.

$280 \times \frac{5}{8} = \frac{\overset{35}{\cancel{280}}}{1} \times \frac{5}{\underset{1}{\cancel{8}}} = \frac{175}{1} = 175$

6. **(1) $200** 12.5% is $\frac{1}{8}$. Solve by multiplying by $\frac{1}{8}$ or dividing by 8. $1,600 ÷ 8 = $200

Lesson 13

All items relate to the skill of application.

GED Practice: Lesson 13 (Pages 121–125)

1. **15** 0.03 × 500 = 15
2. **$142.50** 0.15 × $950 = $142.50
3. **64.8** 0.90 × 72 = 64.8
4. **119** 0.85 × 140 = 119
5. **$275** 1.25 × $220 = $275
6. **276** 1.50 × 184 = 276
7. **60** 0.75 × 80 = 60
8. **$11** 0.055 × $200 = $11
9. **$33.75** 0.0675 × $500 = $33.75
10. **$1.25** 0.025 × $50 = $1.25
11. **$48** 0.04 × $1,200 = $48
12. **343** 0.70 × 490 = 343
13. **$11** 0.55 × $20 = $11
14. **$38.40** 0.40 × $96 = $38.40
15. **315** 2.10 × 150 = 315
16. **$0.75** 0.015 × $50 = $0.75
17. **40** $\frac{1}{3} \times \frac{120}{1} = 40$
18. **$240** $\frac{2}{3} \times \frac{360}{1} = \$240$
19. **(3) 117** Find 36% of 325. 325 × 0.36 = 117
20. **(2) 0.30($2,080)** You need to find 30% of $2,080. Change 30% to a decimal (0.30) and multiply by the base, $2,080.
21. **(5) 1,600** Find 125% of 1,280. 1,280 × 1.25 = 1,600
22. **(3) $16.25** Find 2.5% of $650. $650 × 0.025 = $16.25
23. **(5) $3,250** Find 5% of $65,000. $65,000 × 0.05 = $3,250
24. **(2) $18.60** Find 6% of $310. 310 × 0.06 = 18.6
25. **(4) $9.00** Find 20% of $45. $45 × 0.20 = $9.00
26. **(3) $864.00** Find 9% of $9,600. $9,600 × 0.09 = $864.00
27. **15%** 123 ÷ 820 = 0.15 = 15%
28. **20%** 125 ÷ 625 = 0.2 = 20%
29. **45%** $112.50 ÷ $250 = 0.45 = 45%
30. **2%** $3.50 ÷ $175.00 = 0.02 = 2%
31. **140%** Since the part ($252) is greater than the base ($180), you know the percent will be greater than 100%. $252 ÷ $180 = 1.4 = 140%
32. **4.5% or $4\frac{1}{2}\%$** 225 ÷ 5,000 = 0.045 = 4.5%
33. **75%** $82.50 ÷ $110.00 = 0.75 = 75%
34. **180%** Since the part ($72.00) is greater than the base ($40.00), you know the percent will be greater than 100%. $72 ÷ $40 = 1.8 or 180%
35. **15%** Subtract: $1,725 − $1,500 = $225. Divide by the original amount: $225 ÷ $1,500 = 0.15. Convert to a percent: 0.15 = 15%
36. **12%** Subtract: $582.40 − $520.00 = $62.40. Divide by the original amount: $62.40 ÷ $520 = 0.12. Convert to a percent: 0.12 = 12%.
37. 25% Subtract: 15 − 12 = 3. Divide by the original amount: 3 ÷ 12 = 0.25. Convert to a percent: 0.25 = 25%.
38. **50%** Subtract: 525 − 350 = 175. Divide by the original amount: 175 ÷ 350 = 0.5. Convert to a percent: 0.5 = 50%.
39. **75%** Subtract: 280 − 70 = 210. Divide by the original amount: 210 ÷ 280 = 0.75. Convert to a percent: 0.75 = 75%.
40. **5%** Subtract: $1,200 − $1,140 = $60. Divide by the original amount: $60 ÷ $1,200 = 0.05. Convert to a percent: 0.05 = 5%.
41. **20%** Subtract: $11.00 − $8.80 = $2.20. Divide by the original amount: $2.20 ÷ $11 = 0.2. Convert to a percent: 0.2 = 20%.

42. 18% Subtract: 6,000 − 4,920 = 1,080. Divide by the original amount: 1,080 ÷ 6,000 = 0.18. Convert to a percent: 0.18 = 18%.

43. (4) 20% To find the percent of decrease, subtract: $90 − $72 = $18. Divide by the original price: $18 ÷ $90 = 0.2 = 20%.

44. (5) (567 − 540) ÷ 540 You need to find the percent of increase. First find the difference between the two amounts; then divide by the original amount. Only option (5) shows these operations in the correct sequence.

45. (4) 85% This question asks "119 is what percent of 140?" Divide the part (119) by the whole (140): 119 ÷ 140 = 0.85 = 85%.

46. (1) 40% Divide the part (10) by the whole (25): 10 ÷ 25 = 0.4 = 40%.

47. (4) 4% You need to find the percent of increase. Subtract: $1,508 − $1,450 = $58. Divide by the original amount, and convert to a percent: $58 ÷ $1,450 = 0.04 = 4%.

48. (2) 60% You need to find the percent of increase. Subtract: $1,008 − $630 = $378. Divide by the original amount, and convert to a percent: $378 ÷ $630 = 0.6 = 60%.

GED Practice: Solving Word Problems (Page 127)

1. (3) 852 Find 10% of 8,520 by moving the decimal point one place to the left: 8,520. = 852

2. (4) $3.90 Find 10% by moving the decimal point one place to the left: $26.00 = $2.60 Divide $2.60 by 2 to find 5%: $2.60 ÷ 2 = $1.30. Add 10% and 5%: $2.60 + $1.30 = $3.90.

3. (5) $7.50 Find 10% by moving the decimal point one place to the left: $150.00 = $15 Divide by 2 to find 5%: $15.00 ÷ 2 = $7.50.

4. (4) $1.32 Find 10% by moving the decimal point one place to the left: $8.80 = $0.88. Divide $0.88 by 2 to find 5%: $0.88 ÷ 2 = $0.44. Add 10% and 5%: $0.88 + $0.44 = $1.32.

5. (1) $14.00 Find 10% by moving the decimal point one place to the left: $70.00 = $7 Multiply by 2 to find 20%: $7.00 × 2 = $14.00.

6. (3) $300 Find 10% by moving the decimal point one place to the left: $1,500. = $150 Multiply by 2 to find 20%: $150 × 2 = $300.

7. (2) $1.20 Find 10% by moving the decimal point one place to the left: $24.00 = $2.40 Divide $2.40 by 2 to find 5%: $2.40 ÷ 2 = $1.20.

8. (2) 540 Find 10% by moving the decimal point one place to the left: 1,800. = 180. Multiply 180 by 3 to find 30%: 180 × 3 = 540.

Lesson 14

All items relate to the skill of application.

GED Practice: Lesson 14 (Pages 129–133)

Items 1 through 18 require you to divide the part by the rate to find the whole.

1. $90 $54 ÷ 0.6 = $90

2. $5 $3.75 ÷ 0.75 = $5

3. 64 9.6 ÷ 0.15 = 64

4. 8,000 720 ÷ 0.09 = 8,000

5. $450 $157.50 ÷ 0.35 = $450

6. $32 $1.92 ÷ 0.06 = $32

7. 500 85 ÷ 0.17 = 500

8. $112 $26.88 ÷ 0.24 = $112

9. 550 495 ÷ 0.90 = 550

10. $60 $62.40 ÷ 1.04 = $60

11. 300 810 ÷ 2.7 = 300

12. 180 207 ÷ 1.15 = 180

13. $20 $0.76 ÷ 0.038 = $20

14. 160 36 ÷ 0.225 = 160

15. $12,940 $679.35 ÷ 0.0525 = $12,940

16. $50 $1.25 ÷ 0.025 = $50

17. 9,000 $6{,}000 \div \frac{2}{3} = \frac{\overset{3{,}000}{\cancel{6{,}000}}}{1} \times \frac{3}{\underset{1}{\cancel{2}}} = 9{,}000$

18. $719.73 $\$239.91 \div \frac{1}{3} = \frac{239.91}{1} \times \frac{3}{1} = \$719.73$

UNIT 1

19. **(3) $9,375** Divide the part ($7,500) by the rate (80%) to find the base: $7,500 ÷ 0.8 = $9,375

20. **(3) $23,700** Divide the part ($3,555) by the rate (15%) to find the base: $3,555 ÷ 0.15 = $23,700

21. **(4) $140** Divide the part ($98) by the rate (70%) to find the base: $98 ÷ 0.7 = $140

22. **(5) 22** Divide the part (178) by the rate (89%) to find the base: 178 ÷ 0.89 = 200. Then subtract: 200 − 178 = 22. You need to find the difference between the base and the part: total members − number voting for Elena = number that did not vote for Elena.

23. **(4) 621** Divide the part (207) by the rate ($33\frac{1}{3}$%). Since $33\frac{1}{3}$% equals $\frac{1}{3}$, divide by the fraction $\frac{1}{3}$ instead.
$207 \div \frac{1}{3} = \frac{207}{1} \times \frac{3}{1} = \frac{621}{1} = 621.$

24. **(2) 558** First find the base by dividing 42 by 7%: 42 ÷ 0.07 = 600. Then subtract to find the difference between the base (total appliances) and the part (defective appliances) to find the number not defective: 600 − 42 = 558.

25. **(4) $3.75** Divide the part ($0.75) by the rate (20%) to find the base: 0.75 ÷ 0.2 = $3.75

26. **(3) 300** Divide the part (9) by the rate (3%) to find the base: 9 ÷ 0.03 = 300

Items 27–46 use the following formula:

p × r × t = i

27. **$300** $1,250 × 0.12 × 2 = $300

28. **$228** $2,400 × 0.095 × 1 = $228

29. **$54** $900 × 0.12 × 0.5 = $54 (6 months = $\frac{6}{12} = \frac{1}{2}$ or 0.5)

30. **$50** $4,000 × 0.05 × 0.25 = $50 (3 months = $\frac{3}{12} = \frac{1}{4}$ or 0.25)

31. **$176** $3,200 × 0.055 × 1 = $176

32. **$260** $500 × 0.08 × 6.5 = $260

33. **$103.50** $2,300 × 0.09 × 0.5 = $103.50 (6 months = $\frac{6}{12} = \frac{1}{2}$ or 0.5)

34. **$15** $800 × 0.075 × 0.25 = $15 (3 months = $\frac{3}{12} = \frac{1}{4}$ or 0.25)

35. **$630** First find the interest: $600 × 0.1 × 0.5 = $30 (6 months = $\frac{6}{12} = \frac{1}{2}$ or 0.5); then add the principal and interest: $600 + $30 = $630.

36. **$5,724** First find the interest: $5,400 × 0.08 × 0.75 = $324 (9 months = $\frac{9}{12} = \frac{3}{4}$ or .75); then add the principal and interest: $5,400 + $324 = $5,724.

37. **$12,800** First find the interest: $8,000 × 0.12 × 5 = $4,800; then add the principal and interest: $8,000 + $4,800 = $12,800.

38. **$306** First find the interest: $300 × 0.06 × $\frac{1}{3}$ = $6 $\left(4 \text{ months} = \frac{4}{12} = \frac{1}{3}\right)$; then add the principal and interest: $300 + $6 = $306.

39. **(3) $3,075** Multiply: $10,000 × 0.1025 × 3 = $3,075

40. **(1) ($750)(0.03)(0.5) + $750** Multiply the principal, $750, by the rate, 0.03, by the time, which can be expressed as $\frac{1}{2}$ or 0.5. Then add the principal to the interest.

41. **(3) $84** Multiply: $600 × 0.14 × 1 = $84 12 months is 1 year.

42. **(4) $4,278** Multiply: $3,450 × 0.12 × 2 = $828 Add: $828 + $3,450 = $4,278

43. **(1) $112.50** Multiply: $1,000 × 0.075 × 1.5 = $112.50

44. **(2) $940.50** Multiply: $900 × 0.18 × $\frac{1}{4}$ (or 0.25) = $40.50. Add $40.50 + 900 = $940.50

45. **(3) $562.50** Multiply: $2,500 × 0.075 × 3 = $562.50

46. **(1) $101.89** $1,235 × 0.165 × 0.5 = $101.89 (6 months = $\frac{6}{12} = \frac{1}{2}$ or 0.5)

GED Practice: Solving Word Problems (Page 135)

1. **(4) $17.00** Find the amount of the discount: $20 × 0.15 = $3. Subtract to find the sale price: $20 − $3 = $17.

2. **(3) $9.73** Find the first raise and add to find his new wage:
$9 × 0.06 = $0.54 $9.00 + $0.54 = $9.54.
Using the new wage, find the second raise:
$9.54 × 0.02 = $0.1908 rounds to $0.19.
$9.54 + $0.19 = $9.73.

3. **(3) 200%** Find the difference in the two amounts: $4,500 − $1,500 = $3,000. Divide by the original amount:
$3,000 ÷ $1,500 = 2.00 = 200%.

4. **(2) 0.4(30) + 30** You need to find the increase by multiplying 0.4 by 30. Then add the increase to the original amount.

5. **(1) $24** Find the amount of the down payment: $160 × 0.1 = $16. Subtract the down payment from the cost of the chair: $160 − $16 = $144. Divide by the number of payments: $144 ÷ 6 = $24.

6. **(2) 336** Find the number to be laid off first: 1,400 × 0.05 = 70. Subtract from the original: 1,400 − 70 = 1,330. Find the number in the second layoff: 1,330 × 0.20 = 266. Add the number of employees in both layoffs:
70 + 266 = 336.

GED Mini-Test: Lessons 12–14 (Pages 136–137)

All items relate to the skill of application.

1. **(2) 16 ÷ 0.8** You need to solve for base. Divide the part, 16, by the rate 80% (0.8).

2. **(4) 5%** Find the difference between the two numbers to find how many people had free tickets: 42,000 − 39,900 = 2,100. Divide by the total attendance:
2,100 ÷ 42,000 = 0.05 = 5%.

3. **(4) 736** Multiply the rate (40%) by the base (1,840): 0.4 × 1,840 = 736.

4. **(4) 2,082** You know the result will be greater than the base, 1,735, because the rate is more than 100%. Multiply the base by the rate:
1,735 × 1.2 = 2,082.

5. **(5) $1.17** Multiply the base ($19.49) by the rate (6%) and round to the nearest whole cent.
$19.49 × 0.06 = $1.1694; $1.1694 rounds to $1.17.

6. **(2) $726.80** Find the interest using i = prt.
Multiply: $690 × 0.08 × $\frac{2}{3}$ = 36.8. Notice that 8 months or $\frac{8}{12}$ of a year reduces to $\frac{2}{3}$. Add the interest to the principal to find the total amount to be paid back:
$36.80 + $690.00 = $726.80.

7. **(3) $404.40** First find the amount of commission he has earned during the week by multiplying: $5,320 × 0.045 = $239.40. Add the fixed salary, $165, and the commission, $239.40: $165 + $239.40 = $404.40.

8. **(5) 0.065($15.60) + $15.60** The multiplication step in the expression comes first and allows you to find the sales tax by multiplying the rate by the base, the purchase. The sales tax is then added to the purchase, or the base.

9. **(2) $1,260** Use the formula: i = prt. Multiply: $3,500 × 0.18 × 2 = $1,260.

10. **(3) $42.10** Find the amount of tax: Multiply $40 by 5$\frac{1}{4}$%. $40 × 0.0525 = $2.10. Add the tax to the price: $40 + $2.10 = $42.10.

11. **(3) $500** Subtract 25% from 100% to get the rate that represents the sale price of the TV (100% − 25% = 75%). Divide the part, $375, by the rate, 0.75, to get the original price of the TV. $375 ÷ 0.75 = $500.

12. **(1) $3,832.50** Find how much her investment will earn using the formula: i = prt.
$3,500 × 0.095 × 1 = $332.50. Add the interest to the principal:
$332.50 + $3,500 = $3,832.50.

13. **(3) $236.40** Find the amount of the discount by multiplying the regular price by the discount rate: $394 × 0.4 = $157.60. Subtract the amount of the discount from the regular price to find the sale price:
$394 − $157.60 = $236.40.

14. **(4) $20** Solve for base. Divide the part ($12) by the rate (60%): $12 ÷ 0.6 = $20.

GED Cumulative Review: Arithmetic (Pages 138–144)

All items relate to the skill of application.

1. **(2) 6** Subtract the amount he has already saved from the amount he needs to save:
$600 − $420 = $180. Divide by 30:
$180 ÷ $30 = 6.

2. **(3) 101.6** Multiply the length of one side by 4, the number of sides: 25.4 × 4 = 101.6.

3. **(4) 12(26) − 59** First find the number of pens on hand at the beginning by multiplying 26 (the number of boxes) by 12 (the number of pens in each box). To find the number used subtract 59 from the result.

UNIT 1

4. **(3) $18\frac{3}{4}$** Add the hours worked each day.

$$\begin{aligned} 2\tfrac{3}{4} &= 2\tfrac{3}{4} \\ 4\tfrac{1}{2} &= 4\tfrac{2}{4} \\ 3\tfrac{1}{2} &= 3\tfrac{2}{4} \\ 5\tfrac{1}{4} &= 5\tfrac{1}{4} \\ +2\tfrac{3}{4} &= +2\tfrac{3}{4} \\ \hline & 16\tfrac{11}{4} = 16 + 2\tfrac{3}{4} = 18\tfrac{3}{4} \end{aligned}$$

5. **(2) $940.52** Find 7% (the rate) of $13,436 (the base) by multiplying: $13,436 × 0.07 = $940.52.

6. **(5) Not enough information is given.** You need to know the percent of the voters who voted for the winning candidate or the total number of registered voters who voted. There could be more than one way to find the answer if you had more information. The percent of voters who voted does not relate to the question.

7. **(5) $38.25** Use base × rate = part to find the amount of the discount: $45 × 0.15 = $6.75. Subtract the amount of the discount from $45: $45 − $6.75 = $38.25.

8. **(5) 0.025** Since the options are in decimals, solve the problem using decimals. Convert $\frac{1}{10}$ to a decimal: $\frac{1}{10} = 0.1$. Subtract from 0.125: 0.125 − 0.1 = 0.025.

9. **(4) 18** Multiply 3 by 3 to find the number of meetings over the 3-week period. Multiply the result, 9, by 2 to find the total number of hours: 9 × 2 = 18.

10. **(1) $1.10(30 − 25) + $18** The first step is to find out how many words more than 25 Zoe has in her ad: (30 − 25). Multiply the difference by $1.10. Then add $18, the cost for the first 25 words.

11. **(3) 31** Subtract $18, the charge for the first 25 words, from $25: $25 − $18 = $7. Divide $7 by $1.10 to find how many words over 25 Sally can afford. Round the answer to the nearest whole word. She can buy approximately 6 more words. Add to find the maximum number of words: 25 + 6 = 31.

12. **(2) $45.90** Subtract 25 from 55: 55 − 25 = 30. Multiply 30 by $1.10 and add the result to $18: 30 × $1.10 = $33; $33 + $18 = $51. Since Jesse's ad has over 50 words, find the 10% discount by multiplying $51 by 10%: $51 × 0.1 = $5.10. Subtract the discount: $51.00 − $5.10 = $45.90.

13. **(4) $7,100** Use the formula: $i = prt$. $5,000 × 0.14 × 3 = 2,100. Add the principal and interest to find the total owed: $5,000 + $2,100 = $7,100.

14. **(3) 38.5** Multiply 3.5 by 11: 11 × 3.5 = 38.5.

15. **(4) $497.47** There are two ways to solve the problem. #1—Subtract the advertising expenses from the money taken in, then divide by 4: $2,118.84 − $128.96 = $1,989.88 ÷ 4 = $497.47
#2—Divide both the figures by 4 and subtract the results: $2,118.84 ÷ 4 = $529.71
$128.96 ÷ 4 = $32.24
$529.71 − $32.24 = $497.47

16. **(2) 46%** Divide the part, 138, by the base, 300, to find the rate: 138 ÷ 300 = 0.46 = 46%.

17. **(5) 1,132.7** Subtract 8,867.3 from 10,000:

$$\begin{array}{r} 10{,}000.0 \\ -\ 8{,}867.3 \\ \hline 1{,}132.7 \end{array}$$

18. **(2) 9** Multiply $4\frac{1}{2}$ by 120; then divide by 60 to find the number of bolts. You may find it easier to change $4\frac{1}{2}$ to 4.5 before multiplying:
120 × 4.5 = 540
540 ÷ 60 = 9

19. **(3) 48** Divide: $6 \div \frac{1}{8} = \frac{6}{1} \times \frac{8}{1} = \frac{48}{1} = 48$.

20. **(2) 2,952** Multiply: 246 × 12 = 2,952.

21. **(3) $\frac{1}{3}$ cup** Multiply: $\frac{\overset{1}{\cancel{2}}}{3} \times \frac{1}{\underset{1}{\cancel{2}}} = \frac{1}{3}$.

22. **(2) $12\frac{1}{8}$** Add the lengths:

$$\begin{aligned} 3\tfrac{1}{2} &= 3\tfrac{4}{8} \\ 5\tfrac{1}{4} &= 5\tfrac{2}{8} \\ +3\tfrac{3}{8} &= 3\tfrac{3}{8} \\ \hline & 11\tfrac{9}{8} = 11 + 1\tfrac{1}{8} = 12\tfrac{1}{8}. \end{aligned}$$

23. **(5) $\frac{1}{8}$** Subtract $\frac{3}{4}$ from the whole can (1).
$1 - \frac{3}{4} = \frac{1}{4}$. Multiply: $\frac{1}{2} \times \frac{1}{4} = \frac{1}{8}$.

24. **(5) 32** Multiply the distance by 2 to find the number of miles to work and back:
3.2 × 2 = 6.4 Multiply by 5: 6.4 × 5 = 32.

25. **(3) $147** The pool was split 5 ways:
Ben + 4 others = 1 + 4 = 5. Divide the total by 5: $735 ÷ 5 = $147.

26. **(2) $\frac{1}{12}$** Subtract $\frac{3}{4}$ from 1 to find what fraction of his space Phil uses for his samples: $1 - \frac{3}{4} = \frac{1}{4}$. Multiply $\frac{1}{4}$ by $\frac{1}{3}$ to find what fraction of the company's space Phil uses for samples: $\frac{1}{4} \times \frac{1}{3} = \frac{1}{12}$.

27. **(5) $44.52** Find the price of a case by multiplying: $1.59 × 8 = $12.72. Multiply the price of a case by $3\frac{1}{2}$. The problem can be worked more quickly if you change $3\frac{1}{2}$ to 3.5. $12.72 × 3.5 = $44.52.

28. **(1) 2** Multiply to find the total number of slices: 16 × 4 = 64. Divide by the number of guests: 64 ÷ 32 = 2.

29. **(3) $79\frac{3}{4}$** Multiply $14\frac{1}{2}$ by $5\frac{1}{2}$. You may want to use decimals instead of fractions. Both methods are shown:
$14\frac{1}{2} \times 5\frac{1}{2} = \frac{29}{2} \times \frac{11}{2} = \frac{319}{4} = 79\frac{3}{4}$
$14.5 \times 5.5 = 79.75 = 79\frac{3}{4}$

30. **(2) $2,053.50** Use the formula $i = prt$.
$7,400 × 0.0925 × 3 = $2,053.50.

31. **(3) 465** Subtract Katalina's mileage from Vu's mileage: 701 − 236 = 465.

32. **(4) 28%** Find the total number of miles driven: 756 + 236 + 524 + 483 + 701 = 2,700. The result is the base. The miles driven by Hector is the part. Solve for rate by dividing the part by the base:
756 ÷ 2,700 = 0.28 = 28%.

33. **(1) $28.50** Find the total regular price (the base). Multiply: 250 × $0.12 = $30. Find the 5% discount. Multiply: $30 × 0.05 = $1.50. Subtract the discount from the base amount: $30 − $1.50 = $28.50. You could also use "mental math" to find the discount: 10% of $30 is $3; 5% is $\frac{1}{2}$ of 10%: $3.00 ÷ 2 = $1.50.

34. **(2) 45,000** Multiply: 7,500 × 6 = 45,000.

35. **(5) $\frac{4}{5}$** Subtract: 35 − 7 = 28. Write a fraction and reduce: $\frac{28}{35} \div \frac{7}{7} = \frac{4}{5}$.

36. **(4) $124.46**
Subtract: $149.45 − $24.99 = $124.46.

37. **(1) $2,625** Use the formula: i = prt.
Find the interest: $2,500 × 0.025 × 2 = $125
Add the interest and the principal:
$125 + $2,500 = $2,625.

38. **(5) 8%** Find the difference between the two amounts: $378 − $350 = $28; then divide by the original amount: $28 ÷ $350 = 0.08 = 8%.

39. **(5) Not enough information is given.** You need to know the base or whole amount (Kyle's earnings for the year) to find the bonus (the part). You only know one of the three elements (the rate).

40. **(5) Not enough information is given.** There is no way to find the actual number of stereos Minh sold on a specific day from the total and the length of time. You can calculate only the average number he sold each day.

41. **(1) The charge will be the same.** In Lot A at 75¢ for 4 hours, the cost would be $3.00, but there is a $5.00 minimum charge. In Lot B, $1.25 × 4 is $5.00, so the cost is the same at both lots.

42. **(4) 2($33) + 6($5)** He has to pay the insurance amount, $33, for both months; multiply $33 by 2. Then he has to pay $5 for each of the 6 visits; multiply $5 by 6. Add the two results.

43. **(2) $249 ÷ (100% − 25%)** You have the part, the discount price ($249) and the rate of the discount (25%). You can find the rate that represents the part by subtracting 25% from 100%. The difference (75%) represents the rate. Divide the part ($249) by the rate (75%) to find the base. You would multiply the discount rate (25%) by the original price to find the discount price.

44. **(2) $729 ÷ 0.04** You have the part and the rate. Divide the part by the rate to find the base.

45. **(3) $142.50** Add $50 + $140 = $190. Multiply by $\frac{3}{4}$ (or 0.75): $190 × 0.75 = $142.50.

46. **(1) $3\frac{1}{2}$** Multiply: $1\frac{3}{4} \times 2 = \frac{7}{\not{4}_2} \times \frac{\not{2}^1}{1} = \frac{7}{2} = 3\frac{1}{2}$.

47. **(4) 20** There is $\frac{2}{3}$ of a cup of rice in a serving. Multiply 30 by $\frac{2}{3}$: $30 \times \frac{2}{3} = 20$.

48. **(2) $1\frac{1}{2}\left(\frac{1}{2}\right)$** Multiply the amount of uncooked rice $\left(\frac{1}{2}\right)$ by $1\frac{1}{2}$.

49. **(1) $134.39** Add the amounts:
$69.50 + $39.99 + $24.90 = $134.39.

UNIT 1

50. **(3) 16** Add the amounts and round to the nearest whole pound: 3.32 + 2.45 + 4.67 + 2.25 + 3.08 = 15.77, rounds to 16.

51. **(4) $19\frac{1}{2}$** Multiply $4\frac{7}{8}$ by 4: $4\frac{7}{8} \times 4 = \frac{39}{\cancel{8}_2} \times \frac{\cancel{4}^1}{1} = \frac{39}{2} = 19\frac{1}{2}$ yd.

Lesson 15

All items relate to the skill of application.

GED Practice: Lesson 15 (Pages 148–151)

1. $\frac{4}{3}$ $\frac{16}{12} \div \frac{4}{4} = \frac{4}{3}$
2. $\frac{1}{3}$ $\frac{15}{45} \div \frac{15}{15} = \frac{1}{3}$
3. $\frac{1}{6}$ $\frac{6}{36} \div \frac{6}{6} = \frac{1}{6}$
4. $\frac{2}{5}$ $\frac{14}{35} \div \frac{7}{7} = \frac{2}{5}$
5. $\frac{1}{4}$ $\frac{9}{36} \div \frac{9}{9} = \frac{1}{4}$
6. $\frac{5}{2}$ $\frac{25}{10} \div \frac{5}{5} = \frac{5}{2}$
7. $\frac{2}{3}$ $\frac{24}{36} \div \frac{12}{12} = \frac{2}{3}$
8. $\frac{4}{3}$ $\frac{12}{9} \div \frac{3}{3} = \frac{4}{3}$
9. $\frac{17}{20}$ $\frac{85}{100} \div \frac{5}{5} = \frac{17}{20}$
10. $\frac{1}{3}$ $\frac{18}{54} \div \frac{18}{18} = \frac{1}{3}$
11. $\frac{3}{1}$ $\frac{18}{6} \div \frac{6}{6} = \frac{3}{1}$
12. $\frac{5}{1}$ $\frac{35}{7} \div \frac{7}{7} = \frac{5}{1}$
13. $\frac{4}{5}$ $\frac{80}{100} \div \frac{20}{20} = \frac{4}{5}$
14. $\frac{7}{25}$ $\frac{14}{50} \div \frac{2}{2} = \frac{7}{25}$
15. $\frac{13}{25}$ $\frac{26}{50} \div \frac{2}{2} = \frac{13}{25}$
16. **80 miles per hour** Divide 400 by 5.
17. **17 cents per pound** Divide 85 by 5.
18. **$18.75 per hour** Divide 225 by 12.
19. **4 calories per gram** Divide 112 by 28.
20. **64 people per team** Divide 512 by 8.
21. **225 oranges per bag** Divide 5,400 by 24.
22. **64.5 feet per second** Divide 1,032 by 16.
23. **$0.15 oz.** Divide $2.40 by 16.
24. **10** $2 \times 15 = 30$; $30 \div 3 = 10$
25. **6** $14 \times 12 = 168$; $168 \div 28 = 6$
26. **6** $20 \times 3 = 60$; $60 \div 10 = 6$
27. **7** $18 \times 3.5 = 63$; $63 \div 9 = 7$
28. **14** $4.2 \times 10 = 42$; $42 \div 3 = 14$
29. **8** $5 \times 24 = 120$; $120 \div 15 = 8$
30. **30** $15 \times 24 = 360$; $360 \div 12 = 30$
31. **3** $6 \times 7 = 42$; $42 \div 14 = 3$
32. **3** $11.5 \times 6 = 69$; $69 \div 23 = 3$
33. **32** $7 \times 16 = 112$; $112 \div 3.5 = 32$
34. **7** $4.9 \times 10 = 49$; $49 \div 7 = 7$
35. **6** $15 \times 3.2 = 48$; $48 \div 8 = 6$
36. **10** $3 \times 6 = 18$; $18 \div 1.8 = 10$
37. **1** $5 \times 1.2 = 6$; $6 \div 6 = 1$
38. **20** $6 \times 7 = 42$; $42 \div 2.1 = 20$
39. **Range: 65 to 100; Mean: 87; Median: 85** The range is the lowest number to the highest: 65 to 100. To find the mean, add the scores and divide by 7: 609 ÷ 7 = 87. To find the median, arrange the scores in order and find the middle score: 65, 80, 85, <u>85</u>, 94, 100, 100.
40. **Range: $199.48 to $245.82; Mean: $221.66; Median: $219.82** The range is the lowest number to the highest: $199.48 to $245.82. To find the mean, add the amounts and divide by 5 and find the middle amount: $1,108.30 ÷ 5 = $221.66. To find the median, arrange the amounts in order: $199.48, $215.35, <u>$219.82</u>, $227.83, $245.82.
41. **Range: 60.8°F to 69.0°F; Mean: 64.3°F; Median: 64.8°F** The range is the lowest number to the highest: 60.8°F to 69.0°F. To find mean, add the amounts and divide by 7: 450.1 ÷ 7 = 64.3. To find the median, arrange the numbers in order and find the middle temperature: 60.8°F, 61.1°F, 61.5°F, <u>64.8°F</u>, 65.6°F, 67.3°F, 69.0°F
42. **Range: 0.04 to 0.06; Mean: 0.051; Median: 0.052** The range is the lowest number to the highest: 0.04 to 0.06. To find the mean, add the amounts and divide by 5: 0.255 ÷ 5 = 0.051. To find the median, arrange the numbers in order and find the middle weight: 0.04, 0.048, <u>0.052</u>, 0.055, 0.06.
43. **Range: 128 to 215; Mean: 163; Median: 154.5** The range is the lowest number to the highest: 128 to 215. To find the mean, add the amounts and divide by 4: 652 ÷ 4 = 163. To find the median, arrange the numbers in order and find the middle figure: 128, 135, 174, 215. Average the two middle numbers: 135 + 174 = 309 ÷ 2 = 154.5.

44. **Range: 12 to 150; Mean: 65; Median: 35** The range is the lowest number to the highest: 12 to 150. To find the mean, add the amounts and divide by 7: $455 \div 7 = 65$. To find the median, arrange the numbers in order and find the middle number: 12, 24, 27, 35, 87, 120, 150.

45. **Range: 35 to 67; Mean: 52; Median: 52** The range is the lowest number to the highest: 35 to 67. To find the mean, add the amounts and divide by 5: $260 \div 5 = 52$. To find the median, arrange the numbers in order and find the middle time: 35, 41, 52, 65, 67.

46. **(2) $\frac{3}{5}$** Write the ratio and reduce: $\frac{\text{Friday}}{\text{Both Days}} = \frac{9}{9+6} = \frac{9}{15} = \frac{3}{5}$

47. **(4) 55** $\frac{12}{5} = \frac{132}{?}$
Cross multiply: $5 \times 132 = 660$.
Divide by the remaining number: $660 \div 12 = 55$.

48. **(2) 9** $\frac{8}{2} = \frac{36}{?}$
Cross multiply: $2 \times 36 = 72$.
Divide by the remaining number: $72 \div 8 = 9$.

49. **(3) 1,050** $\frac{315}{3} = \frac{?}{10}$
Cross multiply: $315 \times 10 = 3{,}150$.
Divide by the remaining number: $3{,}150 \div 3 = 1{,}050$.

50. **(3) 0.25** $\frac{\text{favorable outcomes}}{\text{total possible outcomes}} = \frac{1}{4} = 0.25$

51. **(5) $\frac{3}{4}$** $\frac{\text{favorable outcomes}}{\text{total possible outcomes}} = \frac{3}{4}$

52. **(2) $\frac{1}{5}$**
$5 + 8 + 10 + 2 = 25$ possible outcomes.
5 of the 25 are Model A.
Reduce the fraction: $\frac{5}{25} \div \frac{5}{5} = \frac{1}{5}$.

53. **(4) Model D** $0.08 = \frac{8}{100}$
Set up a proportion: $\frac{8}{100} = \frac{?}{25}$.
Cross multiply: $8 \times 25 = 200$.
Divide: $200 \div 100 = 2$.
There are 2 Model D mixers, so Model D has a probability of 0.08 of being the one selected.

54. **(3) Model C** Since the total possible outcomes is the same for each model, compare the numbers of the different models. Model C has the greatest number of chances of being chosen: 10 out of 25.

55. **(5) 0.4** Model C has a 10 out of 25 probability of being chosen: $\frac{10}{25} \div \frac{5}{5} = \frac{2}{5} = 0.4$.

GED Practice: Solving Word Problems (Page 153)

1. **(4) 50%** Find the difference ($\$1{,}800 - \$1{,}200 = \$600$), and set up the proportion: $\frac{\$600}{\$1{,}200} = \frac{?}{100}$
Cross multiply: $\$600 \times 100 = \$60{,}000$.
Divide by the remaining number: $\$60{,}000 \div \$1{,}200 = 50$.

2. **(1) \$7** $\frac{?}{\$35} = \frac{20}{100}$
Cross multiply: $\$35 \times 20 = \$700$. Divide by the remaining number: $\$700 \div 100 = \$7$.

3. **(4) \$562.50** $\frac{?}{\$625} = \frac{90}{100}$
Cross multiply: $\$625 \times 90 = \$56{,}250$.
Divide by the remaining number: $\$562.50 \div \$100 = \$562.50$.

4. **(3) \$33.60** $\frac{?}{\$1{,}344} = \frac{2.5}{100}$
Cross multiply: $\$1{,}344 \times 2.5 = \$3{,}360$.
Divide by the remaining number: $\$3{,}360 \div 100 = \$33.60$.

5. **(2) 70** $\frac{56}{?} = \frac{80}{100}$
Cross multiply: $56 \times 100 = 5{,}600$.
Divide by the remaining number: $5{,}600 \div 80 = 70$.

6. **(1) 40%** $\frac{18}{45} = \frac{?}{100}$
Cross multiply: $18 \times 100 = 1{,}800$.
Divide by the remaining number: $1{,}800 \div 45 = 40$.

7. **(3) 25%** Find the difference: $\$112.80 - \$84.60 = \$28.20$, set up a proportion and solve: $\frac{\$28.20}{\$112.80} = \frac{?}{100}$
Cross multiply: $\$28.20 \times 100 = \$2{,}820$.
Divide by the remaining number: $\$2{,}820 \div \$112.80 = 25$.

8. **(5) 375** $\frac{300}{?} = \frac{80}{100}$
Cross multiply: $300 \times 100 = 30{,}000$.
Divide by the remaining number: $30{,}000 \div 80 = 375$.

GED Mini-Test: Lesson 15 (Pages 154–156)

All items relate to the skill of application.

1. **(3) 7.5** Of the 10 quarts in the paint mixture ($2 + 3 + 5 = 10$), 3 were green.
Set up a proportion: $\frac{3}{10} = \frac{?}{25}$
Cross multiply: $3 \times 25 = 75$
Divide: $75 \div 100 = 7.5$

2. **(4) 8** The baseball team has won 16 out of 20 (16 + 4) games. Set up a proportion: $\frac{16}{20} = \frac{?}{30}$
Cross multiply: 16 × 30 = 480
Divide: 480 ÷ 20 = 24 They need to win 24 to keep the same winning ratio. Find the difference between 24 and 16 and see how many more they have to win. 24 − 16 = 8

3. **(2) Only the mean is $142.** To find out which statement is true, find the mean and the median of the amounts. Mean—Add the amounts:
$120 + $120 + $170 + $140 + $160 = $710
Divide by 5, the number of amounts:
$710 ÷ 5 = $142
Median—Arrange the amounts in order:
$120, $120, $140, $160, $170
The median or middle amount is $140. Only the mean is $142.

4. **(1) $\frac{1}{3}$** There are 12 cars and 4 of them are red. Set up a ratio and reduce. $\frac{4}{12} = \frac{1}{3}$
There is a 1 in 3 chance of choosing a red car.

5. **(2) $\frac{1}{3}$** Now there are 15 cars on the lot. Five of the cars are yellow (2 before + 3 added = 5 cars). Set up a ratio and reduce: $\frac{5}{15} = \frac{1}{3}$

6. **(2) $\frac{5}{8}$** Find the total number of votes:
150 + 56 + 34 = 240.

Set up a ratio and reduce: $\frac{150}{240} = \frac{5}{8}$

7. **(3) 1,800**

Set up a proportion and solve: $\frac{2}{5} = \frac{?}{4{,}500}$

Cross multiply: 2 × 4,500 = 9,000
Divide: 9,000 ÷ 5 = 1,800

8. **(3) between 42 minutes and 46 minutes**
Find the total of the times:
45 + 35 + 30 + 42 + 3(55) =
45 + 35 + 30 + 42 + 165 = 317
Divide by 7: 317 ÷ 7 = 45.29

9. **(2) $\frac{1}{4}$** $\frac{\text{Team A wins}}{\text{Team C wins}} = \frac{3}{12}$ reduce $= \frac{1}{4}$

10. **(5) Team C with $\frac{12}{14}$** Find the ratio of wins to games played for each team and compare the fractions.
Team A $= \frac{3}{14}$ Team B $= \frac{5}{14}$ Team C $= \frac{12}{14}$
This is the greatest of the three fractions.

11. **(4) 15** Set up a proportion and solve:
$\frac{\text{Games Won}}{\text{Games Played}}$ $\frac{5}{14} = \frac{?}{42}$
Cross multiply: 5 × 42 = 210
Divide: 210 ÷ 14 = 15

12. **(2) 1,600**
Set up a proportion and solve: $\frac{4}{10} = \frac{640}{?}$
Cross multiply: 10 × 640 = 6,400
Divide: 6,400 ÷ 4 = 1,600

13. **(4) 49.5** Multiply 3,300 by 5 to find how many characters will be on the 5 pages:
3,300 × 5 = 16,500.
Set up a proportion and solve: $\frac{1{,}000}{3} = \frac{16{,}500}{?}$
Cross multiply: 3 × 16,500 = 49,500
Divide: 49,500 ÷ 1,000 = 49.5

14. **(4) $\frac{1}{2}$** Of the 6 numbers, 3 numbers are even numbers. Set up a ratio and reduce: $\frac{3}{6} = \frac{1}{2}$.
There is a 1 in 2 chance of rolling an even number.

15. **(2) $\frac{2}{3}$** Of the 6 numbers, 4 numbers are less than 5. Set up a ratio and reduce: $\frac{4}{6} = \frac{2}{3}$. There is a 2 in 3 chance of rolling a number less than 5.

16. **(3) $\frac{1}{3}$** The numbers 4 or 5 make up 2 of the 6 numbers. Set up a ratio and reduce: $\frac{2}{6} = \frac{1}{3}$.
There is a 1 in 3 chance of rolling a 4 or a 5.

17. **(2) 385 miles** Set up a proportion and solve:
$\frac{4}{220} = \frac{7}{?}$
Cross multiply: 7 × 220 = 1,540
Divide: 1,540 ÷ 4 = 385 miles

18. **(3) $23,600** Find the total of the amounts:
$21,200 + $24,500 + $25,100 = $70,800
Divide by 3: $70,800 ÷ 3 = $23,600

19. **(4) 90 miles** Set up a proportion and solve:
$\frac{\frac{1}{2}}{15} = \frac{3}{?}$
Cross multiply: 3 × 15 = 45
Divide: $45 \div \frac{1}{2} = 45 \times 2 = 90$

20. **(4) $50** Set up a proportion and solve:
$\frac{\$35}{7} = \frac{?}{10}$
Cross multiply: $35 × 10 = $350
Divide: $350 ÷ 7 = $50

21. **(4) $140** Find the amount he saves out of every $20: $20 − $15 = $5.
Set up a proportion and solve: $\frac{\$5}{\$20} = \frac{?}{\$560}$
Cross multiply: $5 × $560 = $2,800
Divide: $2,800 ÷ $20 = $140

Lesson 16

All items relate to the skill of application.

GED Practice: Lesson 16 (Pages 162–163)

1. **10 feet 6 inches** $3\frac{1}{2}$ yds. $= \frac{7}{2}$ yds.
   $\frac{7}{2}\cancel{\text{yds.}} \times \frac{36\text{ in.}}{1\cancel{\text{yd.}}} = 126\text{ in.} \times \frac{1\text{ ft.}}{12\cancel{\text{in.}}} = 10\text{ ft. }6\text{ in.}$

2. **240 seconds** $4\cancel{\text{min.}} \times \frac{60\text{ sec.}}{\cancel{\text{min.}}} = 240\text{ sec.}$

3. **9 pints** $18\cancel{\text{cups}} \times \frac{1\text{ pint}}{2\cancel{\text{cups}}} = 9\text{ pints}$

4. **$3\frac{3}{4}$ tons** $7{,}500\cancel{\text{lb.}} \times \frac{1\text{ ton}}{2{,}000\cancel{\text{lb.}}} = 3\frac{3}{4}\text{ tons}$

5. **140 minutes** $2\frac{1}{3}$ hr. $= \frac{7}{3}$ hr.
   $\frac{7}{\cancelto{1}{3}}\cancel{\text{hr.}} \times \frac{\cancelto{20}{60}\text{ min.}}{1\cancel{\text{hr.}}} = 140\text{ min.}$

6. **9 cups** $\cancelto{9}{72}\cancel{\text{fl. oz.}} \times \frac{1\text{ cup}}{\cancelto{1}{8}\cancel{\text{fl. oz.}}} = 9\text{ cups}$

7. **23 quarts** $5\frac{3}{4}$ gal. $= \frac{23}{4}$ gal.
   $\frac{23}{\cancelto{1}{4}}\cancel{\text{gal.}} \times \frac{\cancelto{1}{4}\text{ qt.}}{1\cancel{\text{gal.}}} = 23\text{ qt.}$

8. **$2\frac{1}{4}$ yards** $6\text{ ft. }9\text{ in.} = 6\cancel{\text{ft.}} \times \frac{12\text{ in.}}{\cancel{\text{ft.}}} + 9\text{ in.}$
   $= 72\text{ in.} + 9\text{ in.} = 81\text{ in.}$
   $\cancelto{9}{81}\cancel{\text{in.}} \times \frac{1\text{ yd.}}{\cancelto{4}{36}\cancel{\text{in.}}} = \frac{9}{4}\text{ yd.} = 2\frac{1}{4}\text{ yd.}$

9. **3 feet** $3\cancel{\text{ft.}} \times \frac{12\text{ in.}}{\cancel{\text{ft.}}} = 36\text{ in.} > 30\text{ in.}$

10. **13 pints** $1\frac{1}{2}$ gal. $= \frac{3}{2}$ gal.
    $\frac{3}{\cancel{2}}\cancel{\text{gal.}} \times \frac{\cancelto{2}{4}\cancel{\text{qt.}}}{1\cancel{\text{gal.}}} \times \frac{2\text{ pt.}}{\cancel{\text{qt.}}} = 12\text{ pt.} < 13\text{ pt.}$

11. **$3\frac{1}{4}$ hours** $3\frac{1}{4}$ hr. $= \frac{13}{4}$ hr.
    $\frac{13}{\cancelto{1}{4}}\cancel{\text{hr.}} \times \frac{\cancelto{15}{60}\text{ min.}}{1\cancel{\text{hr.}}} = 195\text{ min.} < 200\text{ min.}$

12. **$4\frac{1}{4}$ pounds** $4\frac{1}{4}$ lb. $= \frac{17}{4}$ lb.
    $\frac{17}{\cancelto{1}{4}}\cancel{\text{lb.}} \times \frac{\cancelto{4}{16}\text{ oz.}}{\cancel{\text{lb.}}} = 68\text{ oz.} < 75\text{ oz.}$

13. **6 gallons 1 quart**

3 gal. 3 qt.	5 gal.
2 gal. 2 qt.	1 gal. 1 qt.
5 gal. 5 qt.	6 gal. 1 qt.

    5 qt. = 1 gal. 1 qt.

14. **16 ft. 10 in.**

	8 ft.	8 in.
	3 ft.	5 in.
+	4 ft.	9 in.
	15 ft.	~~22 in.~~

    $22\cancel{\text{in.}} \times \frac{1\text{ ft.}}{12\cancel{\text{in.}}} = 1\text{ ft. }10\text{ in.}$

	15 ft.	
+	1 ft.	10 in.
	16 ft.	10 in.

15. **9 min. 15 sec.** $\frac{1}{\cancelto{1}{2}}\cancel{\text{hr.}} \times \frac{\cancelto{30}{60}\text{ min.}}{1\cancel{\text{hr.}}} = 30\text{ min.}$

$\frac{1}{2}$ hr. =	$\cancelto{29}{30}$ min.	$\cancelto{5\,10}{60}$ $\cancelto{0}{00}$ sec.
	−20 min.	45 sec.
	9 min.	15 sec.

16. **7 in.**

$\cancelto{1}{2}$ ft.	$\cancelto{14}{2}$ in.
−1 ft.	7 in.
	7 in.

17. **16 lb. 4 oz.**

3 lb.	4 oz.
×5	
15 lb.	~~20 oz.~~

    $20\cancel{\text{oz.}} \times \frac{1\text{ lb.}}{16\cancel{\text{oz.}}} = 1\text{ lb. }4\text{ oz.}$

	15 lb.	
+	1 lb.	4 oz.
	16 lb.	4 oz.

18. **18 yd. 2 ft.**

2 yd.	1 ft.
×8	
16 yd.	8 ft.

    $8\cancel{\text{ft.}} \times \frac{1\text{ yd.}}{3\cancel{\text{ft.}}} = 2\text{ yd. }2\text{ ft.}$

	16 yd.	
+	2 yd.	2 ft.
	18 yd.	2 ft.

19. **8 inches** Convert 1 yd. 4 in. to inches.
    $1\cancel{\text{yd.}} \times \frac{36\text{ in.}}{1\cancel{\text{yd.}}} + 4\text{ in.} = 40\text{ in.} \div 5 = 8\text{ in.}$

20. **3 quarts** Convert 3 gal. to quarts.
    $3\cancel{\text{gal.}} \times \frac{4\text{ qt.}}{1\cancel{\text{gal.}}} = 12\text{ qt.} \div 4 = 3\text{ qt.}$

21. **(3) 10**
    $2\cancel{\text{gal.}} \times \frac{4\cancel{\text{qt.}}}{1\cancel{\text{gal.}}} \times \frac{2\text{ pints}}{1\cancel{\text{qt.}}} =$

16 pints	made
− 6 pints	to keep
10 pints	to give away

22. **(1) 12**
    $3\frac{1}{4}\text{ yd.} = 3\text{ yd.} + \frac{1}{\cancelto{1}{4}}\cancel{\text{yd.}} \times \frac{\cancelto{9}{36}\text{ in.}}{1\cancel{\text{yd.}}} =$

$\cancelto{2}{3}$ yd.	$\cancelto{3}{0}$ ft.	9 in.
−2 yd.	2 ft.	9 in.
	1 ft.	

    $1\cancel{\text{ft.}} \times \frac{12\text{ in.}}{1\cancel{\text{ft.}}} = 12\text{ in.}$

23. **661 centimeters**

75 cm.
126 cm.
460 cm.
661 cm.

24. **(3) $41\frac{1}{4}$** There are two methods:
    #1: $2\text{ ft. }9\text{ in.} = 2\text{ ft.} + \cancelto{3}{9}\cancel{\text{in.}} \times \frac{1\text{ ft.}}{\cancelto{4}{12}\cancel{\text{in.}}} = 2\frac{3}{4}\text{ ft.} = \frac{11}{4}\text{ ft.}$
    $\frac{11}{4}\text{ ft.} \times 15 = \frac{165}{4}\text{ ft.} = 41\frac{1}{4}\text{ ft.}$
    #2

2 ft.	9 in.
×15	
30 ft.	~~135~~ in.

    $\cancelto{45}{135}\cancel{\text{in.}} \times \frac{1\text{ft.}}{\cancelto{4}{12}\cancel{\text{in.}}} = 11\frac{1}{4}\text{ ft.}$

30 ft.
$+11\frac{1}{4}$ ft.
$41\frac{1}{4}$ ft.

UNIT 1

25. (3) $8\frac{1}{2}$

$$\begin{array}{ll} 3 \text{ tn.} & 750 \text{ lb.} \\ 1 \text{ tn.} & 150 \text{ lb.} \\ 4 \text{ tn.} & 100 \text{ lb.} \\ \hline 8 \text{ tn.} & 1{,}000 \text{ lb.} \end{array}$$

$\overset{1}{\cancel{1{,}000 \text{ lb.}}} \times \frac{1 \text{ tn.}}{\underset{2}{\cancel{2{,}000 \text{ lb.}}}} = \frac{1}{2} \text{ tn.}$

$$\begin{array}{r} 8 \text{ tn.} \\ +\frac{1}{2} \text{ tn.} \\ \hline 8\frac{1}{2} \text{ tn.} \end{array}$$

26. (3) 9 $3\frac{3}{4}$ hr. $= \frac{15}{4}$ hr.

$\frac{15}{\underset{1}{\cancel{4}}} \cancel{\text{hr.}} \times \frac{\overset{15}{\cancel{60}} \text{ min.}}{\cancel{\text{hr.}}} = 225$ min.

$\frac{1 \text{ cabinet}}{25 \text{ min.}} = \frac{x}{225 \text{ min.}}$

$\frac{1 \text{ cabinet} \times \overset{9}{\cancel{225 \text{ min.}}}}{\cancel{25 \text{ min.}}} = 9$ cabinets

27. (2) 10 There are two ways to solve this problem.

#1 Find the number of servings in 24 oz (1 pkg) and multiply by 5 packages.

1 serving $= 1\frac{1}{2}$ cups $= \frac{3 \cancel{\text{ cups}}}{\cancel{2} \text{ servings}} \times \frac{\overset{4}{\cancel{8}} \text{ oz.}}{\cancel{1 \text{ cup}}} = \frac{12 \text{ oz.}}{\text{serving}}$

$\frac{24 \text{ oz.}}{1 \text{ pkg}} \div \frac{12 \text{ oz.}}{\text{serving}} = \frac{\overset{2}{\cancel{24 \text{ oz.}}}}{1 \text{ pkg.}} \times \frac{1 \text{ serving}}{\underset{1}{\cancel{12 \text{ oz.}}}} = \frac{2 \text{ servings}}{1 \cancel{\text{ pkg.}}} \times 5 \cancel{\text{ pkg.}} = 10$ servings

#2 Find the total number of ounces in 5 pkg and divide by the number of ounces in 1 serving.

$\frac{24 \text{ oz.}}{1 \cancel{\text{ pkg.}}} \times 5 \cancel{\text{ pkg.}} = 120 \text{ oz.} \div \frac{12 \text{ oz.}}{\text{serving}} =$

$\overset{10}{\cancel{120}} \cancel{\text{ oz.}} \times \frac{1 \text{ serving}}{\underset{1}{\cancel{12 \text{ oz.}}}} = 10$ servings

28. (5) 250 Set up a proportion:

$\frac{60}{15} = \frac{1{,}000}{x}$

Cross multiply: $15 \times 1{,}000 = 15{,}000$
Divide: $15{,}000 \div 60 = 250$

GED Practice: Solving Word Problems (Page 165)

1. (2) 38 ft. 8 in. Estimate by rounding: 4 ft. 10 in. rounds to 5 ft.; $5 \times 8 = 40$ Option (2) is closest.

Solve:

$$\begin{array}{l} \quad 4 \text{ ft. } 10 \text{ in.} \\ \times 8 \\ \hline 32 \text{ ft. } 80 \text{ in.} = 32 \text{ ft.} + 6 \text{ ft. } 8 \text{ in.} = \\ 38 \text{ ft. } 8 \text{ in.} \end{array}$$

2. (3) 4 ft. 6 in. Estimate: 5 ft. 8 in. rounds to 6 ft. 1 ft. 2 in. rounds to 1 ft. Subtract: $6 - 1 = 5$. Options (3) and (4) are equally close to the estimate. Solve the problem:

$$\begin{array}{r} 5 \text{ ft. } 8 \text{ in.} \\ -1 \text{ ft. } 2 \text{ in.} \\ \hline 4 \text{ ft. } 6 \text{ in.} \end{array}$$

3. (3) 5 gallons 1 quart Estimate: 1 gallon 3 quarts rounds to 2 gallons. Multiply: $2 \times 3 = 6$. Option (4) is closest to the estimate. Solve the problem:

$$\begin{array}{l} \quad 1 \text{ gal. } 3 \text{ qts.} \\ \times 3 \\ \hline 3 \text{ gal. } 9 \text{ qts.} = 3 \text{ gal.} + 2 \text{ gal. } 1 \text{ qt.} = \\ 5 \text{ gal. } 1 \text{ qt.} \end{array}$$

4. (4) 19.5 Estimate by rounding: 5.8 km rounds to 6 km, 6.5 km rounds to 7 km, 7.2 km rounds to 7 km. Add $6 + 7 + 7 = 20$.

Solve:

$$\begin{array}{r} 5.8 \\ 6.5 \\ +\ 7.2 \\ \hline 19.5 \end{array}$$

5. (2) 5 Estimate: 4 yards 6 inches rounds to 4 yards ($\frac{1}{2}$ yd. is 18 inches). $21 \div 4$ is 5 r1. To solve change 6 inches to a fraction. 6 inches $= \frac{6}{36}$ or $\frac{1}{6}$ of a yard. Divide 21 by $4\frac{1}{6}$.

6. (5) 13 tons 1,600 pounds Estimate: 2 tons 1,500 lb. rounds to 3 tons, 4 tons 300 lb. rounds to 4 tons, 6 tons 1,800 lb. rounds to 7 tons. Add: $3 + 4 + 7 = 14$ tons. The best choice is (5). Now solve the problem:

$$\begin{array}{ll} \ \ 2 \text{ tn.} & 1{,}500 \text{ lb.} \\ \ \ 4 \text{ tn.} & \ \ \ 300 \text{ lb.} \\ +6 \text{ tn.} & 1{,}800 \text{ lb.} \\ \hline 12 \text{ tn.} & 3{,}600 \text{ lb.} = 12 \text{ tn.} + 1 \text{ tn. } 1{,}600 \text{ lb.} \\ & \qquad\qquad = 13 \text{ tn. } 1{,}600 \text{ lb.} \end{array}$$

7. (3) 30 Estimate: 4.2 rounds to 4. $126 \div 4 = 31\frac{1}{2}$. The best answer is (3).

Solve:

$$\begin{array}{r} 3\,0 \\ 4.2\overline{)126.0} \\ 126 \\ \hline \end{array}$$

8. (5) $6\frac{3}{4}$ Estimate: $2\frac{1}{4}$ is about 2. Multiply: $3 \times 2 = 6$.

Solve: $3 \times 2\frac{1}{4} = \frac{3}{1} \times \frac{9}{4} = \frac{27}{4} = 6\frac{3}{4}$

Lesson 17

All items relate to the skill of application.

GED Practice: Lesson 17 (Pages 170–173)

1. Rectangle A

Perimeter $= 2l + 2w$
$= 2 \times 12 + 2 \times 7$
$= 24 + 14 =$ **38 inches**

Area $= l \times w = 12 \times 7$
$=$ **84 square inches**

2. **Square B**
Perimeter $= 4s$
$= 4 \times 15 =$ **60 inches**
Area $= l \times w$
$= 15 \times 15 =$ **225 square inches**

3. **Triangle C**
Perimeter $= a + b + c$
$= 10 + 15 + 18 =$ **43 inches**
Area $= \frac{1}{2}b \times h = \frac{1}{2} \times 18 \times 8$
= 72 square inches

4. **Triangle D**
Perimeter $= a + b + c$
$= 10 + 10 + 10 =$ **30 inches**
Area $= \frac{1}{2}b \times h = \frac{1}{2} \times 10 \times 8.7$
= 43.5 square inches

5. **Parallelogram E**
Perimeter $= a + b + c + d$ (or $2l + 2w$)
$= 6 + 12 + 6 + 12$ or
$(2 \times 6 + 2 \times 12)$
= 36 inches
Area $= b \times h = 12 \times 4$
= 48 square inches

6. **Circle F**
Circumference $= \pi d$ or $2\pi r$
$= 2 \times 3.14 \times 5$
= 31.4 inches
Area $= \pi r^2 = 3.14 \times 5 \times 5$
= 78.5 square inches

7. **(3) 67**
$P = 2l + 2w$
$= 2 \times 25 + 2 \times 8.5$
$= 50 + 17$
$= 67$ feet

8. **(4) 212.5**
$A = lw$
$= 25 \times 8.5$
$= 212.5$ square feet

9. **(4) 80** The shaded portion is a triangle. Use the formula for finding the area of a triangle. The height of the parallelogram is also the height of the triangle.
$A = \frac{1}{2}bh$
$= \frac{1}{2} \times 16 \times 10$
$= 8 \times 10$
$= 80$ square feet

10. **(1) 135**
$A = \frac{1}{2}bh$
$= \frac{1}{2} \times 18 \times 15$
$= 9 \times 15$
$= 135$ square inches

11. **(5) Not enough information is given.** You need the height of a triangle to find the area.

12. **(4) 68** Add the five sides: $14.3 + 14.3 + 13.5 + 13.5 + 12.4 = 68$ centimeters.

13. **(5) 21**
$A = s^2$
$= 4.6 \times 4.6$
$= 21.16$ which rounds to 21 square meters

14. **(5) 64**
$A = bh$
$= 16 \times 4$
$= 64$ square inches

15. **(4) 1,260** Find the area of one section:
$A = lw$
$= 35 \times 18$
$= 630$ square feet
Multiply the area by 2 (the number of sections): $630 \times 2 = 1{,}260$ square feet.

16. **(4) 153.9**
$A = \pi r^2$
$= 3.14 \times 7 \times 7$
$= 3.14 \times 49$
$= 153.86$, which rounds to 153.9 square meters

17. **(5) 800**
$P = 2l + 2w$
$= 2 \times 250 + 2 \times 150$
$= 500 + 300$
$= 800$ feet

18. **(4) 9,600**
$A = lw$
$= 120 \times 80$
$= 9{,}600$ square feet

19. **(3) 27,900** Subtract the area of the cement section from the total area of the park.
Cement Area: 9,600 square feet (see item 18)
Park Area:
$A = lw$
$= 250 \times 150$
$= 37{,}500$ square feet
Subtract: $37{,}500 - 9{,}600 = 27{,}900$.

20. **(5) All three circles are the same size.** If Circle A has a diameter of 7 inches, it has a radius of $3\frac{1}{2}$ inches and is the same as Circle B ($7 \times \frac{1}{2} = 3\frac{1}{2}$). If Circle C has a circumference of 22 inches, it must have a radius of 7 inches, ($C = \pi d$; $22 = \frac{22}{7} \times d$; $d = 7$) which makes it the same size as the other circles.

21. **(3) 31**
$C = \pi d$
$= 3.14 \times 10$
$= 31.4$ which rounds to 31 centimeters

22. **(4) 30** The diameter of the smaller circle is the radius of the larger circle. Double the radius to find the diameter: $15 \times 2 = 30$.

23. **(2) 177** Divide 15 by 2 to find the radius of the smaller circle: $15 \div 2 = 7\frac{1}{2}$ or 7.5.
$A = \pi r^2$
$= 3.14 \times 7.5 \times 7.5$
$= 3.14 \times 56.25$
$= 176.625$ which rounds to 177 square meters

24. **(3) 530** Subtract the area of the smaller circle from the area of the larger circle to find the area of the shaded portion. See the explanation for item 23 to find the area of the smaller circle.
Larger circle:
$A = \pi r^2$
$= 3.14 \times 15 \times 15$
$= 3.14 \times 225$
$= 706.5$ square meters
Subtract: $706.500 - 176.625 = 529.875$, which rounds to 530 square meters.

25. **(5) 201**
$A = \pi r^2$
$= 3.14 \times 8 \times 8$
$= 3.14 \times 64$
$= 200.96$, which rounds to 201 square inches

26. **(2) 122** Find the area of the inner circle and subtract from the area of the circle with a radius of 8 (see explanation for item 25).
$A = \pi r^2$
$= 3.14 \times 5 \times 5$
$= 3.14 \times 25$
$= 78.5$ square inches
Subtract: $200.96 - 78.5 = 122.46$ square inches.

GED Mini-Test: Lessons 16 and 17 (Pages 174–176)

All items refer to the skill of application.

1. **(4) 72** Find the perimeter of one flower bed:
$P = 4s = 4 \times 6$ feet $= 24$ feet.
Multiply by 3 to find the perimeter of 3 beds:
24 feet $\times 3 = 72$ feet.

2. **(5) 22** Find the area of one flower bed:
$A = s^2 = 6 \times 6 = 36$ square feet.
Multiply by 3 to find the total area of the three beds: $36 \times 3 = 108$ square feet.
Divide by 5 (the number of square feet one pint will cover): $108 \div 5 = 21$ r3.
He will need 22 pints.

3. **(3) 31** The area of one flower bed is 36 square feet (see the explanation for item 2).
Each bush will need 5 square feet.
Subtract: $36 - 5 = 31$ square feet.

4. **(2) 324** Find the circumference of a whole circle with a radius of 63. Double the radius (63 ft.) to get the diameter.
$C = \pi d = 3.14 \times 126 = 395.64$
Divide the result by 2 to find the length of the curved line. $395.64 \div 2 = 197.82$
The length of the straight line is the diameter: $2(63) = 126$.
Add it to the length of the curved line:
$126 + 197.82 = 323.82$ which rounds to 324.

5. **(4) 12 × 18** The area of a parallelogram is found with the formula: $A = bh$. The base in the drawing is 18 ft. and the height is 12 ft.

6. **(1) 56** Use the formula to find the perimeter of a rectangle:
$P = 2l + 2w$
$= 2 \times 25 + 2 \times 3$
$= 50 + 6$
$= 56$ feet

25 ft.
3 ft.

7. **(4) 108** Use the formula for finding the area of a parallelogram:
$A = bh$
$= 18 \times 6$
$= 108$ square meters

8. **(4) 36** Use the formula for finding the perimeter of a triangle:
$P = a + b + c$
$= 14 + 10 + 12$
$= 36$ feet

UNIT 1

9. **(1) 56** Use the formula for finding the area of a triangle:
$A = \frac{1}{2}bh$
$= \frac{1}{2} \times 14 \times 8$
$= 7 \times 8$
$= 56$ square feet

10. **(5) Not enough information is given.** You need to know both the length and the width to find the area of a rectangle. You only know the length.

11. **(3) 332 feet** You are asked to find the perimeter of the track. If the two half-circle portions were joined you would have one whole circle. Find the circumference of that circle. Then add the lengths of the rectangle. The diameter is 2 times the radius.
$C = \pi d$
$= 3.14 \times 2(21) = 3.14 \times 42$
$= 131.88$
$131.88 + 100 + 100 = 331.88$ which rounds to 332.

12. **(4) 5,585 square feet** Find the area of the circle (see the explanation for item 11) and the area of the rectangle portion. Add the two amounts together.
$A = \pi r^2$
$= 3.14 \times 21 \times 21$
$= 3.14 \times 441$
$= 1{,}384.74$ square feet
$A = lw$
$= 100 \times (21 + 21) = 100 \times 42$
$= 4{,}200$ square feet
$1{,}384.74 + 4{,}200 = 5{,}584.74$ which rounds to 5,585.

13. **(2) 98** Add the lengths of the sides:
$8 + 22 + 15 + 23 + 30 = 98$ cm.

14. **(3) 360** Find the perimeter of the pool:
$P = 2l + 2w = 200 + 70 = 270$ feet.
Multiply by 12 to find the number of inches in the perimeter:
12 inches/foot $\times$ 270 feet = 3,240 inches.
Divide by 9 (the number of inches in 1 brick):
$3{,}240 \div 9 = 360$ bricks.

15. **(1) 12** Find the area of the floor:
$A = lw = 40 \times 60 = 2{,}400$ square feet.
Divide by 200 (the number of square feet covered by 1 gallon):
$2{,}400 \div 200 = 12$ gallons.

16. **(3) 696** Find the area of the two rectangles. Subtract the area of the smaller rectangle from the area of the larger rectangle.
$A = lw = 30 \times 40 = 1{,}200$ square feet.
$A = lw = 28 \times 18 = 504$ square feet.
$1{,}200 - 504 = 696$ square feet

17. **(4) 48** Find the perimeter of both rectangles. Subtract the smaller number from the larger number.
$P = 2l + 2w = 60 + 80 = 140$ feet
$P = 2l + 2w = 56 + 36 = 92$ feet
$140 - 92 = 48$ feet

18. **(4) 188** Use the formula for finding the circumference of a circle:
$C = \pi d = 3.14 \times 2(30) = 3.14 \times 60 = 188.4$, which rounds to 188 meters.

19. **(1) 78** Find the missing lengths. Then add all the lengths to find the perimeter. The longer missing length is $12 + 9 = 21$ feet. The shorter missing length is $18 - 12 = 6$ feet.
The perimeter is $18 + 9 + 6 + 12 + 12 + 21 =$ 78 feet.

20. **(2) 306** Find the area of each part. Add the two amounts.
$A = lw = 12 \times 12 = 144$ square feet
$A = lw = 18 \times 9 = 162$ square feet.
$144 + 162 = 306$ square feet

Lesson 18

All items relate to the skill of application.

GED Practice: Lesson 18 (Pages 179–181)

1. **(3) February and April** The two shortest bars are labeled F and A for February and April.

2. **(4) between $45 and $50 million** The bar for January reaches more than halfway to the bar marked $50. Remember that the scale is shown in millions of dollars. The bar represents approximately $48 million.

3. **(2) March** $40 million is the approximate average. March's sales were nearest the average.

4. **(1) Federal tax** The Federal tax sector is the largest and has the highest percent at 30%.

5. **(4) 30%** Add 12% (health insurance) and 18% (life insurance): 12% + 18% = 30%.

6. **(5) $250** Add the percents for state and federal taxes: 20% + 30% = 50%. Find 50% of $500. $500 × 0.5 = $250. You could also divide $500 by 2 because you know 50% is $\frac{1}{2}$.

7. **(5) Not enough information is given.** The whole that the graph represents is the total deductions. You are not told what percent the deductions are of the employee's earnings. Since you do not know the relationship of the deductions to the earnings, you cannot solve the problem.

8. **(5) 4 P.M.** Find the highest point on the line. The highest temperature occurred at 4 P.M.

9. **(2) 10 A.M. to 12 noon** The space between each pair of vertical lines is a span of 2 hours. Examine each section that shows a rise in temperature. The steepest rise was between 10 A.M. and noon. In that period the temperature rose from 55 degrees to 65 degrees.

10. **(3) 6** Find the temperature at 4 P.M.: approximately 68 degrees. Find the temperature at 6 P.M.: approximately 62 degrees. Subtract: 68 − 62 = 6 degrees. Even if your reading of the graph was slightly different, the best estimate is 6 degrees.

11. **(2) 11 A.M.** The graph line first crosses the 60 degree line between 10 A.M. and noon. You are looking for the first time. The line crosses the horizontal 60 degree line halfway between 10 A.M. and noon. 11 A.M. is the best estimate.

12. **(3) between 20 and 25 degrees** The lowest temperature (approximately 45 degrees) occurred between 6 and 8 A.M. The highest temperature, recorded at 4 P.M., is between 65 and 70 degrees—about 68 degrees. Subtract: 68 − 45 = 23 degrees.

13. **(2) $33.4 billion** Make sure you are looking in the 95 column on the row marked excise taxes. The amount is $33.4 billion. Remember the numbers on the chart represent billions of dollars.

14. **(2) $468.8 billion** Add the amounts paid by corporations and individuals for 1994: 104.8 + 364.0 = 468.8

15. **(3) 13%** Find the difference between the social security amounts from 1994 to 1995: 307.4 − 273.2 = 34.2. Divide by the original amount and round to the nearest whole percent: 34.2 ÷ 273.2 is approximately 13%.

16. **(4) $1,600,000,000** Subtract: 23.8 − 22.2 = 1.6
Remember the answer is in billions of dollars: 1.6 × $1,000,000,000 = $1,600,000,000.

GED Practice: Solving Word Problems (Page 183)

To answer items 1 to 4, make a table. For the following table, the amounts for rugs and drapes have been entered.

Address	Rent	Parking	Deposit	Lease Fee	Key Fee	Rugs	Drapes
Mulberry	$565	$30	$580	$60		$450	$250
Parke	$485		$970		$50	$450	
Meridan	$525	$25	$1000	$50			$250
Tenth	$495		$990	$49.50			

1. **(4) $5,820** Multiply $485 by 12 (the number of months in a year).

2. **(3) Meridan Street** From the table you can see that the $1,000 deposit is the highest.

3. **(2) $1,935** Add all the costs on the table for the Mulberry apartment: $565 + $30 + $580 + $60 + $450 + $250 =$1,935.

4. **(2) $29.50**
Add the Tenth Avenue costs:
$495 + $990 + $49.50 =$1,534.50
Add the Parke Boulevard costs:
$485 + $970 + $50 =$1,505.
Find the difference:
$1,534.50 − $1505.00 = $29.50.

GED Mini-Test: Lesson 18 (Pages 184–185)

All items relate to the skill of application.

1. **(1) October, November, and December** The actual rainfall is the right-hand bar for each month. The right-hand bar is lower than the left-hand bar (the normal rainfall) for only the first three months.

2. **(2) 3** Look only at the right-hand bar for each month. This bar is higher than the 5.0 line for December, January, and February.

3. **(1) 2** Look only at the right-hand bar for each month. This bar is lower than the 4.0 line for October and March.

4. **(4) 4.4** The right-hand bar for January is more than halfway to the line above 6.0, which would be 7.0. Estimate that the bar reaches to about 6.8 inches. The right-hand bar for March reaches over 2.0, but less than halfway to 3.0. Estimate that this second bar reaches to 2.4. Subtract: 6.8 − 2.4 = 4.4 inches.

5. **(4) 6.0 inches** To find the average normal rainfall, add the estimated normal amounts for the three months and divide by 3:
November is about 5.5
December is about 6.5
January is about 6.2
18.2 ÷ 3 = 6.06 inches.

6. **(3) 2.7** The right-hand bar for November is halfway between the lines for 4.0 and 5.0, which is 4.5. The right-hand bar for October is about $\frac{4}{5}$ of the length from the bottom line, which would be 1.0 to the line for 2.0. Estimate that the bar reaches 1.8. Subtract: 4.5 − 1.8 = 2.7

7. **(3) from 8,700 to 10,000** The lowest point was in 1989 at about 8.7 thousand, or 8,700. The highest point was in 1980 at about 10 thousand, or 10,000.

8. **(3) four** The direction changes from up to down in 1980, from down to up again in 1983, from up to down again in 1987, and back up in 1989 for a total of four changes.

9. **(5) 1995 to 1996** You have to evaluate each of the choices by looking at the graph. Of the answer choices, the line moves up for only 1995 to 1996.

10. **(4) 7%** The miles driven for 1989 is 8.9 thousand. The amount for 1987 is 9.5 thousand. Subtract: 9.5 − 8.9 = 0.6, and divide by the original number. 0.6 divided by 8.9 is 0.067. Round to the nearest whole percent, 0.7 or 7%.

11. **(4) Business, Legal, and Professional occupations** Find $\frac{1}{2}$ of the percent figure for Retail occupations: $\frac{1}{2}$ of 42 is 21. Business, Legal, and Professional occupations has the closest percent figure with 22%.

GED Cumulative Review: Measurement (Pages 186–192)

All items relate to the skill of application.

1. **(4) 22%** According to the graph, the difference in subscriptions from 1994 to 1995 was 10,000: 45,000 − 35,000 = 10,000. Divide by the original amount: 10,000 ÷ 45,000 is $22\frac{2}{9}$% or about 22%.

2. **(5) 1992** Subscriptions for 1992 were between 30 and 35 thousand, which is closer to 30,500 than any of the other bars.

3. **(4) 8,000** According to the graph, the number of subscriptions for 1994 is about 45,000 and for 1993 about 37,000. Subtract to find the difference: 45,000 − 37,000 = 8,000.

4. **(3) 19,250** The subscriptions for 1989 were about 35,000. Find the number of new subscriptions by finding 45% of 35,000: 35,000 × 0.45 = 15,750. Subtract to find the number of renewals:
35,000 − 15,750 = 19,250.

5. **(4) Team C is farthest from its goal.** Change the fractions to decimals for easier comparisons and evaluate each answer choice:
A $\frac{2}{3} = 0.66\frac{2}{3}$
B $\frac{1}{3} = 0.33\frac{1}{3}$
C $\frac{3}{10} = 0.30$
D $\frac{1}{2} = 0.50$
E $\frac{3}{8} = 0.37\frac{1}{2}$
Team C is farthest from its goal because it has reached the smallest portion of its goal.

6. **(5) 17.8** The top line of the table shows the distance from Eagle's Nest to Morning Peak is 5.6 miles. Reading on the bottom line of the table, from Morning Peak to Rock Face is 3.3 miles. Add the two figures: 5.6 + 3.3 = 8.9 Multiply by 2 to find the mileage going and coming back: 8.9 × 2 = 17.8.

7. **(1) about $3\frac{1}{2}$ times as long as the shortest** The shortest distance is from Eagle's Nest to Rock Face: 2.1 miles. The longest distance from Rock Face to Crow's Point, 7.2 miles, is about $3\frac{1}{2}$ times the shortest distance:
2.1 × 3.5 = 7.35.

8. **(4) about 6 hr.** Set up a proportion to find the time it took him to go from Rock Face to Crow's Point, a distance of 7.2 miles.
$\frac{1.8 \text{ miles}}{1 \text{ hour}} = \frac{7.2 \text{ miles}}{x}$
Cross-multiply: 7.2 × 1 = 7.2.
Divide: 7.2 ÷ 1.8 = 4 hours.
It took 4 hours for the first part of the hike. Now find the time it took for the distance from Crow's Point to Eagle's Nest, 3.8 miles.
$\frac{2 \text{ miles}}{1 \text{ hour}} = \frac{3.8 \text{ miles}}{x}$
Cross-multiply: 3.8 × 1 = 3.8.
Divide: 3.8 ÷ 2 = 1.9 hours.
Add: 4 hours + 1.9 hours = 5.9 hours or about 6 hours.

9. **(2) Her reading rate has doubled.** Set up both speeds as proportions to find how many words she can read in a minute.

$\frac{820}{4} = \frac{205}{1}; \frac{4,100}{10} = \frac{410}{1}$

At the beginning of the course, she could read 205 words in one minute. Now she can read twice that amount in one minute. $205 \times 2 = 410$. Option (4) is incorrect because there is no way to know how fast she will read in 4 weeks.

10. **(4) 860** Use the formula $P = 2l + 2w$ to find the area of the rectangle.
$P = 2 \times 310 + 2 \times 120$
$= 620 + 240$
$= 860$

11. **(3) 496** Find the area of the driving range:
$A = lw$
$= 310 \times 120$
$= 37,200$
Divide by 75 to find the number of bags of seed needed: $37,200 \div 75 = 496$.

12. **(2) 5,600**
$A = bh$
$= 80 \times 70$
$= 5,600$ square centimeters
The base does not have to be the longest side.

13. **(5) 320** Add the sides:
$90 + 70 + 90 + 70 = 320$ cm.

14. **(1) 10:10** Add the time needed for the projects:
20 min. + 3(20) min. + 30 min. = 20 min. + 60 min. + 30 min. = 110 min. or 1 hr. 50 min.
Subtract from 12 o'clock:
12 noon = 11 hr. 60 min.
− 1 hr. 50 min.
10 hr. 10 min.
Jerry should start at 10:10 A.M.

15. **(5) 1,920** Set up a proportion and solve:
$\frac{60 \text{ survivors}}{100 \text{ cases}} = \frac{x}{3,200 \text{ cases}}$
Cross multiply: $60 \times 3,200 = 192,000$.
Divide: $192,000 \div 100 = 1,920$.

16. **(4) $18,825** Add and divide by 5:
$16,907
$17,756
$19,174
$19,558
+$20,730
$94,125
$94,125 ÷ 5 = $18,825.

17. **(3) $18,465** Arrange the 6 amounts in order from lowest to highest:
$16,907, $17,204, $17,756, $19,174, $19,558, $20,730. Since there is an even number of amounts, average the middle two:
$17,756 + $19,174 = $36,930 ÷ 2 = $18,465.

18. **(5) 10 to 12 months** The change during this period was 2 kg., more than any other change.

19. **(1) 2 to 4 months** Determine the change for each period from the graph:
0 to 2 months: 1.5 kg.
2 to 4: 1.5 kg.
4 to 6: 1.25 kg.
6 to 8: 1.00 kg.
8 to 10: 0.5 kg.
10 to 12: 2.0 kg.

20. **(5) 1.5 kg.** The Hierro baby weighed 12 kg. at 12 months. Subtract: $12 - 10.5 = 1.5$ kg.

21. **(3) 8 to 10 months** The baby gained only 0.5 kg. during this period, the smallest gain shown on the graph.

22. **(2) 40** Find the total number of sales:
$21 + 42 + 34 + 45 + 51 + 47 = 240$.
Divide by 6: $240 \div 6 = 40$.

23. **(2) 98,000**
Set up a proportion: $\frac{3 \text{ days}}{19,600} = \frac{15 \text{ days}}{x}$.
Cross multiply: $19,600 \times 15 = 294,000$.
Divide: $294,000 \div 3 = 98,000$.

24. **(2) 5 ft. 4 in.**
$P = 4s$
$= 4 \times 1$ ft. 4 in. $= 4 \times 1$ ft. $+ 4 \times 4$ in.
= 4 ft. 16 in.
= 5 ft. 4 in.

25. **(2) 270**
$A = \frac{1}{2}bh$
$= \frac{1}{2} \times 45 \times 12$
$= 22\frac{1}{2} \times 12$
$= 270$ square feet

26. **(3) 17 + 35 + 45** $P = a + b + c$. Only option (3) shows the addition of the three sides.

27. **(4) 0.24** Total number of marbles is:
$12 + 28 + 10 = 50$. Set up a ratio and divide to convert to a decimal:
$\frac{12 \text{ red}}{50 \text{ marbles}} = 0.24$.

28. **(4) 25**
$C = \pi d$
$= 3.14 \times 8$
$= 25.12$ feet rounds to 25 feet.

29. (4) They will fit with 4 inches left over.
Multiply 35 inches by 4: $35 \times 4 = 140$ inches
Convert to feet and inches:
$140 \div 12 = 11$ ft. 8 in.
Compare to 12 feet: 12 ft. − 11 ft. 8 in. = 4 in.

30. (3) 12 If they won 2 out of 3, they lost 1 out of 3. Set up a proportion to find the total games played.
$\frac{1 \text{ lost}}{3 \text{ played}} = \frac{6 \text{ lost}}{x}$
Cross multiply: $3 \times 6 = 18$ games played.
Divide: $18 \div 1 = 18$. Subtract 6 losses from 18 games played = 12 games won.

31. (3) $7\frac{1}{2}$ in. by $10\frac{1}{2}$ in.
Multiply both measurements by 2.
$3\frac{3}{4} \times 2 = 6\frac{6}{4} = 7\frac{1}{2}$ inches
$5\frac{1}{4} \times 2 = 10\frac{2}{4} = 10\frac{1}{2}$ inches

32. (3) 1 ft. 2 in.
Convert and subtract: 2 yards = $\overset{5}{\cancel{6}}$ ft. $\overset{12 \text{ in.}}{}$

$$\begin{array}{r} -4 \text{ ft. } 10 \text{ in.} \\ \hline 1 \text{ ft. } \ \ 2 \text{ in.} \end{array}$$

33. (3) $\frac{1}{50}$ Set up a ratio and reduce: $\frac{4}{200} = \frac{1}{50}$.

34. (4) $253.95 Add the first three amounts for January through June:
$89.36 + $90.12 + $74.47 = $253.95.

35. (2) $79.39 Arrange the amounts in order:
$59.76, $63.15, <u>$74.47</u>, <u>$84.31</u>, $89.36, $90.12.
Since there is an even number of amounts, average the middle two:
$74.47 + $84.31 = $158.78 ÷ 2 = $79.39.

36. (1) $76.86 The "average" is the "mean." To find a mean, add the amounts:
$89.36 + $90.12 + $74.47 + $63.15 +
$59.76 + $84.31 = $461.17
and divide by 6, the number of amounts:
$461.17 ÷6 = $76.86.

37. (5) Not enough information is given. You need to know the total number of employees working for the company to create the ratio.

38. (4) more than $20 Set up a proportion and solve: $\frac{\$5.75}{1.25 \text{ lb.}} = \frac{x}{5 \text{ lb.}}$
Cross multiply: $5.75 ×5 = $28.75.
Divide $28.75 ÷ 1.25 = $23.

39. (3) 2,520 Subtract the area of the triangle from the area of the rectangle.
Rectangle Area $= lw$
$= 80 \times 42$
$= 3{,}360$ square inches
Triangle Area $= \frac{1}{2}bh$
$= \frac{1}{2} \times 40 \times 42$
$= 20 \times 42$
$= 840$ square inches
$3{,}360 - 840 = 2{,}520$ square inches

40. (4) 83 Add the scores and divide by 15:
$100 + 100 + 98 + 95 + 92 + 87 + 84 + 84 + 84 + 80 + 76 + 72 + 68 + 67 + 60 = 1{,}247$.
$1{,}247 \div 15 = 83\frac{2}{15}$ rounds to 83.

41. (3) 84 The scores are already in order. The middle score is the 8th score: 84.

42. (5) 13 First, find the number of hours from 8 A.M. to 5:45 P.M. There are 4 hours before noon, and $5\frac{3}{4}$ hours after noon: $4 + 5\frac{3}{4} = 9\frac{3}{4}$.
Divide by $\frac{3}{4}$: $9\frac{3}{4} \div \frac{3}{4} = \frac{39}{4} \div \frac{3}{4} = \frac{\overset{13}{\cancel{39}}}{\underset{1}{\cancel{4}}} \times \frac{\overset{1}{\cancel{4}}}{\underset{1}{\cancel{3}}} = 13$.

43. (1) 2 feet 3 inches Subtract 2 ft. 3 in. from $1\frac{1}{2}$ yd.

$$\begin{array}{r} 1\frac{1}{2} \text{ yd.} = 1 \text{ yd. } 1 \text{ ft. } 6 \text{ in.} \\ -2 \text{ ft. } 3 \text{ in.} \\ \hline 2 \text{ ft. } 3 \text{ in.} \end{array}$$

(Note: 1 yd. = 3 ft.; $\frac{1}{2}$ yd. = $\frac{3}{2}$ ft. = $1\frac{1}{2}$ ft.
$1\frac{1}{2}$ ft. = 1 ft. 6 in.).

44. (1) 150
$A = \frac{1}{2}bh$
$= \frac{1}{2} \times 20 \times 15$
$= 10 \times 15$
$= 150$ square inches.

45. (3) 206 Add the scores and divide by 3:
$220 + 186 + 212 = 618$
$618 \div 3 = 206$.

UNIT 2: ALGEBRA

Lesson 19

All items relate to the skill of application.

GED Practice: Lesson 19 (Pages 198–199)

The absolute value is the number value without the sign.

1. 6

2. 7

3. 14.6

4. $\frac{2}{3}$

5. 0 The absolute value of 0 is 0.

Use positive numbers for ideas of upward, gain, and increase. Use negative numbers for ideas of downward, loss, and decrease.

6. $+45$
7. $-\frac{1}{4}$
8. -10
9. $-\$200$
10. $+10$
11. $+30$

See addition and subtraction rules on page 196.

12. $+12$
13. -16
14. -1
15. $+3$
16. -10
17. -6
18. -2
19. $+1\frac{1}{2}$
20. -7
21. $+9$
22. $+12$
23. $+5$
24. -6
25. $+5$
26. -8.7

See multiplication/division rules on page 197.

27. -6
28. -5
29. $+28$
30. 0
31. -15
32. $+8$
33. 0
34. $+4$
35. -16
36. -5
37. $+4$
38. $+14$
39. -3
40. -5
41. $+7$
42. -14

43. **$+38$** Begin by multiplying: $4 \times 8 = 32$. Then add: $6 + 32 = 38$.

44. **$+10$** The numerator is -10 and the denominator is -1. $-10 \div -1 = +10$.

45. **-30** The expression inside the first parentheses is -36. Inside the second parentheses is -6.
$-36 - (-6) = -36 + 6 = -30$.

46. **-1** Begin inside the parentheses with $4 \times (-2) = -8$. Next add $-3 + -8 = -11$. The expression becomes $10 + -11 = -1$.

47. **-37** Begin by multiplying 4×3. Then $-25 - 12 = -37$.

48. **-2** The numerator is -10 and the denominator is $+5$. $10 \div +5 = -2$.

49. **-25** Begin inside the parentheses with 4×8. Next add $32 + (-1) = 31$. The expression becomes $6 - 31 = -25$.

50. **-38** The value of the expression inside the first parentheses is -20. The value of the second parentheses is -18.
Add: $-20 + -18 = -38$.

GED Practice: Lesson 19 (Page 201)

1. $\mathbf{2x + 3y}$
$5x - 3x = 2x$; $1y + 2y = 3y$

2. $\mathbf{10x - 2}$
$8x + 2x = 10x$; $3 - 5 = -2$

3. $\mathbf{4a - 2b}$
$a + a + 2a = 4a$; $6b - 8b = -2b$

4. $\mathbf{14 - 11x}$
$-7x - 4x = -11x$

5. $\mathbf{5z + 5y - 3}$
$4z + z = 5z$

6. $\mathbf{-2x + 2y}$
$3x - 5x = -2x$; $12y - 10y = 2y$

7. **0**
$+2 - (+5) + (+3)$
$+2 + (-5) + (+3) = 0$

8. **$+8$**
$(+3) + (+5) + (+5) - (+5)$
$(+3) + (+5) + (+5) + (-5) = +8$

9. **$+3$**
$(-1) + (-4) + (+8) = +3$

10. **$+6$**
$(+2) - (-0.5) - (-0.5) - (-3)$
$(+2) + (+0.5) + (+0.5) + (+3) = +6$

11. **$+17$**
$5(3) + 2 - 4 + 2(2)$
$15 \quad + 2 - 4 + 4 \quad = +17$

12. **-24**
$3(-2)(4) + 2(-2) + 4$
$-24 \quad + (-4) \quad + 4 = -24$

13. **-12**
$\frac{6(-3)(2)(-1)}{-3} = \frac{36}{-3} = -12$

14. **$+2$**
$\frac{4 + 5(2)}{2(4) - 1} = \frac{4 + 10}{8 - 1} = \frac{14}{7} = +2$

UNIT 2

In the answers to Items 15–22, x is used as the unknown number or the variable. Any other letter can be used as well.

15. $x - 2$

16. $\frac{x}{7}$

17. $2x + (-4)$

18. $3x - 2x$

19. $3x - 9$

20. $\frac{4x}{3}$

21. $11 + x - 2$

22. $19 - 5 + x - 2$

LESSON 20

All items relate to the skill of application.

GED Practice: Lesson 20 (Pages 206–207)

1. 19 $x - 15 = 4$; $x = 19$

2. 10 $x - 7 = 3$; $x = 10$

3. 6 $2x = 12$; $x = 6$

4. 7 $-6x = -42$; $x = 7$

5. 10 $x + 12 = 22$; $x = 10$

6. −2 $x - 8 = -10$; $x = -2$

7. −6 $-3x = 18$; $x = -6$

8. −9 $5x = -45$; $x = -9$

9. 5 $6x + 7 = 37$; $6x = 30$; $x = 5$

10. 5 $4x + 5x - 10 = 35$; $9x - 10 = 35$; $9x = 45$; $x = 5$

11. −2 $3x - 6x + 2 = -4x$; $-3x + 2 = -4x$; $2 = -1x$; $-2 = x$

12. 1 $6 - x + 12 = 10x + 7$; $18 - x = 10x + 7$; $18 = 11x + 7$; $11 = 11x$; $1 = x$

13. −1 $5x + 7 - 4x = 6$; $x + 7 = 6$; $x = -1$

14. −4 $9x + 6x - 12x = -7x + 2x - 12 + 5x$; $3x = -12$; $x = -4$

15. 4 $7x + 3 = 31$; $7x = 28$; $x = 4$

16. 12 $3x - 8 = 28$; $3x = 36$; $x = 12$

17. 1 $8x + 6 = 5x + 9$; $3x + 6 = 9$; $3x = 3$; $x = 1$

18. 5 $11x - 10 = 8x + 5$; $3x - 10 = 5$; $3x = 15$; $x = 5$

19. 1 $-2x - 4 = 4x - 10$; $-2x = 4x - 6$; $-6x = -6$; $x = 1$

20. −4 $5x + 8 = x - 8$; $4x + 8 = -8$; $4x = -16$; $x = -4$

21. 7 $11x - 12 = 9x + 2$; $2x - 12 = 2$; $2x = 14$; $x = 7$

22. 14 $5(x + 1) = 75$; $5x + 5 = 75$; $5x = 70$; $x = 14$

23. 8 $5(x - 7) = 5$; $5x - 35 = 5$; $5x = 40$; $x = 8$

24. 3 $6(2 + x) = 5x + 15$; $12 + 6x = 5x + 15$; $12 + x = 15$; $x = 3$

25. $x + 12 = 25$, **so** $x = 13$

26. $x - 4 = 11$, **so** $x = 15$

27. $12x = 60$, **so** $x = 5$

28. $\frac{x}{10} = -2$, **so** $x = -20$

29. $7x - 8 = 5x$; $2x = 8$; $x = 4$

30. $8(x + (-5)) = x + 2$; $8x + (-40) = x + 2$; $7x = 42$; $x = 6$

31. $2x - 6 = 24$; $2x = 30$; $x = 15$

32. $8x + 10 = 3x$; $5x = -10$; $x = -2$

33. $x + 20 = 5x$; $-4x = -20$; $x = 5$

34. $x + 8 = 19$, **so** $x = 11$

35. $21 - x = 10$. **Here you find** $-x = -11$, **so divide each side by** -1 **to find** $x = +11$.

36. $-8x = -32$, **so** $x = 4$

37. $\frac{x}{-6} = 4$, **so** $x = -24$

UNIT 2

38. $2x + 7 = x + 10$
$x = 3$

39. $2x + 2 = 20$
$2x = 18$
$x = 9$

40. $(8 + x) + 12 = 3x$
$x + 20 = 3x$
$-2x = -20$
$x = 10$

41. $5x = x - 12$
$4x = -12$
$x = -3$

42. $4x = x - 12$
$3x = -12$
$x = -4$

GED Practice: Solving Word Problems (Pages 210–211)

1. **(1) 2** Let x represent the number.
$5(9 + x) - 15 = 40$
$45 + 5x - 15 = 40$
$30 + 5x = 40$
$5x = 10$
$x = 2$

2. **(3) 24** Let x represent the lesser number. The expression $x + 1$ represents the next consecutive number. Write the equation.
$x + x + 1 = 49$
$2x + 1 = 49$
$2x = 48$
$x = 24$
The lesser number is 24.

3. **(3) $53 - 21 - 18 = b$** Use the formula to solve for b and substitute the values.
$P = a + b + c$
$P - a - c = b$
$53 - 21 - 18 = b$
or
$P = a + b + c$
$53 = 21 + b + 18$
$53 - 21 - 18 = b$

4. **(2) 87** Let x represent the number of employees who work in management. Let $3x + 12$ represent the number who work in production.
$x + 3x + 12 = 360$
$4x = 348$
$x = 87$

5. **(1) $\frac{30 - 12}{2} = l$**
Use the formula to solve for l and substitute the values.
$P = 2l + 2w$
$P - 2w = 2l$
$\frac{P - 2w}{2} = l$
$\frac{30 - 2(6)}{2} = l$
$\frac{30 - 12}{2} = l$

6. **(2) \$300** Let x represent her husband's earnings. Bonnie's earnings can be written as $2x - \$150$.
$x + 2x - \$150 = \$750$
$3x - \$150 = \$750$
$3x = \$900$
$x = \$300$

7. **(4) 5** $8x - 5 = 7x$
$x = 5$

8. **(4) 39** Let x represent the number Frank did, and $x + 12$ the number that Tom did.
$x + x + 12 = 66$
$2x + 12 = 66$
$2x = 54$
$x = 27$
Tom did 39 pushups.
$(27 + 12 = 39)$.

9. **(3) \$18** Let x represent the fine for the first ticket and $2x$ the fine for the second ticket.
$x + 2x = 54$
$3x = 54$
$x = 18$
The fine for the first ticket is \$18.

10. **(3) $2(x + 19) = 5x + 19$**
Make a chart. Let x represent the son's age.

	George	Son
Now	$5x$	x
In 19 years	$5x + 19$	$x + 19$

Write the equation: multiply 2 times the son's age in 19 years, and equate it with George's age in 19 years.

UNIT 2

11. (1) 2
Make a chart.
Let x represent Diana's age.

	Nora	Diana
Now	$x + 4$	x
2 years from now	$(x + 4) + 2$	$x + 2$

Set up the equation so that 2 times Diana's age in 2 years equals Nora's age in 2 years.

$$2(x + 2) = (x + 4) + 2$$
$$2x + 4 = x + 6$$
$$x + 4 = 6$$
$$x = 2$$

Diana is now 2 years old and Nora is 6 years old.

12. (3) 425 miles per hour
Substitute the known values and solve for rate.

$$d = rt$$
$$2{,}125 = r5$$
$$\frac{2{,}125}{5} = r$$
$$425 = r$$

13. (4) $17.40
Substitute the known values and solve for r, the unit cost.

$$c = nr$$
$$\$104.40 = 6r$$
$$\frac{\$104.40}{6} = r$$
$$\$17.40 = r$$

14. (5) 100 degrees Substitute the known values and solve for C.

$$F = \frac{9}{5}C + 32$$
$$212 = \frac{9}{5}C + 32$$
$$180 = \frac{9}{5}C$$
$$\frac{\overset{20}{\cancel{180}}}{1} \times \frac{5}{\underset{1}{\cancel{9}}} = C$$
$$100 = C$$

Lesson 21

All items relate to the skill of application.

GED Practice: Lesson 21 (Pages 214–215)

For Items 1–6, substitute each of the seven values into the given inequality. If a true statement results, the number being substituted is a solution.

1. All of the numbers except 3 are solutions, because when any of them is substituted for x, a true statement results.
2. All of them except 2 and 3 are solutions.
3. Only -3 and -2 are solutions.
4. All numbers less than 0 are solutions.
5. 1, 2 and 3 are the only solutions.
6. 1, 2 and 3 are the only solutions.
7. 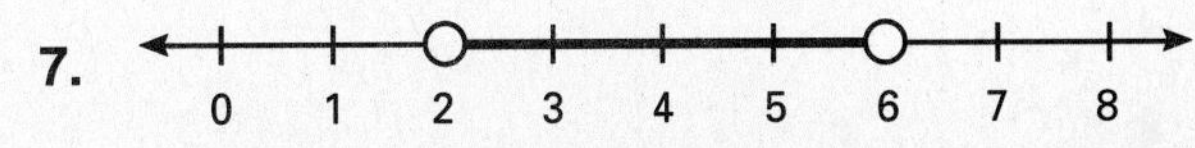

8. 3 4 5 6 7 8 9 10 11
9. −4 −3 −2 −1 0 1 2 3 4
10. −6 −5 −4 −3 −2 −1 0 1 2
11. $x < 1$ An open circle at 1 and the arrow pointing left means $x < 1$.
12. $x \leq -2$ The arrow points left: less than. The solid dot at -2 means it is included in the solution: less than or equal to -2.
13. $x \geq 0$ The arrow points right. The dot is solid.
14. $x > 1$ The arrow points right. The dot is open.
15. $x < 8$
$$3x - 7 < 2x + 1$$
$$3x - 2x - 7 < 1$$
$$x - 7 < 1$$
$$x < 8$$
16. $x > -1$
$$5x + 2 > 4x + 1$$
$$5x + 2 - 4x > 1$$
$$x + 2 > 1$$
$$x > 1 - 2$$
$$x > -1$$
17. $x < 2$
$$6x - 4 < 3x + 2$$
$$6x - 4 - 3x < 2$$
$$3x - 4 < 2$$
$$3x < 2 + 4$$
$$3x < 6$$
$$x < 2$$
18. $x < 4$
$$3(x + 1) > x + 4x - 5$$
$$3x + 3 > 5x - 5$$
$$3x + 3 - 5x > -5$$
$$-2x + 3 > -5$$
$$-2x > -5 - 3$$
$$-2x > -8$$
$$\frac{-2}{-2} < \frac{-8}{-2}$$
$$x < 4$$
Note the inequality reversal because of multiplication by a negative number.

UNIT 2

19. $x < 2$

$$5 + 8(x - 2) < x + 3$$
$$5 + 8x - 16 < x + 3$$
$$8x - 11 < x + 3$$
$$8x - 11 - x < 3$$
$$7x - 11 < 3$$
$$7x < 14$$
$$x < 2$$

20. $x > -7$

$$x + 12 < 5(x + 8)$$
$$x + 12 < 5x + 40$$
$$x + 12 - 5x < 40$$
$$-4x + 12 < 40$$
$$-4x < 28$$
$$\frac{-4x}{-4} > \frac{28}{-4}$$
$$x > -7$$

Note the inequality reversal because of division by a negative number.

21. $x > -17$

$$2x + (4 - 3x) < 21$$
$$2x + 4 - 3x < 21$$
$$-x + 4 < 21$$
$$-x < 17$$
$$x > -17$$

22. $x > -5$

$$7x - 3x - x < 3x + 2x + 10$$
$$3x < 5x + 10$$
$$-2x < 10$$
$$x > -5$$

Note the inequality reversal because of division by a negative number.

23. **(2)** $x < 4$ Set up an inequality and solve:

$$5x + 6 < 4x + 10$$
$$5x + 6 - 4x < 10$$
$$x + 6 < 10$$
$$x < 10 - 6$$
$$x < 4$$

24. **(4) 26** Use the formula for the perimeter of a triangle: $a + b + c = P$. Set up an inequality and solve:

$$21 + 18 + x \leq 65$$
$$39 + x \leq 65$$
$$x \leq 26$$

UNIT 2

GED Practice: Lesson 21 (Page 217)

1. $5(x + 6)$ Divide both terms by 5.
2. $3(2y + 5)$ Divide both terms by 3.
3. $2(4x - 1)$ Divide both terms by 2.
4. $2(2z - 7)$ Divide both terms by 2.
5. $b(b + 9)$ Divide both terms by b.
6. $y(y + 3)$ Divide both terms by y.
7. $2x(x + 2)$ Divide both terms by $2x$.
8. $3x(x + 3)$ Divide both terms by $3x$.
9. $y(7y - 1)$ Divide both terms by y.
10. $2x(2x + 1)$ Divide both terms by $2x$.

11. $(x + 4)(x + 5)$
Check: $(x + 4)(x + 5)$
$= x^2 + 5x + 4x + 20$
$= x^2 + 9x + 20$

12. $(x - 2)(x - 3)$
Check: $(x - 2)(x - 3)$
$= x^2 - 3x - 2x + 6$
$= x^2 - 5x + 6$

13. $(x + 6)(x - 1)$
Check: $(x + 6)(x - 1)$
$= x^2 - x + 6x - 6$
$= x^2 + 5x - 6$

14. $(x - 7)(x + 4)$
Check: $(x - 7)(x + 4)$
$= x^2 + 4x - 7x - 28$
$= x^2 - 3x - 28$

15. $(x + 1)(x - 3)$
Check: $(x + 1)(x - 3)$
$= x^2 - 3x + x - 3$
$= x^2 + 2x - 3$

16. $(x + 6)(x + 2)$
Check: $(x + 6)(x + 2)$
$= x^2 + 2x + 6x + 12$
$= x^2 + 8x + 12$

17. $(x - 3)(x - 4)$
Check: $(x - 3)(x - 4)$
$= x^2 - 3x - 4x + 12$
$= x^2 - 7x + 12$

18. $(x + 8)(x - 1)$
Check: $(x + 8)(x - 1)$
$= x^2 + 8x - x - 8$
$= x^2 + 7x - 8$

19. $(x - 2)(x + 5)$
Check: $(x - 2)(x + 5)$
$= x^2 - 2x + 5x - 10$
$= x^2 + 3x - 10$

20. $(x + 7)(x + 3)$
Check: $(x + 7)(x + 3)$
$= x^2 + 7x + 3x + 21$
$= x^2 + 10x + 21$

21. $(x - 8)(x - 5)$
Check: $(x - 8)(x - 5)$
$= x^2 - 5x - 8x + 40$
$= x^2 - 13x + 40$

22. $(x - 4)(x + 3)$
Check: $(x - 4)(x + 3)$
$= x^2 + 3x - 4x - 12$
$= x^2 - x - 12$

23. $(x + 2)(x - 10)$
Check: $(x + 2)(x - 10)$
$= x^2 - 10x + 2x - 20$
$= x^2 - 8x - 20$

24. $(x - 9)(x - 2)$
Check: $(x - 9)(x - 2)$
$= x^2 - 2x - 9x + 18$
$= x^2 - 11x + 18$

25. $(x + 5)(x - 11)$
Check: $(x + 5)(x - 11)$
$= x^2 - 11x + 5x - 55$
$= x^2 - 6x - 55$

26. $(x + 12)(x + 4)$
Check: $(x + 12)(x + 4)$
$= x^2 + 4x + 12x + 48$
$= x^2 + 16x + 48$

27. $(x + 9)(x - 2)$
Check: $(x + 9)(x - 2)$
$= x^2 + 9x - 2x - 18$
$= x^2 + 7x - 18$

28. $(x - 4)(x - 6)$
Check: $(x - 4)(x - 6)$
$= x^2 - 4x - 6x + 24$
$= x^2 - 10x + 24$

29. $(x + 5)(x + 5)$
Check: $(x + 5)(x + 5)$
$= x^2 + 5x + 5x + 25$
$= x^2 + 10x + 25$

30. $(x - 7)(x + 1)$
Check: $(x - 7)(x + 1)$
$= x^2 - 7x + x - 7$
$= x^2 - 6x - 7$

GED Practice: Lesson 21 (Pages 222–223)

1. **D: (1,1), E: (2,4.5), F: (5,4)**

2. **A: (−3,3), B: (−5,1)**

3. **C: (0,3), K: (0,−3), L: (0,2), M: (0,5)**

4. **D: (1,1)**

5. **F: (5,4), G: (5,0), H: (2,0), K: (0,−3)**

6. **Points I, L**

7. **Points A, B, C, E, M**

8. **Points A, B, C, D, H, I, J, K, L**

9. **Point E**

10. **Points A, B, C**

11. **Slope = −1** Use the points (1,0) and (0,1). Find the slope using the slope formula.
$$m = \frac{y_2 - y_1}{x_2 - x_1} = \frac{1 - 0}{0 - 1} = \frac{1}{-1} = -1$$

12. **Slope = $\frac{3}{2}$** Use the points (−2,1) and (0,4). Find the slope using the slope formula.
$$m = \frac{y_2 - y_1}{x_2 - x_1} = \frac{4 - 1}{0 - (-2)} = \frac{3}{2} = 1\frac{1}{2}$$

13. **Slope = 1** Use the points (2,1) and (0,−1). Find the slope using the slope formula.
$$m = \frac{y_2 - y_1}{x_2 - x_1} = \frac{-1 - 1}{0 - 2} = \frac{-2}{-2} = 1$$

14. **(2) $x - y = -1$** Substitute the three values into each equation until all three make the equation true. Only option (2) is correct.

15. **(3) $y = -1$** Substitute 2 for x in the equation:
$$\begin{aligned} 3x - 4y &= 10 \\ 3(2) - 4y &= 10 \\ 6 - 4y &= 10 \\ -4y &= 4 \\ y &= -1 \end{aligned}$$

16. **(5) A, B, and C** Substitute 2 for x and 1 for y in each equation. The pair (2,1) is a solution for each equation.

17. **(1) (−3,−4)** Substitute each value into the equation:
$(-3) - (-4) = (-3) + (+4) = 1$ (true)
$(-3) - (-2) = (-3) + (+2) = -1$ (false)
$(-1) - (0) = -1$ (false)
$(0) - (+1) = 0 + (-1) = -1$ (false)
$(+1) - (-2) = (+1) + (+2) = 3$ (false)
Therefore, only option (1) is correct.

18. **(2) B** Substitute −4 for x and 2 for y in the equations:
$2(-4) - 3(2) = -8 - 6 = -14$
Because $-14 < 5$, it is found in the graph of B.

19. **(5) Neither B nor C is a solution.** Substitute each point value into the equation.
$2x - y = 1$
For A: (3,5) $6 - 5 \stackrel{?}{=} 1$ (true)
For B: (1,−3) $2 + 3 \stackrel{?}{=} 1$ (false)
For C: (0,1) $0 - 1 \stackrel{?}{=} 1$ (false)
Therefore, option (5) is correct.

20. **(2) −2** Substitute the known value $y = 1$ into the equation and solve for x.
$$\begin{aligned} -4x + 7y &= 15 \\ -4x + 7(1) &= 15 \\ -4x + 7 &= 15 \\ -4x &= 8 \\ x &= -2 \end{aligned}$$

21. **(4) $x + y = 2$** Every point on the line will make this equation true. Pick any point, such as (0,2) and substitute the values into the equation. Only option (4) gives a true statement: $0 + 2 = 2$.

UNIT 2

Lesson 22

All items relate to the skill of application.

GED Practice: Lesson 22 (Pages 226–229)

1. **16** $2^4 = 2 \times 2 \times 2 \times 2 = 16$
2. **1** $1^3 = 1 \times 1 \times 1 = 1$
3. **27** $3^3 = 3 \times 3 \times 3 = 27$
4. **25** $5^2 = 5 \times 5 = 25$
5. **1,024** $4^5 = 4 \times 4 \times 4 \times 4 \times 4 = 1{,}024$
6. **81** $3^4 = 3 \times 3 \times 3 \times 3 = 81$
7. **125** $5^3 = 5 \times 5 \times 5 = 125$
8. **64** $2^6 = 2 \times 2 \times 2 \times 2 \times 2 \times 2 = 64$
9. **9** $3^2 = 3 \times 3 = 9$
10. **36** $6^2 = 6 \times 6 = 36$
11. **625** $5^4 = 5 \times 5 \times 5 \times 5 = 625$
12. **32** $2^5 = 2 \times 2 \times 2 \times 2 \times 2 = 32$
13. **10,000** $10^4 = 10 \times 10 \times 10 \times 10 = 10{,}000$
14. **64** $8^2 = 8 \times 8 = 64$
15. **64** $4^3 = 4 \times 4 \times 4 = 64$
16. **8** $2^3 = 2 \times 2 \times 2 = 8$
17. **second** $10^2 = 10 \times 10 = 100$
18. **6** $6^2 = 6 \times 6 = 36$
19. **5**
20. **32** $2^5 = 2 \times 2 \times 2 \times 2 \times 2 = 32$
21. **sixth** $10^6 = 1{,}000{,}000$ (6 zeros)
22. **6,000** $6 \times 1{,}000 = 6{,}000$
23. **300** $3 \times 100 = 300$
24. **800** $8 \times 10 \times 10 = 8 \times 100 = 800$
25. **40** $4 \times 10 = 40$
26. **560,000** $56 \times 10{,}000 = 560{,}000$
27. **12** $2 \times 2 + 2 \times 2 \times 2 = 4 + 8 = 12$
28. **13** $3 \times 3 + 2 \times 2 = 9 + 4 = 13$
29. **100** $5 \times 5 \times 4 = 25 \times 4 = 100$
30. **586** $7 \times 7 \times 7 + 3 \times 3 \times 3 \times 3 \times 3 = 7 \times 49 + 3 \times 9 \times 9 = 343 + 243 = 586$
31. **15** $5 \times 5 - 3 \times 3 - 1 \times 1 = 25 - 9 - 1 = 15$
32. **3** $2 \times 2 \times 2 \times 2 \times 2 - 5 \times 5 - 2 \times 2 =$ $32 - 25 - 4 = 3$
33. **29** $4 \times 4 + 3 \times 3 + 2 \times 2 = 16 + 9 + 4 = 29$
34. **1** Any number to the zero power is 1.
35. **5**
36. **0.0001** $\frac{1}{10 \times 10 \times 10 \times 10} = \frac{1}{10{,}000} = 0.0001$
37. **0.01** $\frac{1}{10 \times 10} = \frac{1}{100} = 0.01$
38. **1** Any number to the zero power is 1.
39. **6,300** $6 \times 1{,}000 + 3 \times 100 = 6{,}000 + 300 = 6{,}300$
40. **350** $4 \times 100 - 5 \times 10 = 400 - 50 = 350$
41. **28,000** $3 \times 10{,}000 - 2 \times 1{,}000 = 30{,}000 - 2{,}000 = 28{,}000$
42. **504,000** $5 \times 100{,}000 + 4 \times 1{,}000 = 500{,}000 + 4{,}000 = 504{,}000$
43. **3,087** $3.1 \times 1{,}000 - 1.3 \times 10 = 3{,}100 - 13 = 3{,}087$
44. **6,750** $6.25 \times 1{,}000 + 5 \times 100 = 6{,}250 + 500 = 6{,}750$
45. **19,000** $2.5 \times 10{,}000 - 6 \times 1{,}000 = 25{,}000 - 6{,}000 = 19{,}000$
46. **220** $1.75 \times 100 + 4.5 \times 10 = 175 + 45 = 220$
47. $\mathbf{\frac{1}{4}}$ $\frac{1}{2} \times \frac{1}{2} = \frac{1}{4}$
48. $\mathbf{\frac{4}{9}}$ $\frac{2}{3} \times \frac{2}{3} = \frac{4}{9}$
49. $\mathbf{\frac{1}{9}}$ $\frac{1}{3} \times \frac{1}{3} = \frac{1}{9}$
50. $\mathbf{\frac{1}{8}}$ $\frac{1}{2} \times \frac{1}{2} \times \frac{1}{2} = \frac{1}{8}$
51. $\mathbf{\frac{1}{64}}$ $\frac{1}{4} \times \frac{1}{4} \times \frac{1}{4} = \frac{1}{64}$
52. $\mathbf{\frac{1}{125}}$ $\frac{1}{5} \times \frac{1}{5} \times \frac{1}{5} = \frac{1}{125}$
53. **0.04** $0.2 \times 0.2 = 0.04$
54. **0.001** $0.1 \times 0.1 \times 0.1 = 0.001$
55. **0.064** $0.4 \times 0.4 \times 0.4 = 0.064$
56. **0.0025** $0.05 \times 0.05 = 0.0025$
57. **0.000064** $0.04 \times 0.04 \times 0.04 = 0.000064$

58. 0.000008
$0.02 \times 0.02 \times 0.02 = 0.000008$

59. 0.41

60. 6.2 × 0.1

61. 0.18

62. 6.1×10^{-2}

63. 4.65 × 0.01

64. 6 centimeters
Since $6^2 = 36$, the length of each side is 6 centimeters.

65. 10 feet
Since $10^2 = 100$, the length of each side is 10 feet.

66. 4 yards Since $4^2 = 16$, the length of each side is 4 yards.

67. 7 inches Since $7^2 = 49$, the length of each side is 7 inches.

68. 9 meters
Since $9^2 = 81$, the length of each side is 9 meters.

69. 8 centimeters
Since $8^2 = 64$, the length of each side is 8 centimeters.

70. 4 $4 \times 4 = 16$

71. 0 $0 \times 0 = 0$

72. 10 $10 \times 10 = 100$

73. 9 $9 \times 9 = 81$

74. 7 $7 \times 7 = 49$

75. 11 $11 \times 11 = 121$

76. 13
You know $12 \times 12 = 144$.
Try $13 \times 13 = 169$.

77. 1 $1 \times 1 = 1$

78. 12 $12 \times 12 = 144$

79. (3) 865,000 mi. Add a string of zeros to 8.65 and move the decimal point 5 places to the right.
8.65000.0000

80. (4) Only $3^4 = 64$ is false.
$3^4 = 81$, not 64, so $3^4 = 64$ is false. Both $8^2 = 64$ and $4^3 = 64$ $(4 \times 4 \times 4)$ are true.

81. (3) 4 and 5 You know the square of 4 is 16, and the square of 5 is 25. So, the square root of 22 must fall between 4 and 5.

82. (1) 4.7×10^{-1}, 2.34×10^2, 5.2×10^2
Compare the numbers and arrange them in order:
$4.7 \times 10^{-1} = \frac{4.7 \times 1}{10} = 0.47$
$2.34 \times 10^2 =$
$2.34 \times 100 = 234$
$5.2 \times 10^2 =$
$5.2 \times 100 = 520$

83. (4) 10^6
Multiply:
$10^2 \times 10^4 = 100 \times 10{,}000 = 1{,}000{,}000 = 10^6$
(OR: $10^2 \times 10^4 = 10^{2+4} = 10^6$)

84. (3) 3^4 is the least and 2^7 is the greatest.
Find the value of each and then compare the values:
$10^2 = 10 \times 10 = 100$
$2^7 = 2 \times 2 \times 2 \times 2 \times 2 \times 2 \times 2 = 128$
$3^4 = 3 \times 3 \times 3 \times 3 = 81$

85. (4) They are both wrong, since the expression equals 1.
Recall that $3^2 = 3 \times 3 = 9$ and $2^3 = 2 \times 2 \times 2 = 8$. Therefore, $9 - 8 = 1$. Option (4) is the only correct answer.

86. (5) Multiply by 2, and then multiply by 2 again. Because $12 - 10 = 2$, you need to multiply by 2 two more times:
$2^{12} = 2^{10} \times 2 \times 2$

UNIT 2

GED Practice: Solving Word Problems (Page 231)

1. **(4) 14, 15, and 16** Add the numbers in the options. Only option (4) totals 45: $14 + 15 + 16 = 45$.

2. **(5) 300** Eva drove 100 miles less than Juan and 200 miles less than Pam. Go through the answer options to find an option that equals 1,200 when you add it to a number 100 more than the option to a number 200 more than the option. Only option (5) works.
$300 + 400 + 500 = 1{,}200$

3. **(3) 47 and 48** Add the numbers in the options. Only option (3) totals 95: $47 + 48 = 95$.

4. **(2) 8, 9, 10, and 11** Add the numbers for each answer option. Only option (2) totals 38: $8 + 9 + 10 + 11 = 38$.

5. **(3) 14 and 16** Add the number of days in the options. Only option (3) totals 30: $14 + 16 = 30$.

6. **(1) 9, 11, and 13** Add the numbers in the options. Only option (1) totals 33: $9 + 11 + 13 = 33$.

7. **(2) 44 and 49** Add the scores in the options. Only option (2) totals 93: $44 + 49 = 93$.

8. **(2) 31, 33, and 35** Add the numbers in the options. Options (2), (3), and (4) total 99. Option (3) contains two even numbers. The numbers in option (4) are not consecutive.

GED Mini-Test: Lessons 19–22 (Pages 232–233)

All items relate to the skill of application.

1. **(1) A and B** Perform each operation:
   A: $(-2) + (-7) = -9$
   B: $(-6) + (+8) = +2$
   C: $(-3) - (-4) = +1$
   D: $(+4) - (+10) = -6$
   The point farthest to the left is -9, and the point farthest to the right is $+2$.

2. **(1) $\frac{2}{x} - 9x$** The "product of 9 and x" can be written $9x$. The "quotient of 2 and x" is 2 divided by x or $\frac{2}{x}$. The statement says the product $9x$ is subtracted from $\frac{2}{x}$. Write the expression $\frac{2}{x} - 9x$.

3. **(3) $\frac{1}{2}$** Find the coordinates of points A and B on the line: Point A $(-3,-2)$, Point B $(3,1)$. Use the formula for slope:
   $$m = \frac{y_2 - y_1}{x_2 - x_1} = \frac{1-(-2)}{3-(-3)} = \frac{3}{6} = \frac{1}{2}$$

4. **(2) $x = -3$**
   $2(2x + 3) = -3(x + 5)$
   $4x + 6 = -3x - 15$
   $7x = -21$
   $x = -3$

5. **(1) 17** Let x represent Joe's number. "Three more than the number I'm thinking of is 30" can be written $3 + x = 30$.
   Solve for Joe's number: $3 + x = 30$
   $x = 27$
   "Then 10 less than your number is the number __________."
   Subtract 10 from Joe's number: $27 - 10 = 17$.

6. **(1) $(x - 4) - 3$**

	Coop's Age in Months	Carni's Age in Months
Now	1 x	2 $x - 4$
3 months ago	3 $x - 3$	4 $(x - 4) - 3$

   Box 1 is x, Coop's age now. Box 2 is $x - 4$ because Carni must be 4 months younger. To complete these ages 3 months ago, subtract 3 from each expression.

7. **(3) $(-6) + (+13)$** The first arrow (closest to the number line) shows a move of 6 units in a negative direction (to the left). The second arrow shows a move of 13 units in a positive direction (to the right). Check your answer by adding: $(-6) + (+13) = +7$. The second row ends at $+7$ so the answer is correct.

8. **(4) $3(x - 11) = 2x - 11$**
   Create a chart:

	Becky's Age	Her Mother's Age
Now	x	$2x$
11 Years Ago	$x - 11$	$2x - 11$

   Eleven years ago, her mother was three times as old as Becky. Set up an equation.
   $3(x - 11) = 2x - 11$

9. **(4) 15 and 16** Let x represent the first number and $x + 1$ the second number. Set up an equation: When 13 is subtracted from the sum of the two numbers the answer is 18.
   Solve $x + x + 1 - 13 = 18$
   $2x - 12 = 18$
   $2x = 30$
   $x = 15; x + 1 = 16$
   The first number is 15; the second number is 16.

10. **(3) $(x - 6)(x - 6)$** The factors of 36 are: 1 and 36, 2 and 18, 3 and 12, 4 and 9, and 6 and 6. The number part of the middle term must be -12. The last pair of factors total 12. Both 6s are subtracted from x because the middle term must be negative ($-6 + -6 = -12$) and the final term must be positive ($-6 \times -6 = +36$).

11. **(4) $x = 9$**
   $-2(3 + x) - 4 = -3x - 1$
   $-6 - 2x - 4 = -3x - 1$
   $-10 - 2x = -3x - 1$
   $-10 + x = -1$
   $x = 9$

12. **(1) $x > -2$**
   $5x + 2 < 6x + 3x + 10$
   $5x + 2 < 9x + 10$
   $-4x < 8$
   $x > -2$
   Reverse the sign when you divide both sides by -4.

UNIT 2

13. **(3)** 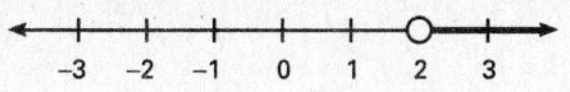

Solve: $x - 3 < -1$
$x < 2$
The graph must have an empty circle at 2 and point to the left.

14. **(5) 2.4×10^{-3}** 0.0024
The decimal is moved three places to the left when a number is multiplied by 10^{-3}.

Cumulative Review: Unit 2: Algebra (Pages 234–240)

All items relate to the skill of application.

1. **(3) 0.001** $10^{-3} = \frac{1}{10 \times 10 \times 10} = 0.001$

2. **(5) $(x - 3)(x - 5)$**
The factors of 15 are $15 \cdot 1$, $-15 \cdot -1$, $5 \cdot 3$, and $-5 \cdot -3$. Only -5 and -3 sum to -8.
Check: $(x - 3)(x - 5)$
$= x^2 - 5x - 3x + 15$
$= x^2 - 8x + 15$.

3. **(5) C only** Try the pair in each equation. Only equation C is true. $-1 + -2 = -3$

4. **(3) $4 + (-7) + 4 = 1$** The bottom arrow moves 4 units to the right (a positive direction). The second arrow moves 7 units to the left (a negative direction). Finally the top arrow moves 4 units back to the right. Only option (3) shows these 3 moves.

5. **(3) $-x - 5$** $2 - (x + 7) = 2 - x - 7 = -x - 5$

6. **(2) −19** $4x - 2y + xy =$
$4(-1) - 2(5) + (-1)(5) =$
$(-4) - 10 + (-5) = -19$

7. **(3) 21** Multiply both terms within the parentheses by 7: $7 \times 3 = 21$.

8. **(1) $\frac{1}{2}$**
$-6(x + 1) + 4 = 8x - 9$
$-6x - 6 + 4 = 8x - 9$
$-6x - 2 = 8x - 9$
$-14x = -7$
$x = \frac{1}{2}$

9. **(1) $x < -1$**
$3x + 7 < -2x + 2$
$5x + 7 < 2$
$5x < -5$
$x < -1$

10. **(2) Two less than x is the same as 7 more than the quotient of x and 4.** Compare each statement with the equation in the problem. Remember, "quotient" means divided by.

11. **(4) $(-6 + y) - (-6x)$** "Product" means multiply. The product of -6 and x is $-6x$. "Sum" means add. The sum of -6 and y is $-6 + y$. Only option (4) shows $-6x$ subtracted from $-6 + y$.

12. **(5) $\frac{x}{6} - 2$** The quotient of x and 6 is x divided by 6 or $\frac{x}{6}$. Two less means to subtract 2 from the quotient.

13. **(3) $4x - 3$**
$7x + 5 - 3x - 8$
$4x + 5 - 8$
$4x - 3$

14. **(4) $P = 16x + 4y$** Total the number of x's and y's in the drawing:
$x + x + x + x + x + x + 5x + 5x = 16x$
$y + y + y + y = 4y$.

15. **(4) $x \le 2$**
$5x + 3 \le 17 - 2x$
$7x + 3 \le 17$
$7x \le 14$
$x \le 2$

16. **(2) $x - 12$**
Divide $5x - 60$ by 5.
$\frac{5x - 60}{5} = x - 12$

17. **(2) 1 and 8** Let x represent the first number. The second number can be written $8x$.
$8x - x = 2 + 5x$
$7x = 2 + 5x$
$2x = 2$
$x = 1$
The first number is 1, and the second number is 8.

18. **(5) All are false.** Evaluate each statement. The symbol $<$ means less than. None of the statements is true.

19. **(5) $x > 1$**
$6 - 5x < 7x - 6$
$6 - 12x < -6$
$-12x < -12$
$x > 1$

20. **(2) +7** The series of plays can be written:
$+7 - 9 - 2 + 1 + 10$
Add the numbers:
$7 - 9 - 2 + 1 + 10 = 7$.

21. (2) A and C Evaluate each statement:
A. $4 + (1 - 4) = 4 - 3 = 1$
B. $(7 - 3) + 1 = 4 + 1 = 5$
C. $6 - (3 + 2) = 6 - 5 = 1$
D. $8 - (1 - 2) = 8 + 1 = 9$
A and C have the same value.

22. (1) −6
$\frac{-12}{(4 + (-2))} = \frac{-12}{2} = 6$

23. (3) 4^5 Evaluate each of the options: $2^5 = 32$, $4^3 = 64$, $4^5 = 1{,}024$, $5^4 = 625$, $6^3 = 216$. Only option (3) is greater than 1,000.

24. (4) 9.8×10^6 9800000. You would need to move the decimal point 6 places to the left, so 9.8×10^6 is the correct expression.

25. (2) $3x - 2y = -5$ Evaluate each equation using the coordinates (−1,1). Only option (2) is a true statement using these values.
$3x - 2y = -5$
$3(-1) - 2(1) = -5$
$-3 - 2 = -5$

26. (4) (0,−2) and (5,0) Start by substituting 0 for x and solving for y:
$2(0) - 5y = 10$
$-5y = 10$
$y = -2$. One pair of coordinates is (0,−2). Only options (2) and (4) include this pair. Now substitute 5 for x and solve for y:
$2(5) - 5y = 10$
$10 - 5y = 10$
$-5y = 0$
$y = 0$. The second pair is (5,0). Option (4) is correct.

27. (3) $x^2 + 7x - 18$
$(x + 9)(x - 2) = x^2 - 2x + 9x - 18 = x^2 + 7x - 18$

28. (1) $x \leq -2$ The direction to the left shows "less than." Since the dot is filled in, you know the correct symbol is $\leq$.

29. (5) $x > \frac{-10}{7}$ You need to divide both sides by −7. Dividing or multiplying by a negative number changes the direction of the sign.

30. (4) 20 Let x equal the number. Write an equation:
$x - 10 = \frac{x}{2}$. Multiply both sides by 2 to get rid of the fraction:
$2x - 20 = x$
$-20 = -x$
$20 = x$

31. (3) $4x(2x + 3)$ Divide by $4x$: $\frac{8x^2 + 12x}{4x} = 2x + 3$.

32. (3) negative x-value and positive y-value The intersection of the two lines is the only point that belongs to both lines. The intersection point is in the upper left quadrant. It would have a negative x-value because it is left of the y-axis and a positive y-value because it is above the x-axis.

33. (1) A Line A rises more rapidly than Line B. Line C has a negative slope.

34. (3) C Moving left to right, Line C goes in a downward direction. It has a negative slope.

35. (5) $(x + 1)(x - 2)$ The factors of −2 are $-1 \cdot 2$ and $1 \cdot -2$. Adding the last pair results in a negative middle term.
Check: $(x + 1)(x - 2) = x^2 - 2x + x - 2 = x^2 - x - 2$.

36. (1) $1 < x < 4$ Only the section between +1 and +4 is shaded. The shaded area is greater than 1 and less than 4.

37. (3) 6 and 7 The square of 6 is 36, and the square of 7 is 49, so the square root of 45 must be between 6 and 7.

38. (2) 17 Let x represent Caroline's age. $2x - 1$ can represent Bill's age. Write an equation and solve:
$x + 2x - 1 = 26$
$3x - 1 = 26$
$3x = 27$
$x = 9$
Bill is 17. $2(9) - 1 = 17$

39. (2) 0
$3^2 + 2^4 - 5^2$
$9 + 16 - 25 = 0$

40. (4) $x = 12$
$5x - 24 = 3x - x + 12$
$5x - 24 = 2x + 12$
$3x = 36$
$x = 12$

41. (1) $-4x = -5x + 2 + 8$ The product of a number and −4 becomes $-4x$; is 8 more than 2 becomes $= 2 + 8$; added to −5 times a number becomes $= -5x + 2 + 8$. Therefore, $-4x = -5x + 2 + 8$.

42. (2) $6x - 12$ Multiply 6 by both terms in the parentheses: $6(x - 2) = 6x - 12$.

UNIT 2

43. **(2) $x = -5$**
$12 + 3x = x + 2$
$2x = -10$
$x = -5$

44. **(5) $(x + 4)(x - 6)$** The possible factors of -24 are 4×-6, -4×6, 3×-8, -3×8, 2×-12, -2×12, 1×-24, and -1×24. Using the first pair, a middle term of -2 can result. $(-6 + 4 = -2)$. Check: $(x + 4)(x - 6) = x^2 - 6x + 4x - 24 = x^2 - 2x - 24$.

45. **(1) A only** Evaluate each statement: 25 is greater than 10. Statement A is true. $8^2 = 64$ so statement B is false. $4^2 = 16$ so statement C is false.

46. **(4) All are equal.** Evaluate each expression.
A. $10 - (2)(2) = 10 - 4 = 6$
B. $2 + \frac{8}{2} = 2 + 4 = 6$
C. $3 - \frac{12}{-4} = 3 - (-3) = 6$

47. **(3) C, G** The positive x-values are to the right of the y-axis. The negative y-values are below the x-axis. Only C and G are in the lower right quadrant.

48. **(3) I and K** The points on the x-axis have a y-value of 0. Only I and K are on the x-axis.

49. **(3) C** You could either graph the line or substitute the coordinates of each point into the equation. To graph the line, substitute any value for x and solve for y. If $x = 2$, $x - y = 8$; $2 - y = 8$. Then $y = -6$. This gives you the first pair of coordinates. Repeat with a different value for x. Graph the two pairs of coordinates and draw a line through them. Point C is on the line. To prove that point C is on the line, substitute $(6,-2)$, the coordinates of point C, and solve the statement: $x - y = 8$; $6 - (-2) = 8$. Point C is on the line.

50. **(5) L** See the explanation for item 49 to graph the line. Use the coordinates for L $(0,-2)$ to check the statement: $x + 2y = -4$; $0 + 2(-2) = -4$.

51. **(1) $\frac{1}{4}$** The coordinates for point I are $(-5,0)$. The coordinates for point E are $(-1,1)$.
Use the formula: $m = \frac{y_2 - y_1}{x_2 - x_1}$

$$m = \frac{1 - 0}{-1 - -5} = \frac{1}{4}$$

52. **(4) -1** The coordinates for point J are $(0,4)$. The coordinates for point K are $(4,0)$.
Use the formula: $m = \frac{y_2 - y_1}{x_2 - x_1}$

$$m = \frac{0 - 4}{4 - 0} = \frac{-4}{4} = -1$$

53. **(5) $x + 6x + x + 6x > 110$** Let x represent the width and $6x$ represent the length. The perimeter can be written $x + 6x + x + 6x$. This expression must be greater than 110.

54. **(2) $x \geq 3$**
$2x + 5 \leq 4x - 1$
$-2x \leq -6$
$x \geq 3$

55. **(1) $3 + 2x = -x$**

Twice a number	$2x$
added to 3	$3 + 2x$
is equal to the negative of the number.	$3 + 2x = -x$

56. **(5) -27**
$-\frac{1}{3}x = 9$
$(-3)\left(-\frac{1}{3}x\right) = 9(-3)$
$x = -27$

57. **(4) $x^2 - 15x + 54$**
$(x - 6)(x - 9) = x^2 - 6x - 9x + 54 = x^2 - 15x + 54$

58. **(4) 8 and 9** The square of 8 is 64, and the square of 9 is 81, so the square root of 70 must be between 8 and 9.

59. **(2) $(x - 3)(x + 4)$** The possible factors of -12 are (3×-4), (-3×4), (2×-6), (-2×6), (1×-12), (-1×12). Using the second pair, a middle term of $+1$ can result. $(-3 + 4 = +1)$ Check: $(x - 3)(x + 4) = x^2 - 3x + 4x - 12 = x^2 + x - 12$.

60. **(3) $-2x + 3y = -6$** Evaluate each equation using the coordinates $(0,-2)$. Only option (3) is a true statement using these values.
$-2x + 3y = -6$
$-2(0) + 3(-2) = -6$
$0 - 6 = -6$

61. **(2) 4.51×10^5** 4.51000 You would need to move the decimal point 5 places to the left, so 4.51×10^5 is the correct expression.

62. **(1) 5^3** Evaluate each of the options: $5^3 = 125$, $5^4 = 625$, $4^6 = 4{,}096$, $6^4 = 1{,}296$, $4^5 = 1{,}024$. Only option (1) is less than 500.

63. (3) $x = 4$

$3x + 12 = 6x - x + 4$
$3x + 12 = 5x + 4$
$8 = 2x$
$4 = x$

64. (3) $5x - 15$ Multiply 5 by both terms in the parentheses: $5(x - 3) = 5x - 15$.

UNIT 3: GEOMETRY
Lesson 23

All items relate to the skill of application.

GED Practice: Lesson 23 (Pages 248–249)

1. **Rectangular solid**
$V = l \times w \times h$
$= 15 \times 8 \times 6$
$=$ **720 cubic inches**

2. **Cube**
$V = s \times s \times s$
$= 5.1 \times 5.1 \times 5.1$
$=$ **132.651 cubic inches**

3. **Cylinder**
$V =$ Area of base $\times h$
$= 18 \times 10$
$=$ **180 cubic feet**

4. **Pyramid**
$V = \frac{1}{3} \times$ Area of base $\times h$
$= \frac{1}{3} \times 35 \times 9$
$=$ **105 cubic yards**

5. **Pyramid**
$V = \frac{1}{3} \times$ Area of base $\times h$
$= \frac{1}{3} \times 22\frac{1}{2} \times 6$
$=$ **45 cubic yards**

6. **Cone**
$V = \frac{1}{3} \times$ Area of base $\times h$
$= \frac{1}{3} \times 12.5 \times 6$
$=$ **25 cubic yards**

7. **254 cubic feet**
$V = \text{A} \times h$
$= 28.26 \times 9$
$=$ **254 cubic feet**

8. **377 cubic inches**
$V = \frac{1}{3} \times$ Area of base $\times h$
$= \frac{1}{3} \times 113.04 \times 10$
$=$ **377 cubic inches**

9. **15 cubic yards**
$V = \frac{1}{3} \times$ Area of base $\times h$
$= \frac{1}{3} \times 6 \times 7\frac{1}{2}$
$=$ **15 cubic yards**

10. **(3) 1,440**
$V = 15 \times 24 \times 4 = 1{,}440$ cu. in.

11. **(4) 8** $V = 2 \times 2 \times 2 = 8$ cu. yd.

12. **(4) 12,500**
$V = 100 \times 25 \times 5 = 12{,}500$ cu. ft.

13. **(5) 576** $V = 12 \times 8 \times 6 = 576$ cu. ft.

14. **(3) 54.6** $V = 9.1 \times 6 = 54.6$ cu. ft.

15. **(3) 33.6** $V = \frac{1}{3} \times 12.6 \times 8 = 33.6$ cu. ft.

16. **(2) $37\frac{1}{2}$** $V = \frac{1}{3} \times 9 \times 12\frac{1}{2} = 37\frac{1}{2}$ cu. ft.

17. **(3) 25** $V = \frac{1}{3} \times 12.56 \times 6 =$
$25.12 = 25$ cu. in. (rounded)

GED Practice: Solving Word Problems (Page 251)

1. **(2) $7\frac{1}{2}$** Convert 33 inches to feet: $\frac{33}{12} = 2\frac{3}{4}$ ft.
$P = 2l + 2w$
$= 2(2\frac{3}{4}) + 2(1)$
$= 5\frac{1}{2} + 2$
$= 7\frac{1}{2}$ ft.

2. **(3) 396** Convert 1 foot to inches: $1 \times 12 = 12$ in.
$A = lw$
$= 33(12)$
$= 396$ sq. in.

3. **(4) 4,748** Convert $3\frac{1}{2}$ feet to inches:
$3\frac{1}{2} \times 12 = 42$ inches. Find the area of the base: The radius is $\frac{1}{2}$ the diameter.
$A = \pi r^2$
$= 3.14(6^2)$
$= 113.04$ sq. in.
Find the volume of the cylinder:
$V = Ah$
$= 113.04(42)$
$= 4{,}747.68$ which rounds to 4,748 cu. in.

4. **(4) between 6 and 7**
Convert the width and height to feet:
1 ft. 6 in. $= 1\frac{6}{12} = 1\frac{1}{2}$ ft.
1 ft. 9 in. $= 1\frac{9}{12} = 1\frac{3}{4}$ ft.
Find the volume of the box:
$V = lwh$
$= 2\frac{1}{2}(1\frac{1}{2})(1\frac{3}{4})$
$= \frac{5}{2} \times \frac{3}{2} \times \frac{7}{4}$
$= \frac{105}{16}$
$= 6\frac{9}{16}$

5. **(3) 15,000** Convert the base measurement to centimeters:
$2 \times 100 \frac{\text{cm.}}{\text{m.}} = 200$ cm.
$A = \frac{1}{2}bh$
$= \frac{1}{2}(200)(150)$
$= 15{,}000$ sq. cm.

6. **(3) 96** Convert 1 ft. 4 in. to inches: $(1 \times 12) + 4 = 16$ in. Find the volume of the cone:
$V = \frac{1}{3} \times$ Area of base $\times h$
$= \frac{1}{3} \times 16 \times 18$
$= 96$

Lesson 24

All items relate to the skill of application.

GED Practice: Lesson 24 (Pages 255–257)

1. **Obtuse** Angles that measure between 90° and 180° are obtuse angles, so an angle of measure 150° is obtuse.

2. **Acute** An angle that measures between 0° and 90° is an acute angle, so an angle of measure 35° is acute.

3. **Right** An angle that measures exactly 90° is a right angle.

4. **Straight** An angle that measures exactly 180° is a straight angle.

5. **Acute**

6. **Obtuse**

7. **Obtuse**

8. **Acute**

9. **∠2 and ∠4** Adjacent angles share a common vertex and a common ray. There are two angles adjacent to, or touching, ∠3, namely ∠2 and ∠4.

10. **∠5** is the vertical angle to ∠2. These two angles are made up of the same two intersecting lines. They are opposite angles.

11. **∠3** is the vertical angle to ∠6.

12. **∠4** ∠4 is the vertical angle to 1.

13. **∠6 and ∠2** Both ∠6 and ∠2 are adjacent to ∠1.

14. **∠6 and ∠4** Both ∠6 and ∠4 are adjacent to ∠5.

15. **∠2** is congruent to ∠5 because vertical angles are congruent.

16. **∠1** is congruent to ∠4 because vertical angles are congruent.

17. **∠6** is congruent to ∠3 because vertical angles are congruent.

18. **42°** $\angle BXC$ and $\angle AXB$ are complementary. That is, their sum is 90°. Therefore, $90° - 48° = 42° = m\angle BXC$.

19. **90°** $\angle AXB$ and $\angle BXC$ are complementary. These two angles form $\angle AXC$. Therefore, $\angle AXC$ is 90° and $\angle AXD$ is a straight angle of 180°. Since
$\angle AXC + \angle DXC = \angle AXD$
$90° + \angle DXC = 180°$
$\angle DXC = 90°$

20. **132°** $\angle BXD = \angle BXC + \angle DXC$
$42° + 90° = 132°$

21. **$\angle CXA$** Supplementary angles sum to 180° or a straight angle. Therefore, $\angle DXC$ and $\angle CXA$ are supplementary.

22. **35°** $\angle ZXY$ and $\angle YXQ$ are complementary—their sum is 90°. $90° - 55° = 35° = m\angle ZXY$

23. **145°** $\angle ZXR$ and $\angle ZXY$ are supplementary.
$m\angle ZXY = 90° - 55° = 35°$
$m\angle ZXR + m\angle ZXY = 180°$
$m\angle ZXR = 180° - m\angle ZXY = 180° - 35° = 145°$

24. **125°**
$m\angle QXR + m\angle QXY = 180°$
$m\angle QXR = 180° - m\angle QXY = 180° - 55° = 125°$

25. **270°** The sum of the measures of the reflex angle and $\angle QXZ$ is 360°, a complete circle. Therefore, the measure of the reflex angle is $360° - 90° = 270°$.

26. **∠2, ∠3, ∠6, ∠7**

27. $\angle 7$

28. $\angle 1, \angle 4, \angle 5, \angle 8$

29. $\angle 1$

30. $\angle 3$

31. $\angle 5$

32. $\angle 6$

33. $\angle 4$

34. $\angle 7$

35. $\angle 4$

36. **(4) 6** Each of the four angles, 1, 2, 4, and 5, are acute angles. But the angle made up of the adjacent angles 1 and 2 is an acute angle, and so is the angle made up of the adjacent angles 4 and 5.

37. **(1) $\angle A$ is complementary to $\angle B$** The sum of the measures of $\angle A$ and $\angle B$ is $28° + 62°$, or $90°$. Angles which sum to $90°$ are complementary.

38. **(2) 50°** To say that two angles are complementary means that their measures sum to $90°$: $\angle M + \angle R = 90°$. Since the measure of $\angle M$ is $40°$, $90° - 40°$, or $50°$, is the measure of $\angle R$.

39. **(2) $\angle 5$** From the figure we see that both $\angle 2$ and $\angle 5$ are right angles, because $\angle 1$ is a right angle. $\angle XOZ$ is also a right angle. Therefore, any two of these four angles, $\angle 1$, $\angle 2$, $\angle 5$, and $\angle XOZ$ are supplementary. However, $\angle 1$ and $\angle XOZ$ are both adjacent to $\angle 2$. Therefore, of these possible angles, only $\angle 5$ fits the requirement that it not be adjacent to $\angle 2$.

40. **(4) 155°** $\angle WOY$ is made up of $\angle 4$ and $\angle 5$: $m\angle WOY = m\angle 4 + m\angle 5$. Since $\angle 3$ and $\angle 4$ make up a $90°$ angle and the measure of $\angle 3$ is $25°$, $\angle 4$ must be $90° - 25° = 65°$. Since $\angle 5$ is a right angle, the measure of $\angle 5$ is $90°$. Add $m\angle 4$ to $m\angle 5$: $65° + 90° = 155° = m\angle WOY$.

41. **(2) a right angle** $\angle XOZ$ is a right angle. Its supplement must also be a right angle because "supplement" means the sum of the angles is $180°$. Options (1) and (3) are eliminated because a right angle can be neither acute nor obtuse. Option (4) is not a necessary condition, and option (5) would not make an angle large enough to be supplementary.

42. **(5) $m\angle 1 + m\angle 7 = 180°$** $\angle 1$ and $\angle 8$ are alternate exterior angles, therefore $m\angle 1 = m\angle 8$. $\angle 8$ and $\angle 7$ are supplementary angles, therefore $m\angle 8 + m\angle 7 = 180°$. Since $m\angle 1 = m\angle 8$, then $m\angle 1 + m\angle 7 = 180°$.

43. **(5) $\angle Q$ and $\angle R$ have the same degree measure.** Option (5) is true because the meaning of "congruent" is that the angles "have the same degree of measure." Vertical angles are always congruent to each other and could be both acute, both obtuse, supplementary, or complementary. However, none of the options (1) through (4) is necessarily true.

44. **(2) $m\angle 2 = 100°$** $\angle 4$ and $\angle 2$ are vertical angles, so they have the same measure. Option (1) is eliminated because $\angle 4$ and $\angle 2$ are not complementary. Option (3) is eliminated because $\angle 1$ and $\angle 4$ are supplementary, so $m\angle 1 = 180° - 100° = 80°$. Option (4) is eliminated because $m\angle 1 = m\angle 3 = 180° - 100° = 80°$, and $80° + 80° = 160°$. Option (5) is eliminated because $\angle 4$ and $\angle 2$ are not supplementary. $m\angle 4 + m\angle 2 = 200°$.

45. **(1) $m\angle 2 + m\angle 5 = 180°$** $\angle 2$ and $\angle 6$ are corresponding angles, therefore $m\angle 2 = m\angle 6$. $\angle 6$ and $\angle 5$ are supplementary angles, therefore $m\angle 6 + m\angle 5 = 180°$. Since $m\angle 2 = m\angle 6$, then $m\angle 2 + m\angle 5 = 180°$.

46. **(3) 120°** $\angle 4$ and $\angle UZX$ are supplementary angles, so $m\angle 4 + m\angle UZX = 180°$, and $m\angle UZX = 180° - 60° = 120°$.

47. **(1) 30°** $\angle XZY$ and $\angle 4$ are complementary angles, so $m\angle XZY + m\angle 4 = 90°$, and $m\angle XZY = 90° - 60° = 30°$.

GED Practice: Solving Word Problems (Page 259)

1. **(3) 443** Draw a diagram:
$l = 30$ ft.
$w = \frac{1}{2}l = \frac{1}{2}(30) = 15$ ft.
Find the radius of the statue:

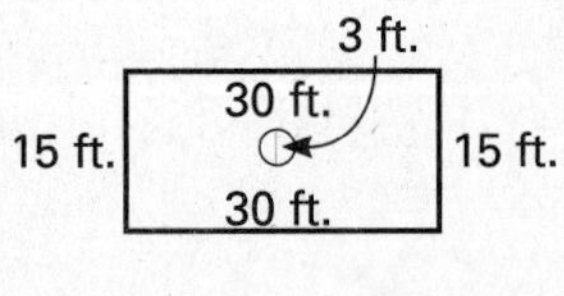

$r = \frac{1}{2}d = \frac{1}{2}(3) = 1.5$ ft.
Find the area of the statue:
$A = \pi r^2 = 3.14(1.5^2) = 7.065$ sq. ft.
Find the area of the garden:
$A = lw = 30(15) = 450$ sq. ft.
Find the difference: $450 - 7.065 = 442.935$ which rounds to 443 sq. ft.

UNIT 3

2. **(5) 104** Draw a diagram:

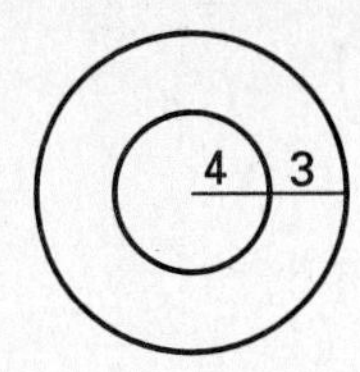

You can see that the radius of the pool and walkway (the outer circle) is 7 ft.
Find the area of the pool and walkway combined: $A = \pi r^2 = 3.14(7^2) = 153.86$ sq. ft.
Find the area of the pool:
$A = \pi r^2 = 3.14(4^2) = 50.24$ sq. ft.
Find the difference: $153.86 - 50.24 = 103.62$ which rounds to 104 sq. ft.

3. **(2) $67\frac{1}{2}$** Find the area:

15 ft.
6 ft. h = $4\frac{1}{2}$ft. 6 ft.
15 ft.

$A = bh = 15\left(4\frac{1}{2}\right) = 67\frac{1}{2}$

4. **(3) 80** Draw a diagram:

28 ft.
12 ft. 12 ft.
28 ft.

The width $= \frac{1}{2}(28) - 2 = 12$ ft.
Find the perimeter of the figure:
$P = 2l + 2w = 2(28) + 2(12) = 56 + 24 = 80$ ft.

5. **(3) 17,500** Draw a diagram:

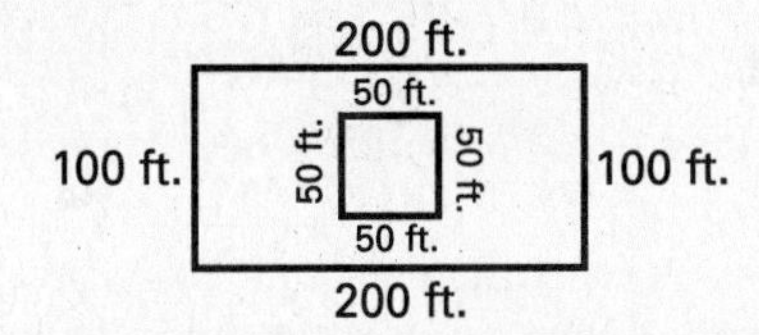

Divide the perimeter of the square blacktop area by 4 to find the length of one side:
$200 \div 4 = 50$ ft.
Find the area of the entire playground:
$A = lw = 200(100) = 20{,}000$ sq. ft.
Find the area of the blacktop portion:
$A = s^2 = 50^2 = 2{,}500$ sq. ft.
Find the difference:
$20{,}000 - 2{,}500 = 17{,}500$ sq. ft.

6. **(2) 128** Draw a diagram: The base is square, therefore all sides are 4 in. From the diagram it is easy to see that you are solving for the volume of a rectangular container.
$V = lwh = 4(4)(8) =$ 128 cu. in.

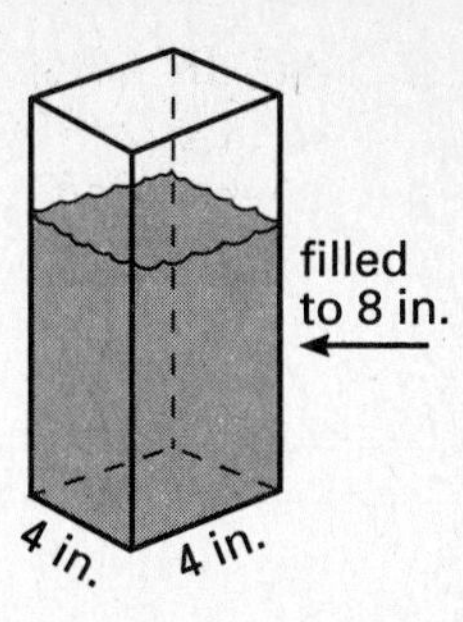

Lesson 25

All items relate to the skill of application.

GED Practice: Lesson 25 (Pages 262–263)

1. **$\triangle ABC$** Each side has the same length, 10. (Note: This triangle could also be named $\triangle ACB$, $\triangle BCA$, $\triangle BAC$, $\triangle CBA$, or $\triangle CAB$.)

2. **$\triangle ABD$** Two of its sides have the same length, 13. (The equilateral triangle, $\triangle ABC$, is a special case of an isosceles triangle.)

3. **$\triangle ACD$ and $\triangle BCD$** In each triangle, there is one angle that is obtuse, that is, greater than 90°.

4. **$\triangle ABE$, $\triangle ABC$, and $\triangle ABD$** All the angles are acute angles, that is, less than 90°.

5. **$\triangle ABE$, $\triangle ACD$, and $\triangle BCD$** No sides are equal.

6. **$\triangle ABE$**

7. **$\triangle BDC$ and $\triangle CDE$**

8. **$\triangle ACB$ and $\triangle BCE$**

9. **35°** The measures of all three angles must sum to 180°: $m\angle A + m\angle ABE + m\angle E = 180°$. Since $m\angle A = 55°$ and $m\angle ABE = 90°$, then $m\angle E = 180° - 55° - 90° = 35°$.

10. **35°** In $\triangle ABC$ we know that $\angle ACB = 90°$ and $m\angle A = 55°$, which sums to 145°. Then $m\angle ABC = 180° - 145° = 35°$.

11. **55°** In $\triangle DCE$, $m\angle CDE = 90°$ and $m\angle E = 35°$ (see item 9), which sum to 125°. Then, $m\angle DCE = 180° - 125° = 55°$.

12. **55°** In $\triangle BCE$, $m\angle BCE = 90°$ and $\angle E = 35°$ (item 9), which sum to 125°. Then, $m\angle CBD = 180° - 125° = 55°$. Also, since $\angle ABE = 90°$ and $\angle ABC = 35°$ (from item 10), then $\angle CBD = 90° - 35° = 55°$.

13. **35°** In $\triangle BCD$, $m\angle BDC = 90°$ and $m\angle CBD = 55°$ (item 12), which sum to 145°. Then $\angle BCD = 180° - 145° = 35°$. Also, since $m\angle BCE = 90°$ and $m\angle DCE = 55°$ (item 11), $m\angle BCD = 90° - 55° = 35°$.

14. All four triangles are **similar, right** triangles: the angles in all four triangles measure 35°, 55°, and 90°.

15. The figure could be a **rhombus**, or a **square**. Since a rhombus is a special **parallelogram** and a square is a special **rectangle,** the figure could also be called one of those other figures.

16. The figure could be a **square** or a **rectangle**. However, since a square is a special **rhombus** and a rectangle is a special **parallelogram,** the figure could also be called one of these.

17. The figure could be a **parallelogram,** a **square,** a **rectangle** or a **rhombus**. A trapezoid has one pair of parallel lines.

18. The figure is a **trapezoid**.

19. The figure could be a **parallelogram,** a **rhombus,** or a **trapezoid.**

20. The figure is a **trapezoid**.

21. **(3) between 15° and 20°** Let x be the smaller angle in the triangle. Then $4x$ is the measure of the other acute angle. The third angle is known to be 90°. The sum of these three angles is 180°. We can write this equation:
$x + 4x + 90 = 180°$
$5x = 90°$
$x = 18°$

22. **(5) parallelogram** The figure has four sides, so option (1) is eliminated. It has no right angles, so option (2) is eliminated. At least one pair of opposite sides in a trapezoid are not equal, so option (3) is eliminated. A rhombus has all four sides equal, so option (4) is eliminated. A parallelogram can have sides as listed, so option (5) is the correct answer.

23. **(5) There are exactly three right triangles.** There are no equilateral, obtuse, or acute triangles in the figure. But the three right triangles are $\triangle ABC$, $\triangle ADB$, and $\triangle BDC$. For each, the right angle of the triangles is listed in the middle.

24. **(1) isosceles** The triangle has two sides that are the same length. This is the definition of an isosceles triangle.

25. **(2) $(x) + (x + 15°) + (2x) = 180°$** Let x be the smallest angle in the triangle. Then the measure of the second is $x + 15°$. Since the measure of the third is twice the measure of the smallest, its measure is $2x$. The sum of these three expressions must equal 180°, and that is the basis for the equation.

26. **(4) 6** They are $\triangle ADB$, $\triangle DEB$, $\triangle ECB$, $\triangle AEB$, $\triangle DCB$, and $\triangle ACB$.

27. **(4) 4** A triangle with two sides of the same length is isosceles:
$\triangle AEB$ has two sides of length 6.
$\triangle CED$ has two sides of length 3.
$\triangle ACD$ has two sides of length 5.
$\triangle CDB$ has two sides of length 5.

GED Practice: Solving Word Problems (Page 265)

1. **(2) $\angle ABD$ and $\angle DBC$ are supplementary angles.** $m\angle ABD + m\angle DBC = m\angle ABC$
Straight angles measure 180°, so
$m\angle ABC = 180°$.
$m\angle ABD + m\angle DBC = 180°$
When the sum of two angles is 180°, the angles are supplementary.

2. **(3) Angle 3 is congruent to angle 7.**
$\angle 1$ and $\angle 3$ form a straight line, so they are supplementary and their measures must add up to 180°. The same is true of $\angle 5$ and $\angle 7$.
Thus: $m\angle 1 + m\angle 3 = m\angle 5 + m\angle 7$
Since we know from the problem that:
$m\angle 1 = m\angle 5$, then $m\angle 3 = m\angle 7$.
Since angles 1 and 5 have the same measure, we can subtract that value from both sides.

3. **(2) $x + (x - 12°) = 90°$** The larger angle is represented by x. So the other angle must be 12° less than that, $x - 12°$. They are complementary angles, so their sum is 90°.

4. **(3) The sum of the measure of $\angle A$, $\angle B$, and $\angle ACB$ is 180°.** Option (1) is ruled out even though it is true, because this is not the reason for their sum being 100°. Option (2) is incorrect because adjacent angles must have the same vertex. Option (4) is true, but it is not the reason for the sum of $\angle A$ and $\angle B$ being 100°. Option (5) is incorrect because there are no such things as "adjacent interior angles."

5. **(5) $x + (3x + 10°) = 180°$** The smaller angle is x, so the larger must be 10° added to $3x$, or $3x + 10°$. The two angles are supplementary, so $x + (3x + 10°) = 180°$.

UNIT 3

6. **(5) $m\angle 4 = m\angle 3$ because of the "corresponding angles" rule.** This is the only false statement because $m\angle 3$ is known to be 80° and $m\angle 4$ is 100°. They are not equal.

Lesson 26

All items relate to the skill of application.

GED Practice: Lesson 26 (Congruence) (Pages 267–269)

1. **$\angle A$ corresponds to $\angle C$.**
2. **$\angle ABE$ corresponds to $\angle CBD$.**
3. **$\angle E$ corresponds to $\angle D$.**
4. **$\overline{BA}$ corresponds to $\overline{BC}$.**
5. **$\overline{AE}$ corresponds to $\overline{CD}$.**
6. **$\overline{BE}$ corresponds to $\overline{BD}$.**
7. **congruent** In lesson 25, you learned that a triangle with all three sides congruent is an equilateral triangle. $\triangle ABC$ is an equilateral triangle because all three sides are congruent. In lesson 25, you learned that the angles of an equilateral triangle all measure 60°. Each of the angles in $\triangle ABC$ measures 60°. $\triangle ABC$ and $\triangle DEF$ have congruent corresponding angles: $\angle A$ is congruent to $\angle D$ and $\angle B$ is congruent to $\angle E$. Also side $\overline{AB}$ and the side $\overline{DE}$ are marked as congruent corresponding sides. $\triangle ABC$ and $\triangle DEF$ are congruent because two angles and the side between them are congruent (ASA).
8. **not congruent** The information shown indicates only one of the three angles in $\triangle GHI$, $\angle I$ is equal to one of the three angles in $\triangle JKL$, $\angle L$, and only one side of $\triangle GHI$; $\overline{GI}$ is equal to one of three sides in $\triangle JKL$, side $\overline{JL}$. For the two triangles to be congruent, information must show that three corresponding sides are congruent (SSS), or that two sides and the angle between them are congruent (SAS), or that two angles and the side between them are congruent (ASA).
9. **not congruent** The information shown indicates only that three corresponding angles are congruent. For the two triangles to be congruent, information must show that three corresponding sides are congruent (SSS), or that two sides and the angle between them are congruent (SAS), or that two angles and the side between them are congruent (ASA).
10. **congruent** The two triangles are congruent because two sides and the angle between them are congruent (SAS).
11. **congruent** Three sides are congruent (SSS).
12. **not congruent** The information shown indicates only that one set of corresponding angles is congruent. For the two triangles to be congruent, information must show that three corresponding sides are congruent (SSS), or that two sides and the angle between them are congruent (SAS), or that two angles and the side between them are congruent (ASA).
13. **(2) 70°** $\angle A = 180° - (60° + 50°)$
    $\angle A = 70°$
    $\angle A = \angle D$
    $\angle D = 70°$
14. **(2) 7** $\angle J = 180° - (90° + 30°)$
    $\angle J = 60°$
    $\angle J = \angle G$
    $\angle H = 180° - (90° + 60°)$
    $\angle H = 30°$
    $\angle H = \angle K$
    $\overline{HI} = \overline{KL}$
    $\overline{HI} = 7$
    $\overline{KL} = 7$
15. **(2) 30°** $\angle S = \angle V$
16. **(3) 60°** $\angle U = 180° - (90° + 30°)$
    $\angle U = 60°$
17. **(3) 55°** $\angle L = \angle I$
    $\angle I = 55°$
    $\angle L = 55°$
18. **(5) Not enough information is given.** The information shown indicates only that one set of angles is congruent and one set of sides is congruent. To decide whether the triangles are congruent so that the missing angle measurement can be found, information must show that three corresponding sides are congruent (SSS), or that two sides and the angle between them are congruent (SAS), or that two angles and the side between them are congruent (ASA).
19. **(3) 4.5** $\angle A = \angle D$
    $\overline{AB} \cong \overline{DE}$
    $\overline{AC} \cong \overline{DF}$
    $\overline{BC}$ and $\overline{EF}$ must be congruent, so $\overline{EF} = 4.5$.
20. **(3) 35°** $\angle A = \angle C$
    $\angle C = 180° - (55° + 90°)$
    $\angle C = 35°$
    $\angle A = 35°$

21. (3) 55° $\angle ADB = \angle CDB$
$\angle CDB = 55°$
$\angle ADB = 55°$

22. **(5) Not enough information is given.** The information shown indicates only that one set of corresponding angles is congruent and one set of corresponding sides is congruent. To decide whether the triangles are congruent so that the missing side length can be found, information must show that three corresponding sides are congruent (SSS), or that two sides and the angle between them are congruent (SAS), or that two angles and the side between them are congruent (ASA).

GED Practice: Lesson 26 (Similarity) (Pages 272–273)

All items relate to the skill of application.

All answers in items 1 to 9 depend on matching the corresponding parts of the two triangles. Note that $\angle A$ in the first triangle matches with $\angle GBD$ in the second triangle, $\angle E$ in the first triangle matches $\angle G$ in the second, and $\angle ACE$ in the first triangle matches with $\angle D$ in the second.

1. $\angle GBD$
2. $\angle BDG$
3. $\angle DGB$
4. $\overline{BD}$
5. $\overline{BG}$
6. $\overline{EF}$
7. $\overline{DG}$
8. $\overline{BD}$
9. $\overline{BG}$

10. $\overline{AC} \cong \overline{BC}$ because in an isosceles triangle two sides are congruent. $\angle B \cong \angle A$ because the base angles in an isosceles triangle are congruent. $\overline{AY} \cong \overline{BY}$ because in an isosceles triangle the line forming the right angle bisects (cuts into two equal parts) the base. $\angle ACY \cong \angle BCY$, because in an isosceles triangle, this line bisects the vertex angle.

11. $\overline{XC} \cong \overline{CZ}$
$\angle CXY \cong \angle CZY$
$\overline{XY} \cong \overline{YZ}$
$\angle XCY \cong \angle ZCY$
All three congruent parts are based on the properties of an isosceles triangle (see item 10).

12. **14.5 ft.** The shadows are the same length. Therefore, the objects are also the same length.

13. **22 ft.** Write $\frac{6}{9} = \frac{x}{33}$. So $x = \frac{6 \times 33}{9} = 22$

14. **35 ft.** Write $\frac{5}{3} = \frac{x}{21}$. So $x = \frac{5 \cdot 21}{3} = 35$

15. **4 ft.** Write $\frac{x}{8} = \frac{20}{40}$. So $x = \frac{20 \cdot 8}{40} = 4$

16. **48 ft.** Write $\frac{5}{12} = \frac{20}{x}$. So $x = \frac{12 \cdot 20}{5} = 48$

17. **(2) 7** $\frac{4}{12} = \frac{x}{21}$
$12x = 4(21)$
$12x = 84$
$x = 7$

18. **(5) 20** $\frac{6}{10} = \frac{12}{x}$
$10(12) = 6x$
$120 = 6x$
$x = 20$

19. **(4) 36** $\frac{6}{5} = \frac{x}{30}$
$5x = 6(30)$
$5x = 180$
$x = 36$

20. **(3) 100** $\frac{60}{90} = \frac{x}{150}$
$90x = 60(150)$
$90x = 9{,}000$
$x = 100$

21. **(3) 16** $\frac{8}{7} = \frac{x}{14}$
$7x = 8(14)$
$7x = 112$
$x = 16$

GED Practice: Solving Word Problems (Page 275)

1. **(2) 150** Set up a proportion:
$\frac{1 \text{ in.}}{40 \text{ mi.}} = \frac{3\frac{3}{4} \text{ in}}{x \text{ mi.}}$
$1x = 40\left(3\frac{3}{4}\right)$
$x = 150$ mi.

2. **(4) 580** Add the distances on the map between the cities:
7 in. + $2\frac{1}{2}$ in. + 5 in. = $14\frac{1}{2}$ in.
Set up a proportion:
$\frac{1 \text{ in.}}{40 \text{ mi.}} = \frac{14\frac{1}{2} \text{ in.}}{x \text{ mi.}}$
$1x = 40\left(14\frac{1}{2}\right)$
$x = 580$ mi.

3. **(4) Scale: 1 in. = 20 mi.** Set up the ratio in the form of inches to miles.
Then $\frac{2.5 \text{ inches}}{50}$ as $\frac{1 \text{ inch}}{x \text{ miles}}$
Find that $x = 50 \div 2.5$, or $x = 20$.

4. **(5) 20 ft. and 25 ft.** The ratio of the short sides is $\frac{12}{15}$. All pairs of corresponding sides will be in this same ratio. So $\frac{12}{15} = \frac{16}{x}$ for the middle sides. Here $x = (15 \times 16) \div 12$, so $x = 20$. For the long sides, $\frac{12}{15} = \frac{20}{x}$. Here $x = (15 \times 20) \div 12$, so $x = 25$.

5. **(3) Yes, the bookshelves fit exactly into this space.** Set up the ratio in the form of inches to feet. Then
$\frac{1 \text{ inch}}{8 \text{ feet}} = \frac{1\frac{3}{4} \text{ inches}}{x \text{ feet}}$
Solve for x to find that $x = 14$ feet.

UNIT 3

6. **(4) 6.3 mi.** Set up the ratio in the form of inches to miles. Then $\frac{1}{1.8} = \frac{3.5}{x}$. So $x = 1.8 \times 3.5 = 6.3$.

Lesson 27

All items relate to the skill of application.

GED Practice: Lesson 27 (Pages 278–279)

In items 1–8, we compare the entries in the "$a^2 + b^2$" column with those in the "c^2" column. If they are equal, the triangle is a right triangle.

	a^2	b^2	$a^2 + b^2$	c^2	Right Triangle?
1.	4	9	13	16	No
2.	9	16	25	25	Yes
3.	4	4	8	9	No
4.	9	9	18	25	No
5.	25	144	169	169	Yes
6.	36	64	100	100	Yes
7.	49	576	625	625	Yes
8.	625	3,600	4,225	4,225	Yes

9. **26 ft.** The figure forms a right triangle with legs of 10 and 24. The unknown side is x. Then $x^2 = 10^2 + 24^2 = 100 + 576 = 676$. A few trials with numbers near 25 show that $26^2 = 676$. The wire is 26 feet long.

10. **4 ft.** Now the triangle has legs of 18 and 24. $x^2 = 18^2 + 24^2 = 324 + 576 = 900$. $30^2 = 900$, so the new wire is 30 feet long, or 4 feet longer than before.

11. **65 miles** The triangle has legs of 39 and 52 miles.
$c^2 = a^2 + b^2$
$= 39^2 + 52^2$
$= 1{,}521 + 2{,}704$
$= 4{,}225$
The square root of 4,225 is 65, so it is 65 miles from Point C to Point B.

12. **72 miles** Add to find the distance from A to D: $11 + 39 = 50$. The triangle now has legs of 50 and 52 miles.
$c^2 = a^2 + b^2$
$= 50^2 + 52^2$
$= 2{,}500 + 2{,}704$
$= 5{,}204$
The square root of 5,204 is between 72 and 73, but it is closer to 72. It is 72 miles from Point D to Point B.

13. **(4) $\sqrt{400 - 36}$** Because this is a right triangle, the short leg squared plus the longer leg squared equals the hypotenuse squared. Here, $6^2 + x^2 = 20^2$
$36 + x^2 = 400$
$x^2 = 400 - 36$
To find the length of the longer leg, x, find the square root of 400 – 36, or $\sqrt{400 - 36}$.

14. **(3) triangle with sides of 5, 12, and 13** In any right triangle, the square of the short leg plus the square of the longer leg will equal the square of the hypotenuse.
Test each choice:
(1) $16 + 25 \neq 36$
(2) $25 + 49 \neq 81$
(3) $25 + 144 = 169$
(4) $36 + 64 \neq 121$
(5) $49 + 81 \neq 144$

15. **(2) 26** Focus on $\triangle ABC$. It is a right triangle with its right angle at C. The legs are 10 and 24, and the hypotenuse is $\overline{AB}$, the length we seek. $(\overline{AB})^2 = 10^2 + 24^2 = 100 + 576 = 676$. Now see which choice is closest: $27^2 = 729$ and $26^2 = 676$. We need go no further; 26 is exactly the distance.

16. **(4) 8**
$c^2 = a^2 + b^2$
$10^2 = 6^2 + b^2$
$100 = 36 + b^2$
$64 = b^2$
The square root of 64 is 8.

17. **(3) Between 95 and 105 ft.** $\triangle ABC$ is a right triangle with its right angle at C. Leg $\overline{AC}$ is 60, and leg $\overline{BC} = 80$. The length of $\overline{AB}$, the hypotenuse, is $\sqrt{60^2 + 80^2}$. $60^2 = 3{,}600$ and $80^2 = 6{,}400$, so $\sqrt{60^2 + 80^2} = \sqrt{10{,}000}$, which is 100.

18. **(4) Find 15^2 and 8^2 and subtract, then find the square root of the answer.** In this right triangle the ladder length, 15, is the hypotenuse, and the short leg is the distance from the building, 8. The unknown distance is x. So $8^2 + x^2 = 15^2$, so $x^2 = 15^2 - 8^2$. To find x, take the square root of this expression.

19. **(3) 18**
$c^2 = a^2 + b^2$
$30^2 = 24^2 + b^2$
$900 = 576 + b^2$
$324 = b^2$
The square root of 324 is 18.

UNIT 3

GED Practice: Solving Word Problems (Page 281)

For items 1–7, use the shortcuts to estimate an answer, and then solve by trial and error.

1. **(3) 55**
2. **(3) 51**
3. **(3) 26**
4. **(4) 85**
5. **(3) 56**
6. **(4) 68**
7. **(2) 35**
8. **(1) $\sqrt{100}$, $\sqrt{144}$, 13** The square root of 100 is 10. The square root of 144 is 12. Only option (1) is in order from least to greatest: 10, 12, 13.
9. **(2) 49^2, 2,300, 45^2** The square of 49 is 2,401. The square of 45 is 2,025. Only option (2) is in order from greatest to least: 2,401, 2,300, 2,025.

GED Mini-Test: Lessons 23–27 (Pages 282–283)

All items relate to the skill of application.

1. **(4) 216**
   $V = s^3$
   $= 6^3$
   $= 216$ cu. cm.
2. **(4) from 500 to 575 cu. in.**
   $V = \pi r^2 h$
   $= 3.14\,(5^2)\,(7)$
   $= 3.14(25)(7)$
   $= 549.5$ cu. in.
3. **(2) 12** Convert 2 yards to 6 feet.
   $A = lw$
   $= 6(2)$
   $= 12$ sq. ft.
4. **(3) 255**
   $A = \frac{1}{2}bh$
   $= \frac{1}{2}(30)(17)$
   $= 255$ sq. cm.
5. **(3) equilateral** The triangle has three equal sides which is the definition of an equilateral triangle.
6. **(2) 60°** Let $2x = m\angle A$ and $x = m\angle B$. Supplementary angles sum to 180°.
   $x + 2x = 180°$
   $3x = 180°$
   $x = 60°$
7. **(1) They are alternate exterior angles.** Angles 4 and 5 are on alternate sides of the transversal t and are on the outside of the parallel lines p and m.
8. **(5) $m\angle 3 = 135°$** Since $m\angle 8$ is 45°, $m\angle 4$ must be 135°. Since $\angle 4$ and $\angle 3$ are supplementary, $m\angle 3$ must be 45° also.
9. **(4) $\angle 7$** $\angle 7$ is on the same side of the transversal and is in the same position in regard to the parallel lines.
10. **(2) 22** Set up a proportion:
    $$\frac{\text{shadow of post}}{\text{actual post}} = \frac{\text{shadow of pole}}{\text{actual pole}}$$
    $$\frac{4\frac{1}{2}}{3} = \frac{33}{x}$$
    $$4\tfrac{1}{2}x = 99$$
    $$x = 22 \text{ ft.}$$
11. **(4) 30 mi.** Set up a proportion:
    $$\frac{\text{map distance}}{\text{actual distance}} = \frac{\text{map scale}}{\text{actual scale}}$$
    $$\frac{\frac{3}{4}}{x} = \frac{1}{40}$$
    $$x = (\tfrac{3}{4})(40) = 30 \text{ mi.}$$
12. **(3) 13 ft.** Use the Pythagorean Theorem:
    $c^2 = a^2 + b^2$
    $c^2 = 5^2 + 12^2$
    $c^2 = 25 + 144$
    $c^2 = 169$
    $c = 13$ ft.
13. **(4) 23 ft.** Use the Pythagorean Theorem:
    $c^2 = a^2 + b^2$
    $c^2 = 7^2 + 22^2$
    $c^2 = 49 + 484$
    $c^2 = 533$
    c is about 23 ft. $23 \times 23 = 529$.

Lesson 28

All items relate to the skill of application.

GED Practice: Lesson 28 (Pages 286–287)

1. **132 sq. ft.** Think of the figure as a rectangle and a triangle. Find the area of the rectangle.
   $A = lw$
   $= 8(4)$
   $= 32$ sq. ft.
   Find the area of the triangle.
   $A = \frac{1}{2}bh$
   $= \frac{1}{2}(10)(20)$
   $= 100$ sq. ft.
   Add the two areas:
   Area $= 32 + 100 =$ **132 sq. ft.**

UNIT 3

2. **304.5 sq. in.** Think of the figure as two triangles and a rectangle. Find the area of each element and add all the results.
Left triangle:
$A = \frac{1}{2}bh$
$= \frac{1}{2}(14)(18)$
$= 126$ sq. in.
Rectangle:
$A = lw$
$= 14(9)$
$= 126$ sq. in.
Right triangle:
$A = \frac{1}{2}bh$
$= \frac{1}{2}(14)(7.5)$
$= 52.5$ sq. in.
Area = 126 + 126 + 52.5 = **304.5 sq. in.**

3. **208 sq. in.** Think of the figure as three rectangles. Find the area of each rectangle and add the results. The missing length of the middle rectangle is 20 – (3 + 5) = 12.
Left rectangle:
$A = lw$
$= 14(3)$
$= 42$ sq. in.
Middle rectangle:
$A = lw$
$= 12(8)$
$= 96$ sq. in.
Right rectangle:
$A = lw$
$= 14(5)$
$= 70$ sq. in.
Area = 42 + 96 + 70
= **208 sq. in.**

4. **172 sq. cm.** Think of the figure as a parallelogram and a triangle. Find the area of each element and add the results.
Parallelogram:
$A = bh$
$= 12(9)$
$= 108$ sq. cm.
Triangle:
$A = \frac{1}{2}bh$
$= \frac{1}{2}(16)(8)$
$= 64$ sq. cm.
Area = 108 + 64 = **172 sq. cm.**

5. **28.56 sq. in.** Find the area of the circle and the area of the square and add the results. Remember, the radius of the circle is half the diameter.
Circle:
$A = \pi r^2$
$= 3.14(2^2)$
$= 3.14(4)$
$= 12.56$ sq. in.
Square:
$A = s^2$
$= 4^2$
$= 16$ sq. in.
Area: 12.56 + 16 = **28.56 sq. in.**

6. **270 cu. in.** Find the volume of the rectangular solid and the volume of the pyramid. Add the results.
Rectangular solid:
$V = l \times w \times h$
$= 6 \times 5 \times 6$
$= 180$ cu. in.
Pyramid:
$V = \frac{1}{3} \times (\text{Area of base}) \times h$
$= \frac{1}{3} \times (6 \times 5) \times 9$
$= 90$ cu. in.
Volume = 180 + 90 = **270 cu. in.**

7. **263.9 cu. ft.** The area of the base (A) of the cylinder is also the area of the base of the cone. Find the volume of the cylinder and the volume of the cone and add the results.
Cylinder:
$V = A \times h$
$= 27.3(8)$
$= 218.4$ cu. ft.
Cone:
$V = \frac{1}{3}(A \times h)$
$= \frac{1}{3}(27.3)(5)$
$= 45.5$ cu. ft.
Volume = 218.4 + 45.5 = **263.9 cu. ft.**

8. **414 cu. ft.** Think of the figure as two rectangular solids. Find the volume of each and add the results. The missing height for the lower solid is 18 – 6, or 12 ft. The missing length for the upper solid is 9 – 4, or 5 feet.
Lower solid:
$V = lwh$
$= 9(3)(12)$
$= 324$ cu. ft.
Upper solid:
$V = lwh$
$= 6(5)(3)$
$= 90$ cu. ft.
Volume = 324 + 90 = **414 cu. ft.**

9. **84.5 cu. ft.** The area of the base (A) of the cylinder (6.5 sq. ft.) is also the area of the base of the cone.
Cone:
$V = \frac{1}{3} \times A \times h$
$= \frac{1}{3} \times 6.5 \times 12$
$= 26$ cu. ft.
Cylinder:
$V = A \times h$
$= 6.5 \times 9$
$= 58.5$ cu. ft.
Volume = 26 + 58.5 = **84.5 cu. ft.**

10. **1,260 cu. in.** Think of the figure as three rectangular solids. The lower has a height of 18 – 6 – 6, or 6 in.
Lower solid:
$V = lwh$
$= 15(7)(6)$
$= 630$ cu. in.
The middle solid has a length of 15 – 5, or 10 in.
Middle solid:
$V = lwh$
$= 10(7)(6)$
$= 420$ cu. in.
The upper solid has a length of 15 – 5 – 5, or 5 in.
Upper solid:
$V = lwh$
$= 5(7)(6)$
$= 210$ cu. in.
Volume = 630 + 420 + 210 = **1,260 cu. in.**

11. **(1) 1,242** Think of the figure as two rectangles. Find the area of each and add the results.
Top:
$A = lw$
$= 54(15)$
$= 810$ sq. in.
Bottom:
$A = lw$
$= 24(18)$
$= 432$ sq. in.
Area = 810 + 432 = **1,242 sq. in.**

12. **(5) 165** The area of the base of the cylinder is also the area of the base of the cone. Find the two volumes and add.
Cylinder:
$V = A \times h$
$= 9 \times 14$
$= 126$ cu. ft.
Cone:
$V = \frac{1}{3} \times A \times h$
$= \frac{1}{3} \times 9 \times 13$
$= 39$ cu. ft.
Volume = 126 + 39 = **165 cu. ft.**

13. **(3) 33** Think of the figure as a rectangle and two congruent triangles.
Rectangle:
$A = lw$
$= 8(3)$
$= 24$ sq. cm.
Triangle:
$A = \frac{1}{2}bh$
$= \frac{1}{2}(4.5)(2)$
$= 4.5$ sq. cm.
Two triangles:
$A = 2(4.5)$
$= 9$ sq. cm.
Area = 24 + 9 = **33 sq. cm.**

14. **(5) 164** Think of the figure as two parallelograms and a square.
Parallelogram:
$A = bh$
$= 16(4)$
$= 64$ sq. ft.
Two parallelograms:
$A = 2(64)$
$= 128$ sq. ft.
Square:
$A = s^2$
$= 6^2$
$= 36$ sq. ft.
Area = 128 + 36 = **164 sq. ft.**

15. **(3) 51** Think of the figure as two pyramids. The two pyramids have the same base area (A). Find the base area.
$A = s^2$
$= 3^2$
$= 9$ sq. cm.
Left pyramid:
$V = \frac{1}{3} \times A \times h$
$= \frac{1}{3} \times 9 \times 12$
$= 36$ cu. cm.
Right pyramid:
$V = \frac{1}{3} \times A \times h$
$= \frac{1}{3} \times 9 \times 5$
$= 15$ cu. cm.
Volume = 36 + 15 = **51 cu. cm.**

UNIT 3

16. (4) 800 Think of the figure as a rectangle and a triangle.
Rectangle:
$A = lw$
$= 30(20)$
$= 600$ sq. in.
Triangle:
$A = \frac{1}{2}bh$
$= \frac{1}{2}(20)(20)$
$= 200$ sq. in.
Area $= 600 + 200 =$ **800 sq. in.**

GED Practice: Solving Word Problems (Pages 290–291)

1. 816 sq. ft. The area of the walkway is the difference between the area of the larger rectangle and the area of the garden.
Larger rectangle:
$A = lw$
$= 48(32)$
$= 1{,}536$ sq. ft.
Garden:
$A = lw$
$= 36(20)$
$= 720$ sq. ft.
$1{,}536 - 720 =$ **816** sq. ft.

2. 8 boxes Find the volume of the carton and the volume of 1 box. Divide the volume of the carton by the volume of 1 box to find the number of boxes that will fit into a carton. First convert the feet into inches.

Carton:	Box:
2 ft. $= 12 \times 2 = 24$ in.	1 ft. $= 12$ in.
$V = lwh$	$V = lwh$
$= 24(8)(8)$	$= 12(8)(2)$
$= 1{,}536$ cu. in.	$= 192$ cu. in.

$1{,}536 \div 192 =$ **8 boxes**

3. 408 strips Find the perimeter of the pool in inches. First find the missing measures. Convert the feet in the perimeter to inches. Divide the number of inches in the perimeter by 6, the number of inches in 1 strip, to find the number of strips.
The left side of the pool is $30 + 12 = 42$ ft. The bottom missing dimension is $60 - 18 = 42$ ft.
Perimeter =
$60 + 30 + 42 + 12 + 18 + 42 = 204$ ft.
$12 \times 204 = 2{,}448$ in.
$2{,}448 \div 6 =$ **408 strips**

4. 26 cu. ft. Find the volume of the refrigerator in feet. Subtract the volume advertised to find the volume taken up by the parts. First convert the inches to feet.
66 in. $= 66 \div 12 = 5\frac{1}{2}$ ft.
36 in. $= 36 \div 12 = 3$ ft.
32 in. $= 32 \div 12 = 2\frac{2}{3}$ ft.
$V = lwh$
$= 3 \times 2\frac{2}{3} \times 5\frac{1}{2}$
$= \overset{1}{\cancel{3}} \times \frac{\overset{4}{\cancel{8}}}{\underset{1}{\cancel{3}}} \times \frac{11}{\underset{1}{\cancel{2}}}$
$= 44$ cu. ft.
$44 - 18 =$ **26 cu. ft.**

5. 792 cu. ft. Separate the lawn into two rectangles. Find the volume of the rectangular solids, which have heights of 6 inches, or $\frac{1}{2}$ ft. Add the volumes. The length of the top rectangle is $18 + 30 = 48$ ft. The width is 24 ft.

Top rectangle:	Bottom rectangle:
$V = lwh$	$V = lwh$
$= 48 \times \overset{12}{\cancel{24}} \times \frac{1}{\underset{1}{\cancel{2}}}$	$= 24 \times \overset{9}{\cancel{18}} \times \frac{1}{\underset{1}{\cancel{2}}}$
$= 576$ cu. ft.	$= 216$ cu. ft.

$576 + 216 =$ **792 cu. ft.**

6. 32 bundles Find the area of the roof in square yards. Divide by 3 to find the number of bundles needed.
Area of half the roof:
$A = lw$
$= 8 \times 6$
$= 48$ sq. yd.
Area of whole roof $= 2 \times 48 = 96$ sq. yd.
$96 \div 3 =$ **32 bundles**

7. (1) 208 Find the volume of the cylinder and subtract the volume of the cubic toy. The difference is the amount of space to be filled with packing.
Cylinder:
$V = A \times h$
$= 80 \times 9$
$= 720$ cu. in.
Cubic toy:
$V = s^3$
$= 8 \times 8 \times 8$
$= 512$ cu. in.
$720 - 512 =$ **208 cu in.**

8. (2) $197.55 Find the area of the floor. Divide the area by 300 to find how many gallons are needed. Multiply the number of gallons by the cost per gallon.
$A = lw$
$= 60 \times 45$
$= 2{,}700$ sq. ft.
$2{,}700 \div 300 = 9$ gallons
$9 \times \$21.95 =$ **$197.55**

UNIT 3

9. **(3) 6** Find the volume of the container and convert the number to cubic yards. Divide 128 cubic yards by the amount each container holds.

$V = lwh$

$= 12 \times 8 \times 6$

$= 576$ cu. ft.

$576 \div 27 = 21\frac{1}{3}$ cu. yd.

$128 \div 21\frac{1}{3} =$ **6 containers**

10. **(3) $429.84** Find the area of the floor and convert to square inches. Divide by 36 square inches—the area of 1 tile. Multiply the number of the tiles by $1.99.

$A = lw$

$= 9 \times 6$

$= 54$ sq. ft.

54 sq. ft. $= 144 \times 54 = 7{,}776$ sq. in.

$7{,}776 \div 36 = 216$ tiles

$216 \times \$1.99 =$ **$429.84**

11. **(5) 15** Find the volume of the rectangular solid formed by the rectangular driveway with 3 inches of concrete. First convert all measures to yards.

64 ft. $= \frac{64}{3}$ yd.; 24 ft. $= 8$ yd.; 3 in. $= \frac{3}{36} = \frac{1}{12}$ yd.

$V = lwh$

$= \frac{64}{3} \times \overset{2}{\cancel{8}} \times \frac{1}{\underset{3}{\cancel{12}}}$

$= \frac{128}{9} = \mathbf{14\frac{2}{9}}$ **cu. yd.**

Shawn needs to order **15 cu. yd.**, since he must round up to have enough concrete.

12. **(1) 82** Find the perimeter of the playground (rectangle), first changing the dimensions from feet to yards. Subtract 6 feet (2 yards) for the gate. The difference is the number of yards of fencing needed.

90 ft. = 30 yd.; 36 ft. = 12 yd.

$P = 2l + 2w$

$= 2(30) + 2(12)$

$= 60 + 24$

$= 84$ yd.

$84 - 2 =$ **82 yd.**

Lesson 29

All items relate to the skill of application.

GED Practice: Area Relationships (Pages 294–295)

1. **24 inches** Find the area of the parallelogram:

$A = bh$

$= 32(18)$

$= 576$ sq. in.

Use the formula for the area of a square:

$A = s^2$

$576 = s^2$

$\sqrt{576} = s$

$24 = s$

2. **18 feet** Find the area of the rectangle:

$A = lw$

$= 15(12)$

$= 180$ sq. ft.

Use the formula for area of a triangle:

$A = \frac{1}{2}bh$

$180 = \frac{1}{\underset{1}{\cancel{2}}}(\overset{10}{\cancel{20}})h$

$180 = 10h$

$\frac{180}{10} = h$

$18 = h$

3. **11.5 centimeters** Find the area of the triangle:

$A = \frac{1}{2}bh$

$= \frac{1}{2}(12.5)(9.2)$

$= 57.5$ sq. cm.

Use the formula for area of a parallelogram:

$A = bh$

$57.5 = b(5)$

$\frac{57.5}{5} = b\frac{(5)}{5}$

$11.5 = b$

4. **28 yards** Find the area of the rectangle:

$A = lw$

$= 49(16)$

$= 784$ sq. yd.

Use the formula for area of a square:

$A = s^2$

$784 = s^2$

$\sqrt{784} = s$

$28 = s$

5. **(3) 20 inches** Find the area of the rectangle:

$A = lw$

$= 16(25)$

$= 400$ sq. in.

Use the formula for area of a square:

$A = s^2$

$400 = s^2$

$\sqrt{400} = s$

$20 = s$

6. **(1) 3.0 centimeters** Find the area of the triangle:

$A = \frac{1}{2}bh$

$= \frac{1}{2}(12)(10.5)$

$= \frac{1}{\cancel{2}_1}(\cancel{12}^6)(10.5)$

$= 63$ sq. cm.

Use the formula for the area of a parallelogram:

$A = bh$

$63 = 21h$

$\frac{63}{21} = h$

$3 = h$

7. **(2) 5 feet** Find the area of the rectangle:

$A = lw$

$= \frac{1}{2}(90)$

$= 45$ sq. ft.

Use the formula for the area of a triangle:

$A = \frac{1}{2}bh$

$45 = \frac{1}{2}(18)h$

$45 = 9h$

$5 = h$

8. **(5) Not enough information is given.** You need to know the measure of the base of the triangle.

9. **(4) 32 centimeters** Find the area of the rectangle:

$A = lw$

$= 16(64)$

$= 1{,}024$ sq. cm.

Use the formula for the area of a square:

$A = s^2$

$1{,}024 = s^2$

$\sqrt{1{,}024} = s$

$32 = s$

10. **(5) 10.5 meters** Find the area of the triangle:

$A = \frac{1}{2}bh$

$= \frac{1}{2}(28.7)(12)$

$= 172.2$ sq. m.

Use the formula for the area of a rectangle:

$A = lw$

$172.2 = 16.4w$

$\frac{172.2}{16.4} = \frac{16.4w}{16.4}$

$10.5 = w$

GED Practice: Solving Word Problems (Page 297)

1. **(1) 6** Let w stand for the width. The length is $3w$. Use the formula for the perimeter of a rectangle to write an equation.

$P = 2l + 2w$

$48 = 2(3w) + 2w$

$48 = 8w$

$\frac{48}{8} = \frac{8w}{8}$

$6 = w$

Check: $P = (2 \times 6) + (2 \times 18) = 12 + 36 = 48$

2. **(5) 22** Let w stand for the width. The length is $2w$. Use the formula for area of a rectangle to write an equation.

$A = lw$

$242 = 2w(w)$

$\frac{242}{2} = \frac{2w(w)}{2}$

$121 = w^2$

$\sqrt{121} = w$

$11 = w$

Length $= 2w = 22$

Check: $A = 22 \times 11 = 242$

3. **(2) 92.5** Let l stand for the length. The width is $l - 10$. Use the formula for perimeter to write an equation.

$P = 2l + 2w$

$350 = 2l + 2(l - 10)$

$350 = 2l + 2l - 20$

$350 = 4l - 20$

$370 = 4l$

$\frac{370}{4} = \frac{4l}{4}$

$92.5 = l$

Check: $P = 2(92.5) + 2(82.5) =$
$185 + 165 = 350$

4. **(5) 12** Let s stand for the side of the square. Use the formulas for area of a square and area of a rectangle to write an equation.

Area of rectangle $= \frac{1}{2}$(Area of square)

$9 \times 8 = \frac{1}{2}s^2$

$2 \times 9 \times 8 = 2 \times \frac{1}{2}s^2$

$2 \times 72 = s^2$

$144 = s^2$

$\sqrt{144} = s$

$12 = s$

Check: $A = 12 \times 12 = 144$; $\frac{1}{2} \times 144 = 72 = 9 \times 8$

UNIT 3

5. **(1) 14** Let b stand for the base of the triangle. The height is $4b$. Use the formula for the area of a triangle to write an equation.

$A = \frac{1}{2}bh$

$392 = \frac{1}{2}b(4b)$

$392 = 2b^2$

$196 = b^2$

$\sqrt{196} = b$

$14 = b$

Check: $A = 14 \times 4 \times 14 \times \frac{1}{2} = 392$

6. **(2) 21** Let h stand for the height of the triangle. Write an equation using the area formulas.

Area of triangle = Area of parallelogram

$\frac{1}{2} \times 24 \times h = 12 \times 21$

$12h = 12 \times 21$

$h = 21$

Check: $A = \frac{24 \times 21}{2} = 12 \times 21$

7. **(5) 15** Let l stand for the length. The width is $\frac{1}{2}l$. Use the perimeter formula to write an equation.

$P = 2l + 2w$

$45 = 2l + 2(\frac{1}{2}l)$

$45 = 3l$

$\frac{45}{3} = \frac{3l}{3}$

$15 = l$

Check: $P = (2 \times 15) + (2 \times 7.5) =$
$30 + 15 = 45$

8. **(3) 27** Let s stand for the side of the square. Use the perimeter formulas to write an equation.

Perimeter of square = Perimeter of rectangle + 20

$4s = 2l + 2w + 20$

$4s = 2(28) + 2(16) + 20$

$4s = 56 + 32 + 20$

$4s = 108$

$s = 27$

Check: $P = 4s = 4 \times 27 = 108;$
$P = 2l + 2w + 20 = 56 + 32 + 20 = 108$

GED Mini-Test: Lessons 28 and 29 (Pages 298–299)

All items relate to the skill of application.

1. **(3) 18** Find the area of the rectangle. $A = lw = 40.5 \times 8 = 324$ sq. cm. Use the formula for area of a square: $A = s^2 = 324;$ $s = \sqrt{324} = 18$

2. **(5) $282.24** Convert feet to inches: 18 ft. = $18 \times 12 = 216$ in.; 9 ft. = $9 \times 12 = 108$ in. Find the area of the room in square inches: $A = lw = 216 \times 108 = 23{,}328$ sq. in. Divide by the number of square inches in a tile, 81, to find the number of tiles: $23{,}328 \div 81 = 288$. Multiply the number of tiles by $0.98 to find the cost: $288 \times \$0.98 = \$282.24$

3. **(3) 144** Think of the figure as three congruent triangles. Find the area of one triangle and multiply by 3. Convert feet to inches. $A = \frac{1}{2}bh = \frac{1}{2} \times 8 \times 12 = 48$ sq. in. $48 \times 3 = 144$ sq. in.

4. **(4) 84** Let w stand for the width. Then the length is $4w$. Use the formula for the perimeter of a rectangle to write an equation:

$2l + 2w = P$

$2(4w) + 2w = 210$

$10w = 210$

$\frac{10w}{10} = \frac{210}{10}$

$w = 21$

$l = 4 \times 21 = 84$

Check: $P = (2 \times 84) + (2 \times 21) =$
$168 + 42 = 210$

5. **(1) 160** Find the volume of the rectangular solid 20 ft. by 16 ft. by 6 inches. Convert the inches to feet: 6 in. = $\frac{1}{2}$ ft.

$V = lwh = 20 \times 16 \times \frac{1}{2} = 160.$

6. **(2) 6** Find the area of the rectangle: $A = lw = 6 \times 3.5 = 21$ sq. in. Use the formula for area of a triangle: $A = \frac{1}{2}bh = 21$; $\frac{1}{2} \times 7 \times h = 21$; $\overset{1}{\cancel{2}} \times \frac{1}{\underset{1}{\cancel{2}}} \times 7 \times h = 2 \times 21$; $7 \times h = 42$; $h = 6$

7. **(5) 549.5** Find the volume of the cylinder and the volume of the cone and add the results. The area of the base of the cone is equal to the area of the base of the cylinder.

Cylinder:

$V = Ah$

$= \pi(5^2)(5)$

$= 392.5$ cu. cm.

Cone:

$V = \frac{1}{3}Ah$

$= \frac{1}{\underset{1}{\cancel{3}}}\pi(5^2)(\overset{2}{\cancel{6}})$

$= 2 \times 78.5$

$= 157$ cu. cm.

$392.5 + 157 = 549.5$ cu. cm.

UNIT 3

8. **(3) 254** Find the area of the larger circle, which includes the walk, and find the area of the pool. Subtract the area of the pool from the area of the larger circle. The radius of the pool is 12 feet; the radius of the larger circle is $12 + 3 = 15$ feet.
Larger circle:
$A = \pi r^2$
$= 3.14(15)^2$
$= 706.5$ sq. ft.
Pool:
$A = \pi r^2$
$= 3.14(12)^2$
$= 452.16$
$706.5 - 452.16 = 254.34$, which rounds to 254 sq. ft.

9. **(3) 90** Find the area of the rectangle and the area of the two congruent triangles and add the results.
Rectangle:
$A = lw$
$= 10 \times 6$
$= 60$ ft.
One triangle:
$A = \frac{1}{2}bh$
$= \frac{1}{2}(6)(5)$
$= 15$ ft.
Area of two triangles $= 15 + 15 = 30$ ft.
Total area $= 60 + 30 = 90$ ft.

10. **(2) 9** Let l stand for the length. Then the width is $l - 5$. Use the formula for perimeter to write an equation:
$2l + 2w = P$
$2l + 2(l - 5) = 26$
$4l - 10 = 26$
$4l = 36$
$l = 9$
Check: $P = 2(9) + 2(9–5) = 18 + 8 = 26$

11. **(3) 175** Find the volume of the rectangular solid and the volume of the pyramid and add the results.

Rectangular solid:	Pyramid:
$V = lwh$	$V = \frac{1}{3}Ah$
$= 5 \times 5 \times 6$	$= \frac{1}{\cancel{3}_1}(5 \times 5)(\cancel{3}^1)$
$= 150$ cu. in.	$= 25$ cu. in.
	$150 + 25 = 175$ cu. in.

12. **(2) 8** Let w stand for the width. Then the length is $3w$. Use the area formula to write an equation:
$lw = A$
$w(3w) = 192$
$3w^2 = 192$
$w^2 = 64$
$w = \sqrt{64}$
$w = 8$

13. **(3) $137** Find the area of the lawn. Divide the area by 150 to find how many pounds of seed are needed. Multiply the result by $6.85 to find the total cost. $A = lw = 100 \times 30 =$ 3,000 sq. ft.; $3,000 \div 150 = 20$ pounds of seed. $20 \times \$6.85 = \$137$

Cumulative Review: Unit 3: Geometry (Pages 300–308)

All items relate to the skill of application.

1. **(3) ∠*HBE*** A straight angle forms a straight line.

2. **(2) ∠*EAG*** Vertical angles are opposite one another.

3. **(1) ∠*CBH*** Two angles that form a straight line are also supplementary angles.

4. **(4) ∠*GCF*** Vertical angles are opposite one another.

5. **(5) 5 ft. 3 in.** Set up a proportion:
$\frac{3\frac{1}{2}}{x} = \frac{4}{6}$
$4x = 21$
$x = 5\frac{1}{4}$ ft. = 5 ft. 3 in.

6. **(3) 20 m.** Set up a proportion:
$\frac{4}{4.8} = \frac{x}{24}$
$4.8x = 96$
$x = 20$ m.

7. **(4) 92** Find the sum of all the sides: $12 + 18 + 1.5 + 12 + 17 + 12 + 1.5 + 18 = 92$.

8. **(3) ∠*AHG*** The angles are alternate exterior angles.

9. **(4) △*ABD*** There is no pair of congruent sides in the triangle.

10. **(3) $c^2 = a^2 + b^2$** This is the Pythagorean Theorem.

UNIT 3

11. **(3) 13** Use the Pythagorean Theorem twice. First find $\overline{BD}$, then find $\overline{AB}$.

$c^2 = a^2 + b^2$ $\quad$ $c^2 = a^2 + b^2$
$\overline{BD}^2 = 3^2 + 4^2$ $\quad$ $\overline{AB}^2 = 12^2 + 5^2$
$= 9 + 16$ $\quad$ $= 144 + 25$
$= 25$ $\quad$ $= 169$
$\overline{BD} = 5$ $\quad$ $\overline{AB} = 13$

12. **(5)** $\pi = \frac{C}{r}$ This formula gives $C = \pi r$, which is false.

13. **(2) 6π cm.** $C = 2\pi r = (2)(\pi)(3) = 6\pi$

14. **(3) 9π sq. in.** $A = \pi r^2 = (\pi)(3)(3) = 9\pi$

15. **(2) ∠*ZXY* and ∠*BZD* are corresponding angles that are congruent.** Lines *m* and *p* are parallel and $\overline{XB}$ is a transversal intersecting both of them. ∠*ZXY* and ∠*BZD* are then corresponding angles, which are always congruent.

16. **(1) ∠*ZYX* and ∠*CZA* are corresponding angles.** See the explanation for item 15. Option (2) is clearly false, and options (3), (4), and (5) do not make clear which angles are involved.

17. **(1) 2.5 ft.** Using the volume formula $V = lwh$ we have $19.5 = 5.2 \times w \times 1.5$, or $19.5 = 7.8w$. Divide 19.5 by 7.8 to find that $w = 2.5$ ft.

18. **(5) Not enough information is given.** While ∠*AED* appears to be obtuse, there is not enough information given to be able to come to that conclusion. Remember: appearance is not enough to reach a conclusion.

19. **(1) *m*∠*ABC* = 55°** The sum of the angles of a triangle is 180° and △*ABC* is isosceles because two sides are equal. Then, $m\angle C = m\angle B$ because the angles opposite the congruent sides in an isosceles triangle are equal.
$m\angle A + m\angle B + m\angle C = 180°$
$70° + m\angle B + m\angle B = 180°$
$2m\angle B = 180° - 70°$
$m\angle B = 55°$
Remember that angles can be named using the vertex only or the vertex as the middle letter.

20. **(4) 13** Use the Pythagorean Theorem: (the wire is the hypotenuse)
$c^2 = a^2 + b^2$
$c^2 = 5^2 + 12^2$
$c^2 = 25 + 144$
$c^2 = 169$
$c = 13$

21. **(1) 16 + 2π** The perimeter of the rectangular part is $6 + 4 + 6 = 16$. The circumference of the semi-circle is $\frac{1}{2}\pi(4) = 2\pi$. The perimeter of the whole figure is found by adding these perimeters together.

22. **(3) Box Z has the greatest volume.** Multiply length times width times height to find the volume of each box.
Box X: $7.6 \times 5 \times 10 = 380$ cu. in.
Box Y: $8 \times 4.5 \times 10 = 360$ cu. in.
Box Z: $7 \times 5.5 \times 10 = 385$ cu. in.

23. **(1) isosceles** The triangle has two sides that are the same length. This is the definition of an isosceles triangle.

24. **(5) right** $9^2 + 12^2 = 81 + 144 = 225 = 15^2$

25. **(2) △*TUR*** ∠*U* is a right angle. △*TUR* is the only option which contains ∠*U*.

26. **(3) a scalene triangle** No sides of the triangle are equal.

27. **(3) three** △*RUV*, △*RUT*, and △*RUS* are right triangles.

28. **(4) > 6 < 7 in. high** The volume formula is $\frac{1}{3} \times A \times h$, where $A = 3$ in. $\times$ 3 in. $= 9$ sq. in. and h is unknown.
$21 = \frac{1}{3} \times 9 \times h = 21 = 3h$. The height is 7 inches. The tallest bottle that can fit is $> 6 < 7$.

29. **(4) $\overline{AC}$ and $\overline{DC}$** When two figures are congruent or similar, then the smallest angle of one "corresponds" to the smallest angle of the other, and the longest side of one "corresponds" to the longest side of the other. To identify corresponding sides we simply match up the parts of the figures that are the same (or proportional in the case of similar figures). In option (4) $\overline{AC}$ is the longest side, the hypotenuse, of one triangle, but $\overline{DC}$ is not the hypotenuse of the other.

30. **(3) $\overline{AD} \cong \overline{BE}$** The clue here is that we want the proof to be by SSS (side-side-side). The only pair of corresponding sides that are not known to be congruent are $\overline{AD}$ in △*ADC* and $\overline{BE}$ in △*BEC*.

31. **(2) $\overline{AE} \cong \overline{BD}$** Two pairs of corresponding sides of these triangles are known to be congruent from the figure. The only missing sides are $\overline{AE}$ in △*ACE* and $\overline{BD}$ in △*BCD*.

UNIT 3

32. **(4) $\triangle CED$** The problem here is "seeing" the triangle named and identifying one of the same shape.

33. **(2) $\overline{AC} \cong \overline{AE}$** We can only assume what is shown in the figure and what is told specifically. Thus we cannot assume that $\overline{AB} \cong \overline{BC} \cong \overline{CD}$, nor that any pair of corresponding sides are congruent. $\overline{AC} \cong \overline{AE}$ because they are congruent sides of the known isosceles triangle.

34. **(3) 1,400 sq. ft.**
Use the formula:
$V = \frac{1}{3}Ah$
$82{,}600 = \frac{1}{3}A(177)$
$82{,}600 = 59A$
$1{,}400 = A$

35. **(3) 16** $16 \times 16 = 256$

36. **(5) Angle 1 is an obtuse angle.**

37. **(3) $m\angle 1 = m\angle 2$** $\angle 1$ and $\angle 2$ are vertical angles. Therefore their measures are equal.

38. **(2) $\angle 1$ and $\angle 2$ are vertical angles.** Angles 1 and 2 are the opposite angles formed by two intersecting lines. This is the definition of vertical angles.

39. **(4) Triangle PQR is an equilateral triangle.** We know $m\angle 3 = m\angle 4$. (This is given in the information about the figure.) If $m\angle 3 = 60°$, then $m\angle 4 = 60°$. Since the sum of the angles of a triangle equals 180°, then $m\angle 5$ must be 60° also. Since the three angles of $\triangle PQR$ are equal, $\triangle PQR$ must be an equilateral triangle.

40. **(2) $\frac{\overline{AB}}{\overline{AC}} = \frac{\overline{FB}}{\overline{GC}}$** Corresponding sides of similar triangles have equal ratios.

41. **(4) 120** $\triangle MON$ and $\triangle POQ$ are similar isosceles triangles. Therefore, the corresponding sides are similar.
$\frac{x}{24} = \frac{150}{30}$
$30x = 3{,}600$
$x = 120$

42. **(2) 17** $17 \times 17 = 289$

43. **(3) 75°** The sum of the angles is 180°.
Let $m\angle C = x$.
$25° + 80° + x = 180°$
$x = 75°$

44. **(4) the supplement of $\angle BAC$** $\overline{AB}$ is parallel to $\overline{DE}$. $\overline{AD}$ transverses $\overline{AB}$ and $\overline{DE}$. Interior angles on the same side of a transversal are supplementary.

45. **(1) 105°** $\angle BCD$ is supplementary to $\angle ACB$ which is 75° (see item 40). $180° - 75° = 105°$.

46. **(3) 1,500 cu. in.** Use the formula: $V = \frac{1}{3}Ah$
$= \frac{1}{3}(4.5)(5)$
$= 7.5$ cu. in.
This is the volume of a single candle. Multiply by 200 to find that 1,500 cubic inches of wax are needed to make 200 candles.

47. **(1) $\sqrt{529}$** $23 \times 23 = 529$

48. **(1) $\angle C \cong \angle D$** In triangles that are similar, the corresponding angles are congruent. None of the other choices shows corresponding parts.

49. **(4) 16** $\overline{DF}$ and $\overline{BC}$ are corresponding parts, and so their ratio is the same as all other ratios of corresponding parts. Use the ratio of $\overline{DB}$ to $\overline{CA}$, which is 9 to 12.
Therefore, $\frac{\overline{DF}}{\overline{CB}} = \frac{9}{12}$.
Since $\overline{DF} = 12$, we have $\frac{12}{\overline{CB}} = \frac{9}{12}$.
$\overline{CB} = (12 \times 12) \div 9 = 16$.

50. **(1) 12** Use the Pythagorean Theorem:
$c^2 = a^2 + b^2$
$15^2 = 9^2 + b^2$
$225 = 81 + b^2$
$144 = b^2$
$12 = b$

51. **(5) $\frac{5}{8} = \frac{x}{40}$** At any given time, the ratio of all objects to their shadows is the same. Set up the ratios in the same order:
$\frac{\text{object height}}{\text{shadow length}}$
$\frac{\text{fence}}{\text{shadow}} = \frac{\text{flagpole}}{\text{shadow}}$
$\frac{5}{8} = \frac{x}{40}$

52. **(3) about 30 ft. long** Use the ratio of object length to shadow length: $\frac{5}{8} = \frac{18}{x}$, where x is the length of the shadow. Solve: $x = (8 \times 18) \div 5$, so x is exactly $28\frac{4}{5}$ ft., which is about 30 feet.

53. **(5) 140 sq. ft.** Find the dimensions of the kitchen: map scale is 1 in. = 2 ft.
$\frac{1 \text{ in.}}{5 \text{ in.}} = \frac{2 \text{ ft.}}{w}$ $w = 10$ ft.
$\frac{1 \text{ in.}}{7 \text{ in.}} = \frac{2 \text{ ft.}}{l}$ $l = 14$ ft.
The area is 10×14 or 140 sq. ft.

UNIT 3

54. **(2) 8 yd.** Let l stand for the length. Then $w = \frac{1}{4}l$. Use the perimeter formula.

$P = 2l + 2w$
$20 = 2l + 2(\frac{1}{4}l)$
$20 = 2l + \frac{1}{2}l$
$20 = \frac{5}{2}l$
$2(20) = 2(\frac{5}{2}l)$
$40 = 5l$
$8 = l$

55. **(1) $\frac{2}{3}$ cu. in.** Find the volume of each cone using the formula $V = \frac{1}{3}$ (area of base) × height. Sample B is 24 cu. in. Sample A is $23\frac{1}{3}$ cu. in. (The information about the length of the side is not needed.) Subtract these two volumes to find that the difference is $\frac{2}{3}$ cu. in. So Sample B, the larger cone, holds $\frac{2}{3}$ cu. in. more of the ice cream mixture.

56. **(1) ∠6 and ∠2** Angle 6 is supplementary to ∠8. So, ∠6 = 180° – 50° = 130°. Since l and m are parallel lines intersected by a transversal, ∠2 is congruent to ∠6. Thus, ∠6 and ∠2 both measure 130°.

57. **(1) 50 sq. ft.** $A = \pi r^2 = 3.14(4)^2 = 50.24$, which rounds to 50 sq. ft.

58. **(3) $2\frac{2}{3}$ in.** Use the proportion:

$\frac{1}{1.5} = \frac{x}{4}$
$1.5x = 4$
$x = 2\frac{2}{3}$ in.

59. **(4) 29** Use the Pythagorean Theorem to find the diagonal: (the diagonal is the hypotenuse)

$c^2 = a^2 + b^2$
$c^2 = 15^2 + 25^2$
$c^2 = 225 + 625$
$c^2 = 850$
c is approximately 29 inches, since $29 \times 29 = 841$.

60. **(2) $12\frac{1}{2}$** Use the Pythagorean Theorem:

$c^2 = a^2 + b^2$
$c^2 = (7\frac{1}{2})^2 + 10^2$
$c^2 = 56\frac{1}{4} + 100$
$c^2 = 156\frac{1}{4}$
$c = 12\frac{1}{2}$

61. **(2) 15 degrees** A right triangle has a 90° angle and two acute angles. The sum of the angles in any triangle is 180°. Set up an equation, where x equals the measure of the smaller angle and $5x$ equals the measure of the larger acute angle.

$90° + x + 5x = 180°$
$90° + 6x = 180°$
$6x = 90°$
$x = 15°$

62. **(2) 50** Think of the figure as a rectangle and two half-circles. Find the area of the rectangle and the area of the whole circle and add the results.

Rectangle:
$A = lw$
$= 6 \times 5$
$= 30$ sq. ft.
Circle:
$A = \pi r^2$
$= 3.14(2.5)2$
$= 19.625$
$30 + 19.625 = 49.625$, which rounds to 50 sq. ft.

63. **(4) 16 yards** Find the area of the parallelogram. Then use the formula for the area of a square. $A = lw$

$32 \times 8 = 256$ sq. yd.
$A = s^2 = 256$; $s = \sqrt{256} = 16$ yd.

64. **(4) 12** Let w stand for the width. The length is $w + 6$. Use the formula for perimeter of a rectangle to write an equation.

$P = 2l + 2w$
$60 = 2(w + 6) + 2w$
$60 = 2w + 12 + 2w$
$60 = 4w + 12$
$48 = 4w$
$w = 12$ in.

Posttest: (Pages 311–319)

All items relate to the skill of application.

1. **(5) $18\frac{2}{3}$** Multiply $\frac{4\frac{2}{3}\text{ yd.}}{\text{cover}} \times 4$ ~~covers~~ $= \frac{14}{3} \times 4 = \frac{56}{3} = 18\frac{2}{3}$.

2. **(5) The maximum frequency of Model B is 100 times as great as that of Model A.** 3×10^3 is 3,000. 3×10^5 is 300,000. $\frac{300{,}000}{3{,}000} = 100$.

3. **(4) $502.50** Add the amounts she collected: $1,900 + $2,350 + $5,800 = $10,050; multiply by 5% (0.05): $10,050 × 0.05 = $502.50.

4. **(1) 20(5) + 12(9)** Multiply the number of students taking a particular lesson by the lesson price. Add the two products.

POSTTEST

5. **(3) 1999** To find the number of years needed to build 80 planes at a rate of 16 planes/year, divide: 80 planes ÷ 16 planes/year = 5 years. Add: 1994 + 5 = 1999.

6. **(4) 900** Multiply 1,200 by 75%: 1,200 × 0.75 = 900.

7. **(2) B** Start at 0 and count 5 to the left on the number line.

8. **(2) 32%** Add the rates for taxes withheld and taxes owed: 24% + 3% + 4% + 1% = 32%.

9. **(3) 2.5%** Divide the amount of disability tax by her taxable income. Change the decimal to a percent. \$475 ÷ \$19,000 = 0.025 = 2.5%.

10. **(4) between 5 and 6 hours** Use the formula: $d = rt$. Solve for time (t) by dividing distance (d) 305 by rate (r) 55. Estimate $\frac{305}{55}$ is less than $\frac{300}{50}$ or 6.

11. **(3) 3 and 4** Estimate the square root of 14. The square of 3 is 9 and the square of 4 is 16. The square root of 14 falls between 3 and 4.

12. **(4) C and F** Cliff has worked $4\frac{1}{2}\left(2 + 2\frac{1}{2}\right)$ hours of the 6 hours available. He has $1\frac{1}{2}$ hours left to work. The only combination among the options on the list that is less than $1\frac{1}{2}$ is C and F. $\left(\frac{3}{4} + \frac{1}{2} = 1\frac{1}{4}\right)$.

13. **(5) Not enough information is given.** You need to know the weight of the tomatoes in pounds, not the number of tomatoes: $\frac{79¢}{\text{pound}} \times 6$ tomatoes—the units are NOT the same, so you cannot perform this operation.

14. **(2) 15%** Divide 12 by the total number of employees (80): 12 ÷ 80 = 0.15 = 15%.

15. **(4) $\frac{1}{10}$** The ratio of buyers to total employees is $\frac{8}{80}$, which reduces to $\frac{1}{10}$.

16. **(3) 150** Use a proportion:
$\frac{1\text{ in.}}{40\text{ mi.}} = \frac{3\frac{3}{4}\text{ in.}}{x} = 3\frac{3}{4} \times 40$
$= \frac{15}{4} \times 40 = 150.$

17. **(2) \$112.50** Multiply \$90 by 1.25.

18. **(5) (3.14)(4)²(14)** Use the formula: $V = \pi r^2 h$. If the diameter is 8 inches, the radius r is 4 inches.

19. **(5) 6(4) + 2(3)** Multiply the number of patterns of each type by the cost for each type: $\cancel{\text{patterns}} \times \frac{\$}{\cancel{\text{pattern}}}$. Then add the two products.

20. **(4) $\frac{2}{3}$** Divide $\frac{1}{2}$ hour by 45 $\frac{\text{minutes}}{\text{lawn}}$ $\left(\frac{3}{4}\text{ hour}\right)$.
$\frac{1}{2}$ hour ÷ $\frac{3}{4}$hour/lawn = $\frac{1}{2}$ hour × $\frac{4}{3} = \frac{2}{3}$.

21. **(4) 1,600** Multiply:
$\frac{200\text{ square feet}}{\cancel{\text{gallon}}} \times 8\ \cancel{\text{gallons}} = 1{,}600$ square feet.

22. **(1) 2.062493 ×10⁶** Move the decimal 6 places to the right to multiply by 10^6.

23. **(4) 64%** 46% + 18% = 64%.

24. **(5) 76%** 100% − (15% + 9%) = 76%.

25. **(1) $\frac{72 + 66 + 74 + 68}{4}$** Add the scores and divide by the number of scores.

26. **(5) \$480,000**
100% − 4% = 96%. Multiply: \$500,000 × 0.96 = \$480,000 or \$500,000 × 4% = \$500,000 × .04 = 20,000
\$500,000 − \$20,000 = \$480,000.

27. **(3) $\frac{1}{30}$** Write a ratio and reduce: $\frac{5}{150} = \frac{1}{30}$.

28. **(2) 201** Use the formula: $A = \pi r^2$. Since the diameter is 16 feet, the radius of the platform is 8 feet. So, 3.14 × 8 × 8 = 201 (rounded).

29. **(1) 180** Multiply: $\frac{12\text{ plants}}{\cancel{\text{row}}} \times 15\ \cancel{\text{rows}} = 180$ plants.

30. **(3) 60** Find the area of Lot B: $A = lw$ = (90)(40) = 3,600. The area of Lot A is $A = s^2$. Find the square root of the area (3,600) to find the length of the sides of Lot A: 6 × 6 = 36; since 3,600 ends in 2 zeros, multiply 6 × 10 = 60.

31. **(2) 26%** Divide:
18 calories ÷ $\frac{70\text{ calories}}{1\text{ serving}}$ =
18 $\cancel{\text{calories}}$ × $\frac{1\text{ serving}}{70\ \cancel{\text{calories}}}$ =
$\frac{.257\text{ calories}}{1}$ = 26% of the calories are from fat.

32. **(3) \$423.87** Multiply to find the sales tax $\left(6\frac{1}{2}\% = 0.065\right)$ (\$398 × 0.065 = \$25.87) and add to the price of the desk (\$398 + \$25.87 = \$423.87).

33. **(4) 0.04(\$500) + \$500** Find the increase by multiplying the cost by 4%. Then add the increase to the original cost.

34. (5) Not enough information is given. Use the formula for volume of a rectangular container: $V = lwh$. You have the length and volume, but you need the height to solve for the width.

35. (3) $5x + 5y$ Combine the x's $(6x - x = 5x)$; combine the y's $(2y + 3y = 5y)$.

36. (5) \$7.56 Convert $3\frac{1}{2}$ dozen.

$\frac{7}{2}$ ~~dozen~~ $\times$ $\frac{\overset{6}{\cancel{12}} \text{ oranges}}{\cancel{\text{dozen}}}$ = 42 oranges.

$\frac{6}{\$1.08} = \frac{42}{x}$

$6x = 42(\$1.08)$

$x = \frac{42(\$1.08)}{6} = \$7.56$

37. (1) 93 $15\%(620) = 93$

38. (4) 60

$p = 3(4)(3 + 2)$

$p = 12(5)$

$p = 60$

39. (1) $1\frac{1}{4}x = 25$ Set up a proportion:

$\frac{5 \text{ miles}}{1\frac{1}{4} \text{ hrs}} = \frac{x \text{ miles}}{5 \text{ hours}}$

40. (5) $74 - 2(22) = 2w$ Use the formula:

$p = 2l + 2w$

$74 = 2(22) + 2w$

So $74 - 2(22) = 2w$

41. (3) \$153.60 Multiply: $\frac{\$12.80}{\cancel{\text{month}}} \times 12$ ~~months.~~

42. (1) $3m - n$ Combine the m's $(2m + m = 3m)$; combine the n's $(2n - 3n = -n)$.

43. (2) p – \$24.95 – \$18.65 Subtract both purchases from the amount of the gift certificate.

44. (1) 3 $3z < 12$

$z < 4$

45. (5) 110° Straight angle $BAD = 180°$. Therefore, $180° - 70° = 110°$.

46. (3) 135° The sum of the interior angles of a triangle is 180°.

$m\angle x + 25° + 20° = 180°$

$m\angle x = 135°$

47. (1) 72 ft. Set up a proportion:

$\frac{\text{pole}}{\text{shadow}} = \frac{\text{building}}{\text{shadow}}$

$\frac{3}{5} = \frac{x}{20}$

$5x = 360$

$x = 72$

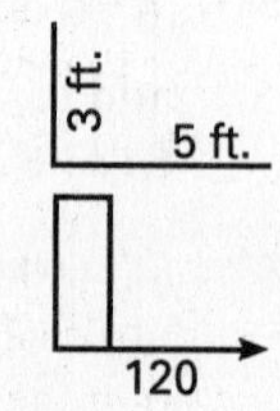

48. (2) 12 Find the sum of the given sides $(16 + 15 + 20 + 9 = 60)$ and subtract from the perimeter, 72: $(72 - 60 = 12)$.

49. (4) 15 ft. Draw a diagram. Set up a proportion.

$\frac{8}{6} = \frac{(8 + 12)}{x}$

$8x = 120$

$x = 15$

50. (4) 8 The radius of the large circle is the diameter of the small circle. $(32 \div 2 = 16)$. Divide 16 by 2 to find the radius of the small circle.

51. (5) 46 Add all the sides. The leftmost side is 8. The right inner side is 5. The base is $10(3 + 4 + 3)$. Thus: $3 + 5 + 4 + 5 + 3 + 8 + 10 + 8 = 46$.

52. (5) 20 ft. Use the Pythagorean Theorem. The wire is the hypotenuse, or c^2 in the formula.

$a^2 + b^2 = c^2$

$12^2 + 16^2 = c^2$

$144 + 256 = c^2$

$400 = c^2$

$20 = c$

53. (3) side *FH* $\angle FDE$ corresponds to $\angle GHF$ because they have the same measure. If the triangles were oriented by placing points D and H together, sides DF and FH would lie on the same line.

54. (4) 50 Similar triangles have proportionate sides. Set up a proportion using any corresponding pair of sides.

$\frac{FE}{FG} = \frac{DE}{GH}$

$\frac{40}{80} = \frac{x}{100}$

$50 = x$

55. (1) $a^2 + 9^2 = 15^2$ Use the Pythagorean Theorem on the formula page. The longest side (15) is the hypotenuse (c).

56. (4) 90° The sum of the internal angles of a triangle is 180°.

$m\angle 1 + m\angle 2 + m\angle 3 = 180°$

$55° + 35° + m\angle 3 = 180°$

$m\angle 3 = 180° - 35° - 55°$

$m\angle 3 = 90°$.

Simulated GED Test (Pages 322–331)

All items relate to the skill of application.

1. **(3) $37.20** Add the three items ($15.80 + $19.20 + $11.50 = $46.50), find the discount ($46.50 × 0.20 = $9.30), and subtract the discount from the original price ($46.50 – $9.30 = $37.20).

2. **(3) 20(10.75)** Multiply the total hours worked (20) by the rate ($10.75).

3. **(1) –4** Start at 0 and count to the left on the line to the point *B*.

4. **(2) *p* – $0.50 – $1.00** Subtract the amount of the coupon discount ($0.50) and the rebate amount ($1.00) from the original price.

5. **(4) $2,600** Let x = Mark's earnings and $2x$ = Rosa's earnings. Set up an equation:
$$2x + x = 3{,}900$$
$$3x = 3{,}900$$
$$x = 1{,}300$$
$$2x = \$2{,}600$$

6. **(3) 27** Let x = the losing team's points and $x + 13$ = the winning team's total.
Set up an equation:
$$x + x + 13 = 41$$
$$2x + 13 = 41$$
$$2x = 28$$
$$x = 14;$$
$$x + 13 = 27.$$

7. **(1) 25°** The sum of the internal angles of a triangle is 180°. Add the two angles you have been given (37° + 118° = 155°) and subtract from 180° (180° – 155 ° = 25°).

8. **(2) 15($25.80) + 12($41.76)** Multiply the number of subscriptions by the price in each case, then add.

9. **(2) 20** Use the formula $V = lwh$. $10 \times 4 \times \frac{1}{2} = 20$.

10. **(1) 20** Forty men preferred ham and 20 preferred tuna. Find the difference (40 – 20 = 20).

11. **(5) Not enough information is given.** You need to know how many people will order the specials. The graph shows only the <u>preference</u> of items in relation to one another.

12. **(3) 6** Use this formula:
$$m = \frac{y_2 - y_1}{x_2 - x_1}$$
$$2 = \frac{y_2 - 2}{3 - 1}$$
$$2 = \frac{y_2 - 2}{2}$$
$$4 = y_2 - 2$$
$$6 = y_2$$

13. **(2) $18,189.60** Find the amount of the raise ($17,160 × 0.06 = $1,029.60) and add it to the original salary ($17,160 + $1,029.60 = $18,189.60)

14. **(5) $\frac{3}{4}$** Set up a ratio and reduce:
$$\frac{\text{large}}{\text{total}} = \frac{150}{200} = \frac{3}{4}.$$

15. **(5) $13\frac{1}{2}$** Multiply $3\frac{3}{8}$ by 4.

16. **(5) 120** Substitute and solve.
$$x = 3a(b^2 - 8)$$
$$x = 3(5)(4^2 - 8)$$
$$x = 15(16 - 8)$$
$$x = 15(8) = 120$$

17. **(2) 1.8592×10^4** Moving the decimal point 4 places to the right means multiplying by 10^4.

18. **(3) $1,261.60** You can solve the problem using either of these equations:
$$1{,}660(100\% - 24\%) = x$$
$$1{,}660 - (1{,}660)(.24) = x$$

19. **(4) 12** Substitute and solve.
$$4x^3 - 2y$$
$$4(2^3) - (2)(10) =$$
$$4(8) - 20 =$$
$$32 - 20 = 12$$

20. **(3) 10** Divide $22\frac{1}{2}$ by $2\frac{1}{4}$.

21. **(5) the length of the shadow of the street sign** You need the other shadow to set up a proportion.
$$\frac{\text{height of sign}}{\text{shadow of sign}} = \frac{\text{shadow of sign}}{\text{shadow of building}}$$

22. **(3) 45 and 45** The sum of internal angles of a triangle is 180°. Subtract the measure of the right angle (90°) from 180° (180 – 90 = 90). An isosceles triangle must have two equal angles. Divide 90 by 2 to find the measure of the two angles.

23. **(3) $41.30** Multiply: $2,065 × 0.02.

24. **(1) 4** Solve for y. Cross multiply. $5 > y$. Only option (1) satisfies this equation.

25. **(2) $N = \frac{7}{4}(92)$** Set up a proportion and solve for N:

$$\frac{4 \text{ hours}}{92 \text{ monitors}} = \frac{7 \text{ hours}}{N}$$
$$4N = (7)(92)$$
$$N = \frac{(7)(92)}{4} \text{ or}$$
$$N = \frac{7}{4}(92)$$

26. **(2) 20 miles** "Due east" and "due south" form a right triangle. These distances are the legs. Use the Pythagorean formula to solve for the hypotenuse:
$c^2 = a^2 + b^2$
$c^2 = 12^2 + 16^2$
$c^2 = 144 + 256$
$c^2 = 400$
$c = \sqrt{400}$
$c = 20.$

27. **(2) A and E** Add to find out how much Linda has already spent: ($14.50 + $16.80 = $31.30). Subtract to find out how much she has left to spend ($50 − $31.30 = $18.70). Test each option. Only option (2) A and E totals less than $18.70 ($8.95 + $9.45 = $18.40).

28. **(1) $s^3 = 216$** Use the formula for the volume of a cube: $V = s^3$.

29. **(3) $597.53** Add the amounts: $593.60 + $585.92 + $602.35 + $608.25 = $2,390.12. Divide by 4 to average: ($2,390.12 ÷ 4 = $597.53).

30. **(4) 79** In an isosceles triangle the sides opposite the equal angles must be equal. Therefore, side AB and side BC are both 32 inches. The perimeter is $P = a + b + c = 15 + 32 + 32 = 79$.

31. **(5) 64** Let x represent the number of professional golfers. Use this equation:
$x + 2x = 96$
$3x = 96$
$x = 32$
The number of amateurs is $2x = 64$.

32. **(4) $\frac{1,170}{5}$** Divide the total amount to be saved in 5 months by 5 to find the average.

33. **(5) $a^2 + 8^2 = 10^2$** Use the Pythagorean relationship and substitute the values from the drawing.

34. **(2) 3** Divide the amount spent on food ($463) by the amount spent on recreation ($158). Round off to the nearest whole number.

35. **(4) 0.35($15) + $15** The price is 35% more than the cost.
Price = 35% cost + cost
= 0.35($15) + ($15).

36. **(5) $2,695** Find the new cost for materials: ($1,500 × .25) + $1,500 = $1,875. Add the labor and equipment costs ($1,875 + $220 + $600 = $2,695).

37. **(2) $9,500 − (0.20)($9,500)** Subtract the discount (0.20 × $9,500) from the original amount ($9,500).

38. **(3) 58 degrees** Since a right angle measures 90 degrees, angles 1 and 2 must add up to 90 degrees. Subtract the measure of angle 1 from 90° (90° − 32° = 58°).

39. **(2) 30** Set up a proportion.

$$\frac{5 \text{ milligrams}}{30 \text{ pounds}} = \frac{x \text{ milligrams}}{180 \text{ pounds}}$$
$$(5)(180) = 30x$$
$$900 = 30x$$
$$30 = x$$

40. **(3) 4 and 5** The square of 4 is 16 and the square of 5 is 25; therefore, the square root of 24 must be between 4 and 5.

41. **(1) 9** Set up a series of equations:
Stuart + 4 = Michela
Sofia − 3 = Stuart
If Sofia = 8, then
Stuart = 8 − 3 = 5 and Michela = 5 + 4 = 9.

42. **(2) $\frac{1}{2}$** Convert the times to the same units, minutes or hours. Set up a proportion:

$$\frac{90 \text{ min.}}{1 \text{ floor}} = \frac{45 \text{ min.}}{x}$$
$$90x = 45$$
$$x = \frac{1}{2}$$

43. **(4) 125** Use the proportion:

$$\frac{1 \text{ in.}}{50 \text{ mi.}} = \frac{2\frac{1}{2} \text{ in.}}{x}$$
$$x = 2\tfrac{1}{2} \times 50 = 125$$

44. **(3) $3a = 33$** Let a represent Anne's age, $3a$ equals Doris's age (33); therefore $3a = 33$.

45. **(1) $(3.14)(1.5)^2(5)$** Use the formula for the volume of a cylinder $V = \pi r^2 h$. Divide the diameter (3) by 2 to find the radius (3 ÷ 2 = 1.5).

46. **(3) 75.6%** Subtract the percent spent on recruiting and salary (19.3 + 5.1 = 24.4) from 100%. (100% − 24.4% = 75.6%).

47. **(3) 15(9) + 22(7)** Multiply the hours worked by the wage rate for each job, then add.

48. **(1) 5** Use the formula for the volume of a rectangular solid: $V = lwh$. Substitute the known quantities and solve for w.
$900 = (20)(w)(9)$
$900 = 180w$
$5 = w$

49. **(4) 208** Find the area of the 8-by-8 square and the rectangle and add the results. The unmarked length of the rectangle is $8 + 16 = 24$.
Square:
$A = s^2$
$= 8^2$
$= 64$
Rectangle:
$A = lw$
$= 24 \times 6$
$= 144$
Add:
$64 + 144 = 208$

50. **(4) Between 7 and 8 hours** Divide 420 by 55.

51. **(4) 15** Use the proportion $\frac{200 \text{ mi.}}{1 \text{ tank}} = \frac{3{,}000 \text{ mi.}}{x \text{ tanks}}$
$x = \frac{3{,}000}{200} = 15$

52. **(2) The range of Type B sonar is 15 times greater than the range of Type A sonar.**
$\frac{4.8 \times 10^4}{3.2 \times 10^3} = \frac{48}{32} \times \frac{10}{1}$
$= \frac{3}{2} \times 10 = 15$

53. **(2) 12** Set up a proportion:
$\frac{\overline{GH}}{\overline{DE}} = \frac{\overline{FG}}{\overline{DF}}$
Substitute the values you have and solve for the unknown.
$\frac{8}{\overline{DE}} = \frac{12}{(12 + 6)}$
$144 = 12(\overline{DE})$
$12 = \overline{DE}$

54. **(4) 303** Add the distances $(318 + 315 + 320 + 298 + 264 = 1{,}515)$ and divide $(1{,}515 \div 5 = 303)$.

55. **(5) 10.5** Find the area of the triangle. Then use the formula for area of a rectangle.
$A = \frac{1}{2}bh = \frac{1}{2}(28.7)(12) = 172.2$ sq. m.
$A = lw = 172.2$; $16.4w = 172.2$; $w = 10.5$ m.

56. (4) $54 Find the lowest cost for each supply and add:

1 expense register:	$15
50 file folders:	$25
3 notebooks:	$10
6 paper rolls:	$4
	$54

SIMULATED TEST

Glossary

absolute value the distance an integer is from zero on a number line

acute angle an angle that measures less than 90°

addition combining quantities to find a total; the symbol + ("plus") is used in addition
Examples:

$$\begin{array}{r} 300 \\ +478 \\ \hline 778 \end{array} \qquad 12 + 6 = 18$$

adjacent angles angles that have a common vertex and a common ray

algebra the branch of mathematics which describes the relations between quantities

algebraic expression a mathematical expression that uses letters and numbers instead of words

alternate exterior angles angles formed when two parallel lines are cut by a transversal; always outside the parallel lines

alternate interior angles angles formed when two parallel lines are cut by a transversal; always inside the parallel lines

angle a pair of rays extending from a common point

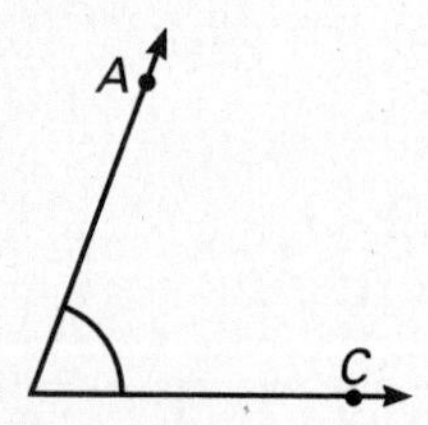

arc part of a circle

area the measure of the surface inside a flat object or figure expressed in square units

area relationships relationships between areas of figures with different shapes

arithmetic using numbers to calculate or compute, usually by adding, subtracting, multiplying, or dividing

average the sum of the data in a list divided by the number of items on that list; the mean

bar graph a graph with bars that represent different numbers

base the whole amount in a percent problem
Example: 30% of 200 = 60
↑
base (points to 200)

base (triangle) the line or plane along which a triangle or figure rests

cancellation the process of reducing factors when multiplying and dividing fractions
Example:

$$\frac{1}{3} \times \frac{3}{2} = \frac{1}{\cancel{3}_{1}} \times \frac{\cancel{3}^{1}}{2} = \frac{1}{2}$$

chart information arranged in columns and rows

circle the curve formed by all the points that are the same distance from one point, called the **center**

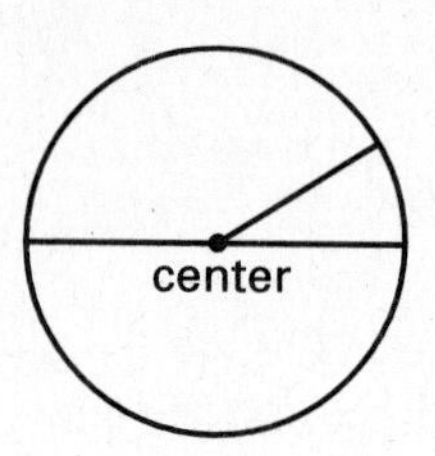

circle graph information shown in a circle that is cut into sections to show the parts that make up the total

circumference the perimeter of, or distance around, a circle

common denominator a number that two or more denominators divide into evenly

comparing (ordering) numbers finding if two numbers are equal (=), if one number is less than (<) another, or if one number is greater than (>) another
Examples:

$14 = 14$ Fourteen equals fourteen.
$17 < 19$ Seventeen is less than nineteen.
$12 > 10$ Twelve is greater than ten.

complementary angles two angles whose sum is 90°

cone a solid figure with a circular base and sides that meet at a point

congruent angles angles that have equal measures

congruent figures figures that are the same shape and size

coordinates two numbers that represent a point on a grid that has a horizontal axis and a vertical axis

coordinate system a set of points on a coordinate grid that has a horizontal axis and a vertical axis

corresponding angles angles that are on the same side of a transversal that cuts two parallel lines; the angles are either both above or both below the two parallel lines and are always equal in measure

cross multiplying a process by which you can tell if two fractions are equal
Example:
$\frac{4}{8} \stackrel{?}{=} \frac{3}{6}$ $4 \times 6 = 24$ $8 \times 3 = 24$ $\frac{4}{8}$ and $\frac{3}{6}$ are equal.

cross-product rule to find the missing number in a proportion, cross multiply and divide the product by the third number

cube a rectangular solid with six square sides

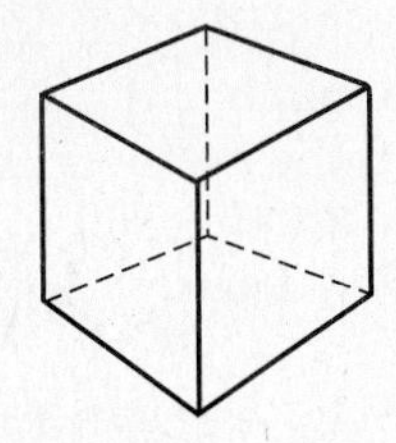

cylinder a solid figure with a circular base and straight sides

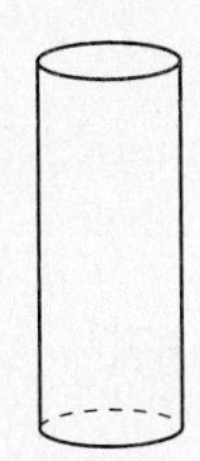

data a list of numbers

decimal a fraction that uses a place-value system to show a part of 10, 100, 1,000, and so on
Examples: 3.25 0.9

degree a unit for measuring angles

denominator the bottom number in a fraction; the number that tells the number of equal parts in the whole object or group

diagonal a line segment drawn between the vertices of two non-adjacent sides of a figure that has four or more straight sides

diameter a line segment drawn through the center of a circle

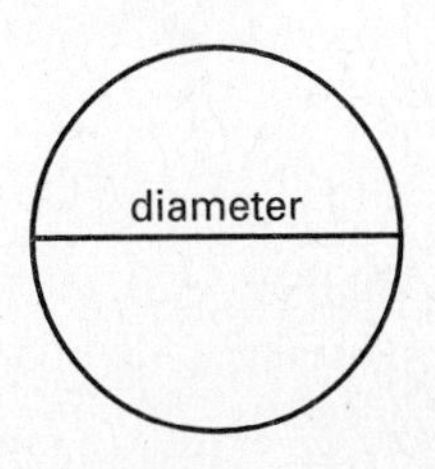

difference the answer in a subtraction problem

digit one of the numbers 0, 1, 2, 3, 4, 5, 6, 7, 8, and 9, used to represent numbers in a place value system

dimension a measurement of length, width, or thickness

division splitting a quantity into equal groups; the symbols ÷ and $\overline{)\ }$ are used in division; a fraction can also be used to show division
Examples:
$34\overline{)68}$ = 2 $12 \div 4 = 3$ $\frac{25}{5} = 5$

estimation finding an approximate amount when an exact answer is not needed

equal fractions fractions that have the same value

equation a statement that says two expressions are equal

equilateral triangle a triangle that has all congruent sides

exponent in an expression, a raised number at the right of a number, which tells how many times the number is used as a factor
Example: $5^3 = 5 \times 5 \times 5 = 125$
↑
exponent

factors numbers or algebraic expressions that are multiplied together

factoring finding the factors of a number or of an algebraic expression

formula an equation in which the letters stand for specific kinds of quantities

fraction a number that shows a part of a whole or a part of a group
Examples: $\frac{1}{4}$ $\frac{1}{2}$

geometry the branch of mathematics which studies the relationship of points, lines, angles, and surfaces of figures in space

hypotenuse in a right triangle, the side opposite the right angle

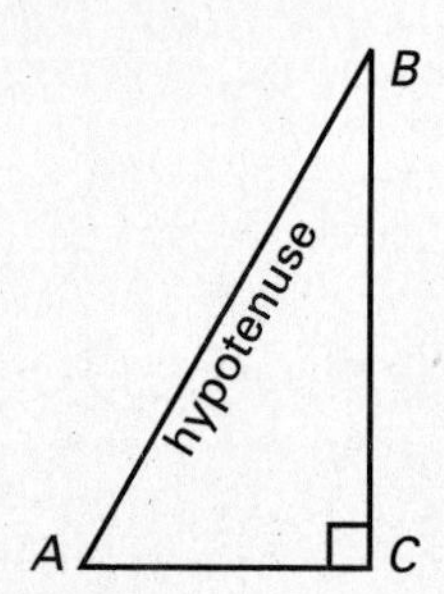

improper fraction a fraction that shows a quantity equal to or greater than 1; the numerator is equal to or greater than the denominator
Examples: $\frac{12}{7}$ $\frac{16}{13}$

indirect measurement a method to find measures when there is no way to actually perform the measurement

integers all whole numbers, both positive and negative

interest a fee charged for using someone else's money expressed as a percent

intersecting lines straight lines that cross

irregular figures figures that are made up of several shapes

isosceles triangle a triangle that has two congruent sides

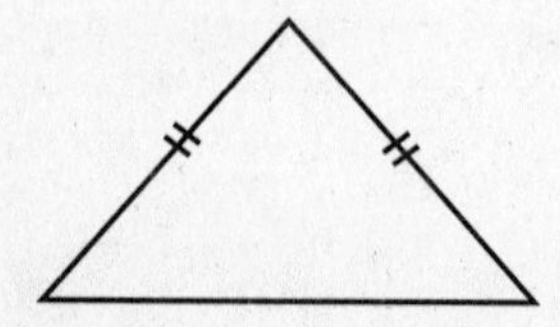

least common denominator the smallest multiple of the denominators of two or more fractions

legs of a triangle in a right triangle, the sides opposite the non-right angles

like fractions fractions that have the same denominator

line a collection of points along a straight path

line graph a graph consisting of lines that connect points representing different amounts to show change

line segment part of a line

lowest terms fraction a fraction in which there is no number other than 1 that will divide evenly into both the numerator and the denominator

mean the sum of the data in a list divided by the number of items on that list; the average

median the middle number in a list of data arranged in order

metric measurement system the measurement system used throughout most of the world; based on the powers of ten

mixed number a number with a whole number part and a proper fraction part, used to show a fraction that is greater than 1

multiple the result of multiplying a whole number by 0, 1, 2, 3, and so on

multiplication joining together a number of quantities to find a total; adding the same number repeatedly; the symbols × ("times"), · (raised dot), and ( ) (parentheses) are used in multiplication
Examples: $\begin{array}{r} 209 \\ \times\ 5 \\ \hline 1{,}045 \end{array}$ $4 \cdot 15 = 60$ $(5)(4) = 20$

negative number a number preceded by a minus sign, indicating a number to the left of zero on a number line; a number less than 0 in value; used to show a decrease, a loss or downward direction

non-adjacent angles angles with a common vertex, but not a common ray

number line line divided into equal segments by points corresponding to integers, fractions, or decimals. Points to the right of 0 are positive; those to the left are negative

numerator the top number in a fraction; the number that tells the number of equal parts you are referring to

obtuse angle an angle that measures more than 90° but less than 180°

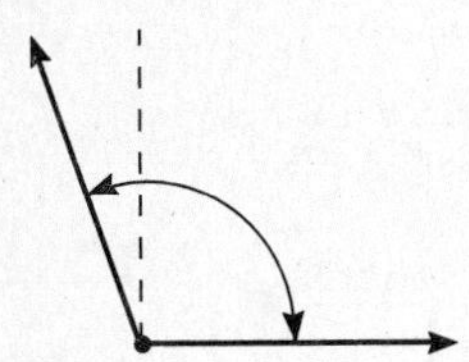

opposite angles the angles that are across from each other when two lines intersect or cross

ordered pair a pair of numbers in parentheses, an x-coordinate followed by a y-coordinate, that names a point on the coordinate grid

ordering arranging a set of integers, fractions or decimals from least to greatest or greatest to least

origin the point of intersection of the x-axis and the y-axis in the coordinate plane

parallel lines two lines on the same plane that do not intersect

parallelogram a flat figure with four sides and opposite sides parallel; opposite sides and opposite angles are equal

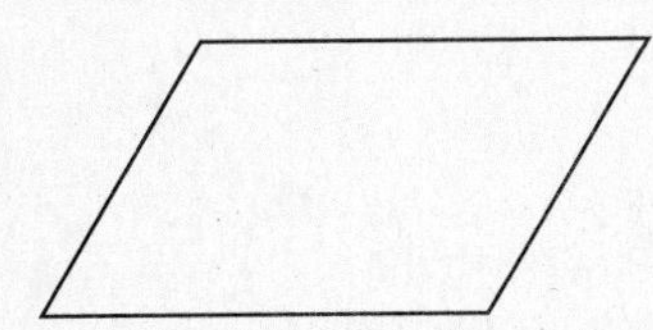

part a piece of the whole or base in percent problems

Example: 30% of 200 = 60
↑
part

percent a way to show part of a whole; the whole is always divided into 100 equal parts

Example: fifty percent = 50% = 50 out of 100

50% of 48 = 24
↑ ↑ ↑
rate **base** **part**

perimeter the measure of the distance around the edge of any flat object or figure

perpendicular two lines that intersect, forming adjacent right angles

pi (π) the constant ratio of the circumference of a circle to the diameter, with the value $\frac{22}{7}$, or approximately 3.14

place value the value of a number determined by its position. Examples: the 5 in 589 has a value of 500; the 5 in 0.05 has a value of $\frac{5}{100}$

plane a set of points that forms a flat surface

point single, exact location often represented by a dot

positive number a number to the right of zero on a number line, sometimes preceded by a plus sign; a number greater than 0 in value; used to show an increase, a gain or upward direction

power a number used to show how many times a number is multiplied by itself

Example: 10^3 ten to the third power

principal an amount of money borrowed or invested

probability a number (whole, fraction, decimal, or ratio) that shows how likely it is that an event will happen

product the answer in a multiplication problem

proper fraction a fraction that shows a quantity less than 1; the numerator is always less than the denominator

proportion an equation that states that two ratios are equal

pyramid a solid figure with a square base and four equal triangular sides that meet at a point

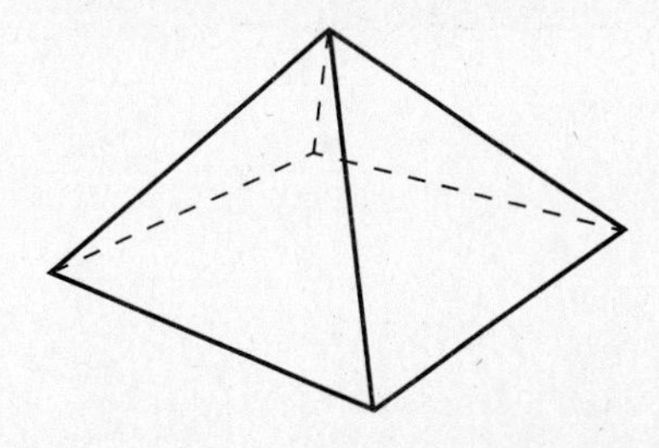

Pythagorean Theorem in a right triangle, the square of the hypotenuse is equal to the sum of the squares of the other two sides

quadrant one-fourth of a coordinate grid, formed by the intersecting axes

quadrilateral a flat figure with four sides

quotient the answer in a division problem

radius a line segment connecting the center of a circle to a point on the circle; the length is $\frac{1}{2}$ the length of the diameter

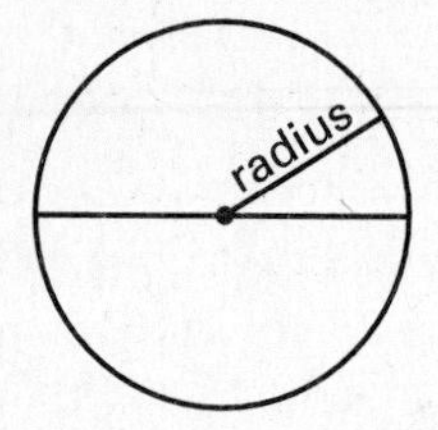

range in a list of data, the spread between the lowest number and the highest number

rate a relationship between quantities of different kinds
Examples: 55 miles per hour,
1 tablet every 4 hours

rate (in percent problems) the relationship of the part to the base. Example: the interest rate on an amount of money borrowed, expressed as a percent

ratio a way of comparing two numbers using division
Examples: 3 out of 4 3:4 $\frac{3}{4}$

ray a part of a line having only one endpoint

rectangle a parallelogram with four right angles

rectangular solid a solid in which all sides are rectangles and all corners are square

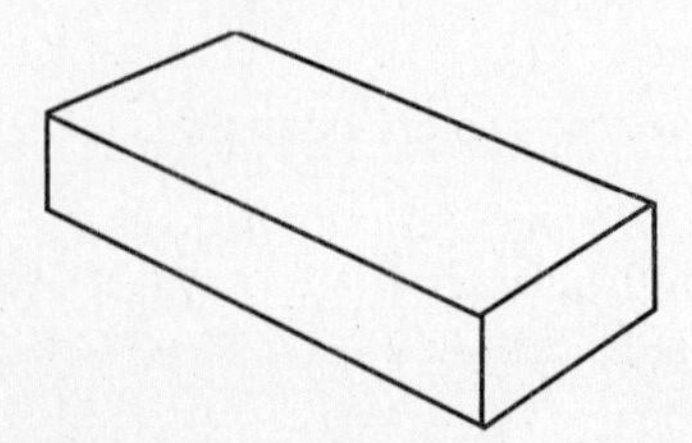

reducing fractions finding an equal fraction with a smaller numerator and denominator

reflex angle an angle that measures more than 180° but less than 360°

remainder the amount left over in a division problem

rhombus a parallelogram with four sides of equal length

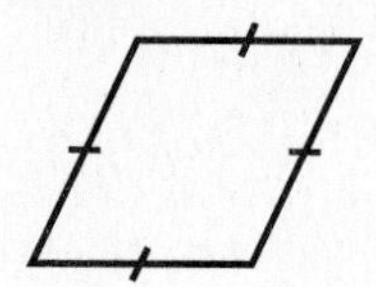

right angle an angle that makes a "square corner" that measures 90°

right triangle a triangle in which one angle is a right angle (90°)

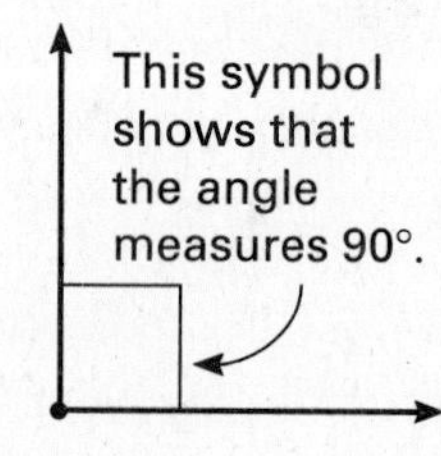

root in an expression with an exponent, the number that is multiplied by itself
Example: **root** $\rightarrow 6^3 \leftarrow$ **exponent**

rounding using an approximation for a number

scale relationship between two sets of measurements

scale drawing a sketch of an object with all distances in proportion to corresponding distances on the actual object; the scale gives the ratio of the sketch measurements to the corresponding measurements on the actual object

scalene triangle a triangle in which no sides are congruent

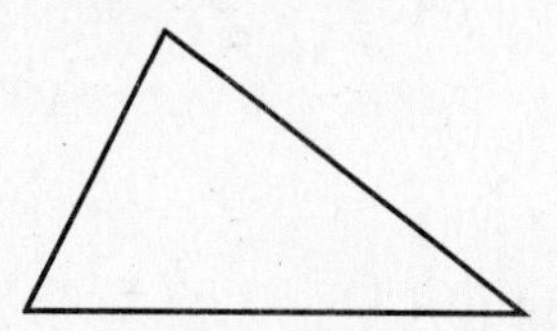

scientific notation a way of writing very large numbers and very small fractions, in which the numbers are expressed as the product of a number between 1 and 10 and a power of 10

Examples: $5.26 \times 10^6 = 5{,}260{,}000$
$3 \times 10^{-4} = 0.0003$

signed numbers positive and negative numbers, used to show quantity, distance, and direction

similar triangles triangles in which the corresponding angles have equal measures and the corresponding sides are in proportion

simplifying an algebraic expression performing all the operations you can within the expression containing a variable

slope a number that measures the steepness of a line

square a rectangle with sides of equal length or a rhombus with right angles

square root a number that when multiplied times itself equals the given number

Example: The square root of 25 is 5.
$\sqrt{25} = 5$ $5^2 = 25$

standard measurement system the measurement system that is used in the United States

straight angle an angle that measures 180°

subtraction taking away an amount from another amount to find the difference; the symbol – ("minus") is used in subtraction

Examples:
$$\begin{array}{r} \$204 \\ -\$167 \\ \hline \$\ 37 \end{array}$$
$25 - 7 = 18$

sum the answer in an addition problem

supplementary angles two angles whose sum is 180°

table information arranged in columns and rows

transversal a line that crosses two or more parallel lines

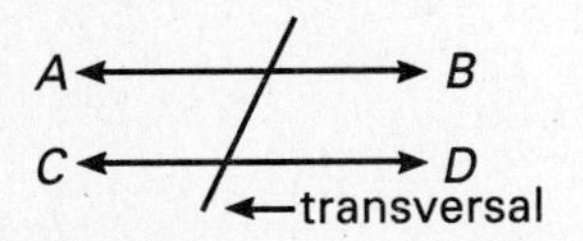

trapezoid a quadrilateral with only one pair of parallel sides

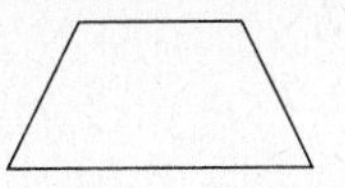

triangle a flat closed figure with three sides and three angles

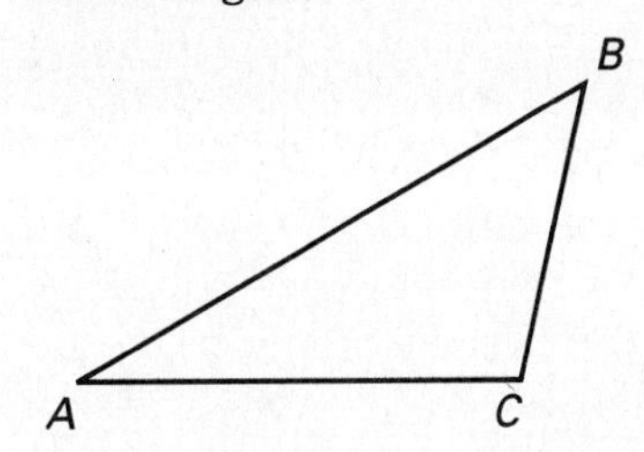

unit rate a ratio with a denominator of 1

unlike fractions fractions with different denominators

variable any letter used to stand for a number

vertex the point in which two or more line segments or sides of a figure meet; the point in which the two rays that form an angle meet

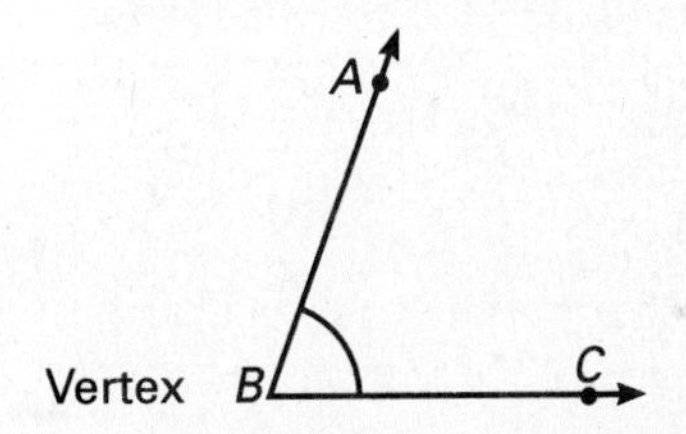

vertical angles the angles that are across from each other when two lines intersect or cross

volume the measure of the amount of space inside a three-dimensional object or figure

whole number any number in the set 0, 1, 2, 3, . . .

***x*-axis** the horizontal axis in the coordinate plane

***x*-coordinate** the first number in an ordered pair, showing distance from the origin along the x-axis

***y*-axis** the vertical axis in the coordinate plane

***y*-coordinate** the second number in an ordered pair, showing distance from the origin along the y-axis

Index

Formulas

Description	Formula
Area *(A)* of a:	
square	$A = s^2$; where s = side
rectangle	$A = lw$; where l = length, w = width
parallelogram	$A = bh$; where b = base, h = height
triangle	$A = \frac{1}{2}bh$; where b = base, h = height
circle	$A = \pi r^2$; where $\pi = 3.14$, r = radius
Perimeter *(P)* of a:	
square	$P = 4s$; where s = side
rectangle	$P = 2l + 2w$; where l = length, w = width
triangle	$P = a + b + c$; where a, b, and c are the sides
circumference *(C)* of a circle	$C = \pi d$; where $\pi = 3.14$, d = diameter
Volume *(V)* of a:	
cube	$V = s^3$; where s = side
rectangular container	$V = lwh$; where l = length, w = width, h = height
cylinder	$V = \pi r^2 h$; where $\pi = 3.14$, r = radius, h = height
Pythagorean relationship	$c^2 = a^2 + b^2$; where c = hypotenuse, a and b are legs of a right triangle
distance *(d)* between two points in a plane	$d = \sqrt{(x_2 - x_1)^2 + (y_2 - y_1)^2}$; where (x_1, y_1) and (x_2, y_2) are two points in a plane
slope of a line *(m)*	$m = \frac{y_2 - y_1}{x_2 - x_1}$; where (x_1, y_1) and (x_2, y_2) are two points in a plane
mean	mean = $\frac{x_1 + x_2 + \ldots + x_n}{n}$; where the x's are the values for which a mean is desired, and n = number of values in the series
median	median = the point in an ordered set of numbers at which half of the numbers are above and half of the numbers are below this value
simple interest *(i)*	$i = prt$; where p = principal, r = rate, t = time
distance *(d)* as function of rate and time	$d = rt$; where r = rate, t = time
total cost *(c)*	$c = nr$; where n = number of units, r = cost per unit

Answer Sheet

GED Mathematics Test

Name: ____________________ **Class:** ____________________ **Date:** ______

○ Pretest ○ Posttest ○ Simulated Test

1 ①②③④⑤	**11** ①②③④⑤	**21** ①②③④⑤	**31** ①②③④⑤	**41** ①②③④⑤	**51** ①②③④⑤
2 ①②③④⑤	**12** ①②③④⑤	**22** ①②③④⑤	**32** ①②③④⑤	**42** ①②③④⑤	**52** ①②③④⑤
3 ①②③④⑤	**13** ①②③④⑤	**23** ①②③④⑤	**33** ①②③④⑤	**43** ①②③④⑤	**53** ①②③④⑤
4 ①②③④⑤	**14** ①②③④⑤	**24** ①②③④⑤	**34** ①②③④⑤	**44** ①②③④⑤	**54** ①②③④⑤
5 ①②③④⑤	**15** ①②③④⑤	**25** ①②③④⑤	**35** ①②③④⑤	**45** ①②③④⑤	**55** ①②③④⑤
6 ①②③④⑤	**16** ①②③④⑤	**26** ①②③④⑤	**36** ①②③④⑤	**46** ①②③④⑤	**56** ①②③④⑤
7 ①②③④⑤	**17** ①②③④⑤	**27** ①②③④⑤	**37** ①②③④⑤	**47** ①②③④⑤	
8 ①②③④⑤	**18** ①②③④⑤	**28** ①②③④⑤	**38** ①②③④⑤	**48** ①②③④⑤	
9 ①②③④⑤	**19** ①②③④⑤	**29** ①②③④⑤	**39** ①②③④⑤	**49** ①②③④⑤	
10 ①②③④⑤	**20** ①②③④⑤	**30** ①②③④⑤	**40** ①②③④⑤	**50** ①②③④⑤	